BETWEEN TWO FIRES

BETWEEN TWO FIRES

A FIRE HISTORY OF CONTEMPORARY AMERICA

Stephen J. Pyne

THE UNIVERSITY OF
ARIZONA PRESS

TUCSON

The University of Arizona Press
www.uapress.arizona.edu

© 2015 The Arizona Board of Regents
All rights reserved. Published 2015

Printed in the United States of America
20 19 18 17 16 15   6 5 4 3 2 1

ISBN-13: 978-0-8165-3182-0 (cloth)
ISBN-13: 978-0-8165-3214-8 (paper)

Cover design by Leigh McDonald
Cover photos: [front] Tree torching near Africa Lake, September 10, 1988, by Jim Peaco; [back] The cumulative fires of 2012 from January 1 through October 31, 2012, courtesy of NASA's Goddard Space Flight Center.

Library of Congress Cataloging-in-Publication Data
Pyne, Stephen J., 1949–, author.
  Between two fires : a fire history of contemporary America / Stephen J. Pyne.
    pages cm
  Includes bibliographical references and index.
  ISBN 978-0-8165-3182-0 (cloth : alk. paper) — ISBN 978-0-8165-3214-8 (pbk. : alk. paper)
  1. Forest fires—United States—Prevention and control—History. 2. Fire ecology—United States. 3. United States. Forest Service—History. I. Title.
  SD421.3.P9586 2015
  634.9—dc23
                                    2015001529

♾ This paper meets the requirements of ANSI/NISO Z39.48-1992 (Permanence of Paper).

To Sonja,

who has seen it all, from spark to conflagration to reburn

and

to Len and Ken and the other Longshots:

"They're missing, Chief. That's all."

And they shall go out from one fire, and another fire shall devour them.

—*Ezekiel 15:7*

CONTENTS

Prologue: Agency on Fire 3
Three Fires, 1961 28
1. Spark 33
   Last Hurrahs, 1967 and 1970 93
2. Hotline 99
   New Normals, 1977 and 1980 177
3. Holding 182
   Lost Fire, 1991 256
4. Slopover 275
   Millennial Fire, 2000 348
5. Blowup 357
   East Meets West, 2007 411
6. Burning Out 418
   Wallowing, 2011 452
   Epilogue: After Action Report 457
   Afterword 467

   Abbreviations 471
   Note on Sources 475
   Notes 477
   Index 525

   Illustrations follow page 262

BETWEEN TWO FIRES

PROLOGUE

AGENCY ON FIRE

You work for the Forest Service, you fight fire.
—CHIEF FORESTER DALE BOSWORTH[1]

In the beginning were the fires.

In America lightning started thousands of fires annually in longleaf and shortleaf pine, in shortgrass prairie, in everglades sawgrass, in pinyon-juniper woodlands, in montane and mixed-conifer forests, in old-growth lodgepole and crinkly chaparral. They clumped with particular fierceness around the Southwest, the Sierra Nevada, and Florida, but they could occur wherever seasonal or decadal rhythms of wetting and drying coincided with dry- or early-season flashes.

For every fire nature started, people started ten or more. These had traditionally clustered around routes of travel and places where people busily converted wildland into hunting grounds, foraging patches, swidden plots, farms, and towns. They burned most easily where fire already existed: in such places humanity's fires overwhelmed nature's. But the torch could burn beyond the range of lightning, and people could slash, crunch, loose livestock, drain, dry, and otherwise modify the landscape and so burn outside nominally natural parameters. By deliberation and default, people burned everywhere they went and tweaked or remade fire's regimes. The more unsettled the land, the more reckless the fires. By whatever cause, the country that became America burned.

Most Americans required fire—needed its power to run hearths and furnaces and wanted access to its ecological alchemy to render the landscapes they inhabited more suitable to their livelihoods. They accepted the occasional bad fire in order to ensure access to good fires. But after industrialization began to alter the fundamentals of controlled burning, they increasingly sought to replace open flame with internal combustion and the fuels of living landscapes with those from fossil ones. In site after site, they replaced flame with sublimated surrogates. When the country established protected wildlands, it projected an industrial model into the wild. People quit lighting fires. They tracked down and quenched fires of any and all origins. And they established institutions to oversee the process.

Of these institutions, the U.S. Forest Service (USFS) became supreme. From its origins, fire control was not just something the agency did among its varied tasks. Perhaps more than anything else, fire was the reason the agency existed at all, and it became a primary index of the agency's success. Its origin stories were mostly forged in flame. By the middle of the 20th century, the Forest Service could be aptly characterized as an agency on fire.

* * *

When John F. Kennedy proclaimed in his presidential inaugural on January 20, 1961, that the torch had passed to a new generation, it is unlikely he had in mind a literal driptorch, but if he had, it is likely that torch would lie in the hands of a firefighter with the U.S. Forest Service. Probably the torch had been designed at a Forest Service equipment center, matched against experiments conducted by a Forest Service fire lab, manufactured to Forest Service specifications from the federal supply service, and stored in a fire cache on a national forest. It would be held by a Forest Service smokechaser, hotshot, smokejumper, helitacker, or militiaman under the direction of a Forest Service fire officer; if it was used to combat a major wildfire, that torchman would operate as part of Large Fire Organization, a Forest Service invention. If the torch was held by someone outside the federal fire establishment, the hands that gripped it probably belonged to a member of a state forestry bureau that cooperated formally with the Forest Service. The torch would be passed according to policies crafted by and overseen by the Forest Service. Possibly the passing would occur under the benevolent eye of Smokey Bear, the national symbol of fire prevention that had become a virtual logo of the Forest Service. If that torchman chatted with others, they would have told stories of past fires, an occupational folklore handed down from generation to generation. Everywhere, at every chain of custody regarding fire, its equipment, its workforce, its purposes, or even knowledge about it, you would see the U.S. Forest Service or its collaborators.

In 1960 the Forest Service was a benign hegemon. It is not simply that the agency administered the largest mass of public lands subject to fire protection or that it controlled the means of production for fire's control but that there were no other institutions to challenge it. Fire use had become a monopoly of the government. No organized civil society existed. Little research existed outside Forest Service sponsorship. Its alliance with professional forestry allowed claims for moral authority and expanded the reach of its fire imperium. Within the United States, the Forest Service was the matrix that connected all the pieces of fire's control. Globally, the United States was a firepower unrivalled by any other country.

It was an astonishing achievement. Fifty years earlier the fabled Great Fires of 1910, climaxing with the Big Blowup in the Northern Rockies, had torn through the fledgling agency and left it exhausted and traumatized. Chief Forester Henry Graves had declared that fire protection was 90 percent of American forestry. During the 1920s Chief Forester William Greeley had squeezed that number down to 75 percent. By 1960 fire claimed only 13 percent of the agency's budget. Instead of driving firefighters into a mine shaft, as the Big Blowup had, the "dragon Devastation," as Gifford Pinchot had once identified it, had itself been muscled back into a cave. In the agency's classification of wildfire, the largest category (Class G) contained fires bigger than 5,000 acres.

But if flame's daily presence had dimmed, if it no longer posed an existential threat to crews and the agency, that is because it no longer needed to flash and blast to shape the quotidian life of the agency. It was embedded in its heritage and virtual identity. It was spliced into the Forest Service's bureaucratic DNA. Everyone in the organization understood something about fire; most had served on firelines or fire camps, and no one doubted that fire control was the first duty and precondition for stewardship over the public lands. "A self-righteous youth in my twenties I thought that my jobs as fire lookout and firefighter gave me a real moral advantage—I told my city friends, 'Look, when I do this kind of work I can really say I'm doing no harm in the world, and am only doing good,'" wrote poet Gary Snyder of the 1950s.[2]

* * *

Fast forward 50 years to the 2010 centennial of the Big Blowup. Former chief Dale Bosworth recalled the lessons branded into his memory by growing up in a Forest Service family, specifically as the son of a fire control officer in Northern California.

Irwin Bosworth had determined to be a forester from the time he was a boy scout. He worked seasonally as a lookout; graduated from the Forestry School at the University of California, Berkeley; got on as a dispatcher on the Lassen

National Forest; then moved up the hierarchy as a staff officer in timber, wildlife, and grazing—and eventually as a district ranger and forest supervisor. Fire had been his entry point, and it remained a constant presence regardless of what bureaucratic title graced his door. Fire control was everyone's job: when big fires struck, he was called out like most of the staff; he rose through the ranks of what the agency deemed its "militia." The senior Bosworth served as a fire boss on overhead teams mobilized for the big burns throughout the West. It was possible for someone assigned to timber to avoid watershed or wildlife issues, but it was not possible to ignore fire. It transcended all other practices; it took priority over all other activities save the safety and safeguarding of human life. During fire season, everyone was on call 24/7. Life in the Forest Service was less a job than a vocation, and its rites, as often as not, revolved around fires that happened or fires that just threatened.[3]

His son recollected how such a world worked. "Like most people in the Forest Service, my dad and his colleagues talked only about the Forest Service when they got together," which meant they talked constantly about fire. His mom, Mary Ellen, would keep the radio tuned to the forest frequency. In the early years, when his dad was away, she staffed the dispatch office. His father took Dale to his first fire, started by a plane crash, when he was eight years old. But fire was a routine of life. Irwin hiked him to the lookout after Sunday dinner, paid him a penny apiece to scrub grunge from hoses dirtied by fire duty, and let him listen to radio traffic when fires broke out. He was told to read Stewart Holbrook's 1943 journalistic survey of historic fires, *Burning an Empire*. He was taken to watch the 1953 firefighting movie (Hollywood's only good one) *Red Skies of Montana*. In 1960 when he turned 17, Dale was hired by the California Department of Forestry to work on an engine crew. His entry into the Forest Service came, as had his father's, through fire.

What he absorbed about fire from that upbringing was a catechism of beliefs that constituted a code of conduct. Fire is bad. Fire is exciting. Nothing good comes from fire. Fire is everyone's job. Fire control is the foundation of land management. You hit fires hard and early. A firefight was a battle between good and evil. Whatever else it aspired to be, the Forest Service was a fire service charged with protecting forests and communities. How it judged itself was also how the public judged it. To call a fire out and have it flare up again bespoke a slovenliness akin to desertion; it was cause for dismissal. Fire control could make or break a forest, a career, or an agency. Looking back in 2010, Dale Bosworth said simply, "You work for the Forest Service, you fight fire."[4]

Fire control could shape a life just as it did the agency. Dale began his Forest Service tenure at Avery, Idaho, close to ground zero of the Big

Blowup. Wherever he subsequently went—to the Powell District on the Lolo National Forest, to the Flathead—he witnessed the lingering vestiges of 1910 and saw equally the scars it left on policy and attitude. Beyond the Great Fires, fire and its most storied landscapes accompanied his time in the Forest Service, as they had his father's. Dale was regional forester for the Northern Rockies when the 2000 firestorm rumbled across the mountains in eerie echo of the 1910 Big Burn. The next year he became chief forester amid an era that featured megafires and budget-crushing suppression costs. When the Forest Service celebrated its centennial in 2005, he looked back and not only saw but knew in his bones how fire had molded the agency. By then it was sucking away over 40 percent of the agency's regular budget. By then, too, Gary Snyder had reversed his belief that in "guarding against forest fires" he had found "Right Occupation." The joke instead was on him "fifty years later as I learn how much the fire suppression ideology was wrong-headed and how it has contributed to our current problems."[5]

Over the course of Dale's career, Forest Service policy toward fire had inverted. The traditional agency response had always been to meet threats with force; the greater the threat, the greater the force. Now fire's absence had become as ecologically significant as its presence. As chief (2001–7), he had declared as an ideal that "we'll light fires where we can and fight them where we must." But in whatever avatar it assumed, fire and the need to control it had always been present, the chisel and hammer that had sculpted the agency as no other task had. When he retired in 2007, with his son a third-generation forest supervisor who had also begun as a journeyman in fire, Dale Bosworth left with the belief that fire would define the Forest Service into the 21st century.

* * *

In 1960 the U.S. Forest Service was the premier wildland fire agency on Earth. Almost no aspect of the American fire scene, save its cities, was unaffected by the agency, and through it virtually every facet was connected to every other.

The national forests were themselves a miniature of America's countryside. They contained sand scrub pine on the coastal plains, hardwoods in the Green Mountains, grasslands in the northern plains, mixed-oak woodlands in Arkansas and Missouri, the recovering cutover pineries of the Lake States, the sky islands of the Southwest and Great Basin, the big timber of the Northwest, the mixed-conifer forests of the Sierra Nevada, and the crinkly chaparral of Southern California. The agency administered both boreal taiga and tropical rainforest and everything in between. When, in 1959, the nation had gotten bigger by reaching across North America to incorporate Alaska as the 49th state and then halfway across the Pacific to absorb Hawaii as the 50th, the Forest Service was already present in both areas. It even sponsored a forestry experiment station in Puerto Rico. Under regulation, those lands were logged, grazed, dammed, mined, researched, camped in, skied on, and hiked over. They held summer cottages and lodges. They had tree farms and primitive areas. They permitted rights-of-way for roads and transmission lines. They were a parallel, public version of the private landscapes of America. In 1960, in addition to 30,000 timber contracts and 17,000 grazing permits, a survey identified 56,000 special-use permits for 110 purposes—all of which, in some fashion, the agency reconciled.[6]

Each use had its partisans, and in some cases, its poets. In 1960 Wallace Stegner, seconded temporarily from Stanford University to the Department of the Interior, wrote his celebrated wilderness letter in which he argued for the preserved wild as part of America's "geography of hope." Stegner's appeal was a harbinger of what was to come. As social and economic demands intensified, as the agency shifted from a primarily custodial role to one producing market commodities, as the contrasts sharpened between clear-cut slopes and pristine backcountry, and as population and housing booms threatened to squeeze land and resources, the Forest Service sought a firmer legislative mandate to justify its complexly stirred cauldron of uses. In 1960 it got its wish with passage of the Multiple Use-Sustained Yield Act, which essentially codified what the agency was doing and wished to do.

The country seemed big enough to accommodate both wild and working landscapes along with others that its citizens might imagine, and the carefully calibrated mechanisms by which the Forest Service balanced local and national needs seemed adequate to assimilate those needs. The agency could take on additional duties and bring newly purposed lands under its direction, much as the country was doing with its freshly admitted states.

* * *

Always there was fire. In 1960 Herbert Kaufman published a classic study of administrative behavior, *The Forest Ranger*. He selected five of the agency's 792 districts for special attention, districts distributed among Virginia, South Carolina, Michigan, Colorado, and the Pacific Northwest. None were noted fire forests, yet when Kaufman described the routine duties of his subjects, he devoted eight and a half pages to fire, four to timber, two to recreation, one to range, and one and a half to all the others. Whatever the newly enacted Multiple-Use Act stated, the practical reality was fire control.[7]

As much as fire protection shaped the Forest Service, so the Forest Service shaped the national system. Over the years, fire had become a monopoly of the state, something government sought to control in the name of public safety. Fire wars, like World War II and Korea and the Cold War with its looming nukes, reinforced the perception that fire was hostile and threatening. Yet whatever the danger, all eyes looked to the U.S. Forest Service. With its vast estate, its routine experience with fire, its nearly exclusive mastery of fire research, its web of interagency agreements, and its capacity to actually fight fire, the Forest Service easily became the responsible agency for coping with the nation's fire problems and threats. By the early 1960s an ever-expanding legislative mandate had made it the indispensable institution, the agency on fire, the prime mover that turned the gears of all the others whose missions might somehow intersect with fire.

With 193 million acres under its jurisdiction, and more added from time to time, the U.S. Forest Service was the nation's largest landholder until Alaskan statehood tipped the scales toward the Department of the Interior's Bureau of Land Management. The bulk of USFS holdings were in the West, but even where it was a secondary presence, it influenced the way fire appeared (or did not) on the scene. In 1911, following the Big Blowup, Congress had enacted the Weeks Act, which allowed for the Forest Service to expand the national forests by purchase and to encourage the states to join by a program of grants. The new forests rose out of the stumps and ashes of abused and abandoned lands. Those grants-in-aid targeted fire control. In 1924 the Clarke-McNary Act expanded the range of cooperation. The 1949 Northeastern States Fire Compact, an aftershock of the disastrous 1947 fires, offered another interstate model. The 1955 Reciprocal Protection Act authorized the Forest Service to enter into agreements with *any* fire organization for "mutual aid, with or without reimbursement," and allowed for "emergency assistance in the vicinity of agency facilities in extinguishing fire when no agreement exists." Beginning in 1956 special fire conferences targeted the southern states, rallying political enthusiasm for cooperative fire. In 1959 a conference in Lexington encouraged Kentucky to fund the program its legislature had passed the year before, which would extend coverage to 50 of the state's 120 counties, "about one-tenth of all the remaining unprotected forest land in the United States." By 1960 the national program, a complex mutual-aid pact, was nearly complete; the last state, Arizona, joined in 1965.[8]

The Forest Service's Division of State and Private Forestry became a conduit for promulgating common policies, equipment, knowledge, and standards to state foresters who in turn typically became fire wardens for their states. Most of the nation's fires burned on private land or on lands under county or state jurisdiction; overwhelmingly, they burned in places not incorporated into any formal protection system. A federal and state condominium sought to reconcile national goals and monies with local needs and peculiarities. Federal assistance, channeled through the Forest Service, allowed something like national standards to spread without direct federal control. Large as the federal estate might be, private and state lands were larger, and it was the states that steadily quenched fire out of America's vernacular landscapes. America's fire protection empire, like much of Britain's imperium, operated through indirect rule.

In 1942, through the Clarke-McNary program and as a wartime measure, the secretary of agriculture assigned rural fire responsibilities to state and federal agencies. Such programs often continued in the postwar era, at first as an expedient and then as part of an emerging national security state. The Federal Excess Property Program sent war-surplus tractors, trucks, planes, jeeps, and even half-tracks to state forestry bureaus. The Federal Civil Defense Act of 1950 built on Clarke-McNary to furnish a template for rural fire protection under the nominal direction of the Forest Service as civilian defense moved from a wartime emergency to a permanent feature of American life. In 1955 Congress authorized the Reciprocal Fire Protection Act, which expanded assistance beyond the states to any entity that needed it during an emergency. By 1958 these exercises had led to a National Plan for Civil Defense and Defense Mobilization and would subsequently underwrite programs to assist rural communities and volunteer fire departments.[9]

When the UN Food and Agriculture Organization sought to promote forest fire protection, its texts were written by former fire officers from the Forest Service. Even the national fire prevention campaign, symbolized by Smokey Bear, operated under the cooperative fire aegis erected by Clarke-McNary (the industry counterpart, Keep America Green, was a pale shadow). The future built on the past: the more responsibilities the Forest Service acquired, the more likely it was to receive more. The Forest Service did not seek out these new tasks; they fell to the agency in the way a stone rolls downhill. There was no one else remotely competent enough to do them: no other agency had its reach or firepower.[10]

Its strength did not derive from control over lands alone. Through its warehouses and engineering centers, the USFS designed and stored most of the hardware needed to fight fire on the ground. To the extent that the equipment was unique to fire — not just a shovel but one modified to scrape line, not just an ax but a dual-purpose pulaski, not a simple rake but a duff-ready McLeod—the USFS had developed those tools, established their manufacturing specifications, and ordered them through the federal supply service. After World War II it assisted the mass transfer of war-surplus goods to the fireline, most spectacularly aircraft. As early as 1946, helicopters were flying over California firelines. Air tankers were dropping fire-retarding borate during the 1956 Malibu and Inaja fires. Once that barrier was broken, adaptations cascaded. By 1960 a national fleet was aloft—mostly converted B-25s, TBM Avengers, and PBY Catalinas (all former military aircraft). Some of this hardware was owned by agencies, and some was contracted from private sources, but for nearly all, the Forest Service forged the templates for agreements and through mutual-aid pacts oversaw distribution throughout the country. Almost no company made firefighting tools for the general market, and very few sold the handful that did exist.[11]

What was true for tools was no less true of the men who used them. They were, on the public lands, overwhelmingly public servants. The institutional demographics loosely resembled an army, with a few career officers amid an annual churn of employees who worked for the fire season and temporaries conscripted during campaign fires. Some fire officers were professionals with college degrees and were headed to the ranger corps, but most came up through the ranks. The best of them had acquired an extraordinary if highly personal knowledge of how fire behaved in their locales. Almost everyone in the organization, however, had some fire experience. For many, a season or two smokechasing or working an engine crew had been their point of entry; the seasonal labor pool was a place of probation and a source for recruitment into the ranks of permanent staff. A district's personnel were a reserve force—widely referred to as a "militia"—ready for callout whenever fires struck. Those mass callouts were held together according to a Large Fire Organization, whose principles the Forest Service had hammered out over tough decades.

Fire crews were at heart manual laborers. When they were not suppressing fires or cleaning up afterward, they built fences, thinned timber stands, cleared trails, and stacked slash. They had minimum training: a two- or three-day fire school taught the basics of handtools and fireline construction. They learned on the job. They were furnished hardhats and by the early 1960s also orange fire shirts of flame-resistant Nomex. They were expected to supply their own gloves and boots; most wore denim jeans. Equipment for ear and eye protection was unknown. They often lived on duty for five and a half days a week in remote camps, the relics of the Civilian Conservation Corps (CCC) years. They chased down smokes, built handline where dozers could not go, and manned engines where roads were available. They burned out fuels and mopped up fires. For all the flushes of adrenaline, it was a life of daily hard labor based on stamina and a strong back. Fire camps were kept small (under 300 men) and lean on the theory that no one would want to stay in them longer than necessary. While hardened fire officers were respected for their experience, theirs was not a realm of arcane knowledge. The first national course for fire officers was held in 1958; the first fire behavior course launched in 1961. A national fire training center opened in 1962.

Dedicated fire crews sprang up in places with heavy fire loads or in regions such as California, where fire season moved north and south, or in places where big fires could suddenly overwhelm local capacities. These were gangs that could travel to campaign fires where they were most needed. The idea was formalized in 1939 with the appearance of a 40-man crew on the Siskiyou National Forest, a crew built of field-tested enrollees from CCC camps. The experiment dissolved during the war and then was revived piecemeal, particularly in Southern California, as self-proclaimed hotshot crews. In 1961 the program expanded to involve crews that could cross Forest Service regional borders, thus birthing the interregional fire suppression crew. Where large labor was needed, corps of seasonally available manpower were cultivated, notably among American Indian and Hispanic labor pools; the Southwest Forest Fire Fighters became the most celebrated. Some states (especially California) also used crews of prison inmates as surrogates for the departed CCC. If matters became more desperate, the military could be called in, beginning with the National Guard. To oversee such gangs, the practice evolved of dispatching experienced men to big fires to serve as "overhead," a mobile officer corps that could be mustered when needed but otherwise tended to their regular duties. Some were full-time fire officers; most had acquired fire experience as their careers matured, but they served in timber or range or as members of the managerial class. All in all the fire community was remarkably self-contained; it served the public without involving the public.[12]

Over the course of 50 years following the Big Blowup, the American fire scene under Forest Service auspices had fashioned what was in effect a fire commons. What began as an exercise in mutual aid and coordination became, as the federal largesse swelled and as alternatives faded away, a homogenization of practice under a uniform policy. What first united the agencies was the fact that every place had its bad fires, the unwanted fires that had to be beaten back as quickly as possible, and for these, people needed help from neighbors. The sheer bulk and zealotry of the USFS placed the agency first among equals, and over time, ad hoc alliances gelled into formal institutions. As far back as 1928, the Coolidge Administration, in a bid for efficiency, had established a Forest Protection Board to coordinate among the federal agencies with responsibilities for forested lands. The Forest Service chaired and dominated the group.[13]

Then in 1935, amid the abundance of money, men, and attention granted by the Roosevelt administration, Chief Forester Gus Silcox announced what became known as the Forester's Policy of fire control by the first burning period, which was understood to run from 10 a.m. to 6 p.m. The 10 a.m. policy was promulgated both as an "experiment on a continental scale" and as a universal standard. It was based strictly on fire behavior, on the simple recognition that the sooner a fire was contained, the smaller, less damaging, and less costly it would be. The 10 a.m. policy ignored values at risk just as the Social Security Act, also passed that year, ignored varied needs. A bold bid to break the back of America's fire problem, the decree was also a preposterous goal that assumed the same result from a fire started 100 miles away as a fire started 100 yards from a ranger station, and it was conceivable only amid the cornucopia of New Deal programs—especially the CCC, which fielded a civilian army of firefighters. Few districts could satisfy that goal, let alone the agencies outside the Forest Service (and some never really tried). But the ambition was unequivocal: to stop wildfire.[14]

The fact was, not until means expanded could ends enlarge in a practical way. Not until the New Deal did wildland fire control receive the political will and practical resources to cover the national forests and parks; not until World War II did it truly nationalize in prevention and purpose; and not until the postwar era (as wartime improvisations hardened into a national security state) did fire protection acquire the equipment and organization that allowed it to flourish. By 1960 all the ingredients had alloyed into a fight against the "red menace." What had begun as an obsessive preoccupation by America's founding corps of foresters had morphed into a collective Cold War on fire.

* * *

It was not enough to replace tools forged in backyards with military-grade hardware. The same had to happen with knowledge. So in 1960, in a bold commitment to supplement or replace the personal lore, the Forest Service opened the Northern Forest Fire Laboratory in Missoula, Montana.[15]

The Forest Service had long committed itself to formal research. In 1916 it established a branch of research equal in rank (if not in heft) to the division of State and Private Forestry and to the National Forest System. William

Greeley declared that firefighting was as much a matter of scientific investigation as silviculture was. In 1928 the McSweeney-McNary Act granted the Forest Service an effective monopoly of government-sponsored fire research: what public-land administrators wanted to know about fire they would get from the scientific arm of the USFS. Gradually, research stations and experimental forests sponsored the research their regions required. Almost all dealt with fire's rate of spread or, from the early 1930s, with fire-danger rating. These ratings were triggers for emergency funds to bolster presuppression activity, and they promised a common standard for evaluating performance. The first quantitative version appeared in 1934, and its apparent success surely fed into the adoption of the 10 a.m. policy the next year.

Research lagged during the war, but the wartime mobilization of science showed persuasively how science could advance a fight, and in the postwar era fire research revived, boosted by the ambitions of a national security state in the Cold War that followed. Because postwar surveys of Germany and Japan suggested that the next war would be a fire war, the Pentagon wanted research to identify how such fires might start and how to stop them; because the combatants had atomic weapons, field tests focused on how blast and heat could kindle firestorms; and because the Forest Service was practically the only entity that had experience with large fires, its scientists assisted with the design and execution of those trials. The experience boosted the morale and standing of the agency's researchers while it compelled them to talk with chemists, physicists, engineers, and meteorologists in ways they never had before. Although foresters began as fire's field empiricists, as with their involvement in nuclear tests, so they found themselves entrained into the plume of theory and models that followed.

The Forest Service reorganized its fire research program. The Department of Defense, and after 1950 the Office of Civil Defense, funneled money into programs investigating the behavior of large fires, a critical catalyst. The National Academy of Sciences, National Research Council sponsored a Committee on Fire Research. It held its first "correlation conference," which included the USFS, in 1956. The next year the National Fire Protection Association published a survey that sought to tally the "facilities, personnel and management of agencies engaged in research on fire." In a small way, big science had come to fire.

The reforms seemed larger than they were because research on wildland fire was so tiny. Overwhelmingly, combustion science had focused on urban sites and industrial themes, not on the countryside beyond. Instead, wildland fire had evolved three approaches. One sought to formalize field experience, abstracting the hard-won knowledge of fire officers—an approach aptly characterized by the career of Harry Gisborne in the Northern Rockies. The second focused on statistical analysis of the thousands of fire reports, supplemented by select field trials, an exercise in operations research; the work of S. B. Show and E. I. Kotok in California was the exemplar. The third applied the techniques of physical science to fire phenomenology, of which the efforts of William Fons in California and especially George Byram of the Southeast Experiment Station in Asheville, North Carolina, stand almost alone. All three perspectives shared a common goal: to improve the rapid detection and control of wildfire by understanding how fires started and spread.

By 1959 the embryonic science of wildland fire was ready for consolidation. Appropriately, it came as a triangle. At one apex stood dedicated forest fire labs. With support from the Georgia Forest Research Council and Georgia Forestry Commission, the Forest Service began the process with a dedicated lab in Macon in November 1959. A second lab opened in Missoula in 1960, and a third was established in 1963 in Riverside, California. A second apex was the decision to scrap the eight regional fire-danger rating indexes in favor of a single national system. Two years later an algorithm for fire spread was ready for field testing. The third apex was the publication, for the first time, of a college-level text. Written by Kenneth Davis of the University of Michigan, *Forest Fire: Control and Use* summarized the state of formal learning.

A fundamental fact accented by the publication of Davis's text was that fire belonged within forestry. Davis, a professor of forest management, authored most of the book, McGraw-Hill published it as a contribution to its American Forestry Series, and it was dedicated "to the men of the United States Forest Service, who through pioneering experience and creative research have formed the art and science of forest fire control." Forest fire, Davis observed, "is the last major area of forestry to receive full book-length treatment." Instead, "much of what has been written on fire control exists in manuals, handbooks, and administrative guides prepared by control organizations, for their own use." The material basis for something like a technical field emerged only in the postwar era, and most of it within the previous decade. George Byram wrote two chapters on combustion and fire behavior. W. R. Krumm of the Weather Bureau wrote one on fire meteorology, but he had worked closely with the Forest Service at Missoula, a center for aerial fire control. Keith Arnold, an agency impresario of big-science fire research (and an architect of Operation FireStop in 1954), had originally signed on as coauthor until other duties with the agency forced him to withdraw. Even university science relied on the Forest Service.

The sole academic entities committed to understanding free-burning fire were departments of forestry and range science. While Yale had an old forestry school, and Harvard was bequeathed a donated forest at Petersham, research outside the USFS was conducted at land-grant universities, and almost all the funding came through the Forest Service. In knowledge, as in equipment and crews, the Forest Service ran what outsiders might rightly regard as a cartel.[16]

* * *

If there was collusion within the fire community, it was a conspiracy by voluntary conformity. All practitioners confronted the same problem: wildfire. Much of what unified the American fire scene did not happen by decree or outright manipulation of funds but by a shared sense of collegial values and a perception of how fire fit into the world. Overwhelmingly, fire's practitioners belonged to the fraternity of foresters. When the World Forestry Congress in 1960 staged its fifth meeting (in Seattle), the U.S. Forest Service hosted the event. President Dwight D. Eisenhower delivered the welcoming address.

Forestry mattered, and as an academic discipline and expression of political economics, forestry had long feared and detested fire. A revealing fact is that forestry had matured in what, from the perspective of fire, was an anomalous patch of Earth. The Mediterranean has a cadence of wetting and drying as a defining trait, and the boreal can undergo long waves of drought from time to time, but central Europe knows neither, and so it has no natural basis for fire. To its foresters, landscape fire was something people made, and many people made it in ways that threatened the ambitions of foresters. Over the centuries, herders had burned to freshen pasture, farmers had burned to clear woods and fertilize soil, hunters had burned to improve habitat for wildlife, and miners had burned to strip away ground cover. But forestry could thrive only if its trees were not swept away in an afternoon's flames or if the sites it planted with conifers were not converted by fire to pastures or fields. Controlling fire not only prevented direct damage to woods, it was also a means to regulate those peoples who sought other purposes for those lands. Prussians planted strips of pine like row crops. Colonial British foresters spoke, as an ideal, of felling the wild woods and replanting the land with tree farms.

While that would never happen in America, because most of its landscapes were fire prone and because the human population had long been accustomed to the free use of fire, America's foresters still had Europe's pyrophobia hardwired into their training, and it was that training that justified calling their guild a profession. Forestry had its own society, publications, public-outreach projects, political organs, origin stories, and code of ethics. Yet it saw itself as more than an administrative exercise in applied science and its agents as more than proconsuls for America's western empire. Foresters implicitly claimed special standing as moral philosophers of land use, much like economists in more recent times. They were even permitted to audit themselves. When its crews died in fires, the Forest Service fielded its own boards of review.

The size of Forest Service holdings—10 percent of the national acreage—gave forestry an exaggerated voice in deciding how to manage the country's landed estate. Foresters declared themselves uniquely positioned to exercise judgment over the proper use of lands and especially of the public domain. Their shared values reinforced a culture of conformity within the Forest Service and between the USFS and other federal and state agencies and among private owners of industrial forest lands. No matter their location, foresters became a community of like-minded practitioners who did not need to be told what the common interest was because they had already internalized it. In principle, forestry reached across agencies and countries. Even the National Park Service relied on its in-house foresters for fire knowledge.

Since the excesses of the Gilded Age, it had been the dream of reformers to have the unpopulated lands of America settled in a rational way under the direction of a disinterested corps of engineers who could make decisions without the anarchy of laissez-faire capitalism or the unseemly and chaotic politics of an adolescent democracy. The U.S. Forest Service seemed to embody that ideal of public administration, or so the agency and its dispersed allies believed.

* * *

Not everyone agreed. Even in 1960, America's pyrogeography resembled less a melting pot than a magma chamber chocked with great lumps of unassimilated country rock. On fire policy, in particular, there were dissenters—renegade regions, apostate foresters, prophets pointing to ecological perdition and calling down prescribed fire in righteous wrath. There were places not yet broken to fire suppression. And there were enthusiasts—zealots to officials—who denied the dogmas that underwrote the homogenization of America's firescapes. However much a map might simplify that texture under a common hue of Forest Service green, the grounded reality was a jumble of ownerships full of quirks and anomalies.

What divided the various communities concerned with fire was a pair of very old questions: Was fire exclusion possible, and if it was possible, was it desirable? The dispute—expressed in pretty much those exact words—dated back to the origins of European colonial forestry. When it was first raised in 1870s India, local practitioners argued that fire was inevitable, useful, and necessary, while colonial administrators, academics, and foresters accepted as an article of faith that rational land use was impossible so long as folk burners roamed the landscape. You could not control fire without controlling people, and you could not control people if you did not control fire. In India, South and East Africa, Cyprus, Algeria, Australia, and Canada—over and over the dispute raged like an epizoic fever when foresters advanced the cause for state-sponsored conservation, and everywhere, officialdom decided for its officers. But the issue would not go away. To foresters it was an administrative zombie, the controversy that would not die. Critics claimed fire was inevitable, useful, and necessary. So long as practitioners continued to apply fire, however isolated the enclaves or tiny the band of partisans, the dispute smoldered on, occasionally flaring into public view.

The fever came to the United States in August 1910. Even as the Big Blowup was ripping, a controversy over "light-burning" flared in California. Light-burners argued explicitly against Forest Service fire policy, which they asserted would lead to a buildup of fuels, larger wildfires, forests degraded by insects and fungi, and generally unusable landscapes. They campaigned for the "Indian way" of forest management—as adopted by western settlers, stockmen, and some timber owners—in which routine surface burns cleared away the surface clutter. Proponents were an unlikely mob that included engineers such as George Hoxie, poets and novelists including Joachim Miller and Steward White, and large landowners like T. B. Walker. Unlike foresters, they had no academic discipline to infuse intellectual authority into their case, and unlike the Forest Service, they did not speak with the prerogative of the state. They argued on practical knowledge, what they saw with their own eyes. All understood that the controversy had a deeply political subtext, the legitimacy of government administration of a public domain for the common good—according to the dictates of formal learning. In January 1910 Gifford Pinchot, patriarch of the Forest Service, staged a fight with Secretary of the Interior Richard Ballinger and found himself fired as a result. Then, amid the Big Blowup, Ballinger publicly sided with the light-burners.[17]

For the next dozen years the Forest Service fought bitterly against them. They dismissed the practice as "Paiute forestry," beneath the dignity of an aspiring great power and Enlightened society. They countered casual light-burning with "systematic fire protection," and they promoted model forests and conducted sporadic field trials. All of this showed, as the agency saw it, that fire protection was possible, economical, and vital. Systematic fire protection underpinned not only forest regeneration, thus forestalling an imminent "timber famine," but also the doctrine of state-sponsored conservation. Right-thinking agencies would have to protect society from itself. In 1920 Aldo Leopold, then a ranger in the Southwest, declared that the light-burning propaganda "directly threatened" the whole edifice of Forest Service–led reforms. In 1923 a special commission convened by the California Board of Forestry formally condemned light-burning. Light-burners belonged with circle-squarers and perpetual-motion mechanics.[18]

In the Southeast, however, resistance persisted, and "woodsburning" joined the region's other peculiar institutions. To outsiders the practice seemed as bad for the land as the slovenly hillside plowing that led to unchecked erosion. It was deeply embedded in the culture, passed from generation to generation, but it was no more justified than the region's entrenched racial segregation. Superstition and mindless rituals were countered with science and demonstration plots. Getting fire right mattered economically because the cutover and burned-over landscapes needed to reforest, and it mattered politically because fire protection was the linchpin of cooperative forestry between the national government and the states. Without federal grants-in-aid for fire control, state forestry bureaus would quickly wither.

But the experiments failed. The demonstration plots seemed to show that fire exclusion actually worsened the woods, and when fires did burn, they were uncontrollable and far more damaging in the fire-exclusion plots. The Forest Service stalled publication of the results, partly because the outcomes were so counterintuitive; the results defied centuries of central European learning. Like novice students told to redo an experiment until they got the right result, the trials were repeated, and the results were kept in file cabinets. They were treated as if probably untrue, but if true, then dangerous.

Rather than improving, the southern firescape was deteriorating. A handful of university foresters, such as H. H. Chapman of Yale, and Forest Service rangers, including Inman Eldridge of the Florida National Forest, began arguing for some accommodation with burning. They were joined by range scientists including S. W. Greene, wildlife biologists such as Herbert Stoddard, and timber companies keen to restock abandoned land as pine plantations. All dismissed forestry's appeals to authority in favor of tradition and what they observed for themselves. While some of their articles were published in national journals, most struggled to pass the establishment's editorial gatekeepers.

In 1935, while Chapman was president of the Society of American Foresters, he organized a panel for the annual meeting on "Forest Fire Control in the Coastal Plains Section of the South." A few months earlier Chief Forester Gus Silcox had promulgated the 10 a.m. policy, and the panel granted proponents of controlled burning a national platform from which to broadcast their criticisms. After the panel concluded, E. V. Komarek growled that "this is the first time that censorship on the subject has been removed and we have been told the facts." In 1942 Raymond Conarro of the Forest Service's Southeast experiment station proposed the term *prescribed burning* as a pragmatic compromise between fire's competing fundamentalists—those who demanded either exclusion or unregulated burning. Two years earlier W. J. Cash had written in his classic *The Mind of the South* that the region was "not quite a nation within a nation, but the next thing to it." That was as true for fire as it was for the South's racial caste system.[19]

If the Southeast remained the rebel region, Florida was its most recalcitrant core. Amid a biota that, as Ron Myers once put it, was shaped by how deep the water and how frequent the fire, the only question was whether the fires were wild or tamed. The only way to abolish flame was to convert the landscape to citrus orchards or shopping malls. Here government adapted to fire. In 1932, south of Tallahassee, St. Marks became the first national wildlife refuge to accept burning. In 1943 the first national forest, the Florida, yielded. In 1954 the first national park, Everglades, embraced burning. Despite misgivings, the Florida Division of Forestry converted as well. Similar waivers spread slowly through the region, but the exceptions did not overturn national policy; they were compromises reluctantly made and vigorously quarantined. Still, they gave prescribed fire a place of refuge with the potential to become a point of propagation.[20]

There were other problem regions. Ranchers in the Flint Hills slashing across Kansas and Oklahoma continued traditional burning to promote pasture. Prairie enthusiasts such as John Curtis at the University of Wisconsin Arboretum understood that without burning they could not restore native grasses and forbs. On many Indian reservations, fires flourished because there was scant means to stop them, and on a few reservations, they became a deliberate policy. Harold Weaver of the Bureau of Indian Affairs (BIA) cultivated prescribed fire on the Warm Springs Reservation in Oregon, and when he was transferred to Arizona, he promoted it on the Hualapai and Fort Apache reservations. His successor, Harry Kallander, made the pineries of the White Mountain Apaches a center for controlled burning largely in the indirect service of logging. Officially, the BIA remained wary and kept some distance from what was happening on the ground;

unofficially, perhaps 200,000 acres were burned. In the western United States, the Arizona reservations were, for a couple of decades, the premier practitioners of prescribed fire. The western yellow pine, ponderosa, did for the Southwest what the southern yellow pine, longleaf, had done for the Southeast. In California, Harold Biswell, a range ecologist formerly of the Forest Service and later a professor at the University of California, Berkeley, sought to transfer the lessons of prescribed burning from the Deep South to the Far West. He found enthusiasts among ranchers and eventually amid the parks of the Sierras, but organized forestry opposed his campaign.[21]

There were yet plenty of places without organized fire protection at all or with subpar capacity (Clarke-McNary took decades to complete its build-out). There were places with novel fire problems, or old problems granted new vigor, as postwar America began bulldozing over its once-rural countryside with suburban sprawl. California's population boom made that state as much a poster child for a new form of suppression as Florida was for burning. (In 1959 the Angeles National Forest had suffered through two fatality fires, the Decker and the Woodwardia, the latter at $1.4 million the most costly in USFS history up till that time.) Much of the Great Basin barely had any organized fire protection. And then there was Alaska—two and a half times the size of Texas and prone to off-the-charts spasms of breakout fires. In 1957 the U.S. Forest Service dispatched Charles Hardy of the Intermountain Experiment Station to investigate reports of conflagrations in what would soon become a state (and was recognized as a frontier in the Cold War). He estimated some five million acres had burned that summer in the interior. While the Forest Service held Alaska lands, they looped along the southeast coast; what happened in the interior was the responsibility of the Bureau of Land Management (BLM), then only a decade old and barely able to field crews or muster an engine in the Lower 48. Although the BLM was keen to mount an aerial fire program complete with smokejumpers in the valley of the Yukon River, skeptics wondered whether this was a flying version of the Alaska Fire Control's origins in which a railroad speeder had made weekly treks between Fairbanks and Anchorage. In brief, a national system was a long way from complete, and what existed had its eccentricities and anomalies.[22]

As the experiences of both Florida and the BLM suggest, even among the federal agencies the system was far from robust or even uniform. The BLM had yet to become a significant presence on the American fire scene, which left 253 million acres (one-eighth of the national landmass) with marginal protection. Perhaps the more interesting case was the National Park Service (NPS), which held much less land but had sites of inestimable public interest. The NPS was feudal in its politics—an archipelago of baronies, each authorized by a separate act of Congress for parks and by autonomous presidential decrees for monuments, held mostly by shared purpose and a common ranger corps. Its crown jewels, such as Yellowstone, Glacier, and Grand Canyon, had staff foresters and fire-protection programs. Something like a national program had begun in 1928 when a fire-toughened ranger, John Coffman, transferred from the Forest Service and brought USFS methods with him. Many NPS units, however, were too small or remote to outfit regular fire crews, and not a few superintendents regarded fire as more of a nuisance than a duty; some cash-strapped chief rangers wondered why fires were fought at all. In 1960 the entire NPS system had fewer than six dedicated fire officers on its payroll.[23]

The dissenters loom larger in retrospect than they seemed at the time. Most revolutions fail; their firebrands die out and survive only as printed ashes in archives. The Stoddards, the Komareks, the Weavers, the Kallandars, and the Biswells have branded themselves into history because the fire revolution they kindled not only survived but remade the pyrogeography of America. In 1960, however, they were but a sideshow. In his college text, Kenneth Davis included 2 chapters (of 20) on the "application" of fire. Like fire control, fire use was embedded in forestry; its primary purpose was to burn off logging slash. The national forests were rapidly moving from a custodial role to one of commodity production; most of those suburban houses that were rolling across the countryside were wood framed, and the effluent of industrial logging was becoming a serious menace. The purpose of controlled fire was to assist fire control.

* * *

In 1960 the smart money held that the Forest Service would continue to expand its dominion. It would bind together, if not oversee outright, the creation of a national infrastructure for fire protection. It would balance regional anomalies with national goals. It would make course corrections when necessary, as it had with fire research and prescribed burning, but it would reconcile local fire needs as it had its many districts. It remained the gold standard of fire protection: no fire agency on the planet had more experience or firepower. None so instilled its mission into its staff. The two books published about the service that year celebrated just that special élan.

The agency was, as Herbert Kaufman's book *The Forest Ranger* announced, a paragon of public administration—a model of American federalism, a model of civil service, and a model of probity. "By any criterion," Kaufman concluded, the Forest Service had "vanquished the tendencies toward fragmentation." It had achieved that goal by many means, but ultimately by recruiting and retaining a corps of men with remarkably shared persuasions and similar personalities. "One of the most striking conclusions about the Forest Service is the degree of similarity among the men in it—their love of outdoor life; their pride in the Forest Service; their habit of taking the long view of things; their patience; their acknowledgment of their obligations to the local users of the national forests; their acceptance of the inevitability of conflict growing out of differences among the many users of the national forests, and between the national interest as against local or special interest; their enjoyment of the variety in Ranger district administration as compared with the narrower scope of industrial forestry; their willingness to do more than is legally required of them in order to get their jobs done."[24]

What this syrupy broth boiled down to was a hard crystal of "voluntary conformity." "To overstate the case," ranger decisions are "predetermined. From the Rangers' point of view, they are not obeying orders or responding to cues when they take action on their districts; they are exercising their own initiative. It is not compulsion or inducement or persuasion that moves them; it is their own wills. Speaking figuratively, it would not occur to them that there is any other proper way to run their areas." They did freely what they had to do. So with the Forest Service overall: it did instinctively what the country wanted; its lands and personnel had internalized the values and wishes of the nation, a melting pot of people and places. The Forest Service had accomplished what few organizations had. In the process it became, literally, a textbook exemplar of competent administration in the public service.[25]

The second book, Michael Frome's *Whose Woods These Are*, showed how that sensibility played out in the field. In his grand circuit of the national forests, Frome devoted special attention to "a fiery western weekend." The chapter followed an obligatory summary of the Great Fires of 1910 and was intended to show both the persistent problem of fire and the astonishing measures taken to contain it. The numbers seemed to speak for themselves. Some 6,000 fires burned, 25 of which exceeded 3,000 acres ("each the size of the Gettysburg Battlefield"); the largest bolted to over 35,000 acres. By the end of July, fires had romped over California, Oregon, Washington, Idaho, and Montana. The climax came toward the end of July, in which over the course of a fortnight, a quarter of a million acres burned while the Forest Service mobilized a force of 25,000 men and "together with co-operating states, cities and private landowners, spent $1,000,000 a day in firefighting." Here was the "same old grim war of the Western states," but now its soldiers had "modern means" to fight back, a civilian "army of infantry, artillery and air force." They called in hotshot crews ("the commandos of fire fighting"), American Indian "shock troops," 400 tractors and bulldozers, 600 engines and water tenders, and an "air force of 300 planes and helicopters."[26]

That was the national reckoning. In order to root those numbers in narrative, Frome followed the life cycle of the Saddle Mountain fire on the Bitterroot National Forest. It began as a lightning strike near Lost Trail Pass in beetle-killed lodgepole pine and found a patch of cheatgrass. District Ranger John Hall took two men and drove a pickup loaded "with fire packs, power saw and radio" in the direction of the smoke. They got within two miles before stalling at a rockslide. The fire was two acres in size. Hall ordered a borate bomber, but all were committed elsewhere. By evening some ranchers and highway workers ("old hands in the woods") had joined in the attack, with the fire at about 100 acres. Fire boss Hall ordered a force of 100 men. The weary militia was demobilized. More men were ordered. A large-fire organization took shape, camps and kitchens appeared, and fire crews poured in from around the region, some from the Crow and Cheyenne tribes. On the fourth day Forest Supervisor Harry Anderson could see the smoke from his office "rolling over the mountains and canyons," mushrooming skyward "like the cloud of a small atomic bomb." By the fifth day the fire had blackened 3,000 acres and was battled by 1,600 men—trained firefighters, loggers, ranchers, pickups, and camp flunkies. Bombers were called in. The fireline wended over 21 miles. Then humidity rose, light rain fell, some crews commenced mopup, and others were released from the fire. The Saddle Mountain fire was held to 3,000 acres at a cost of $500,000.[27]

To Michael Frome, the firefight was a triumph: at 3,000 acres the Saddle Mountain burn was a thousand times smaller than the Big Burn of 1910. Here was the Forest Service at its most effective and dramatic. As his national tour circled back to Southern California, where it had begun, Frome contemplated the litany of big fires that had over the past decade rumbled across the nation's forests, ending with the 1959 Refugio Canyon fire on the Los Padres National Forest and the heartrending spectacle of a citizenry, through the Penny Pines Plantations, attempting to replant burned hillsides with trees.[28]

That was the agency in the round. To distill that sensibility further, consider the career of Lynn Biddison, who in 1960 accepted the fire control officer (FCO) position on the Cleveland National Forest. He already had 17 years of on-the-line firefighting experience and was a third-generation Forest Service fire officer. His grandfather had homesteaded near the Angeles National Forest and had became a forest guard at Bouquet Canyon, his father had worked up the ranks through the CCC fire program to become assistant FCO on the Angeles, and Lynn had begun work as a firefighter in 1943 at age 16.[29]

Before he retired, Lynn Biddison had worked in nearly every position in the fire organization. In 1950 he was a crew foreman of the Chilao Hotshots in their second year, and he acceded to superintendent the next year. Then he supervised an inmate crew. He pioneered helijumping, the Southern California equivalent of smokejumping. Later, he was the Region 5 representative to the first national fire-behavior training course in Missoula in 1958. While on the Cleveland, he established the first standing forest-overhead teams, and he himself joined interregional teams. In 1968 he carried the California methods to the Southwest Region as regional fire director.

What he knew he learned early. His bosses were tough, direct, old-school bulls-of-the-woods and extraordinary teachers. "They were firm, they were fair, they knew what they wanted, and they knew their limitations. Their style was, 'This is the way we're going to do it, we will do it right, and do it now.'" You did your job. To illustrate, he recalls the 1952 Meadows fire on a Mt. San Gorgonio ridge at 10,000 feet amid Santa Ana winds. The district ranger pointed the fire out to them, and the Chilao Hotshots hiked in. They remained for 11 days. They had one blanket for every two crewmen, so they dug pits where they could light fires for cooking and sleeping. They had little food. Every few days a pack string would bring in water and rations. It was late October and "cold, cold, cold." They stayed with the burn until it was dead out.

Thirteen years later Biddison returned to California as regional fire director. When mandatory retirement forced him out, he left with the exhortation to return to tried-and-true basics. That meant never having a fire, once contained, escape. It meant instilling a sense of urgency, critiquing actions on every fire regardless of size, and boring in and bearing down on standards, because high goals and hard work sparked pride. It meant "if the fire runs out, DO NOT GIVE UP—back up and start again." Or simply, "FIGHT FIRE AGGRESSIVELY." In 1998 he distilled the lessons of his long career into one simple "fact that never changes: The safest and least costly fires are the ones that receive strong initial attack and are suppressed while still small."[30]

Of course California was special, and those who had come up its fire ranks like Lynn Biddison saw American fire as through a (smoked) glass, darkly. By the early 1960s the state was distorting the national fire establishment as much as its demographics and economy were distorting national trends. Southern California alone siphoned off nearly a quarter of the national forest system budget for fire control. By the end of the century, California would, by an order of magnitude, lead the nation in firefighter fatalities and burned houses. In 1961 Merle S. Lowden, soon to be director of fire control for the Forest Service, admitted the region's fire problems were "very great." Agency inspections, audits, congressional studies, and "investigations by outside sources and research authorities" all confirmed California's unique and destabilizing status. But its power in training a generation of fire officers could hardly be underestimated. For them suppression was not just a doctrine. It was a way of life.[31]

* * *

In 1960 the Forest Service saw itself as administratively decentralized, geographically diverse, intellectually rational, and politically inclusive. It saw itself, that is, as a cameo of the country: it was the best the nation had to offer. There were plenty of observers to agree with that judgment.

The public could be forgiven for identifying the Forest Service with fire. Fire control was, most visibly, what the Forest Service did, and it was through the Forest Service that something like systematic fire protection had spread across the country. In its capacity to set policy, symbolized by general acquiescence to the 10 a.m. policy; in its unmatched control over the apparatus of firefighting, from equipment caches to overhead teams; in its virtual monopoly over research, with two dedicated fire labs and another in the works and its hands on the spigot of fire-science funding; in its defining role among American fire institutions; in its alliance with the guild of foresters, which extended its reach to private industry and other realms of natural resource management; and in its control over the most valuable block of the national estate, the 193 million acres of the national forest system, the Forest Service stood supreme. By 1960 it had come close to completing the agenda scratched out 50 years earlier in the ashes of the Big Blowup.

Not least on the list, the fire organization enjoyed a certain autonomy because of its peculiar funding. The Division of Fire Control held an annual budget that bound it to its host agency, subject to competition with other needs, the wishes of administrators, and the eccentric rhythms of legislatures. But since 1908 the Forest Service had had authorization to overspend that budget to meet fire emergencies, with supplemental appropriations allotted after the season. In 1910 it spent nearly $1 million fighting the Big Blowup. In 1935, following the invention of a credible fire-danger rating system and in an attempt to bolster the 10 a.m. policy, it extended that option to include emergency *pre*suppression. Since no one could foretell what next year's fire season would require, the system had an economic logic behind it. Among its unexpected consequences, however, it rewarded forests with lots of fires—the more fires, the more funding—and it kept fire control outside of normal bureaucratic audits. It was as though fire protection was on a permanent war footing. It could fight fires irrespective of any other agency mission. It could pay for itself. By the late 1920s the Department of the Interior acquired a similar power.

But success like that achieved by the Forest Service can have a price. Such organizations work because they embody national goals and boast a workforce that does willingly what they must, but they can, as Kaufman wrote in *Forest Ranger*, "well become infertile on the one hand and unreceptive on the other." They excel at doing what they've always done and can boldly push their tried-and-true methods into new territory. But they are often unwilling to respond to outside pressure and can be unable to reform internally. To outside observers they appear monolithic. To someone inspecting the inner workings they seem full of fissures held together by a weak nuclear force of conviction and camaraderie.[32]

The special strength of the Forest Service—its ability to integrate a wide range of landscapes and purposes—was also a liability. It had to be all things to all people. It had to absorb all the stresses that would soon splinter the American citizenry by race, ethnicity, gender, and class; that sought to set apart lands for singular purposes for wildlife or wilderness, timber or recreation; that pursued goals well outside the ken of technocratic foresters, whose values increasingly put them at odds with American society and whose fraternal cohesiveness made reform from within troublesome.

The future lay with special-cause politics and single-purpose lands. Institutions that managed for one ambition thrived; those that tried to hold the lid on a bubbling diversity of intentions and interest groups faltered and often boiled over. American society began to refashion the landscape after its own new self-image. The mosaic superseded the melting pot as a national metaphor, and the rebel replaced the company man. In *Whose Woods These Are*, Michael Frome noted that Chief Forester Richard McArdle had jokingly requested that "for once, please don't call us the 'dedicated men of the Forest Service.'" A decade later no one would.[33]

A year after Frome chronicled the fire war on Saddle Mountain, the Cold War stiffened as East Germany raised the Berlin Wall. To observers at the time it seemed that both wars might go on for a thousand years. The contestants appeared monolithic, relentless, and impregnable. Yet each collapsed almost overnight. The reign of Forest Service fire control toppled first; its informing doctrine, which seemed almost a law of history, imploded from within. The consequences of its application were too contradictory and the costs too high. Perhaps most surprisingly, even its administering officialdom, so long unyielding to any challenge, though likely prompted by a new generation eager to seize the torch, abruptly lost faith in its precepts.

Three Fires, 1961

Amid the worst western fire season in decades, three burns stand out. The Harlow fire was a frontcountry conflagration along the Sierra foothills of the sort that had for decades characterized the border between the national forests and rural America. The Sleeping Child burn was a backcountry fire, a classic blowup deep in the Northern Rockies. And the Bel Air–Brentwood conflagration was a novelty that rushed through an upscale early-adopter landscape of urban sprawl. All three outbreaks occurred in regions rich in fire culture and well armed for firefights.

What makes them omens was that together they triangulated the evolving future of American fire. The Harlow was the kind of fire that had plagued the maturing frontier, but it fed on what was becoming a shrinking base of rural lands. The backcountry Sleeping Child burn typified the type of fire that would spread as more and more of the backcountry moved into roadless landscapes, outright wilderness, or habitats for endangered species. The exurban Bel Air–Brentwood conflagration was propagated by the winds of postwar sprawl that would scatter a California contagion across most of the nation. Together the three fires define an emerging American fire scene.

Still, they had to bond fire with American culture. It was easy, for a while, to dismiss blowups like Sleeping Child as a throwback to fires not seen for 30 years or as fluke expressions of western violence akin to a Montana grizzly attack. It seemed equally possible to denounce Bel Air–Brentwood as a flaky event, a California quirk like earthquakes and Esalen. But while in 1961 they might appear as outliers, each grew at the expense of a middle, once-rural scene of the sort that underwrote the Harlow burn. Fifty years later such fringe fires had become a horrible new normal, while wildfire blasting through protected backcountry and vulnerable exurbs embedded itself into the national almanac and a national consciousness.

* * *

In mid-July 1961 the Harlow fire ripped through Mariposa County in the Sierra Nevada foothills, killed two people, consumed 90 homes, and raced from ignition to 18,000 acres in two hours. The burn ultimately scorched 43,000 acres. Cecil Metcalf, deputy state forester for the region, declared that in 35 years of firefighting, he had never encountered a blowup comparable to it.[1]

But the Harlow, and the Basin fire that burned nearby on the Sierra National Forest, were brasher and bigger versions of what broke out nearly every year. The Harlow was a replay of others that had stretched along the Mother Lode gold belt for a century. It echoed in more lethal form the 1958 McGee fire that had burst out of ranching country onto national forest lands to become the nation's largest fire for the year. It was, in a sense, a normal fire on a greater-than-normal scale. But like hurricanes that derive their power from warm oceans and dissipate over land, so fires like Harlow acquire strength from particular sustaining landscapes, and as those rural settings and working landscapes in the West would, over the coming 50 years, shrink, what was typical became increasingly rare, and what was an outlier became a norm.

* * *

The Sleeping Child burn on the Bitterroot National Forest east of Darby, Montana, was a classic backcountry conflagration, a set piece interchangeable with fires a century or a millennium before. It rumbled through swaths of lodgepole pine, themselves undoubtedly the offspring of previous burns. Between 1928 and 1932 an epidemic of mountain beetles had left much of the forest in gaunt and ghostly silhouettes. In 1949 powerful gusts turned the surviving trunks into a kaleidoscope of standing and jackstrawed blowdown. Grass, shrubs, and spotty reproduction bonded those patches into a common fuelbed. The only way to cleanse that landscape—to transmute a lump of indigestible biomass into currency useful to nature's economy—was to burn it. On August 4, 1961, lightning started that alchemical process.[2]

The fire kindled near the Sleeping Child Mine. Miners and USFS crews struggled to contain it while it spread across the high country. Everything favored the fire and frustrated the firefighting. The jackstrawed trunks littering the surface exhausted handtool crews and made cutting fireline laborious and tedious. The fire could travel faster than crews and needed little encouragement to leap over lines. On the second day the weather bureau issued a red-flag warning as a dry cold front, like those that had powered blowups from time immemorial, pushed through the mountains. Winds quickened, turned blustery, carried embers aloft, and blew them thousands of feet ahead of the front—and then flipped from the southwest to the northwest as the front passed. What had been a barely controllable flank of the fire broke into several heads, spilling into basins. What had been 400 acres at the end of the first day became 9,000 acres by the second day. The fire burned for the next two weeks.

The Forest Service threw everything it had into the fight. It mustered 1,721 firefighters, 41 bulldozers, four helicopters, countless engines and portable pumps, and whatever airtankers it could draw in. Governor Donald Nutter pledged the state's help. The commandant of the regional U.S. Army post ordered his men and matériel to assist. As frontal systems ebbed and flowed, the fire quieted and then roused. Recognizing that they could not stand in the blowdown to battle a blowup, fire bosses backed off and let the fire do what it had to. By August 14, the worst had passed; control lines flanked the burn at some 28,000 acres. It was the largest regional fire since the 1930s. Suppression had cost $11 million (nearly $85 million in 2013 dollars), at that time the most expensive in USFS history.

The Forest Service mobilized immediately for postfire operations. It salvage-logged some 90 million board feet. It seeded grasses by the ton. It graded 70 miles of new permanent road and punched through another 130 miles of skid trails and low-use vehicle tracks. And it studied the burn's recovery through an elaborate array of transects. The fire taught little new in fire behavior: it happened when all big fires in the Northern Rockies did, and its dynamics were the same as those behind the Great Fires. The blowup was simply a matter of timing between woods and wind and spark. Still, the fire and the agency updated training manuals with a contemporary textbook case.

What the transects recorded was something else. The Sleeping Child burn presented a unique opportunity to research fire ecology in the upper montane and subalpine forest zones of Montana. For the next 21 years USFS researchers documented a "vegetal" succession of fescue, lupine, serviceberry, willow, lodgepole, alder, vaccinium, rose, and assorted other species within a matrix of manipulations that included logging, thinning for poles, fuelwood cutting, Christmas-tree harvesting, terracing, checkdamming, grazing by cattle, streamclearing, seeding and planting, mistletoe control, chemical spraying, and—to facilitate all these projects—a network of roads. Researchers were powerless to halt the rash of tinkering. "Postfire rehabilitation of the high-altitude watershed was considered a management necessity. Immediate salvage logging was required to prevent the loss of a substantial volume of timber. The growth of grasses and forbs provided an opportunity to relieve grazing pressure on cattle allotments elsewhere," and "the obvious overstocking of lodgepole pine seedlings led to intensive efforts to reduce tree densities throughout the burn." The outcome was data from sites far from undisturbed by meddling: by 1973 only two transects "had not been compromised in one or more ways." While the fire reaffirmed fire behavior, the results were too flawed to inform fire ecology. The major contribution, appreciated later, was to confirm the vigor of cone serotiny in lodgepole pine.[3]

In brief, the production of goods trumped the production of knowledge. The fire may have occurred at a remote site, but it was one that lay within the capacious bounds and metes of multiple use. Officials knew what they needed to know about fire. What the Sleeping Child blowup demonstrated was the latent power of fire that ever lurked in the mountains, and the long transect through time showed how the failure to catch a fire early could linger for decades.

* * *

The big fire story in the 1950s was that something new was emerging under the Southern California sun. Large, high-intensity, tough-to-fight fires occur in places fluffed with flashy combustibles, battered by drought, broken up by mountains, alive with seasonal winds, and peppered with ignitions. Along the South Coast it was possible to find big fires that were powered by any one of those factors, and it was common to find two or three factors colluding, but usually it was old-age chaparral and Santa Ana winds together that were the shears that cut across the landscape. Fire officers had learned to back off to a nearby ridgeline or to cede a deep-gorged basin or two or to let the winds drive flames down a topographic sluice box like Topanga Canyon. What changed that calculus was the sudden postwar construction of suburbs and exurbs. In 1959 the National Fire Protection Association published a warning about wooden roofs and conflagration potential and declared Los Angeles's hilltop fringe a "design for disaster." In 1963 Keith Arnold, chief of USFS fire research, estimated that "people and their occupancy of California wildlands had increased the potential for conflagrations between two and ten times over the level of the 1930s."[4]

The new normal announced itself on November 6, 1961, when, amid red-flag winds, a fire of unknown but "accidental" origin took hold in Stone Canyon. Within eight minutes the first chief to arrive declared a major emergency. Engines and airtankers converged and saved numerous houses, though they could not contain the surge of flame. The fire leapt over lines and ridges, hundreds of houses igniting hundreds of other houses, while firebrands sailed freely across the San Diego Freeway and rained down like hail on houses and brush alike. Fire-control efforts could not rely on the traditional techniques of pulling back, backfiring, letting flames dance through jackstrawed patches, or breaking up concentrations of slash ahead of the flaming front—because those caches of combustibles were $1 million houses along Mulholland Drive. The ember swarm feasted on wood- and shake-shingle roofing in Bel Air and Brentwood. At 12:56 p.m. another fire, of "suspicious" origin, started near Topanga Canyon; the hard-pressed Los Angeles Fire Department ceded responsibility to the Los Angeles County Fire Department, which fought the "watershed" fire as a wildland burn. A third fire, "criminally ignited," broke out in Benedict Canyon, but borate drops and a swarm of engines held it. About 3 p.m. the Santa Anas eased and containment tightened; by 8 p.m. the fire had ceased its advance. Resurgent normal winds threatened to push the flames back into areas spared the day before, but their force was weaker and suppression stronger, and the lines held. By 4 a.m. the fire was controlled at 8,560 acres. The entire conflagration had lasted less than 20 hours.[5]

The firefight had exhausted the city—some 85 percent of all available apparatus, 154 engines and 54 other vehicles, were on the line—and had drained its cooperators. Fourteen airtankers had flown on the fires, and engine crews had laid 135 miles of hoselines. During the firefighting, water flowed at over 50,000 gallons per minute for a total of 77.5 million gallons, of which about 90 percent went to the fires. Although some 78 percent of houses within the final perimeter of the fire were saved, the Bel Air–Brentwood burn was the worst fire in Los Angeles history so far and the largest urban conflagration since 1923, when a wildland fire had burned into Berkeley and savaged its wooden roofing. The 1961 season racked up the biggest burned area in California since 1924.

These were impressive numbers, but no less critical was the fact that the avalanche of flame had tumbled through Hollywood's back lots. Bel Air–Brentwood became the first of postwar America's celebrity burns.

CHAPTER ONE

SPARK

Let the word go forth from this time and place, to friend and foe alike, that the torch has been passed to a new generation of Americans.
—PRESIDENT JOHN F. KENNEDY, INAUGURAL ADDRESS, JANUARY 20, 1961

I have spent about half of my life influenced, taught, and educated against fire in nature, and then I have spent the other half of it using fire and trying to understand it.
—E. V. KOMAREK, KEYNOTE ADDRESS, FIRST TALL TIMBERS FIRE
  ECOLOGY CONFERENCE, 1962[1]

America's Great Cultural Revolution on Fire

For books about America in the round, 1962 was a stellar year. John Steinbeck published *Travels with Charley: In Search of America*, his tour by pickup camper, which had started in 1960. In his attempt to "rediscover this monster land," he found that the "grinding terrifying poorness" of the 1930s was gone, replaced by a kind of "wasting disease." The country had become a discontented land, he wrote, lacking "the pressures that make men strong and the anguish that makes men great." Americans were bored, indebted, and hungry for more material toys. Yet underneath it all a combustible brew built "like gases in a corpse." When it explodes, he wrote, "I tremble to think what will be the result." Everywhere he saw "a burning desire to go, to move, to get under way, anyplace, away from Here." Americans sought to move "not toward something but away from something."[2]

At about the same time, Michael Frome published his more thematically restrictive but more far-flung tour of the national forests, which was also a search for America. *Whose Woods These Are* was more celebratory and

33

more devoted to reportage than Steinbeck's novelistic quest for character. Its coda was that "there is no end, no beginning, no neatest of packages." Loggers, ranchers, miners, summer home residents, campers, primitive area partisans—each group was convinced of the absolute right, necessity, and primacy of its cause. The lands themselves were a patchwork, often with checkerboarded ownership, surveyed by a logic that reflected historical opportunity rather than political or environmental principles. Even within the federal government, agencies competed over which should administer such lands according to what tenets. Frome could not resolve these issues and concluded with a tribute to those who worked on the forests, a lineage that harkened back to Gifford Pinchot. "My lasting reaction is deep pride in having them on our side."[3]

While both treks circumnavigated the country, and while Frome repeatedly encountered fire and even devoted a chapter to it, neither saw the sparks in the tallgrass prairie or the Red Hills longleaf pine that were the points of origin for the cultural conflagration to come. And while both authors looked to the big questions as they triangulated from past to present, neither envisioned a fundamental reformation in how Americans related to fire or why it might be necessary. Even Eliot Porter's 1962 photobook for the Sierra Club, *In Wildness Is the Preservation of the World*, celebrated a green nature, not a black one. The fires that caught the public's attention were burning buses in Birmingham or the flames that lofted *Friendship 7* into orbit. If they thought of landscape fires, they would likely follow Frome and focus on the stirring firefights in the West. The future, however, lay not with firefighting but with fire lighting. It lay with the caches of fire folk who knew both what they were moving from and where they wanted to move to.

* * *

Before the year was out, the American fire scene began a wholesale reformation in politics, pyrogeography, and practice. While fire has often served as a metaphor for revolution—the single ember that can spread across whole landscapes, the power of flame to transmute what it burns—revolution has less often returned the favor. One reason is the character of fire itself. It is neither a substance like air or stone nor a bounded entity like a hill or a river, neither a creature nor a printed manifesto. It is a chemical reaction: it has no identity apart from the ingredients and setting that make it possible. It synthesizes that setting; it takes its character from its context. Fire is what its circumstances make it, and as those circumstances change, so does fire.

A fire revolution occurred when the cultural landscape of America changed and accordingly remade fire itself. Active fires broke through old conventions and containment barriers. Fires long lost from the landscape became ecologically significant precisely because they were missing. Novel fires appeared unlike any witnessed in the remembered past. There were tragedy fires, blowup fires, prescribed fires, megafires, confined fires, loosely herded fires, and fires that were merely monitored. There were fires kindled from lightning, accident, arson, and the mandates coded in agency handbooks. The traditional tools for working fire—shovel, ax, rake, torch—yielded to handheld calculators, aerial incendiaries, infrared scanning, and satellite uplinks. There were myriad and sometimes conflicting laws that governed what could and could not happen on the land. Threatened and endangered species became as powerful an influence as air tankers. The reclassification of land into wilderness restrained fire control as significantly as bulldozers had once bolstered it. The landscape of learning shifted, as though by tectonic plates, so that fire was understood and valued differently, could be seen as useful and necessary, its presence a neutral perturbation rather than an agent of irremediable destruction. Like wolves and grizzlies, St. John's wort and Chiricahua Mountain columbine, fire could be seen as belonging.

When the revolution began, control over fire had increasingly gravitated toward a government monopoly administered through a hegemonic Forest Service and a guild of foresters. When it ended, a civil society had emerged to reclaim the torch, and environmental enthusiasms and legislation had reorganized the nation's public estate while a service economy remade its private lands. Ecology now trumped forestry. A demographic upheaval transformed social castes overall and the fire-management workforce specifically. Collectively, a fast cascade of events—what would end in a convulsion of conflagrations—transformed American fire policy and practices as thoroughly as the concurrent Second Vatican Council did the liturgy of Roman Catholicism. At the end of a long decade, or of a shortened two decades that extended from 1962 to 1978, Americans found themselves living differently on their lands, among themselves, and with free-burning fire.

Natural Experiments

On March 1, 1962, the Tall Timbers Research Station (TTRS), located between Tallahassee and the Georgia border, with a co-sponsor in the Florida State University Research Council, hosted the first of what it called a fire ecology conference. To accommodate the invited assemblage, the gathering was staged in the Florida State University College of Education building. But this was neither an academic nor a governmental gathering. On the contrary, it marked the onset of a fire counterculture. It announced an insurgency that led to America's great cultural revolution on fire.

The conference had a manifesto; for the more ardent, it would become the equivalent of the Port Huron Statement from the Students for a Democratic Society. Among those who actually worked with fire, mostly those in the Southeast, opinions were converging about the place of fire in nature and its use by humans. Ed Komarek delivered the keynote address in which he described the cultural schizophrenia by which he spent half his life being taught that fire was bad and the other half using it productively. It was time to announce that understanding in order to reeducate the "public at large, the conservation groups, and the leaders of our educational systems." The Tall Timbers Research Station had been chartered in 1958 to do just that. It was dedicated to the study of fire ecology and fire's application to lands under management. It recognized a "landowner's right" to apply or withhold fire, whether as a private individual or an agency of the government. And it recognized the "right of the public to be adequately and honestly informed" regarding fire's potential usefulness as well as its possible destructiveness.[4]

The Tall Timbers Fire Ecology Conference, the first of a series, sought to provide a common meeting ground not only for private landowners and public agencies but also for "the enlightenment of conservation groups whose energies have too frequently been wasted in a crossfire of misleading and opinionated information." As concurrent pesticide controversies showed, there was a "lack of independent, privately organized, yet public, non-government" research organizations. That was the role TTRS would fill. It would be midwife and megaphone to a civil society for fire. Before the decade was out, it kindled a flame seen round the world.[5]

The founding of Tall Timbers has become part of the folklore of modern fire management. After the Civil War, wealthy northerners moved into the Red Hills region in search of quail hunting plantations akin to those their British counterparts enjoyed for grouse in Scotland. Like the Scots, when they found the bird populations plummeting, they commissioned an inquiry. The Cooperative Quail Investigation (1924–31), with some guidance from the U.S. Biological Survey, hired Herbert Stoddard, an accomplished naturalist who had spent much of his youth in the region. Stoddard concluded that the quail population rose and fell according to the health of its habitat. He determined that the postwar disintegration of the regional economy in the late 19th century had left an ideal mosaic of farms, woods, pastures, and scrub, and he argued that routine burning had maintained that mélange in a form suitable for quail. The land needed more of the right kind of fire. (Oddly, or perhaps not, this was the same conclusion the British Grouse Commission had reached two decades before.)[6]

Those discoveries flew in the face of a rising revulsion by officials and sponsored scientists that the South needed far less fire. Stoddard struggled to get his full convictions published; what began as crystals of insight dissolved, under successive edits and outside reviews, into syrupy prose and then into tinted water. In a formulation that would appeal to American instincts, the *Establishment*—a term minted into modern currency in 1955—had squashed the pioneering individual. Herbert Stoddard became a naturalist avatar of the lone inventor or forgotten artist fighting the system.[7]

Stoddard and several landowners then organized the Cooperative Quail Study Association as a consultancy. E. V. Komarek was hired as an assistant in 1934. The association disbanded in 1943 during the war years, but by then Stoddard had transformed Sherwood Plantation into an experiment station. Komarek had acquired Birdsong Plantation, and Henry Beadle, who with his brother owned Tall Timbers Plantation, had converted from his initial horror at seeing his land burned to a deep conviction that only fire could save it. More landowners signed on, including Komarek and his brother Roy with Greenwood Plantation, and ambitions swelled to engage the landscape as a whole, not simply its game birds. In 1958 discussions among the Beadles, the Komareks, and Stoddard ended with the creation of Tall Timbers Research, Inc. Its charter explicitly identified "fire ecology" as a primary charge, and Tall Timbers Plantation became Tall Timbers Research Station. In 1961 an outside grant allowed for a small laboratory building. What had begun as a personal quest had morphed into an institutional presence. The next year TTRS commenced its fire ecology conferences.[8]

It is tempting to see the insurgency as antiestablishment rebels riding a crest of youthful protest, perhaps the fire community's version of Ken Kesey and his Merry Pranksters in their Further bus but with driptorches. But Tall Timbers was deeply conservative. When the first conference convened, Stoddard was 73 and Komarek 53; their West Coast counterparts, Harold Biswell and Harold Weaver, were 56 and 58. This was a revolution of the elders. The conference opened with a dedication to the pioneers of southern fire—Beadle and Stoddard, of course, but also Roland Harper, H. H. Chapman, and S. W. Greene, representing botany, forestry, and livestock. The presentations followed a long peroration on fire and its history. Henry Beadle retold his conversion experience, and Herbert Stoddard recalled the long arc of his personal fire biography since childhood amid the longleaf forest. E. V. Komarek expanded that reach to the evolutionary history of fire on Earth, to the historical ecology of fire in many biotas, and to the ancient associations of fire and humanity. He concluded by distilling those abstractions into hard crystals of local experience: Tall Timbers and its fraternal plantations. His observations harked back to an older style of the naturalist. There was much to learn from what he termed "natural experiments" to be found in the historic interaction of fire, land, and people.[9]

The goal was restoration—the restoration of fire to its rightful place in the land in opposition to fire abolitionists, the restoration of a landowner's right to burn against the state's effective confiscation of the torch, and the restoration of folk knowledge and personal experience with the use of fire against the strictures of a public-sponsored and reductionist science. This was not a call for an uprising of Red Guards to kindle fires willy-nilly; it appeared radical only because modernity had so destroyed the former world. It was a landscape version of what Jane Jacobs had proposed for urban settings in her 1961 book *The Death and Life of Great American Cities*. As with The Nature Conservancy's (TNC) experiment, it promoted a renewal by returning to the old ways, to practices tried and tested over millennia by people actually living on the land who understood what they were doing. As Stoddard remarked, "You can't possibly give a formula for burning on land you haven't seen." Sound burning required art because "science is never going to solve that." The man burning has "to be a man that knows the woods. He's got to be a woodsman."[10]

Over the years such concrete and practical understandings had been lost to institutions, public agencies, and the abstractions of formal science. Despite protests from academics, Tall Timbers declined to submit papers for peer review and published the proceedings at its own expense under its own editorial control; its founders had had too many ill encounters with censorship, whether deliberate or simply de facto because of intellectual fads and institutional biases. The truth lay in what one saw with one's own eyes.

* * *

In Florida the spark of protest found ample tinder. From the Red Hills to the Everglades, Florida had for 30 years opposed efforts to subject itself to national norms of fire control. Hunting plantations had resisted in the Panhandle, and open-range herders had ignored proscriptions throughout the interior. One by one the federal agencies had conceded—the Biological Survey, the Forest Service, the National Park Service—and by the 1960s Florida's state agencies, every one, rejected the doctrine of fire exclusion in favor of prescribed burning. Speaking for the Florida Forest Service, R. A. Bonninghausen admitted, "We changed with the times, but sometimes with reluctance." Tall Timbers built on that immovable stone and gathered acolytes from across the country (and the world) to leverage their collective insights and experiences to bring down the establishment. In 1962 the odds seemed hopeless. It was, for fire, the equivalent of President Kennedy calling for a landing on the Moon.[11]

But it happened. Tall Timbers provided a sanctuary from which to launch the intellectual insurgency. The published proceedings became a veritable encyclopedia of fire ecology and practice. A decade after they began, the annual fire conferences took on the ambience of a big-tent revival meeting or rock-band road show as they brought fire from Europe, Latin America, Africa, North America, and Australia to Tallahassee while taking Tall Timbers to California, Missoula, Fredericton, Lubbock, and the Pacific Northwest. At the 1974 session, the Forest Service formally confessed its failings and announced its conversion. Komarek and company had achieved their goal—had, along with legislative reforms such as the Wilderness Act, forced the fire establishment to accept fire's role in landscape ecology and management. If David had not slain Goliath, he had at least conscripted him.

The Tall Timbers board decided the conferences were no longer needed, but the fire community rose in protest. The conferences had become indispensable even if other organizations now staged them. Tall Timbers was, or would become, the institutional embodiment of landscape fire management. For a time it housed the administrative headquarters of national fire programs for TNC and the U.S. Fish and Wildlife Service, and the National Park Service stationed personnel there, too. When the U.S. Forest Service closed the Macon Lab, Tall Timbers remained the primary center for regional fire research. An NGO had acquired, in effect, the standing that elsewhere resided in state agencies. It was a huge burden on a private institution, though one that Tall Timbers bore willingly. In 1989 Tall Timbers Plantation was named to the National Register of Historic Places.

By then it had joined the establishment. Tall Timbers had helped make Tallahassee one of the apexes of America's institutional fire triangle. What it never lost, however—what spared it from the usual sclerosis of bureaucratization—was its close connection to the land. Those acres lay a few steps outside the Stoddard Laboratory and the Beadle Homestead that served as its administrative center. It never lost the immutable insight of its founders: landowners not only had to know, they had to do, and it was the doing, the close attachment to the land, that made the knowing possible.

Direct Action

On April 26, 1962, a month after the inaugural Tall Timbers conference, Dr. Don Lawrence, botanist, ecologist, and adviser to TNC, tossed matches into dry bluestem at the Helen Allison Savanna Preserve north of Minneapolis and burned 20 acres of oak and prairie. The practice spread, along with the Conservancy. Fifty years later the tally of lands TNC had control-burned approached two million acres. It reached from patches of upstate New York that were burned to support habitat for Karner blue butterflies and along the Lake Wales Ridge in Florida for scrub jays to coastal grasslands in Texas for Attwater's prairie chicken.[12]

Some revolutions are progressive and aspire to a utopian future radically different from anything known before, and some are conservative, seeking to restore a utopia imagined in the past. Prairie burning belonged with the restorers. From the earliest moments of European contact, explorers had identified free-burning fire with the sweep of the central grasslands. So prominent was the association of fire and prairie that a school of fire art flourished in the 19th century as artists from George Catlin to Charlie Russell painted gorgeous canvases filled with flame and often with fleeing bison, antelope, and people. Naturalists debated whether fire was a cause or consequence of the prairie peninsulas, barrens, savannas, and the grassy Great Plains. Pondering the oak savanna of Wisconsin, Aldo Leopold observed that the unstable frontier between grass and oak shattered when prairie lost its "immemorial ally" of fire, broken by the plow. In 1948, the year Leopold died while trying to defend planted woods from a grassfire, John Curtis at the University of Wisconsin Arboretum ignited experimental burns as part of a prairie restoration project. Increasingly, year by year, the scientific evidence matched the historical: those fires were both inevitable and essential in tallgrass prairies and woody savannas. Remove fire and you send prairie into a downward spiral of brush-choking decadence and ultimate demise. Without fire, such grasslands became woodlands.[13]

Lawrence's April burn, done wholly with volunteers outfitted with rakes, matches, and metal backpack pumps, was a bold bid. A science of fire ecology did not exist by name. Forestry, which controlled virtually all fire research, wanted to promote those revanchist trees, not the resisting grasses and forbs. It was tricky enough for fire civilians to burn autumn leaves much less open landscapes designated as critical habitats. The Conservancy itself was a fledgling, started in 1951 as a nonprofit offshoot of the Ecological Society of America by members interested in "direct action." It first purchased a site in 1955, 60 acres of the Mianus River Gorge in

New York. In 1961 it entered its first partnership with a public agency, the Bureau of Land Management (BLM), to protect old-growth forest in California. That same year, the Conservancy received its first donated conservation easement, six acres of salt marsh in Connecticut. The next year it conducted its first burn. Not until 1965 did the Conservancy have a full-time director, and that thanks to a grant from the Ford Foundation.[14]

Land acquisition, particularly in the Midwest, scaled up through contributions from Katharine Ordway, heir to the 3M Company fortune. The Conservancy found itself doing more burning, and it began hybridizing with existing fire cultures such as those in the Flint Hills. What was being restored was not simply an ecological regimen but a culture of fire, and more, a political order. No Second Amendment assured landowners of the right to bear torches, and over time fire control had passed from civil society to the state. On the Helen Allison Savanna Preserve, landowners began to reclaim that heritage. But it was not until TNC moved into Florida in the mid-1980s that it found in fire a primary tool for "direct action" and an identity as a fire organization that would no doubt have stunned its founders.

Vignettes of Primitive America

The movement began in Yellowstone, which was how Yellowstone liked matters, and as with everything Yellowstone, the action seemed to hinge on its megafauna, specifically, its elk. There were too many. For decades the park had coaxed and cajoled more elk into being by feeding them and by killing predators, and now the elk were eating the park raw. Over the winter of 1961–62, rangers shot 4,283 elk in an effort to cull the herd to something that Yellowstone, vast though it was, could accommodate. The public outcry did for the National Park Service what clear-cutting would do for the Forest Service. Interior Secretary Stewart Udall responded, as administrators instinctively did, by establishing a committee.

The Advisory Board on Wildlife Management was an august group, chaired by A. Starker Leopold, who was a professor at the University of California, Berkeley, and a son of Aldo Leopold. The Sierra Club noted that those who would challenge the board's credentials or conclusions faced a formidable task. The Leopold Report, as it became known, was powerful because, in evaluating methods by which to cope with Yellowstone's elk herd, it based its analysis on a rereading of the ultimate goals and fundamental purposes of national parks. Most readers and commentators quickly forgot its strenuous insistence on active management and its specific recommendations (which ironically included the need for in-park culling) in favor of its rhetorical rechartering of national park purposes.[15]

The First World Conference on National Parks had convened in July 1962 in Seattle, and the Leopold committee accepted its report "as a firm basis for park management." To that report the advisory committee added a healthy dose of American nationalism. In memorable language the report declared that a national park should as its primary goal "represent a vignette of primitive America" and should ensure that "the biotic associations within each park be maintained, or where necessary recreated, as nearly as possible in the condition that prevailed when the area was first visited by the white man." The moment of European contact became a baseline for "naturalness."[16]

The implications of this "seemingly simple aspiration," the report concluded with calculated understatement, were "stupendous." The problem was, the biotas of America's parks were "artifacts, pure and simple." They were the progeny of complex ecological histories, not necessarily patches of primitive America. Among the more spectacular examples the report cited was the western slope of the Sierra Nevada. When the forty-niners had spilled over its crestline, it had boasted a montane forest of large trees widely spaced and routinely burned. By 1963 it displayed a "depressing" vegetative tangle, a "dog-hair thicket of young pines, white fir, incense cedar, and mature brush—a direct function of overprotection from natural ground fires."[17]

That primitive scene needed to be restored. This was a task neither easy nor fully possible but an undertaking that called for active measures informed by scientific research and conducted by a competent corps of Park Service personnel. The mangled fire regime was both a paradigm and an obvious point of departure because fire was the most comprehensive means to reform the habitat that underlay wildlife management. Among possible techniques considered by the Advisory Board on Wildlife Management, "the controlled use of fire is the most 'natural' and much the cheapest and easiest to apply." But so profound was the ecological deviation from historical conditions that fire could not do its proper work—would most likely blow up—until the wildlands that fed it were reconstructed; even chainsaws might be needed. What could emerge at the end was "a reasonable illusion of primitive America." Would such interventions succeed? The Leopold savants would not say. "We cannot offer an answer." They were wildlife biologists, not fire scientists. The necessary skills did not exist. They insisted only that the job "will not be done by passive protection alone."[18]

From the moment it was released on March 3, 1963, to the North American Wildlife and Natural Resources Conference, the Leopold Report was a sensation. Most commentators cherry-picked its striking phrases. They ignored its cautionary warnings about historical complexity, ecological ignorance, and the absence of skilled managers and instead seized on its call for the wild. What excited them most were variants of the phrase "naturalness above all." The transformation was identical to what happened at the same time with the legacy of Aldo Leopold, whose *Sand County Almanac* was subsequently reissued in 1966 and read less for its messages about patiently and humbly restoring debased land than for its championing of a land ethic and its celebration of wild nature. So it happened also with fire.

A simple narrative began to congeal, a narrative of how, with European contact, the natural process of fire had been driven to near extinction along with bison and grizzlies. Here was the dark side of America's story, the national creation myth that told how a civilized Europe had encountered a primitive America and spawned a new society. Just as national parks had been established to preserve the memory of that encounter, so those historic fires had to be reinstated. It was a matter of mythic as much as ecological integrity. Reclaiming fire was less a radical innovation than a restorative act, even a penitential one. The narrative turned on its head what had been considered a legal and moral duty—an obligation to control fire. The charge now was to restore it.

Secretary Udall, then completing a book that told the saga of American environmentalism, *The Quiet Crisis*, received the report enthusiastically. He instructed NPS director Conrad Wirth to "take such steps as appropriate to incorporate the philosophy and the basic findings into the administration of the National Park Service." That the report granted so much space to fire was both a problem and a prescription: there existed no more daring symbol of the commitment to a new order or one potentially more damaging. By comparison, loosing wolves seemed almost domesticated. A single wolf could not transform an entire park in an afternoon; a fire could. Everything might, as ecologists like to proclaim, be connected to everything else, but fire burned everywhere and could be seen by anyone. No one might know whether rangers shot a few elk in the deep snow of a Yellowstone winter, but everyone could see the smoke from a fire lit or left to burn.[19]

* * *

That the National Park Service should be the first federal agency to break ranks had a certain symmetry. The national parks had invented modern wildland firefighting beginning in 1886, when the U.S. Cavalry assumed the administration of Yellowstone and then extended that regimen to the California parks. Now, 80 years later, they led a revolution to devise a replacement.

It seems both odd and inevitable. Begin with the agency's assets to promote so daring a change. The NPS was prepared to split from the Forest Service on fire because of the two agencies' long-running rivalries, notably over scenically choice lands and responsibility for outdoor recreation. The NPS did not share in the fraternal order of foresters. It had long seen itself as distinct in mission and esprit, a chip off the block of American exceptionalism. Historically it had known pockets of light-burners, notably in the Sierra Nevada parks. It had accepted controlled burning at Everglades. And, scattered into hundreds of small units, it simply lacked the heft and infrastructure to match the Forest Service in firefighting; not a few parks relied on neighbors to suppress wildfire. Politically, the national parks were less an integrated system than a daisy chain of semiautonomous fiefdoms, which made discretionary experimentation at local parks possible.

As the separate parks were to the system, so the NPS was to the national infrastructure of fire protection: it could be remarkably self-contained, even self-referential. The National Park Service had more cultural cachet and political clout than the Fish and Wildlife Service, which was quietly expanding prescribed fire along the Gulf Coast, and less anxiety about proving its mettle than the adolescent Bureau of Land Management, eager to take on the Forest Service at its own game. The agency's founding charge to maintain its holdings "unimpaired" for future generations disposed it to see natural events as part of the scene and to let nature take its course. The parks had interest groups from the National Parks Conservation Association to the Sierra Club ready to lobby on its behalf. Apart from shooting elk, the public was willing to grant NPS rangers political space. On most controversial issues the public granted the Park Service wide tolerance.

But those assets could as easily flip into liabilities. It was difficult to scale up what happened in a particular park into a service-wide policy. A failure to hammer fire aggressively could be turned to the old charge that the NPS was weak on defense, that it simply was not up to a tough, gritty job like fire suppression. While its ranger corps did not kowtow to forestry, it had no alternative professional identity to counter forestry's guild; there was no program of study or apprenticeship of technical skills that led someone into status as a park ranger. Like the ranger's uniform, his role had evolved out of its cavalry era. What its ranger corps had were camaraderie and cohesion. (The standard joke was that there were two organizations you never left: the Mafia and the National Park Service.) The United States had no national park organic act, only an act creating a National Park Service and a letter of instructions from Secretary of the Interior Franklin Lane to its first director, Stephen Mather, which was widely regarded as the agency's Magna Carta.

The variety and independence of its units could lead equally to enlightened experimentation or administrative anarchy. It was an arrangement that favored personalities and high-value holdings.

The Leopold Report was, in this sense, a mixed blessing. It gave the agency a new charter, but the NPS, unlike the USFS with the Multiple-Use Act, had not asked for one. It came while Mission 66, a $1 billion investment in infrastructure primarily to support the boom in visitors, was at full throttle; the agency was not interested in other initiatives that might divert attention. Its baron superintendents hated any check on their sovereign powers. If adopted—and Secretary Udall was keen to translate proposals into written policies—the tenets of the Leopold Report would enact a universal standard for the management of natural areas, the agency's true crown jewels. And it would likely compel the National Park Service to intervene in the landscape rather than let nature unfold in its own way. Some of those active measures would be distasteful, both to the NPS and to the public. It meant shooting animals. It meant starting fires.

Yet the Leopold Report also offered an anchor point from which to survive the impending firestorm of environmental reform that would consume the federal land agencies over the next 15 years and for some prove schismatic. A few agencies such as the BLM acquired an organic act for the first time; some, notably the USFS, had their statutory authority rewritten; others, like the Fish and Wildlife Service, were granted fundamental new powers. But they all had to cope with the National Environmental Policy Act and assorted legislation that affected how they did business. By adopting the Leopold Report the NPS avoided those imposed recharterings. It reformed more or less internally, it kept control within its own constituencies, and it even acquired, for the first time, its own research program in the natural sciences. Most especially, the Report bequeathed a working alternative to the strictures of the Wilderness Act. The park as vignette of primitive America granted more freedom to maneuver than a place untrammeled by humans. It left to the agency the discretion over what the phrase actually meant and how to manage it. It expanded and refined the notion of "unimpaired for future generations," a vision the Park Service was comfortable with.

Alone among those new charters, the Leopold Report directly addressed fire's presence and possible uses. Other agencies had to interpret how to adopt new fire practices (and purposes) within their changed contexts. The Wilderness Act, the Endangered Species Act, the Clean Air Act, the National Environmental Policy Act—none included fire's management specifically in their directives. The Leopold Report did. It identified fire's removal as a problem, urged fire's restoration as a solution, and proposed controlled burning as a treatment of choice. Once codified into administrative guidelines in 1968, it left fire's management in the national parks with the National Park Service, and it positioned a fulcrum that allowed the agency to leverage its influence outward rather than being moved by outside pressures. The Report received enthusiastic attention at the next (third) Tall Timbers conference, whose attendees instantly recognized a fellow traveler.

* * *

Principles are easy, practice hard, and policies without money are, as Director George Hartzog observed, "just talk." However favorably situated the Park Service appears in retrospect, the factious agency hesitated, stalled, ignored, and moved fitfully. The concerns the Report addressed, particularly regarding fire, had not bubbled up from the bottom; they were imposed from the top and were better understood intellectually than emotionally. This was a revolution from above. Not all fire officers converted to the new doctrine; after all, many had fashioned their careers by fighting fires. Nor was it obvious how to reinstate fire on the ground. The act more resembled restoring a vanished predator species than it did constructing a new visitor center. It was not simply putting something back that had been lost, because restoring that something would alter the dynamics of everything else. And unless it had the right habitat, fire might turn feral — might misbehave and damage what it was intended to enhance. If the agency was to change course, it needed a proof-of-concept test. It found one in the Sierra parks.[20]

Margaret Mead once observed that successful movements — she had in mind American anthropology under Franz Boas — needed a charismatic patriarch to announce it, a sugar daddy to fund it, and young acolytes to proselytize its message. The fire revolution had all that: Herbert Stoddard (segueing into Ed Komarek), the Tall Timbers endowment, and youthful partisans of burning ready to discard the shackles of failed doctrines and practices. But flaming Florida was too idiosyncratic and easily isolated to shake the national establishment. California could do it, though. It symbolized the hopes and horrors of the 1960s, and it quickly created a West Coast counterpart to the Florida agenda. In Harold Biswell it had its patriarch; in Sequoia-Kings Canyon National Park it had its research station; in a generation of new recruits, particularly University of California, Berkeley, students, who had studied under Leopold and Biswell, it had a corps of enthusiasts who did for fire (though with far greater discipline) what Yosemite's Camp 4 covey did for rock climbing.

Harold Biswell—"Doc" to his students, "Harry the Torch" to his critics—was the linchpin. Like many of the pioneering naturalists of his time, he had grown up on a farm in the Midwest. He earned a doctorate in plant ecology at the University of Nebraska, still aglow with the triumphs of the Grassland Lab, amid the environmental (and for grassland scientists, intellectual) trauma of the Dust Bowl. The Forest Service hired him for its Pacific Southwest Experiment Station at Berkeley, California. In 1940 he transferred to the Southeast Experiment Station at Asheville, North Carolina, where he learned the regional fire scene. He stayed until 1947 when he joined the University of California, Berkeley, faculty; there he remained until his retirement in 1973. He and Starker Leopold became colleagues, co-taught graduate seminars, and reinforced their predilections toward fire. He found landowners in Northern California (including Hoberg's Resort and Teaford Forest, in the heartland of the old light-burning controversy) to allow him to create demonstration plots, but real traction required something that could propagate fire through the public estate of the West. More precisely, it demanded the alliance of a premier research university with a high-visibility federal agency on a landscape of supreme public interest. For fire it just did not get any better than California's giant sequoia groves.[21]

Here external and internal pressures converged. The outside forces were those identified in the Leopold Report. The pressures interior to the parks concerned the paradox that despite intense protection, some of the Park Service's most prized treasures were deteriorating. Most spectacularly, the fabled Big Trees of its Sierra Nevada parks were doing poorly, and the suspicion was rife that people were the reason. The effect of trampling and other accommodations to visitors lay behind the doctoral study that Richard Hartesveldt had conducted in 1962 at Yosemite's Mariposa Grove; when it was completed, Sequoia-Kings Canyon commissioned additional research. Begun in 1963, the studies continued until 1970 (with an extra summer in 1974). After serial progress reports, *The Giant Sequoias of the Sierra Nevada* was submitted in 1971.[22]

It confirmed that people were in fact behind the decline of the giants not simply because of what they did but also what they did not do. The Big Trees needed fire. They could thrive amid frequent burns; most bore scorch scars, and a few boasted fire-excavated cavities. But their cones were semiserotinous, and their seeds germinated best in ashy beds temporarily freed from competitors. Sequoia seedlings survived most exuberantly, in fact, in places that burned intensely. What threatened these patches of Pleistocene megaflora was less root damage by visitors than an altered habitat in which fire-sensitive competitors such as fir and cedar flourished, sequoia reproduction was impossible, and overgrown understories threatened even mature sequoias with fires unlike any they had known. Instead of scurrying around the forest floor like mice, flames could soar upward through the latticed canopy of intrusive trees and incinerate the otherwise fire-immune sequoia crown. If the Big Trees were to survive, the old fire regime would have to return. Advocates argued then, as William Everhart would in 1983, that "those who still want the Park Service to put out fires might ask themselves how the wilderness managed to survive for so many millions of years without rangers." Listening to locals explain how they had "saved" the Big Trees from fire 29 times in the past five years, Gifford Pinchot in 1891 had wryly wondered who had saved them the "other three or four thousand years of their age?"[23]

The sequoia research advanced as the Leopold Report percolated through the NPS. In 1964 Harold Biswell proposed to transfer his demonstrations to Whitaker's Forest, a University of California, Berkeley, experimental site on Redwood Mountain adjacent to Sequoia-Kings Canyon. For the next decade he directed trials with cutting, piling, and other strategies to ease fire back into the groves. The park began similar exercises on its side of the fence, as students of Leopold and Biswell staffed positions and created a cadre of partisans for prescribed burning. Redwood Mountain became an *experimentum crucis* for the fire philosophy urged by the Leopold Report. In 1967 Tall Timbers staged its annual fire ecology conference in California in honor of Biswell and other western pioneers such as Harold Weaver. In October, Park Superintendent John McLaughlin and his staff met with Leopold in Berkeley to quicken a plan for fire's reintroduction (Forest Service researchers from the Pacific Southwest Station might also have been present—the record is unclear). When skepticism threatened to stall the project, when the fire and forestry clique began to pile up qualms and queries, Leopold calmly informed them that the issue was not whether the park would restore fire, but how.[24]

The breakthrough came in 1968. It was a year made notorious by assassinations, riots, social mayhem, and political turmoil throughout the Western world. It also marked the culmination of a quiet revolution for fire. It helped that two fires in Glacier National Park the summer before had forced the NPS to reconsider the limits of suppression. Before the next fire season could begin, the National Park Service published a set of administrative guidelines for natural areas that formally recanted the 10 a.m. policy.

With the right ingredients, gently stirred by modest fire seasons and public enthusiasm, the program launched boldly. In the summer of 1968, Sequoia-Kings Canyon ignited an 800-acre prescribed fire on Rattlesnake

Ridge and allowed a lightning fire on Kennedy Ridge to burn freely. As Bruce Kilgore recalled, there seemed no difference between the two fires, and it appeared "that the simplest way" to reinstate fire would be "to let lightning fires burn." In 1969 Sequoia-Kings Canyon designated 129,331 acres of upper-elevation landscapes (15 percent of the park's holdings) for "let-burns" and deliberately fired 6,186 acres under prescription. Even when one of the kindled fires on Redwood Mountain burned more ferociously than anticipated (or desired), even after a large burn had to be contained with bulldozers, and even after administrators recognized that prescribed fire in the West was expensive (and would probably prove as costly in the long run as a traditional program) and might someday cause public relations blowups, the effort soldiered on.[25]

By then, though, Sequoia-Kings Canyon had passed the torch to Yosemite where, under Robert Barbee and with Biswell as mentor, a similar program gathered steam and earned the approbation of Harold Weaver on an inspection tour. The Park Service had its proof of concept.

* * *

Between the 1967 and 1968 fire seasons, the agency utterly overhauled its administrative policies. A shelf of manuals was condensed down to three slim books, each known by the color of its cover, one for natural areas (Green), one for recreational holdings (Red), and one for historical sites (Blue). The Green Book had 67 core pages that opened with a long preamble of purposes, policies, and principles, then discussed their application according to various topics, and concluded with another 99 pages of appendices that ranged from Lane's 1918 letter to Mather to the procedures for public review of master plans. Rather than specify meticulously what a superintendent ought to do under every imaginable circumstance, it granted extraordinary leeway to adapt the general to the local.

The policy on fire came directly from the Leopold Report. It opened by declaring that "the presence or absence of natural fire within a given habitat is recognized as one of the ecological factors contributing to the perpetuation of plants and animals native to that habitat." Accordingly, it acknowledged that fires resulting from natural causes are "natural phenomena and may be allowed to run their course" within limits, and it approved prescribed burning as a valid substitute for natural fire. Fires that threatened lives, infrastructure, or cultural assets would be suppressed. Forty years after they had been condemned as anathema, light-burning and let-burning were not merely to be tolerated but actually promoted.[26]

The Green Book's fire passages were an attempt to reformulate America's relation to nature. Its sentiments leaped ahead of popular opinion, much as the Civil Rights Act had with racial attitudes. The reform stated an ideal: it did not allocate funds to make it happen or reconstruct Park Service organizational charts or establish a national-level staff to assist, much less specify how to execute the new regime. As Bruce Kilgore observed, the "individual parks were on their own." It took another three years before operational guidelines established working parameters and parks beyond the Sierra Nevada (and of course Everglades) joined in. From then on it was a case of letting a hundred fires bloom.[27]

That bald observation, however, glosses over what was within the agency a tough sell. Not everyone agreed. Those who favored the natural landscape—resource managers, scientists—wanted more fire. Those who had risen through the protection division, which embraced both visitors and landscapes, hesitated. The parks were more overwhelmed by visitors than by fires and more dazzled by the sparkling infrastructure of Mission 66 than dismayed by overgrown woods. The agency drew its managerial caste mostly from its ranger corps, and its rangers rose through the protection division, which increasingly meant servicing visitors. The agency's solution was to partition. Resource Management division would be responsible for fire's restoration and a Protection division for its removal. They were separate and unequal. Resource Management (and prescribed fire) had a small budget. Protection (and fire control) had access to big emergency funds.

That decision established an institutional chasm that the Park Service did not begin to close for another 20 years. The fissure could be finessed in the early years when the workforce was small, when nearly everyone knew everyone else, and when almost everyone had some fire experience or background. It became a fault line as the ranger corps was sucked into the widening maelstrom of law enforcement and, later, after big money poured into the fire program following the 1988 season, a "professionalization" of fire management that isolated it from the rest of the Service. For all its ideological swagger, the fire program depended on personalities—the personalities of superintendents and chief rangers, the personalities of those within a park who had to reconcile differing career paths and institutional purposes. The outcome favored bold superintendents like the progressive McLaughlin, but it allowed equally bold skeptics to stall. The NPS could not do what came so readily to the Forest Service: it could not apply a common standard across a wide spectrum of settings. The Green Book freed parks from simple suppression without imposing a standard appropriate to the new era or without fashioning at a national level the enabling tools they would need. It made fire restoration desirable but not obligatory. Although scientists were catalysts, the Green Book based its doctrine not on science but on a standard of "naturalness." By choosing not to dismantle the old fire-suppression organization, it left the new fire practices without a firm institutional home. The Leopold Report had argued that "controlled burning is the only method that may have extensive application," but when pressed how, actually, to apply fire for restoration, its authors confessed that "we cannot offer an answer."[28]

Firefighting remained with the Protection division while fire lighting migrated into newly invented Resource Management divisions, which absorbed what foresters the parks still retained and added wildlife biologists. In this way fire's management in the parks had two co-serving tribunes alternating their command. Because of emergency funding availability and sheer inertia, the deep power remained with suppression. It had the engines, the crews, the infrastructure, the heritage, and the connections with its counterparts across the park border. The fire restorers, like the fire-restoring parks, were on their own.

Green fire's attractiveness to most observers — its appeal to naturalness — also compromised its ability to use all the tools in the fire cache. There was a clear bias for natural fire and against prescribed fire. Even an advocate like Superintendent McLaughlin wanted the term "prescribed burning" banished in favor of "restoring a natural process." Controlled burning was costly, was not always controlled and specified culpable agents if something went wrong, and was tolerated only as a surrogate for nature's fire. Lightning fire was the true vestal fire on America's virgin lands. The driptorch was a grimy expedient useful only until lightning could reclaim its rightful place. The Tall Timbers agenda built on humanity's long use of fire, the California agenda on fire's ecological antiquity. To skeptics, the Green Book's guidelines looked like *Star Trek*'s prime directive, in which nonintervention was the norm and intervention was allowed only to correct the perturbations caused by past intrusions.

The national parks broke the national unity of fire purpose and practice. The NPS could claim it had no choice — its mandate was to preserve the natural scene, and fire was an indispensable part of that order. The Park Service was not the Forest Service. It did not have a mission to assimilate as many uses as possible. Buried in the Leopold Report was the revealing comment that "purely from the standpoint of how best to achieve the goal of park management, as here defined, unilateral administration directed to a single objective is obviously superior to divided responsibility in which secondary goals . . . are introduced." The old all-purpose fire commons was being broken up and parsed into special uses, each of which would have its own fire protocols. The Green Book commenced that bureaucratic enclosure movement.[29]

Untrammeled by Man

On September 3, 1964, President Lyndon Johnson signed Public Law 88-577, the Wilderness Act, to establish "a National Wilderness Preservation System for the permanent good of the whole people, and for other purposes." The signing climaxed over a century of evolving ideas, 40 years of administrative experimentation with designated primitive areas, and a decade of legislative maneuvering. The act defined wilderness "as an area where the earth and its community of life are untrammeled by man, where man himself is a visitor who does not remain."

Less clearly appreciated at the onset, by enacting a new category of land, the act also created a new category of fire, which in turn would require suitable policies and practices. What made this invented fire different from others at the time is that it came with the force of law. Among the act's other purposes, it proved a legal crowbar with which to pry fire suppression out of its clanking machines and maybe out of the backcountry altogether. "The Wilderness Act says that natural processes should proceed," erstwhile fire officer Bud Moore observed. "In light of that, to put out a fire was almost illegal."[30]

Nowhere does the enabling legislation directly address fire. The federal agencies that held legal wilderness would have to interpret what policies were suitable. A fire community that liked to think with its hands, which sought to reduce fire problems to physics, would have to grapple with metaphysics, with a constantly evolving state of mind interacting with a constantly changing state of nature, for that is what wilderness would mean for them. The interaction between the wild and fire was not a dialectic between constants but between variables. The confusion over ideas, however, was nothing compared with the confusion over practice. Prescribed fire was a choice; fire in wilderness was now an unstated but statutory mandate. This was not a topic that could be tabled, ignored, or voted away. The fires came. The Wilderness Act said they had a right to stay.

* * *

An evolving narrative held that wild land was integral to the American experience. In his 1967 book, *Wilderness and the American Mind,* Roderick

Nash declared that "wilderness was the basic ingredient of American culture. From the raw materials of the physical wilderness, Americans built a civilization. With the idea of wilderness they sought to give their civilization identity and meaning." The struggle against the wild and the campaign to spare some significant fragments of it was America's creation story.[31]

That was the foundation narrative, but as interest grew, the notion of the wild collected other identities as a museum, a gym, a lab, and a refuge. It preserved the past, the land as it was when Western civilization and American pioneers first encountered it. It allowed for outdoor recreation, or more strenuously, for an imagined reenactment of the experiences of explorer and pioneer. It upheld a standard of the natural order, the one ordained by Nature and Nature's God, by which it was possible for modern science to judge modernity's dislocations. And it offered spiritual solace, a sanctuary outside the bustle of "the business of America" that was "business." It also intertwined its tendrils with the trestle of national narrative. That Americans had driven the wild, like its wild bison, nearly to extinction did not dismiss the wild's significance to the country's story because that struggle was itself the narrative arc that joined Plymouth Rock to Half Dome.

Inevitably, critics—academics, mostly—challenged the premise that America had been wild in the sense characterized by the Wilderness Act. They pointed to the long evolution of the very idea, which rendered wilderness less a constant, a land metric like the platinum-iridium alloy kilogram housed under glass in Paris, than a sliding index, like the cost of living. And they hammered home the fact that the place had been populated, and those people had shaped the land as they wished and could. America was not a virgin land but a widowed one, emptied of indigenes by conquest and disease. Wilderness was not an intrinsic characteristic of pre-Columbian America but one fashioned by settlement and invented by an industrial America. Surely there were niches that had escaped the routines of anthropogenic tinkering, but most of precontact America had been a cultural landscape. Its original inhabitants had worked the land. They had burned.

By the time Roderick Nash published a sequel, *The Rights of Nature*, in 1989, wilderness had shifted from something profoundly tethered to culture to the notion of the wild itself, a state that was in its essence outside culture and beyond human control. Wilderness merged with broader concepts of deep ecology and intrinsic-value philosophy. What had started as a saga of settlement metamorphosed into a celebration of the acultural, the ahistorical, and the transcendent. The wilderness movement reached a climax between 1978 and 1980 with new reserves in Alaska, and although more sites were added over the next few years, the preservationist instinct found a more powerful mechanism in the 1973 Endangered Species Act. In brief, the wild no longer fit tidily into America's history or into any human history. In more radical visions it stood apart. It was singular. It was a condition, not a tool, and certainly not one use of land among many. It did not play well with others—that, after all, was the point. It partitioned public land into an apartheid of the wild and the nonwild.

What might appear as contradictions were treated at paradoxes. The core paradox is that the wild is managed: it exists through legal and political acts. Its identity is a matter of definition in court and Congress. People must manage what is nominally autonomous from human meddling. Moreover, the Wilderness Act was the expression of an idea whose time had come: it was an invention of its time and place, which helps explain why the concept has traveled poorly outside America's borders. National parks, biosphere reserves, wildlife refuges, national forests, even private reserves held for public good have all circulated around the globe with relatively modest adaptations and concessions to local norms. Wilderness, though, has not. It tends to be too absolute in its segregation of the natural and the cultural, and it is too closely bound to a particular narrative and moment of American experience. It expressed a distinctive era of American experience in which immigration was held to historic lows and the nation looked inward even as it confronted global crises.

The paradox is that its appeal to transcendence, by severing the ties of blood and soil that underwrite nationalism elsewhere, is ideal for the syncretic, pluralistic, multiethnic society that America became. Those regions that had the least wilderness were those that had their own creation stories, their rooted ties to land, and were less subject to immigration—think places like Texas, Utah, and Indian reservations. In principle, wilderness transcended them all, although in reality it meant extinguishing those local claims in the name of the national good. So, too, it meant that the federal land agencies had to renounce some of their historic claims to how they managed those places.

In the past, when the country announced a new category of public land, it created an agency to manage it, though sometimes this took a while. The forest reserves legally existed for 14 years before the Forest Service took them over. The first national park was designated 44 years before Congress established a National Park Service. Wildlife refuges went through serial reorganizations until the Fish and Wildlife Service brought order. But the Wilderness Act did not endow a National Wilderness Service to administer the National Wilderness Preservation System; the agencies had to absorb that charge within themselves. The closest parallel is the experience with the national monuments, which resided with whatever agency held the site until Executive Order 6166 in 1933 transferred them all to the National Park Service.

Wilderness lands proved much harder to assimilate. They were larger, often highly visible, and, paradoxically, they had costs. They required investments of money, staff, and planning. Trickier, they could clash with other management practices. They were impermeable to multiple use—and to compromise. Agencies chartered to do one task or operate under one set of guidelines had to reform internally to accommodate wilderness. They internalized within agencies the conflicts that the country itself was undergoing; even a single agency had to evolve different policies for the lands under its administration.

* * *

The new mandate said nothing directly about how to handle the most transformative process in most wilderness lands—fire. The Endangered Species Act could translate precepts into concrete practices. In the Southeast it compelled prescribed fire as essential to maintain the longleaf habitat of the endangered red-cockaded woodpecker. In the Northwest it held back burning around sites critical to the northern spotted owl. In wilderness it could proscribe bulldozers but not prescribe burning. Yet neither fire nor wilderness would wait. There was no way to exclude fire, and wilderness needed more of it.

What fire management meant coevolved with notions of wilderness. In Muir's day wilderness meant lands shielded from ax and hoof, more or less synonymous with the emerging notion of a national park. In 1924 Aldo Leopold persuaded the Forest Service to create restricted-use "primitive areas" within the national forest system; the first designated was on the Gila National Forest. The decision expressed Leopold's (and others') sentiments about the value of the wild; it also expressed Forest Service determination to resist future pressures to convert such lands into national parks. Special guidelines ("L-20 regulations") were promulgated to assist administration. In 1927 Robert Marshall argued for strengthening those rules, closing loopholes for logging, grazing, and roading, which resulted in the "U-rules." Unlike the national parks, such areas were set aside by administrative decree, which meant they could also be delisted with a change of mind or administration. As draft legislation for a national wilderness act began percolating, Chief Forester McArdle objected to it precisely because it would remove that discretionary power. By the time of the 1960 Multiple-Use

Act, the Forest Service considered wilderness as one "use" among many, another tile in an intricate land mosaic.[32]

Some aspects of the managed wild were easy: do not build roads, do not clear-cut, do not hunt creatures to extinction. But it was not so obvious how to handle fire. The philosophers, promoters, and prophets of wilderness did not give much thought to how fire might be absorbed into such settings; mostly, they condemned wildfire or ignored it. What obsessed them were the depredations of logging, hunting, and roading. None of the partisans addressed how to manage free-burning fire on the ground. There seemed no point to protect the sanctuaries from ax and auto only to let fire burn them to ash. Yet, clearly, mechanized fire suppression could not continue. Why ban Buicks while boosting bulldozers?

Henry Thoreau famously let a fire escape and burn through the Concord woods. John Muir lashed out at fire as destroying 10 times as much forest as the axe. Aldo Leopold, while a ranger in the Southwest, denounced light-burning as subverting the deeper conservation message of the Forest Service. Later, as a wildlife professor, Leopold softened his position, corresponded with Stoddard, and noted the value of fire in preserving prairie, but he never wrote about the red fire dying on the landscape as he did the green fire in the eyes of the wolf, and he died of a heart attack while fighting a fire on a neighbor's land.

Robert Marshall, another founder of the Wilderness Society, reported from one of his solitary treks in the Northern Rockies that "there were some scenes of desolation that pretty nearly drive an imaginative person crazy. . . . A pessimist would conclude that one summer's fires destroyed more beauty than all the inhabitants of the earth could create in many years, while an optimist would go singing through that blackened, misshapen world rejoicing because the forest will look just as beautiful as before—in two or three centuries. Take your choice." The founders of the wilderness movement chose green over black. "I understand philosophically what you are saying to me," explained David Brower, "but emotionally, I just can't handle black trees." Particularly for those who sought spiritual values, they may well have echoed the Buddha, whose fire sermon equated fire with the messiness and havoc of the world, passions that had to be quenched in order to achieve nirvana and enlightenment. So, it would seem, did partisans of the wild need to quench the ecological passions unleashed by fire and let the landscape achieve its calm climax.[33]

When the Sierra Club published the anthology *Voices for the Wilderness* in 1969, it included essays on wilderness and culture, the American Dream, politics, conservation, human self-interest, human rights, literature, science, plants and animals in natural communities, and human ecology—but not one on fire. There were a few parenthetical comments, noting that lookout towers would not violate wilderness norms and that modern firefighting by such means as smokejumpers could hold fires. (It seemed that fire crews could satisfy the language of visitors who did "not remain.") If anything, the perceived higher values at risk argued for continued fire protection, though the expectation was that it would assume a kinder, gentler form. There was little notion that wilderness would establish a new norm for fire.[34]

A fire denied was a fire postponed and an ecosystem harmed, yet a fire regime was trickier to administer because a free-burning fire could renew or obliterate a wild landscape in an afternoon. The ideal fire-management program might be no program at all, but when flames and smoke could bolt from their preserves, there were risks unlike any associated with letting a wolf range freely. Worse, nature might not be adequate for the task of restoring fire to its historic dimensions. The fires that had gone from the land were not just those nature set and crews suppressed but those people once set and no longer did. It was obvious that mechanized firefighting had no place in wilderness; it was less clear that fire lighting should be banned as well.

Just as, in principle, wilderness proponents had to deny wilderness as a homeland in favor of uninhabited space, so they had to deny anthropogenic fire as a process shaping those landscapes even though removing the human presence might be as powerful as stripping away an alpha predator. Wild places—save the most remote and barren—had coevolved with humanity and its fires. Abolishing anthropogenic fire might be as disruptive as suppressing natural fires. Yet to allow fire officers to kindle fires violated the precept to leave "untrammeled by man" and might be a wedge to crack open those sites to hunters, trailer parks, or casinos—the whole riot of human meddlings that wilderness was intended to exclude. Here was fire management's version of the distinction between legal wilderness and simple wild.

Unlike other challenges, fire could not be shelved by committees, held in contempt of court, or subpoenaed for congressional testimony. It obeyed its own dialectic. It would happen, it would force action. It would compel fire protection to change its strategies, tools, techniques, and conception of itself. Wilderness fire had behind it the power of public opinion and the force of law to compel such changes even if the actual means and ends were vague. No less, it would force theorists of wilderness to accept, if reluctantly and with shameless finessing, that the planet's keystone species for fire might belong. It was easy to ban roads, clear-cutting, and hunting species to extinction; it was not so easy to ban burning. Other expressions of the human presence might be scraped away, but humans as keepers of the flame might not. It was a species thing. Legal wilderness might not re-create a precontact scene so much as create something that had never before existed.

Unsilent Springs and Unquiet Crises

Two more books, one by an outsider and one by an insider, both examining the American environment. The first appeared with the shock of a nuclear explosion. In truth, the image of an obliterating humanity and radioactive fallout drizzling over innocents lies behind much of the impact of Rachel Carson's 1962 *Silent Spring*, with its vision of an Earth poisoned into silence by industrial chemicals. With its publication, an outsider had boldly challenged the system. That same year, even as he was commissioning his Advisory Board on Wildlife Management, Stewart Udall completed the manuscript that became *The Quiet Crisis*. This was an inquiry that narrated the long history of Americans and their lands and came with a foreword by President John F. Kennedy. In a sense, it offered a *Profiles in Courage* for American conservation. Like Carson, Udall called for a new ethic because "modern life is confused by the growing imbalance between the works of man and the works of nature." Both authors were responding to the post–World War II era, what proponents of the Anthropocene have called the Great Acceleration, with its rapid and radical alteration of life both social and natural.[35]

Between them, the two books identified the major poles of what became the environmental movement, one focusing on pollutants and the other on land use. The first inspired a whirlwind of legislation to clean up air and water and to begin leaching away toxins such as lead and arsenic. The second led to a cascade of reforms and new charters for the institutions with stewardship over the public domain. The two themes knotted together in 1970 as Earth Day announced the breadth of social enthusiasm for environmental issues, and passage of the National Environmental Policy Act (NEPA) tethered that public sensibility to political power. Over the next decade, the reforms swept over American society and land like a wave train set in motion from a storm far at sea.[36]

Neither *Silent Spring* nor *The Quiet Crisis* spoke directly about fire. Carson got no further than fire ants, and Udall, Fire Island. Big blowups had less public visibility than nuclear tests in the atmosphere, which were partially banned in 1963. Fires' smoke did not contaminate milk or sicken those downwind like fallout from the Nevada test site. The iconic media image of fire was the Cuyahoga River in Cleveland burning in 1969. In many respects, it was the absence of fire that threatened species with extinction. Still, the wellsprings of environmental concern spilled over into fire policy. The Tall Timbers conferences and the Leopold Report addressed fire head on, but the Wilderness Act and other reforms changed the rules by which fire might be suppressed and started and the character of the agencies that dealt with it. The Environmental Protection Agency, established under NEPA, created standards for air quality and regulated open burning; its requirements for environmental impact statements (or assessments) would influence fuel management schemes or even fire practices on public lands. The NEPA process provided a mechanism by which citizens and NGOs could challenge what had been strictly a government prerogative.

* * *

"In recent times we have begun to cross into an entirely new watershed in the history of the conservation movement in the United States." So spoke Secretary Udall when he addressed the Sierra Club's biennial wilderness conference in 1963. "We are doing so by necessity," he observed, "because the path of land conservation that our government has used for more than half a century is running into a dead end." The old path relied on reserving lands by presidential decree or congressional statute out of the public domain. Those lands were now fully allocated. Further progress required more lands, which could come only through purchase, a transfer from private holdings to public. Although he addressed the issue only indirectly through wilderness, lands could also come by reallocating the existing domain or by redefining the mission of the agencies responsible for their administration. The future held not one path but two.[37]

In 1964 the country took both paths. The Wilderness Act showed how existing lands could be reconstituted. While a wilderness site might remain under the nominal control of the Forest Service, it was no longer subject to Forest Service administrative whims, multiple-use guidelines, or the precepts of forestry. Over the coming years the same experience visited the National Park Service, the Bureau of Land Management, and the Fish and Wildlife Service. Other measures followed, for just as preservation replaced conservation as an informing philosophy, so reformers wanted to reconsider not only the lands under management but also how the agencies managed them. Those acts were like citizen initiatives that directed and restricted legislatures' freedom to act.

The second path, building on the 1962 report of the Outdoor Recreation Resources Review Commission, was the Land and Water Conservation Fund Act, which provided a mechanism to buy land for recreation, parks, and preservation. Among its early targets were seashores, redwood groves, and urban parklands. The Weeks Act of 1911 had authorized the Forest Service to acquire by purchase new national forests, notably in landscapes not in the public domain such as the cutover and burned-over slopes of the southern Appalachians and New England. The Land and Water Conservation Fund extended that option to the agencies of the Department of the Interior. The Bureau of Outdoor Recreation had received, the year before, its organic act. The project was, in a sense, a companion to the Park Service's Mission 66, which was then winding down.

That the country could reserve some patches of its estate to preserve as "untrammeled" while simultaneously purchasing other patches to promote their recreational use suggested that the time had come for an overhaul of the public domain. So in 1964 Congress authorized the Public Land Law Review Commission (PLLRC) to investigate the condition of the national estate. Of the 2.2 billion acres in the United States, the federal government retained 755.3 million acres. The governing legislation was, to believers in rational administration, a shambles of historical happenstance and opportunism. If the federal government were to dispose of some of those holdings, it needed congressional guidelines. If it were to manage the retained lands well, it also needed standards. Lands, agencies, and laws no longer aligned since the last Public Land Commission had completed its task in 1903. In the absence of planning, the public domain produced less revenue than it might, advanced environmental goals more feebly than it should, and often yielded contention rather than cooperation. The public domain needed new laws, and existing laws needed codification.

The PLLRC was not a think-tank study. The commission was a political entity staffed by six members of the House of Representatives, six senators, and six citizens from the public appointed by the president. The commission elected as its chair Wayne Aspinall, a Democratic representative from Colorado. It deliberated for nearly five years. Its conclusion, issued in 1970 as *One Third of the Nation's Land*, laid down a series of "fundamental premises" by which the public domain ought to be governed and a complementary raft of "recommendations." Whether or not its particulars became statutes, the PLLRC bestowed public recognition on the public lands and helped knead them into the politics of the environmental movement. Between the time the commission opened its first hearing and a decade after its report, every public-land agency had its organic act rewritten (or in the case of the NPS, found a surrogate).[38]

* * *

The environmental movement was, in its pith, a revolution in values that suffused and propagated in unpredictable ways. Some reforms acted like fences, barring or mandating particular practices in particular landscapes, as the Wilderness Act did. Others behaved like sheepdogs, aggressively guiding a flock of practices while it moved expediently over the scene (think the Clean Air Act and NEPA). Most simply altered the atmospherics—the climate of opinion and unvoiced but understood assumptions about what was right and appropriate.

Such deep movements could not help but reform fire policy. Fire affected air, water, soil, plants, and animals whether domesticated, wild, or endangered. Everything that altered how Americans related to those environmental features affected the kinds of fires that occurred, just as every modification in fire policy affected airsheds, watersheds, timber berths, habitats, scenic vistas, and whatever uses Americans sought from those lands. How Americans fought, lit, fled from, or ignored fire changed. Since it was the institution most responsible for integrating those varied practices, the Forest Service felt those changes more sharply than anyone else.

In this unfolding saga there was perhaps one constant. At almost every instance, the U.S. Forest Service found itself on the wrong side of history. Its traditional friends among the conservation community turned against it; so eventually did segments of its workforce as it appeared to betray its founding creed. An agency that had long touted itself as the leader of conservation discovered that in an age of preservation, within a culture that, as Samuel Hays put it, valued "beauty, health, and permanence," it had become the nemesis of good land management. To its disbelieving horror the Forest Service found itself morphing into the evil empire of American environmentalism.[39]

The Big Boom

Both Rachel Carson and Stewart Udall addressed their appeals in generational terms. Carson declared that "future generations are unlikely to condone our lack of prudent concern for the integrity of the natural world that supports all life." Observing that "each generation has its own rendezvous with the land," Udall put the burden on his own cohort. "By choice or by default, we will carve out a land legacy for our heirs." It is not likely that many members of the American fire community read both books, but if they did not appreciate the demography of generations, they soon would.[40]

In 1950 the United States had a population of 150 million. By 2010 that number had more than doubled to 309 million. Because the nation's pyrogeography reflects not only its population of fires and their changing profile but also its population of people and their changing demographics, that big population boom mattered. It affected the density of torch bearers and quenchers, how they lived and the landscapes they shaped, and the composition and character of the fire institutions they staffed. Landscape sprawl had behind it a sprawl of people. America's modern era of wildland fire would end with a sustained eruption of big fires, but it began with a demographic explosion.

* * *

The upsurge had three tributaries.

The first was the postwar baby boom, which lofted birth rates upward until the early 1960s, when rates plummeted. The 78 million births in that era became, in a famous simile, the "pig in the python" that would distort every institution as it passed through its generational life cycle. The boomers' maturity defines the era of America's fire revolution. The first boomer turned 18 in 1964, the year of the Wilderness Act and the Public Land Law Review Commission. The first boomer reached Medicare in 2011, as the National Cohesive Strategy for Wildland Fire Management commenced its first phase. But boomers did not mastermind the fire revolution or write its manifestos; the patriarchs, from Komarek to Zahniser, were aged 55 to 60. Nor did baby boomers account for agency reforms in the early years; they did not really come into administrative power until the late 1980s. What they did with fire was what they did to everything else they touched: they added the weight of numbers. They gave heft to whatever trends were under way at the time. They helped push suburbs into exurbs. They made a larger, cheaper workforce possible (it is no accident that the American buildup for the Vietnam war coincided exactly with the entry of the first boomers into the draft). They did not make the "Me Generation"; that label, as journalist Tom Wolfe explained, belonged to the generation between the boomers and their parents. The boomers simply inflated it. So, too, they did not invent the fire revolution. They staffed it, exaggerated it, and made it unwieldy.

The second contributor arrived exactly when the first departed. In 1965 a new Immigration Act opened the country to the largest surge of immigrants in its history. To the trauma of absorbing the boomers was added the task of accommodating an even larger influx of newcomers. By the time the Great Recession spread, it was estimated that 10 percent of Mexicans were living in the United States. Moreover, this influx of peoples corresponded with a comparable influx of foreign goods. Globalization, as it was blandly called, helped shift, as it shocked, the economy.

The third tributary changed how those peoples related to one another and to the political economy of the country. A long wave of civil rights legislation and court decisions opened politics and economy to citizens who had previously been denied equal access. In 1962 the Supreme Court issued its verdict in *Baker v. Carr* with its famous formula, "one man, one vote," which forced reapportionment, while the Equal Pay and Equal Employment acts chipped away at economic discrimination. In 1964 the Civil Rights Act extended equal protection to castes formerly denied it. The Voting Rights Act and affirmative action programs followed a year later. Within a decade administrative decrees and lawsuits would begin turning agency workforces inside out along with the tradition of how firefighters learned their craft, how knowledge about fire was transferred across generations, and how careers in fire might be made or broken. Moves to "diversify" cracked apart once-homogeneous workforces just as the Wilderness Act broke up a roughly homogeneous public domain.

The wilderness movement, in fact, demonstrates the curious alliance of demography and destiny. Its narrative is framed almost exactly by immigration policy. The first Forest Service primitive area, on the Gila, was established in 1924 as the restrictive provisions of a new Immigration Act began to bite; this resulted in the smallest and most selective era of newcomers in American history. Both could be seen as acts of assimilation, of stabilizing a sense of American identity, while the country weathered the Great Depression, World War II, and the onset of the Cold War. The modern era of immigration, the largest by number in American history, commenced with the Immigration and Nationality Act of 1965, passed the year after the Wilderness Act. The subsequent population boom added weight and urgency to an idea that was legislated under very different circumstances.

* * *

No entity felt these pressures more forcefully than the Forest Service. It had the largest staff and was more vulnerable simply by virtue of its numbers. As a national institution in ways not really shared by the National Park Service or Fish and Wildlife Service, both rooted in local settings, what happened in one forest or research station could ramify rapidly throughout the system. So even as Herbert Kaufman was extolling the Forest Service as a model administration, capable of extending a sense of voluntary conformity over its rangers, the sustaining society was moving in the opposite direction, replacing the company man with the rebel, promoting the artist over the engineer, beginning a long trek from self-examination to self-obsession. The Forest Service's coherence made it an easy target.

In 1964 the popular television show *Lassie* moved its setting from rural farms to national forests, reflecting the continuing migration of middle-class Americans to suburbs and people's growing enthusiasm for environmental matters. Yet viewership declined. A ranger-and-his-dog motif could not compete with boy-and-his-dog nostalgia; the country was moving elsewhere. In 1968 ranger Corey Stuart was eased out of the series by means of a two-part episode, "The Holocaust," in which he is injured fighting a forest fire. The fires that mattered that year for the American fire community were those lit by the National Park Service on Redwood Mountain and allowed to burn on Kennedy Ridge in Sequoia-Kings Canyon. But the fires that mesmerized the American public were those that raged in Detroit, the outcome of race riots triggered by the assassination of Martin Luther King Jr.

Hostile Fire, Friendly Fire

The revolution was a revolt of values first and later of practice. Behind those values lay new ideas, and behind those ideas lay new ways of probing, interpreting, and understanding—knowing—fire. The 1960s opened a golden age for fire science.

That realm, too, seemed to mirror the multiuse vision of the Forest Service. Whatever research on fire the federal government needed, the Forest Service did it. Through its control over land, it housed most of the sites suitable for experimentation; through cooperative programs, it unified efforts at universities and industry; and through its capacity to grant political shape to forestry, it joined with state bureaus and other organizations. It staffed experimental stations, built dedicated labs, and virtually monopolized the national investment in fire science. No one else had anything comparable. And more than sheer bulk, the Forest Service harmonized those parts into something like a system.

This was government science for government lands. Unsurprisingly, most of the funded research supported the existing apparatus: it sought to make fire control more powerful and fire administration more precise. But not all American research was in fire behavior, not all was funded from official sources, and not all upheld the establishment order. In 1960 Charles Cooper published a long study in *Ecological Monographs* on the dolorous effects of fire exclusion on the ponderosa pine forests of the Southwest, and the next year he broadened that scope to introduce fire ecology to *Scientific American*. Then Tall Timbers inaugurated its annual conferences. Even as major monies poured in and new labs arose, the old order was dying on its feet.

* * *

The USFS had to overcome what Earle Clapp in 1935 had warned were structural flaws in its research aspirations. "It has not in my judgment been conclusively shown that it is possible, in a bureau such as the Forest Service, primarily administrative in its functions to develop and permanently maintain a strong effective research organization." The need to act overwhelmed the need to know, and officials confused demonstration with true experimentation. When knowledge conflicted with policy, administrators sided with the bureaucracy over its intellectual auditors. Having a de facto monopoly over research was not the same as doing genuine research.[41]

Thirty years later those qualms seemed quaint. Internally and externally fire research, like its favorite topic, was soaring from surface flames into the crowns. An infrastructure arose around three labs at Macon, Missoula, and Riverside. In 1961 a Lake States Fire Conference held in Green Bay, Wisconsin, reaffirmed the need for the Lake States Forest Experimentation Station to commit to "a long-range plan for forest fire research," probably a lab in East Lansing, Michigan. Instead, the region had the Forest Products Laboratory and the state-sponsored Roscommon Equipment Development Center, one connected to the region's aging timber industry and the other relying on its proximity to Detroit's still-thriving manufacturing. The labs all had their odd birthmarks, testimony to local obsessions (the Missoula lab had its origins in Project Skyfire, a quixotic inquiry into lightning suppression), but they were regional clones of a national program; each spoke the same language with a provincial accent. By the mid-1960s perhaps 75 percent of fire research had migrated to the labs; the network of experiment stations absorbed the rest.[42]

Even so, the Forest Service's reach exceeded its grasp. Research also had its equivalent to the mutual-aid pacts and equipment transfers that allowed the agency to extend its influence to the states. The primary mechanism was the McIntire-Stennis Cooperative Forestry Research Program launched in 1962, which could distribute monies to land-grant universities. In principle the act allowed the Department of Agriculture to allocate as much as one-half of the funds appropriated to federal forestry, although nothing like this amount was ever approved. The money of course went to forestry schools, which strengthened the ties that bound the forestry guild.

The land-grant universities became for research what state forestry bureaus were for operations. Whatever investigations into fire the Forest Service did not conduct in-house were effectively outsourced to its cooperators.[43]

The upshot for federal fire research was a program similar to the model proposed by the California university system for the nation's research universities. Only the Canadian Forest Service and the Soviet Union, with its Academy of Science lab at the Sukachev Institute in Krasnoyarsk and the archipelago of research facilities maintained by the forestry ministry, had even an approximation of something comparable.

* * *

The Forest Service used this apparatus to investigate fire topics it had traditionally obsessed over. It wanted to know how fires started in order to better prevent them; how they spread to better stop them; how they damaged woods and watersheds to better ameliorate the consequences; how that risk might be anticipated so it could be countered with presuppression measures; and how they behaved ecologically, or at least in terms of reducing fuels, to better apply controlled fire. (Enthusiasm waned somewhat in the late 1950s after a few prescribed burns bolted away from control lines and into headlines, most notoriously on the Prescott and Shasta national forests.) Research provided the intelligence with which to direct operations and guide agency investments of men and money.

That only reinforced a sense that predicting fire's start and spread was the core function of research and that a fire-danger rating system was the most integrative of research ambitions. When the Forest Service reorganized its fire research projects in 1948, eight regional indexes partitioned the country. By mobilizing the power of modern science, it seemed possible to collapse them into one, founded on first principles, notably fuel moisture. As the Macon lab was being erected, a joint committee of researchers and fire officers convened to discuss the properties of such a system. In June 1959, the Division of Fire Research chartered the project. Two years later the program laid out a basic structure in which fire spread was deemed the most critical of four phases. Spread, in turn, had several indexes by which to characterize ignition, describe propagation in forests and fine fuels, and account for cumulative drying (or "buildup"). Field trials tested the spread phase in 1962 and 1963, and the next year it was adopted for field use and inserted into the Forest Service manual.[44]

Before long the various regions were again adapting to local circumstances, like an artificial language that picks up dialects and slang. The hard-core scientists wanted something "analytical," avoiding statistical or empirical data that introduced various biases. A truly national system demanded that algorithms incorporate the remaining phases—ignition, risk, and energy release. A successor project in 1965 reviewed the scene and, noting the remarkable advances in the understanding of fire weather, heat transfer mechanisms, and the modeling of fire spread, recommended that the Forest Service pursue a full-scale National Fire Danger Rating System (NFDRS), of which the California Wildland System was the closest approximation. With the CWS as a model, another round of meetings led to the articulation of a structure and "philosophy" for a genuinely national approach. By 1970 a first version was ready for field trials in the Southwest. That year one of the great texts of fire science, Mark Schroeder and Charles Buck's *Fire Weather*, brought together the discovered principles and practical conclusions of a decade of high-intensity research.[45]

What underwrote the prospect of an analytical system was a national investment in understanding fire behavior. Something had to translate environmental conditions (such as fine fuel moisture and wind) into the types of fires that occurred. A model, ideally mathematical, based on first principles, not correlations, was the missing link between the parameters that fire officers might gather in the field and the fire spread they could expect. Such knowledge could result only from laboratory experimentation and serious funding. The USFS had the labs; the funding came from several sources, among them the Office of Civil Defense (OCD).

What joined Forest Service labs and outside sponsors were the atomic bomb tests and the National Academy of Sciences-National Research Council (NAS-NRC) Committee on Fire Research. That supplemental interest (and money) was likely enough to catalyze a Forest Service investment that otherwise would not have transcended the early work of Wallace Fons and George Byram. For administrative purposes, the traditional analysis of fire reports could supply the data necessary to design a fire protection complex. In 1957, as part of its rural fire defense mission and as a contribution to a second NRC Fire Research Correlation Conference on Methods of Studying Mass Fires, the USFS prepared tables and maps with which to predict the "maximum extent of spread of mass fires" at various times and places in the United States. In 1960 that work was amplified and its methodology simplified.[46]

That leisurely pace quickened in 1962. With the Cuban missile crisis, with the threat of a thermonuclear exchange that would spawn "firestorms upon firestorms," the Cold War heated up, and the hyperventilated prose of research proposals moved from hypothetical to plausible. The USFS was assigned as point man for a toughened program of Rural Fire Defense, a task most fire officers regarded as an unwelcome diversion from their primary duties and another step toward becoming a national fire service. The OCD committed an unprecedented infusion of research funds. Whatever its internal interests, the Forest Service became the medium for these national interests, the protection of America's citizenry against the fiery consequences of a nuclear war.

Its new labs were in place precisely at the right time to absorb the political and fiscal fallout. As with the mobilization of physical science for World War II, and as mini–Manhattan Projects sprouted in the labs, foresters rubbed shoulders with meteorologists, chemists, physicists, and mechanical engineers. The first rush of hires for the Missoula lab—Richard Rothermel, Hal Anderson, Stanley Hirsch—were refugees from a General Electric team researching an atomic airplane for the Air Force. Fire behavior was now flying far beyond applied forestry. In 1962 the OCD contracted with the Forest Service and United Research Services (URS), a California think tank, to prepare a "mathematical model of mass fire spread compatible with the damage assessment system." The study would include both wildland (rural) and urban settings. The USFS would identify critical parameters, suggest metrics, and generate data. URS would devise the models. Field trials ran from April 1962 to June 1964.[47]

In short order, roughly in lockstep with developments of the NFDRS, the Forest Service received two other contracts from OCD. One, culminating in Project Flambeau (1964–67), led to more extensive field experiments in mass fires. Clive Countryman marshaled the effort, which gave the Riverside lab its first grand project. Field experiments subsequently went international, involving Great Britain, Australia, and Canada under a Tripartite Technical Cooperation Program. The second contract, the National Fire Coordination Study (NFCS, 1964–66), reported on the capabilities of America's diffuse fire protection arrangements to cope with large fires. Such fires were now rambling into exclusive suburbs such as Bel Air–Brentwood, and as race riots boiled over, big fires came into the inner cities. Bud Moore, then in the Washington Office, headed the overall project, while Jack Dieterich shepherded its science.[48]

When it ended, the Forest Service sought other allies, appealing to the American Forest Association, which organized a Task Force on a National Program for Wildfire Control in 1967. Its chair, William E. Towell, wrote for *American Forests* that "the need for a disaster fire plan is national in scope," that it had to embrace all regions and all agencies from the Feds to the states to private protective associations and the forest products industry.

It had to range from the Office of Emergency Planning to volunteer fire departments that would need training to "fight fires on our suburban fringes." Ultimately, the country required "a new awakening" hardened into legislation and funding whose thrust would be to curb the disaster fire "wherever it may strike and *do it now!*"[49]

But whether natural or anthropogenic, whether along the Wasatch Range or in Watts, whether the problem was to predict nuclear-sparked firestorms or fire-danger rating for lightning-struck snags, the core was a working model of fire behavior. That joined everything the Forest Service did in fire protection, whether on its own lands or in cooperation with state forestry bureaus or in its oversight capacity for rural fire defense. Better understanding of some aspects of fire behavior improved understanding of all aspects: fire was a dynamic integration of its parts, from which fire whirls could not readily be parsed from plumes nor forward spread from pyrolysis. Better knowledge of fire behavior meant better fire control. All research contributed.

The political panic passed. The term "mass fire" fell from the literature in favor of the old standby "blowup," and a nationwide fire defense program under the OCD went the way of fallout shelters. Cold War obsessions seem a weird diversion amid a fire revolution, and coordinating for a National Fire Defense Plan ("our Nation's security is partly at stake") belongs with *Dr. Strangelove*, released the same year the survey started. Yet fundamentals are fundamentals. The 50-year run of the fire revolution would, prophetically, end with "firestorms upon firestorms" sweeping the West and a desperate scientific quest to model the dynamics of blowup fires. And virtually every recommendation of the National Fire Coordination Study—from the need for the interchangeability of equipment and crews to a common communication medium to a national assessment of fuels and risks to the ability of rural and urban fire to coordinate on shared fires and not least to an appeal for a National Fire Plan—all happened. Firescope, Landfire, Farsite, the National Interagency Fire-Qualification System, the National Advanced Fire and Resource Institute, a National Cohesive Strategy, and others can all trace their pedigrees to the imperatives of the national security state. In the end, however, the country did not need the fear of mushroom clouds over Chicago and New York. Pyrocumulus clouds towering over the Sierra and the Selway were threat enough.[50]

* * *

The real challenge came from critics who questioned the core premises of Forest Service fire control and the research it sponsored to enact its vision—and even the agency's legitimacy as an oracle on fire. As the OCD tugged the Forest Service toward more intensive investigations into fire behavior and a more expansive reach in national fire defense, it faced a fifth column of insurgents.

It began with the Tall Timbers conferences, which announced a fire world organized on essentially different principles. The Tall Timbers consortium—to grant it a firmer term than it merits—looked toward natural experiments, human history, and a naturalist's pragmatic visions, all of which were a very long shot from hallucinations over nuclear fires, countermeasures forged in labs equipped with wind tunnels, and reorganizations of fire services to better battle hostile fire. The Tall Timbers tribe argued for more fire ecology and less combustion chemistry, more research on how to propagate good fires rather than how to suppress bad ones, and more of the empirical, place-centered knowledge that the architects of the NFDRS had wanted to abstract away in favor of universal algorithms. By self-publishing, Tall Timbers could work around the caltrops strewn over the usual paths of publication in fire science, and by continuing year after year, they championed a cause that would not quietly slip away. Even its conference style was alien. James Agee was stunned when he attended the 1967 Tall Timbers conference at Hoberg's Resort in California. "That was an eye-opener," he said. "People from the South would jump up and declare, 'I believe in prescribed fire.' And somebody else would jump up . . . very much like a revival meeting." That was not how graduates from the University of California, Berkeley, or card-carrying researchers for the Forest Service did science.[51]

If the first Tall Timbers conference questioned the Forest Service's judgment over how to study fire, another publication that year disputed the agency's historic legitimacy to do so. Ashley Schiff's *Fire and Water: Scientific Heresy in the Forest Service*, published by Harvard University Press, traced the uneasy relationship between administrative goals and scientific discoveries. Schiff chronicled the sad saga of official fire research in the Southeast, in which agency politics had either biased the data or squelched results that questioned the program of fire exclusion. The founding fervor had corrupted the agency's claims to disinterested administration. "Thus had evangelism subverted a scientific program, impaired professionalism, violated canons of bureaucratic responsibility, undermined the democratic faith, and threatened the piney woods with ultimate extinction."[52]

In a letter to Biswell that same year, Harold Weaver, long a self-described "admirer of the crusading zeal of the Forest Service and of its apparent esprit de corps," admitted its dark side. Even those who privately agreed with him would not publicly declare their solidarity. "When it comes time to speak up in forestry and advisory meetings, however, they simply say nothing. They haven't dared to." Far from being the paragon of public administration sketched in Kaufman's *Forest Ranger*, the agency depicted in Schiff's *Fire and Water* was a clumsy and unstable alloy of researchers and practitioners that compromised science's role as an independent auditor. Its presumed authority had no historical backing. It got fire in the longleaf almost fatally wrong. There was no reason to think it had gotten fire right elsewhere.[53]

Still, so long as the Forest Service controlled the apparatus of federal fire research, and so long as the forestry guild boxed up other venues of publication, its hegemony endured. That, too, began to change when Secretary of the Interior Stewart Udall requested the National Academy of Sciences to advise and assist toward an "expanded program of natural history research by the National Park Service." The resulting committee thus paralleled the Advisory Board on Wildlife, and like the Leopold Report, its conclusions bore the name of its chair, William J. Robbins. As with fire policy, so with fire research: the national parks established, in principle, an autonomous source of influence.[54]

Issued in August 1963, the Robbins Report sought to redefine the significance of the national parks along scientific criteria as the Leopold Report had along naturalness. Parks were "dynamic biological complexes" in which "evolutionary processes will occur under such control and guidance as seems necessary." Instead of vignettes of primitive America, they were natural laboratories. They were certainly not national forests, and the Park Service could no more outsource its science to the Forest Service than it could the management of its elk. The study referenced and reinforced the themes of the Leopold Report. The deeper conundrum, however, was convincing the Park Service, particularly within the context of Mission 66, that the parks were primarily biological preserves rather than public "pleasuring grounds" and that the agency required natural science rather than relying on local lore and the personal preferences of its rangers, a managerial caste with no necessary technical or scientific training. The agency viewed natural science as part of its charge to interpret the scene to visitors. In its famous trope, the report noted that the agency budget for science equaled the cost of a single campground "comfort station."[55]

The NPS resolved the challenge posed by the Leopold Report by embracing it, but it addressed the Robbins Report by burying it. It released the report in typescript. It established a token Division of Natural Science Studies but granted it little funding or autonomy. Power remained with supervisory rangers or with a ranger subcaste of resource managers. In 1967 director Hartzog coaxed Starker Leopold into serving as chief scientist, an offer Leopold accepted for just a year, during which he remained in Berkeley. When he departed the position, the ebb tide carried a downgraded Office of Natural Science Studies along with it. The National Park Service was not—and did not wish to become—a research institution. It did not consider parks as primarily natural research sites, and it certainly did not want science to restrain its discretion to act as it saw fit. The Franklin Lane letter to Stephen Mather was its guide, not data sets from academics.[56]

While the Robbins Report amplified the number of ecological concerns cited by Leopold, it did not single out fire as a topic for investigation. Instead, in the early years, Park Service scientists found a more congenial reception at Tall Timbers than within their own organization. Still, wildlife biologists began replacing foresters, the old forestry pipeline for fire research constricted, and the American fire community found an alternative perspective and revenue source that had never existed before. With dramatic flair, the Park Service broke with national fire policy; with far less fanfare it renounced establishment fire science. The Leopold Report updated the Lane letter with a vision of vignettes of primitive America. The Robbins Report rationalized that vision into natural ecosystems. Both were eager for natural fire.

Wildland fire research remained deeply ghettoized. In the second Tall Timbers conference, Eugene Odum observed that if people "cannot learn to handle intelligently this relatively simple factor"—fire—then we had "no business attempting to control rainfall or other vastly more complex matters." Yet in scanning the research supported by grant agencies such as the National Science Foundation, he was "impressed with the lack of projects on basic fire ecology." He discounted systemic bias; the core problem was that no one applied. They did not apply because they came out of forestry and had funding from the Forest Service, and the USFS had had since 1928 a monopoly over the subject. By 1969–70, however, even the National Science Foundation identified "fire ecology" as a topic for special consideration. Like a creeping ground burn, fire was slowly feeling its way into new intellectual fuels.[57]

*　*　*

Buried among the NFCS recommendations were appeals for better communications, the development of infrared mapping capability, and new training methods. In the early 1960s the nation's technological prowess was as primed for an inflection as was its science. What happened was a tiny revolution—literally—ultimately based on the miniaturization of electronics. But like the cybernetic model in which a small switch can turn on a dynamo, those tiny pieces had vast consequences.

Information technology came stealthily to wildland fire. Forest Service thinking on equipment still pivoted on war-surplus machinery, from half-tracks to S2F Tracker airplanes. The establishment of a Defense Department on a quasi-permanent war footing meant that "excess equipment" became routinely available. The Arcadia (later San Dimas) equipment and Missoula development centers and the Roscommon Center in Michigan, funded by a consortium of states, specialized in beating such swords into plowshares. And there was also the specialized equipment in aerial fire control and smokejumpers. Up through World War II, the USFS had also staffed a radio lab in Portland, Oregon, to improve and dispatch homemade radios, primarily for fireline use. (No one prior had tested wireless in the forested and mountainous terrains typical of national forests.) Before the war, the Portland facility had served as the main electronics lab of the U.S. government. Then the war came, and electronics moved into the private sector, although typically with government sponsorship. But the hunger for better fireline communications and surveillance endured.[58]

In 1960 *Tiros 1*, the first weather satellite, went into orbit equipped with television cameras and infrared sensors. The next year witnessed the first communication satellite, *Telstar 1*. By 1963 electronics had developed rapidly enough to suggest classroom training simulators (the first manufactured by ITT, which had launched *Telstar*). None had any immediate consequences for fire protection; even infrared mappers required more than a decade to mature. Yet much as background demographics changed the workforce for fire, so background technology altered the way fire agencies conducted science and operations. Tellingly, the launch of Telstar coincided with the second edition of Norbert Weiner's seminal *Cybernetics*, in which he argued that the technologies of control—the methods of sensing and switching—were to a postindustrial future what prime movers had been to the industrial past. The knowledge that was power was coded in transistors.[59]

Early gadgets seemed more like trinkets or toys, minuscule compared with retardant-dropping B-17s and D-8 tractors gouging firelines along ridgetops; but eventually knowledge, or at least its simulacrum, information, packaged in usable forms, became the great integrator of wildland fire. Information acted as the interstitial medium that bolstered and bolted together science, operations, and administration. Instead of retrofitting matériel discarded from the Korean War, information technology promised to invent something new. After all, the fire revolution was not based on bigger planes and bulkier pumpers but on a better understanding of how fire functioned in landscapes and how humans ought to respond. It needed to measure its achievements in bytes per second rather than gallons per minute.

While such reforms lay in the future, the future was approaching faster than anyone at the time of the Leopold and Robbins reports could imagine. In 1965 Gordon Moore of Intel announced his eponymous law.

Interior Fire

The fire empire of the Forest Service slowly crumbled. One after another, portions of the public domain broke away to create autonomous fire organizations. The Park Service was the first to declare independence, but although it loomed large in symbolism, it was tiny in area. It could hold its own with the Forest Service in journals but could not match it in the field. It simply lacked the collective mass and coherence to stand by itself when a big fire roared. By the late 1970s the new Alaska parks altered that calculus, but Alaska was remote, and it argued for a fire strategy not based on overwhelming force. Some other institutional presence had to challenge the Forest Service if the revolution were to lead to a realignment of politics and not just of policy.

That counterforce arose when the Bureau of Land Management assumed its modern character and became the big brother for the bureaus of the Interior Department. The BLM oversaw more land than the USFS while lacking the obligations that bound the Forest Service to the states and that made internal reforms complicated. It was free to develop along a Forest Service model without either the public scrutiny or the internal dissent. Rather, critics argued for its build-out while largely sparing it the challenges of wilderness and logging that convulsed its rival. The Interior agencies found in a bulked-up BLM an ally to check and balance the behemoth in the Department of Agriculture. That arrival proved often ungainly, rarely rational, and ultimately uneven as the young guns sought to take on the old guard.

* * *

Until 1905 the United States had the curious experience of housing its forest reserves in the Department of the Interior and its Bureau of Forestry in the Department of Agriculture. Nearly everyone recognized that Interior's General Land Office (GLO) was a flawed instrument to oversee the reserves; it had been created to move the national estate from the public to the private sector and was ill suited by temperament or talent to manage them in perpetuity.

Then Teddy Roosevelt transferred the reserves to the Bureau of Forestry in the Department of Agriculture. At a stroke the renamed Forest Service became one of the most powerful bureaus in the federal government while Interior saw its landed inheritance (and its political clout) rapidly sold off until the Taylor Grazing Act and a presidential order shut down the process in 1934. The GLO retooled to classify lands and begin some nominal management, leaving most of the unalienated lands otherwise to be overseen by a newly constituted Grazing Service. The Department of the Interior sought, repeatedly and unsuccessfully, to reclaim the national forests. In 1946 an executive order consolidated the Grazing Service, GLO, and other agency scraps into the Bureau of Land Management.[60]

The Department of the Interior had become a holding company for a miscellany of public-land agencies, themselves often a miscellany of sites and purposes. Collectively, the BLM, National Park Service, Biological Survey, and Bureau of Indian Affairs oversaw more lands (and more symbolic landscapes) than the national forest system, including the nation's prized parks, wildlife refuges, and monuments. But they had no collective fire presence. Each agency had its own fire organization; and most parks, refuges, reservations, and monuments had their own apparatus. There was scant common effort even among the parks or refuges. There was almost no departmental heft—no Interior fire presence to match that of the Forest Service, much less its pact with the National Association of State Foresters. On the contrary, the Interior agencies had their own staff foresters and followed the Forest Service's lead. Interior had no central fire office.

Except in select parks, Interior hardly had a fire program beyond local volunteers and the occasional conscript. A primitive protection program began with the creation of statutory grazing districts in 1935–37 and then became a departmental affair (a "force account") in 1942 with the termination of the Civilian Conservation Corps program. From then until 1961, the fire budget for the western states was minimal, particularly regarding personnel and infrastructure, which caused "serious disruption to all other resource management programs during the fire season." When Jack Wilson was hired as a range conservationist by a BLM Wyoming district in 1948, the staff consisted of him (as manager) and a clerk to oversee 20,000 square miles. Fire was a seasonal, ancillary duty, fought by per diem guards, ranchers, nearby USFS personnel, and a 1943 bomb carrier converted to a fire engine. It was a situation "very typical at least of most BLM districts between 1936 and 1955." In fact, Wilson concluded, "I was little different than the GLO ranger in 1897."[61]

Gradually, the agency learned to cope with bigger fires, rising with the general tide of fire-protection improvements in the postwar era. In 1949 Congress authorized a fire-suppression account with allowance for supplemental appropriations in big years. In 1958 it authorized the agency to spend those funds for presuppression work, which allowed a buildup of forces outside programmed budgets. But the fire program had to operate within a badly flawed agency that still lacked a congressional charter and a coherent mission. In 1960, the year Herbert Kaufman praised the Forest Service for its integrity, political scientist Phillip Foss condemned the BLM for its bureaucratic turpitude. If *The Forest Ranger* established the USFS as the gold standard of public service, Foss's *Politics and Grass* named the BLM as its dark side. The BLM became the textbook example of administrative "capture" by its clientele, most notably the livestock industry. That perspective was reinforced in 1966 by Grant McConnell's *Private Power and American Democracy*. Slowly, prodded by the energetic Secretary Udall, Congress issued piecemeal variants of the Multiple Use-Sustained Yield Act for the BLM, culminating in the Classification and Multiple-Use Act of 1964. The goal was to bring the BLM into administrative parity with the USFS.[62]

It was easiest to scale up the fire program. During Udall's tenure the buildup became serious, with a sudden flush of budgeted funds that permitted the hiring of staff and "limited replacement of obsolete and unserviceable World War II military equipment obtained from surplus." But access to emergency accounts for fires meant it could hire crews and equipment as needed, and some of the purchases added to a quasi-permanent infrastructure. By 1963 the overhaul had so progressed that thoughts turned to creating a service center for fire operations somewhere in the Intermountain West. Overall, however, the BLM—and the Department of the Interior, for that matter—depended on the national system created by the Forest Service.[63]

During the 1960s, the Interior's bureaus began to evolve away from the USFS. The Leopold Report redirected the national parks. In 1966 the National Wildlife Refuge System Administration Act consolidated assorted refuges and ranges into a unified National Wildlife Refuge System. Fitfully, Congress endowed the BLM with piecemeal charges that clarified by executive fiat some of the ambiguities of its arbitrary origins. The BLM began to move closer to the multiuse model of the Forest Service even as the Forest Service was beginning a forced withdrawal from it. By 1976, as both agencies received new congressional charters, their trajectories appeared to cross, and they began to approach a penumbra of firepower parity.

* * *

What changed the political geography of American fire was the admission of Alaska. The Forest Service held some high-value timber lands on two national forests, the Chugach and the Tongass, along the southern and southeastern coasts. The immense inland empire, however, fell by default to the BLM. The act of admission allotted 20 years for the state to select lands that Alaska would alienate from the federal estate, a tumultuous era during which combustion—internal and free burning—figured hugely. Oil replumbed the political economy of the state, while wildfire everywhere redefined the federal presence. The ambiguities of the BLM's mission allowed it extraordinary freedom to maneuver. It became serious about fire, and fire protection became its most visible administrative presence.[64]

Alaska was too large to ignore and too remote to absorb. Because the United States had reluctantly purchased it from a Russia equally reluctant to sell it, it did not fit into any administrative model. It did not even become an official territory until 1912. Outside the soggy coastal forests, no organized fire protection existed until 1939 when the Alaskan Fire Control Service (AFCS) began patrolling select roads and rails. For two years it ran the CCC in Alaska, and then the military buildup during World War II did for the AFCS what the CCC did elsewhere. In 1947 the newly constituted BLM absorbed the AFCS into its Division of Forestry. Alaska then compressed into the next 10 years what the Lower 48 had taken 70 years to achieve. The prevailing sense was that it had no option, that as Charles Hardy confirmed, "Proper management of these lands is dependent upon adequate protection." Emergency funding, aerial fire control, a fire-danger rating system, cooperative agreements, a research presence (through the Forest Service), an alliance with national security to keep the air clear across a Cold War border state, a smokejumper base, emergency and native crews, big fires, and defining firefights—one after another, the standard pieces of modern fire protection clicked into place.

There were modifications, of course, because Alaska was different. It was huge, larger than the next two big states, Texas and California, combined. It added 374 million acres, nearly twice the size of the national forest system. When Alaska's national parks were established, they amounted to two-thirds of the entire national park system by area. Fire seasons passed in true boreal boom-and-bust fashion, some with hardly a willow burned, and others exploding over trackless taiga; but on average Alaska added a million burned acres annually to the national total. In 1957 some 500 wildfires romped over an estimated five million acres and turned the Yukon Valley into a cauldron of smoke. Alaska was not so much a statistical anomaly as it was an alternative world. It was an outlier in the sense that a supernova is an outlier among stars.[65]

Significantly, Alaska remained politically under the dominion of Interior. The BLM did here what the Forest Service did elsewhere: it was the gravitational presence that attracted everything around it. Even the State of Alaska contracted with the BLM for fire protection on lands for which it had responsibility. In 1960 the arrangement was formalized with a Cooperative Fire Protection Agreement. But the reverse was equally true. Whatever happened to boost fire protection in Alaska ratcheted up the BLM's fire program. The 1960s were a heady era generally for Interior under the charismatic leadership of Secretary Udall. Moreover, Alaska Fire Control had a long-standing director, Roger R. Robinson, who cultivated the program from its origins through the 1960s and bequeathed what few Alaskan institutions enjoyed, longevity and continuity. The admission of Alaska to statehood rallied national enthusiasm to bring to it formal fire protection. The Alaska fire program fused all these factors into what became its heroic age.[66]

Besides, Alaska had plenty of fires, and it soon became evident that they were not all the result of white transients and careless indigenes. Lightning accounted for most of the burned area by far, a realization that led to a special agreement with the National Weather Service that in turn culminated in an automated lightning-detection system. It also meant that fire protection could not simply patrol roads and villages; it would have to fly into the backcountry and cope with mammoth fires of the sort that the Lower 48 had largely throttled out of existence. The interior was a "green desert," similar to the high desert country of Arizona, Utah, and Nevada, and it would require a fire apparatus similar to theirs. With ample money and the technology of aerial fire control, BLM Alaska pushed its fire program ever farther into the hinterlands, all under the glare of a national spotlight.[67]

The early 1960s were benign, granting the program an opportunity to field-test its emergency fire crews, expand its fleet of aircraft, and work out a program that brought an extra contingent of smokejumpers from Missoula. The BLM believed it could suppress perhaps 75 percent of all Alaskan fires and hold burned acres below 100,000. Then the fires returned. In 1966 about 600,000 acres burned; in 1968 a million; and in 1969 the lid blew off and over four million acres burned. The Swanson and Russian river fires on the Kenai Peninsula alone burned a million acres and became the costliest fire in the nation. Before the season ended the BLM was attempting cloud seeding to suppress lightning and encourage rain. The BLM had not, J. H. Richardson affirmed, been able "to identify anywhere fires can safely be left to burn without serious consequences and high costs." Full-bore fire protection in Alaska had arrived.[68]

Paradoxically, it arrived just as environmental issues were remolding fire programs in the Lower 48 into more subtle forms. Fire officers on the Kenai fires sent bulldozers to punch firelines the size of interstate freeways through a prime moose refuge and over permafrost. Other Alaska fires that year roared amid backcountry of a sort the NPS was allowing fire to free-burn in the name of naturalness. Instead of associating fire with the kind of wreckage produced by the gold rush, which overturned not just the biota but the very soil, a new generation of naturalists was inclined to pair it with wildlife and with the kind of forage that Alaska's megafauna required. Wildfires, one old-time Alaska fire officer wryly observed, have "stirred an ever increasing controversy on whether or not to control them."[69]

It was possible to view Alaska binocularly. Through one lens the Last Frontier presented the chance to repeat the best opportunities of the old days; through the other, it was a chance to avoid their worst consequences. The one argued for new mineral rushes, more private lands, and more fire protection. The other pleaded for parks, refuges, and a forbearance toward free-ranging fire akin to the tolerance for grizzlies and wolves. That contest between those two visions intensified, each coiling around the other, until the Alaska National Interest Lands Conservation Act (ANILCA) passed in 1980. Brooding over the 1969 season, the American Forestry Association concluded that "man once again is upsetting the balance of nature on a last frontier. It is up to man to help right that balance" with better fire control. Between statehood and ANILCA, the Alaska BLM had to absorb those tensions, and as it wrestled with fire, it grew stronger. Like the Forest Service in Pinchot's day, the BLM reckoned that fire control was the best bet for institutionalizing a new agency. Fires could not be ignored, and because they brought in emergency monies, fire control could in principle pay for itself and endow a permanent infrastructure at the same time.[70]

By the mid-1960s the BLM's Alaskan outpost—the fire department for all matters Alaskan—challenged its other more traditional programs. In terms of values protected, the Oregon and California Railroad Revested Lands (O&C Lands) in Oregon's Coast Range overflowed with high-value timber, and mineral leases in the mountain West and on the continental shelf brought in significant revenue. But fire control was the BLM's bid for national recognition and parity with the Forest Service. Most of its landed estate remained with western grazing and was thus prone to burning. In 1962 Congress increased the BLM fire operations budget for grazing lands by 150 percent and in 1963 by 175 percent. The BLM was well on its way as the nouveau riche of the American fire community. The funds were enough that they caught the attention of the Bureau of the Budget, which demanded some administrative order. A task force suggested a fire service center in Boise as a focus for operations in the interior West.[71]

* * *

In August 1964, 32 fires blew across 278,159 acres around Elko, Nevada. Before the smoke dissipated, the governor had declared a state of emergency, and suppression had mustered 2,500 firefighters, 65 aircraft, and 280 trucks and bulldozers, all from 10 states and a dozen agencies bound by mutual aid agreements. To cope with this surge, the BLM national director established a coordinating center in Salt Lake City.[72]

It was not just that the fires exceeded anything the local BLM districts could handle. They lay beyond the capability of the agency. After the ashes cooled, it commissioned studies to investigate options for the future, especially whether it should rely on contract or in-house forces for protection. The recommendation was to build up its own capacity, although the O&C Lands remained under contract with the Forest Service, and cooperative agreements acted as a force multiplier. The BLM, however, was still too new and its holdings too dispersed, and like most of the Interior agencies, it had no common fire corps by which to bulk up its presence. The exception of course was Alaska. And the Alaskans had an idea how to reform the inchoate fire scene on the BLM lands of the Lower 48—treat the Great Basin like the Yukon Valley. Bring the kind of fire suppression organization that worked in Alaska to the Intermountain West. Instead of trying to translate a continental model north, they would transfer the Alaska model south. In 1965 the BLM established a Great Basin Fire Center at Gowan Field, Boise, and staffed a downtown joint fire coordination center with the USFS.[73]

The BLM was big enough that it demanded its own stall in the nation's fire cache. It tilted the scales toward Interior in ways that had never been possible before and thus boosted that part of the fire revolution that sought to limit the power of the Forest Service. Yet it did so by creating a parallel institution, not a radically different one. It challenged the Forest Service by aping its policies, practices, and style. The NPS challenged the USFS on the how and why of fire management. The BLM challenged it by shifting a substantial chunk of the national investment to a parallel institution in another bureau.

Coordinated Fire

In the folklore of the fire revolution, the creation of the Boise Interagency Fire Center (BIFC) stands like Chimney Rock on the Oregon Trail, a landmark on the way to the future. It commemorates the moment when the logic of cooperation prevailed and the fire community inflected permanently into interagency forms. The modern era of fire suppression opened in 1969 beside the tarmac of the Boise International Airport.

This is history written by the victor and long after the war. Like most matters affecting the fire community during the 1960s, reality was more complicated and interesting. BIFC actually did not invent interagency cooperation. The Forest Service had pioneered the concept for decades through mutual aid agreements, compacts with the states, local arrangements with adjoining federal land agencies, and memoranda of understanding between the secretaries of Interior and Agriculture. What shifted was not coordination among firefighting forces or even agreement over policy, because suppression was still the national norm. What changed was the centrality of the Forest Service.

The participants differed in interpreting the shift. For the Forest Service the advent of BIFC marked a strengthening of capabilities in its Intermountain Region, an evolution in mutual aid along existing patterns. For the BLM it signified a dramatic break with the past, the moment when its fire program strode onto the national stage and demanded equal billing. Accordingly, even their official records disagree on dates, intentions, and catalysts.

* * *

The rough chronology is as follows. In 1959 Forest Service fire staff in Region Four proposed a "smokejumper and aircraft facility" for Gowen Field. Within a year the concept had ramped up to become an interregional fire facility similar to those developed elsewhere in the West. The Boise National Forest opened discussions with both the city and the state about acquiring a site, probably through a land exchange. All this followed a familiar formula.

What altered the dynamic was the BLM's expression of interest in a joint air facility. In 1961 the agency had dispatched a contingent of its Alaska crews to Boise to help fight the Squaw Butte burn, a 100,000-acre range fire. The agency realized there would be other such calls, and as investment in its fire program swelled, it sought an administrative center of some sort. The 1964 fires showed the hopelessness of attacking Nevada grass fires with

Alaskan crews and the need for greater pooling of personnel, hardware, and talent within the Intermountain region. The temporary coordination center established in Salt Lake City needed to become permanent. The proposed Boise facility was an obvious solution.[74]

The BLM wanted two outcomes. One was a joint air base, or "Boise Interagency Fire Control Air Center," and the other was a dispatching site, what it chose to call the Great Basin Fire Control Center. It brought in Roger Robinson from Alaska to head the operation, and he brought the Alaska model with him. Preliminary meetings were held in Salt Lake after the 1964 fire season. The value of cooperation was self-evident. A working arrangement, the Western Interagency Fire Coordination Center, was in place for the 1965 season. By this time the Weather Bureau, which had responsibility for fire weather forecasts, was also interested. The next year the Idaho State Land Board worked out an exchange with the BLM for what was being called the Boise Interagency Fire Center.

A memorandum of understanding was signed by the BLM, Weather Bureau, and Forest Service Region 4 in July 1967. Construction commenced under BLM supervision, while the Forest Service prepared detailed plans for running the facility. When the base went operational in 1969, in time to support the Kenai Peninsula fires, the BLM under Robinson maintained the buildings, supervised public affairs and training, and ran its own fire program. The USFS oversaw the smokejumper loft, the infrared aerial scanning program, and its western zone air unit, with its National Fire Radio Cache in the warehouse. The USFS also ran the fire control program of the Boise National Forest. Each agency had its own chief of operations. Soon, however, some useful ambiguities that had helped birth the facility needed sharpening.

A dispute over titles threatened to turn toxic. The BLM wanted their chief to be called "BLM-Director, BIFC" and the Forest Service director the "Forest Service Chief of Operations." The facility could have only one director, it insisted—the man whom "the public can go to for answers." The Forest Service replied that if their man was not also a director, he would "be considered subservient to the BLM man." The two agencies had cooperated in that they shared the same facilities and that dispatchers could chat face-to-face. But if further integration occurred as was planned, then the Forest Service demanded that it be recognized "as a full partner both in fact and in appearance."[75]

There was more than status at stake. For the BLM, BIFC was not just its main presence in the Lower 48 but also its bid to establish national standing. For the Forest Service, BIFC was one regional coordination center among several, and the BLM agreement was with the regional (not the Washington) office, and the local forest staffed it. In 1964 the Forest Service had established a rough national fire coordination center in Alexandria, Virginia. At BIFC the Forest Service contributed the bulk of the high-cost hardware. The brouhaha over titles disguised the real issue, which was the national role of BIFC (and the BLM's claim for recognition) and the evolving relations between the BLM and the USFS. What one side insouciantly called "cooperation" looked to the other like competition. In face-to-face meetings, representatives argued this point for hours with no agreement. Writing to the deputy chief of the National Forest Systems, Merle Lowden, director of fire control, concluded that the arrangement would have to be decided between the Forest Service chief and the BLM director.[76]

The Forest Service had the weight of history and sheer presence behind it. It was happy to cooperate—it had done so for all its history. It was not pleased to jointly operate a facility, particularly one with national aspirations, at which it was subordinate. The BLM had its hands on the levers of what it regarded as the future. It had a decade of dramatic growth behind it, it had functioned in Alaska as the Forest Service did elsewhere, and it saw BIFC as a project in national coordination, not simply a regional one. It wanted nominal directorship of the center as a totem of its achievements and ambitions. Over the allocation of titles and the grander vision of the center, it would not yield.

In the end, they agreed to disagree. The BLM's representative would be called BLM, Director, and BIFC would service the agency's fire needs for the entire West. Its strength would make it a champion of Interior overall. The USFS representative would be called Forest Service, Administrator, and would assist mostly the Intermountain Region, supplementing the agency's other centers. The Weather Bureau would transfer to BIFC responsibilities for its Fire Weather Service in the West. BLM director Boyd Rasmussen and Forest Service chief Edward Cliff signed a memorandum of understanding in May 1969, along with the director of the Weather Bureau, George Cressman. The BLM director was R. R. Robinson. The Forest Service administrator was Howard Ahlskog, the supervisor of the Boise National Forest.

The finesse worked. When USFS chief Cliff spoke at the center's dedication on July 25, 1970, he blandly intoned that his agency was "pleased to join with the Bureau of Land Management in the development of a joint center that will serve the needs of both agencies and through its joint nature provide strength through unity." The issues festered, however. (The warehouse had a line painted down the middle to separate the two agencies' caches.) It would not begin resolving until the Forest Service cracked under the pressures of the 1970 fire season and began its stutter steps toward a radical overthrow of the 10 a.m. policy and the world that policy had informed. The BLM suffered no such strains. It remained in full suppression mode.[77]

The Boise Interagency Fire Center did not reform fire policy. It sought to strengthen fire suppression, not redirect its mission. It added clout to suppression by granting it access to more resources and higher public and political visibility, even bestowing on it an aura of glamour. Nor did it reform practice. The BLM hired people with the same training and values as the Forest Service, aspired to apply the same multiuse model to its lands, and intended to fight fire in the same way. The BLM wanted acknowledgment that it had arrived. The Forest Service saw BIFC as a "unique event" in that it addressed the peculiar political geography of the Great Basin, a circumstance not likely to repeat elsewhere. The BLM saw BIFC as unique in that it marked a shift in the general history of American fire. Each might seek an "ever closer union," but each also imagined itself as head.

Yet BIFC did make visible an emerging new order. The fire revolution was fracturing the hegemony of the U.S. Forest Service. The old imperium was splintering under a rash of independence movements as each agency sought to control its own policy and to establish a workforce capable of applying it. But the country could ill afford a free-for-all disintegration: the pieces that flaked off had to be reassembled into working condominiums. The founding of BIFC showed what that future might look like and the stresses that it would undergo. In the old system all the parties had orbited around the Forest Service. Now the BLM, as champion of the Interior agencies, demanded parity and insisted that the American system be reorganized around two suns, even if one sun in 1970 still far outshone the other.

The Revolt in the Provinces

The fire revolution was bicoastal, some would say bipolar. One pole lay in Florida, the other in California. That bifurcation reflected the general strengthening of the two coasts at the cultural expense of the middle. But the poles themselves were only the pith of broader themes and a wider geography. If this was a revolution from above, it was equally a revolution birthed in the provinces.

The fact is, the United States is more than a federation of states; it is also a confederation of regions. North, South, and West have been the big three, and the dynamic between North and South, one based on free labor and one on slavery, competing over the West, once triggered a civil war. Those regions have persisted, underwritten by broad geographic characteristics and reinforced by centuries of habitation, accents, customs, and distinctive land usage. Where fire was among those distinctive practices, the regions also defined the provincial hearths of American fire.[78]

* * *

From time to time different regions have dominated. For a while the epicenter of problem fires was New England, its narrative full of Dark Days. Then it moved to the Great Plains, which gloriously coincided with a procession of traveling artists who created a genre of painting centered on prairie fires. Then the Lake States burst into prominence like a volcano rising from the woods, as land clearing and railroads fused into an unstable alloy that for over 50 years erupted into lethal conflagrations. When the Forest Service assumed responsibility for much of the Northern Rockies, which blew up soon afterward, not just once but over and over, the USFS took that regional fire as a national norm and crafted responses that it diffused across the country. When logging moved into the Pacific Northwest, gorging on mountains of slash, so did the attention of the American fire community. And when the industry relocated to the Southeast for sawtimber and then pulp, fire politics followed. California and Texas were almost fire nations in themselves. All in all, like other Americans, free-burning fire has seemed happiest on the open road. But during the 1960s two regions dominated, and the dialectic between them defined the discourse of the fire revolution. Florida and California argued, respectively, for prescribed fire and natural fire, while California also argued for new notions of suppression.

Florida was southern fire on steroids. It was as though every regional issue and traditional lore of woodsburning poured down the peninsula, refining as it flowed, until at the Everglades the biota itself purified into a simple dichotomy of fire and water. Florida would burn—the usual saying was that it burned twice every year. Whatever ambition anyone had for the land or for fire, controlled burning would remain the informing practice. If you could not burn it, you could not use it, and if you did not burn it right and often, wildfire would do the job for you. A long folk tradition crystallized into an art form, as much a feature as the piney woods or cracker cattle. Open-range herders scattered matches like dandelion seeds. To the national thesis that systematic fire protection was the foundation of land management, it proposed prescribed burning as an antithesis. No place did it with more institutional panache. Anchored at Tall Timbers, Tallahassee became the Silicon Valley of prescribed fire. The past persisted.

California had broken its past. Its modern history began with the shock of a gold rush, obliterating much of the indigenous population almost overnight, remaking the land as a plague of prospectors and speculators swarmed over, dug up, hydraulicked, burned, and finally settled, even as more population rushes piled up. But little that was old survived long in California; each new thing was soon superseded by a next new thing, and transitions could be abrupt. No state packed such extremes into such proximity: Mt. Whitney, the nation's highest point, rose scarcely 60 air miles from Death Valley, the nation's lowest. That pairing of antitheses became a California style, one without a middle synthesis. And so it proved during the fire revolution.

In the early dialectic between fire lit and fire fought, California chose to fight. Systematic fire protection became California's particular invention for the American fire community. The pressures for outright protection intensified as the state's population doubled decade after decade, and then it rode the great surge of postwar population southward, where it began to interbreed with urban fire traditions. What made California's choice nationally significant was that the state unified around that decision in ways that, added to its general heft, unbalanced the national discourse. What California did with water it did with fire. Its statewide master plan forced Northern and Southern California to fuse: it allowed for the transfer of fire forces from north to south as it did waters. The California fire master plan effectively forced Northern California to accept the protocols of Southern California, which meant an uncompromising fire suppression supported with specialty crews and heavy machinery. Here urban fire services absorbed wildland firefighting, nudging it into an all-hazard model. In Southern California experiments in prescribed fire flared briefly and then ceased. They ended for the Forest Service when the threat of smoke and escaped fires overwhelmed ecological enthusiasms and for the Park Service as it appreciated that fires of any origin encouraged invasive species. The only strategy left standing was suppression. The Southern California model became California's, and California's threatened to become the nation's.

Still, interest in light-burning lingered in niches, a counterculture of fire management, a return-to-nature movement that focused on the Sierra Nevada's most monumental sites and its Big Trees. Here the notion that natural fire might be left to range freely and that controlled fire might prove a benign way to heal a man-made distemper underwent its first beta tests. Although the symbolism loomed as majestically as the High Sierra crest, the land available was modest. The idea roamed more freely than the flames it inspired. It thrived only in institutional reservations, as it were, able to escape the California master plan and the push for conformity. If it were to flourish, it would need mountains to match its fires.

Each paradigm propagated where conditions favored it. The Florida model diffused throughout the Southeast and eventually even into the Caribbean and Mesoamerica. It became a beacon for institutions such as the Fish and Wildlife Service and The Nature Conservancy, which could broadcast the message in patches throughout the country. The California model for natural fire required special but large sites. The vision of wilderness fire needed spacious wildlands, and so it migrated to remote sky islands in the Southwest, the rumpled Northern Rockies, and big-taiga Alaska. The wildland-urban hybrid attached to those places whose landscapes slammed together open land and exurban sprawl.

The fire revolution was not simply a national movement or a coordination among national institutions but also an interaction among regions and the nation at large. A national language of fire there might be, but it would express itself in enduring dialects. In the early 1960s it spoke with three accents—Tallahassean, Missoulian, and Southern Californian, and to make that fire triangle into a pyramid, add the idiom of the tallgrass prairie. It looked to Tallahassee for prescribed fire, to Missoula for the big fire in the backcountry, and to California for suppression.

Mostly, in the early 1960s, like the country at large, it looked to California. What the Forest Service was to wildland fire, California was to the Forest Service. By its sheer bulk and brawn California could set the national agenda. When the revolution came, California alone among the American states contained all the fires that would define the coming decades. In the South Coast it effectively announced the era of exurban conflagrations, in the Sierra Nevada it field-tested natural fire, and amid the Big Trees it transplanted prescribed fire from longleaf savannas. It had immense problems, and it was itself an immense problem for the country. If a practice stumbled in California, as natural fire and prescribed burning did, it could stall the practice elsewhere in the West, except in protected refugia. Equally, if a practice succeeded in California, as hybrid wildland-urban suppression did, it might propagate to places that did not need or want it.

What happened in California rarely stayed in California. A generation of fire officers who had grown up together, or had passed through a formative California tour, accepted the California system as an exemplar and then metamorphosed into a cohort of like-minded souls. Their California years affected them as a tour at Fort Riley had so many of World War II's generals. When he became director of the Division of Fire Control, Lowden began to distribute that brotherhood across the country. Lynn Biddison went to

Albuquerque, John Heilman to Missoula, Dick Millar to San Francisco. Carl Hickerson went to Portland, Bob McBride to Ogden, and Ed Corpe to Atlanta. As the agency moved toward interregional solutions, the California model and mind-set infected the others, step by step, traveling along with its crews, pumpers, and overhead teams.

California magnified the best and worst of the revolution. If Harold Biswell could argue for burning on ranches and parks, using Hoberg's and Whitaker's as proving grounds, regional forester Charlie Connaughton could thunder that he would tolerate "no escape burns." Like a fire declared out that came back, a prescribed fire that blew out of control could end a career. It was a bold fire officer who would challenge that dare, and it helps explain why prescribed fire would have to propagate from Florida. If Sequoia-Kings Canyon and Yosemite could tolerate free-burns in their granitic high country, no place else in the state did. A program of natural fires would have to roam where the backcountry was so large and so remote that it was beyond the reach of California foresters. Where reforms most took root was in hard-core suppression. At its dedication in 1970, BIFC might boast that it pioneered an interagency model, but the real future emerged from that year's fire siege in Southern California, which led to Firescope and the national incident management system.

Those firefights meant casualties, in which California again proved a leader. The 1956 Inaja fatality fire, on the heels of the 1953 Rattlesnake tragedy, led to a special task force, with orders to report to the chief forester, on ways "to keep men from being burned on fires." The upshot was the 10 Standard Fire Fighting Orders, along with an embryonic training program to systematize fireline behavior. Merle Lowden brought Bud Moore from Region 1 to Washington, DC, to head the mission, which then merged with the NFCS survey. Then the Loop fire on the Angeles burned over the El Cariso Hotshots, killing 12, and the next year a complacency-shattering blast of big fires returned to the Northern Rockies. One outcome was to upgrade an ad hoc facility and relocate it to Marana as the National Advanced Resource Technology Center. The solutions to California's fires propagated throughout the country, whether they matched local circumstances or not.

The decade ended as it began with defining fires in California and California's response to them. The Tall Timbers conferences might attract attention by championing prescribed burning, and the National Park Service might boldly restate its policy to restore flame, but the enduring reality of the decade was that the American fire establishment existed because of the U.S. Forest Service. It sustained the other agencies; they did not sustain it. When fire bursts exhausted the resources of one region or agency, the call for help went to the Forest Service. When fires went bad, the USFS mustered review boards to analyze and propose recommendations. Those insights became national norms. To its rangers and fire officers, the agency's challenge was how to do more of what it did best, and its best seemed to lie in California.

* * *

The fire provinces have persisted even as they have metamorphosed under the stress of changing land use and migration. Fifty years after the fire revolution began, over three-quarters of the country's prescribed fire burned in the Southeast, while three states with ample public lands—Alaska, Nevada, and Idaho—accounted for two-thirds of large fires. The epicenter of Great Plains burning remained within the cherty fire ring of the Flint Hills. Mixed fire suppression, edging into an urban template, was fire's answer to an urban out-migration that sprawled heedlessly over the countryside. America still experienced the residual embers of former eras, though in pockets. In 1963 nearly 200,000 acres burned in New Jersey's pine barrens. The next year Long Island burned a tenth of its pine barrens preserve. North Carolina blew up in 1966, burning 120,000 acres and 50 houses.

What connected provinces and wildfire shards into something resembling a functioning whole was the Forest Service. The agency had long prided itself on being a national bureaucracy that was regionally decentralized. In 1960 it boasted 10 regions, each of which enjoyed a degree of autonomy. The southeastern forests could prescribe-burn. The Alaskan coastal forests had few fires. The national forests of California reigned supreme at complex suppression. The Lake States forests were still picking up the pieces within the long historical shadows cast by abusive cutting and conflagrations. Yet through inspection tours, a shared professional identity as foresters, common procedures and standard equipment, and a sense of fire as a collective threat, the USFS could coax, badger, and occasionally order a rude conformity among its factious parts. Even the labs aspired to be both national and regional. In this assumption, that the center and its peripheries could find sufficient common cause to cohere, that centripetal and centrifugal forces could match, the Forest Service was synecdoche for the country it served, its chronicle a cipher for the upheaval and fragmentation to come.

Seismic Sixties

In retrospect it is easy to filter the signals from the noise. At the time it was tricky to know what were background tremors and what were the surface shakings of deeper quakes bent on moving the tectonic plates of American society: its demography, its legal system, its values, its wars, its economy. Year after year something shook the fire establishment—the Tall Timbers insurgency in 1962, the manifesto of the Leopold Report in 1963, the legislative coups of the Wilderness Act and Civil Rights Act in 1964. For a while the Forest Service absorbed their aftershocks into its regimen, as though (if only symbolically) they were fires. It found ways to beat back the bad ones and tweak the less threatening into submission. It continued to do what it had done for half a century by expanding the domain of organized fire protection.

When the 1960s began, the Forest Service's standing as a fire organization had never been higher or more vital. It remade a Rube Goldberg assemblage of fire agencies, departments, labs, caches, and machines into a coherent system. But as the 1960s deepened, a widening gyre of unrest pulled that center apart. The Forest Service tried to double down on its proven methods. Whatever symbolic brush fires might break out, it could cope with the real ones that mattered.

* * *

Few beliefs proved more mistaken than the assumption that the Multiple Use-Sustained Yield Act would cap and stabilize the Forest Service mission. Almost immediately a trickle of environmental legislation became a torrent that washed away much of the agency's administrative discretion and freedom to maneuver.

The Clean Air Act, the Water Quality Act, the Endangered Species Act (of 1969), and others culminating in the National Environmental Policy Act all restricted what the agency could do with its skies, waters, creatures, and woods. The Wild and Scenic Rivers Act shaped its ability to manage select watercourses; the National Trail System Act its capacity to maneuver around routes open to outdoor recreation; the National Historic Preservation Act its response to old structures. And of course, the immovable boulder, the Wilderness Act, relocated large chunks of national forest to the National Wilderness Preservation System. Worse were blows directed at its founding tasks, particularly its ability to manage timber. By the late 1960s, even apart from unfavorable rulings by the federal judiciary, the USFS in the court of public opinion stood convicted of arrogance and incompetence. And the hits kept coming.

The immutable argument for a federal Forest Service had always been to protect the nation's public forests from ruination by feckless logging and burning. Yet on the Monongahela and Bitterroot national forests the agency was condemned for doing itself what it had originally denounced in others. The two sites carried a degree of ripe symbolism that was not lost on foresters. The agency had been established to protect existing stands and where possible, particularly along the Appalachians, to reclaim ruined forests. The Monongahela existed because the Weeks Act had allowed wrecked and abandoned lands to be purchased, converted into national forests, and rehabilitated—exactly the mission the agency had sought. The Bitterroot National Forest had heroically wrestled with fire, and more recently with insects, to spare its majestic wild woods. The Bitterroot, too, had long enjoyed special status as the heartland of the most vast of national forest holdings, as the scene of the Big Blowup, and as the nursery of chief foresters.

Both fell to the ax of the postwar boom. As the Monongahela's woods regrew, they were pruned, and when they had matured sufficiently, they were felled. But it was how they were cut that outraged the populace: they were clear-cut. Though the practice could be self-justified within the forestry guild, the public viewed it as the ecological equivalent of strip-mining. The land seemed to be returning to the state at which the Forest Service had first intervened. As West Virginians rallied in political protest in 1965, in 1967, and again in 1970, the state legislature stepped in. The Forest Service yielded, but too late to save the matter from going to court, where the agency lost control over how and even what it logged. Meanwhile, the Bitterroot controversy was led by former employees, including Guy Brandborg, who rose up in opposition; a special panel, overseen by Dean Arnold Bolle of the University of Montana forestry school, reviewed and criticized current practices. Yet again the agency responded too slowly to avoid a political escalation, this time to the U.S. Senate, where new guidelines were promulgated. Agency foresters seemed unable to understand what they had done wrong. They had only executed with the power of government what professional forestry had always advocated: the conversion of wild woods to cultivated plantations. Clear-cutting the Selways and bulldozing slopes into terraces was transforming wasteland into timber.[79]

That, of course, was the problem. The public wanted those wildlands kept wild, or at least wooded. The deeper cause was forestry's inability to communicate with that public. It assumed the argument had ended with passage of the Multiple Use-Sustained Yield Act, which granted the agency statutory authority to do what it thought best. Significant fractions of the public, however, no longer believed that government foresters knew best, and even former employees charged that the Forest Service had abandoned its founding ethos. Forestry was tarred as an apologist for industry rather than celebrated as guardian of the nation's natural estate; federal foresters seemed to have swapped moral authority for the bureaucratic riches of a timber boom. The more the agency defended its actions by arguing over techniques, the more the public, through Congress, was prepared to question its judgment and limit its discretionary powers. As the decade ended, the Forest Service seemed agape, unable to comprehend the depth of the angst it had aroused, confident that the old methods and appeal to authority would eventually triumph.

Nowhere was this more true than with systematic fire protection, America's contribution to global forestry and state-sponsored conservation. The agency assumed it could absorb the newcomers as it had so often in the past. The National Fire Coordination Study submitted its final reports in 1966 along with a plan for implementation by 1969. Instead it got a blowup in the Northern Rockies, the biggest since 1934. Three years later, with the fires of 1970, it weathered firefighting's equivalent of the Tet Offensive in Vietnam. The Glacier Wall fire was the last major firefight by the National Park Service under the doctrine of full suppression. The 1970 firefights were the last stands by the Forest Service as an autonomous fire agency.

Last Hurrahs, 1967 and 1970

Some fires brand themselves into history because of what they do and some because of how they are remembered. The legacy of the doers is preserved in the landscapes and cityscapes they bequeath; the legacy of the remembered is preserved in the songs and scrolls they inspire. The fires may not even be real. It is hard to believe that the most significant fire of ancient Greece occurred on the Plain of Omalos in Crete, but here, according to Hesiod, the Titans and Olympians fought for supremacy amid flames kindled by Zeus's lightning, and so it is preserved in storied prose.

The fires of 1967 belong with those preserved in documents. The fires of 1970, from Oakland to San Diego, align with those branded into places and agency procedures. Each year had its signature big fire, but it was the seasons themselves in aggregate that the country had to pass through as the United States crossed a threshold between a time when fire could be controlled by the sheer force the Forest Service could rally against it and a time when mind had to supplement muscle.

* * *

The 1961 fires had been tough, with the largest area burned since 1934, but the 1967 fires blew past them. They made a season of superlatives: the first time a general forest closure had been invoked in Region 1 history; the greatest number of lightning fires ever recorded on the Nez Perce National Forest, and well above the 10-year average elsewhere; some 44 project fires, drafting everything the agency could throw at them, along with a callout of National Guard and military support. It was, William Moore wrote for a special NFCS analysis, one of the "most difficult fire control situations" in the history of Region 1 and the Forest Service. The potential fire scene, he declared, was actually "worse than in 1910."[1]

But the memorable, the brute fires, burned outside the national forests. The Trapper Peak and Sundance fires started on lands under the jurisdiction of the Priest Lake Timber Protective Association (PLTPA). They escaped initial attack and then grew explosively when the PLTPA exhausted its suppression funds and hesitated, because it would be liable for the costs, to request aid from the Forest Service. On August 17 President Lyndon Johnson declared north Idaho a disaster area, freeing up federal assistance. On September 1 the Office of Emergency Planning and the governor of Idaho authorized the Forest Service to assume command of the Sundance fire, just as it began a 9-hour, 16-mile, 50,000-acre surge over the mountains. Before the smoke cleared, the Trapper Peak fire had burned 18,000 acres at a suppression cost of $6 million; the Sundance fire had burned 56,000 acres at a cost of $18 million. Two firefighters died.

The contrast between the faltering attack on Trapper Peak and Sundance with the successful initial attacks on nearby national forest fires was striking. Two lessons seemed undeniably obvious. One, an uncompromising offense is the surest defense. "Force enough fast enough" was a dictum on which every agency could improve. Moore, then deputy national director of Fire Control, made an inspection tour and concluded that "the strength, speed and effectiveness of Region 1 initial attack and first reinforcements actions needs study with the objective of identifying all possible means to improve them." In February the Forest Service had allowed for variances in early- and late-season fires, but when the season was still flush, there could be no compromise. The logic of the 10 a.m. policy was irrefutable. The second lesson was that cooperation had to be comprehensive, for one escape fire could undo the work of successful suppression around it. Certainly under wartime conditions split jurisdictions that inhibited the free flow of men and matériel would be lethal. So long as there were vacant lots and oily rags in the corners of the American landscape, and firefighters were held at the border of political entities, fire protection was flawed, and a national system of civil defense was perhaps fatally compromised.[2]

Still, big fires are only big smokes unless they connect with cultural institutions and ideas that can leverage them out of their local particularities. What allowed the 1967 fires to become a stele in American history was a study by the fledgling Missoula lab. The roster of colleagues acknowledged by chief author Hal Anderson was a roll call from fire research's golden age. And although lab director Art Brackebusch cautioned about the limits of the case-study approach, *Sundance Fire: An Analysis of Fire Phenomena* set a standard for large-fire analysis and whetted the aspirations of those who longed to see hard science come to the fireline. The investigation combined metrics from the NFDRS for weather and fuels, mathematical modeling of aerial firebrands, formulas for fire-induced blowdowns, and the melting of the Pack River bridge as a measure of intensity. The study seemed to demonstrate the premise that even for wildfire, among nature's most chaotic phenomena, modern science could leap from wind tunnels to the jet streams pouring over the Selkirk Divide.³

The other informing document emerged out of Glacier National Park. On August 11 dry lightning crossed the Continental Divide and kindled 20 fires in the park; 2 became large. The Flathead fire started near the Apgar and Huckleberry mountains, visible from park headquarters; the Glacier Wall fire started near the main scenic drive leading to the Going-to-the-Sun Road. Both proved tough going. Then a cold front on August 18 pushed them across control lines, and in the case of the Flathead fire, onto the Flathead National Forest. Suddenly, the park had a very visible firefight.

With a regional bust already sucking in every available pulaski and bulldozer, reinforcements were mustered from the Blackfeet Nation and Montana National Guard and flown in from afar—miners and steelworkers from Nevada, Inuit from Alaska, smokejumpers from throughout the Northwest, fire officers from distant NPS outposts. Then another front on August 23 blew over firelines and almost doubled the burn. The Park Service steadily increased its fire suppression forces. Glacier, after all, had been birthed amid the Big Blowup, and since 1936 it had served as a model for fire protection throughout the national park system. When the fires finally stopped—the Flathead at 3,658 acres and the Glacier Wall at 3,109—mopup ended with a political firestorm as in-holders and locals, with former U.S. Senator Burton Wheeler serving as mouthpiece, launched a blistering criticism of Park Service management. Constituents complained to Senator Mike Mansfield, who forwarded concerns to NPS director George Hartzog and interior secretary Udall. In response, superintendent Keith Neilsen organized a major two-day public conference on November 30 to review the fires.⁴

The accounts are detailed, and the Park Service cleverly situated its response with comparisons to past Glacier fires, such as those of 1929, with the general outbreak in the Northern Rockies, which diluted the prime resources available but still held burned area to a fraction of that burned in 1910; and with the determination of the agency to protect its lands against fire. The preamble of a publicly released report stated unequivocally that "fire is today, without a doubt, the greatest threat against the scenic grandeur of our National Parks," and for this reason, "the National Park Service is committed to fight each and every fire that threatens scenic values." Reconstructed narratives then demonstrated how it had hurled smokejumpers and air tankers at the Flathead fire from the onset and had dispatched ground crews to Glacier Wall, who reported the snag fire inaccessible and unlikely to spread (which was true until burning debris dropped down the cliff). Shorty Meneely, FCO for the Flathead, opined that under the conditions, the park was lucky it had not burned over. Glacier officials insisted they had done everything reasonably possible. The park regarded the political brouhaha as the outcome of poor "public relations and press coverage," a problem that perhaps could have been avoided had a regional information officer been assigned.[5]

The final report looked around and back and described with extraordinary precision how fires were fought in 1967. But buried among the open discussions was an exchange that looked to the future as well. Les Gunzel was unable to attend the conference, but he sent comments to be entered into the record. He fretted over the effect of mechanized fire control on the lands the park claimed it wanted to protect unimpaired. He had "always been against bulldozers in the wilderness areas of the National Park Service" and still felt that way. Exceptions were possible; he thought they were justified on the Flathead fire, but he wrote that they "were of no value" at Glacier Wall. He wanted guidelines for their use—rules of engagement in place before and outside of the adrenaline rush of emergency firefighting. Fighting fires might leave far worse scars than the fire.[6]

Amid the era's generalized war on fire, Gunzel's comments were the counterpart to the criticism sparked by the Vietnam War officer who declared in 1968 that it had been necessary to destroy a village in order to save it. Bad fires had to be controlled; so did bad firefighting. If a feeble suppression organization could allow fires to become dangerous, so an unchecked fire program could allow suppression to compromise the values being protected. When that winter the Leopold Report was codified into new administrative guidelines, Gunzel's appeal found a home. Mechanized firefighting by bulldozers would henceforth require approval on a case-by-case basis by the Secretary of the Interior.

For all this, the Forest Service could point to the 1967 season as a success. It had encountered big fires in a region famous for them. What it protected, it held. The problems lay with the PLTPA and Glacier National Park, and both had been saved when the Forest Service brought its vast power and firefighting expertise to bear.

* * *

Jacob Bronowski once remarked that a genius was a man with two great ideas. The genius of the 1970 season was that it held two great fire busts, and each exceeded the new standard set in 1967.

The season's long march began on August 23 when lightning kindled 200 fires around the Wenatchee National Forest. A handful defied control, merged into five giant burns, and romped over 200,000 acres. The largest, the Entiat, burned 122,000 acres—twice the size of the Sundance. For the next 15 days the Forest Service siphoned in 8,500 firefighters and mountains of equipment, then cut, burned out, and eventually held lines. The firefight inspired some of the classic photos of American fire as battlefield: a lumbering B-17 dropping retardant above conifers, grimy Redding Hotshots striding grimly along a blasted fireline, towering pyrocumuli. Suppression costs rose as high as the convection columns, reaching $13 million. Postfire responses included massive salvage logging and replanting. The effort exhausted the Forest Service.[7]

Then the story went south, and a feral season turned rabid. On September 22 an arsonist set a fire in the Oakland Hills that raced over the summit, with grass as an accelerant and frost-killed eucalypts and windfall as fodder. It burned 36 houses to the ground, badly damaged 37 others, and terrified a metropolis that had not known a threat of this kind since 1923. Over the next 12 days, while Santa Ana winds blustered across Southern California, the fires—773 in all, 32 of which became big, and 1 of which, the Laguna, transfigured itself into a 170,000-acre monster—stalked the South Coast and recalibrated what fire meant at a time when California dreaming had become the American dream.[8]

It is "almost impossible to avoid superlatives" in describing the Southern California fire scene, wrote Carl Wilson. The California Department of Forestry (CDF) declared that the serial conflagrations of 1970 were unique in modern times; not since the 1947 fires in Maine had so widespread a disaster affected so vast a civilian population. When the winds finally abated, 580,000 acres had burned, 722 houses lay in ash, and 16 lives had been lost. Expand the chronology a bit, and in the two months after September 15 some 1,260 fires burned over 600,000 acres, destroyed 885 homes, and racked up $233 million in suppression costs and losses. As one analysis observed, this was "more than the cost of the Great Chicago fire." Under the California Fire Disaster Plan the response had rallied virtually every significant fire agency in the state. While a crucial fraction had burned on national forest lands, and control of the outbreak had relied "upon the nationwide strength" of the USFS, the state Office of Emergency Services had quickly assumed control, and CDF had borne the brunt of the responsibilities. Compared with past chronicles, the outbreak seemed an outlier. Compared with the annals to come, it was a baseline.[9]

The state soon organized a Task Force on California's Wildland Fire Problem to analyze and make recommendations. It noted that such eruptions had become an almost annual event, that 118 large fires between

1960 and 1969 had destroyed over 2,000 structures, and that what made the 1970 extravaganza unique in modern times was "the geographical area involved, the total acreage burned, the wildland-urban nature of the fires, the large number of homes completely destroyed, and the large number of agencies, people and equipment involved." Five subcommittees proposed 49 recommendations.[10]

The gist, however, boiled down to two main points. First, the problem of colliding wildland and city was systemic, and effective fire protection required codes, zoning, and other techniques beyond after-the-fact firefighting. Second, when a firefight arose, the axis of California fire protection, the postwar fire disaster plan, proved marvelous in principle but messy in practice. No agency could cope alone: everyone had to fight together or they would all burn separately. But cooperative action demanded more than mutual-aid agreements when a cauldron of polyglot agencies had incompatible hose couplings and hydrants, assorted radio frequencies, differing understandings of strategy and tactics, and even incommensurable names for a vehicle that squirted water. It was not enough to draft men and matériel from everywhere in the state. They had to be able to work together when they stood against the flames.

Those were the California lessons learned. The Forest Service learned another, that it could no longer cope by itself. It was no longer larger than the fires it faced. It needed to cooperate not just to help neighbors and thus protect its borders but also so that neighbors might aid it when the big ones came. The 1967 fires made that year the last in which the U.S. Forest Service was a hegemon in policy; the 1970 fires marked the last time it could act as a hegemon in a firefight. Even with allies, the uneasy feeling gnawed in the gut that this might be a war that could not be won.

CHAPTER TWO

HOTLINE

Fire management is change: It is a change in concept, a change in policy, and a change in action.
—HENRY W. DEBRUIN, DIRECTOR OF FIRE AND AVIATION MANAGEMENT, U.S. FOREST SERVICE (1974)[1]

Why not let forest fires burn? Once again we are going to try, and this time we hope nature won't show us why not.
—CRAIG CHANDLER, DIRECTOR, FOREST FIRE RESEARCH, U.S. FOREST SERVICE (1972)[2]

Let a Hundred Fires Bloom

The political upheaval that stirred in the early 1960s and boiled over in 1968 reached its climax with the resignation of Richard Nixon as president in 1974. By the time the country headed into its bicentennial, it suffered stagflation, watched the demoralized conclusion to its Vietnam misadventure, was caught between oil crises, and endured more turmoil as the civil rights movement expanded. Yet the waves begun by the earlier storms still surged, reaching shore by the late 1970s. The fire revolution was among them.[3]

What the environmental movement had so far legislated it now upgraded: the Endangered Species Act in 1973, an Eastern Wilderness Act in 1975, the Clean Air Act in 1977, the Endangered Wilderness Act in 1978, the Alaska National Interest Lands Conservation Act of 1980, and numerous others. Each statute notched a point of entry for lawsuits, particularly where federal lands were involved, a legal access bolstered by the procedural complexities of the National Environmental Policy Act. Ruling upon ruling followed, each carrying the impact of new legislation. *Environmental Defense Fund v.*

Hardin granted judicial standing to people and other living things who might suffer biological harm from actions such as continued use of DDT. *Sierra Club v. Morton* clarified "standing" for litigation by environmental activists. *West Virginia Division of Izaak Walton League v. Butz* questioned Forest Service harvesting practices. The waves kept coming ashore.

By keeping options open, multiple use had been a useful doctrine for a period of custodial guardianship. But as land moved into industrial production and the public clamored for special, intensified, and exclusive uses, multiple use proved both inept and vulnerable. The multiuse monolith exfoliated apart in giant flakes. Those federal land agencies such as the BLM that lacked organic acts got them. Those that possessed legislative charters had them remade, notably the Forest Service. But all found their discretion clipped and their decisions challenged. Gradually, the external contest between state-sponsored conservation and civil society was ingested and made internal. Agencies—the Forest Service especially—had to defend not only against outside critics but against fifth columns within their own ranks. What began as an organizational rechristening ended in a full-immersion baptism, first by the waters of congressional statute and then by the fires of nature.

* * *

Those same trends held for the institutions of wildland fire. Every congressional rechartering and court check affected how agencies coped with fire, which unlike DDT or dam building was not something they willed into being and hence could unwill. Fire happened with the insouciance of hailstorms and blowdowns. Yet wildland fire has its own narrative as well. The arc is one from mind to hand, as ideas sired in philosophy struggled to realize themselves on firelines. It was not enough to issue manifestos about the glories of natural processes; those beliefs had to become routine practice; they had to move from manuals to operations. During the 1970s the process was repeated, agency by agency, some of which were still under construction, and they had to find ways to coordinate what each did into a working whole. Having toppled Humpty Dumpty off its hegemonic wall, all the country's fire folk struggled to put at least a working simulacrum together again.

Prescribed fire had to escape from niche demo plots, slash piles, and its southern ghetto. Natural fire had to find a larger dominion than the ice- and-granite basins of the High Sierra. Suppression had to toughen its stance against rural fires blasting into new suburbs while easing the force with which it extinguished fires in the backcountry. Like the military reassessing battlefield tactics in an age when tanks replaced cavalry or strategy in an era of nuclear weapons, the American fire community, accustomed to the rhythms of fast attack and big burnouts, had to adapt to new political and technological realities. The struggle would occur along generational as much as institutional lines.

The community turned, hopefully, to the new labs. The 10 a.m. policy had been possible because of the munificent bounty of the New Deal, and it had renewed itself through the postwar surplus in military hardware. Both were means well suited to erecting an infrastructure and building firelines. An era that sought to restore fire demanded a different kind of leverage and found it—or discovered the promise of a solution—in hard science. Information could also be power. The ability to predict a fire's behavior might replace the mechanical muscle of a bulldozer to stop its spread. The capacity to understand how fuels might be rearranged to quell fire in advance of ignition could substitute for mustering phalanxes of hotshots to fight them later. But like policy, knowledge gained in labs or test fires had to be transformed into practice—made to work against sullen shrubs, the untrained muscles of a new workforce, and the skepticism of those whose lifetimes of experience told them the scheme was wrongheaded and maybe ruinous. There was also the bureaucratic caution of those who knew that a fire that went bad could end a career (or sink an agency) and the indifference of the majority who would do what they had always done until otherwise rewarded.

The old orthodoxies fell with remarkable suddenness. Institutions and policies that had seemed immutable dissolved into flux, whether by design or through the disarray of inattention. For some fire officers and a few agencies, the liberation encouraged a fever of new thinking and experimentation. By 1978 the great bunker of the 10 a.m. policy, the Forest Service, had surrendered, ceding the field to a new age of environmental nation building. What few observers appreciated was that those long decades of fire exclusion had created an ecological insurrection that would, on one hand, spiral into ever-larger conflagrations and fire busts that suppression would be unable to contain and, on the other hand, would frustrate attempts to reintroduce fire without unacceptable mistakes or exorbitant costs.

That realization, however, lay in the future. What saved both those who sought a new order based on fire's restoration and those who wanted to maintain the old regime through fire's continued containment was a long passage, almost two decades, with few horrific fires, and those in that bastion of exceptionalism, California. Whatever general malaise afflicted the nation, American nature granted the fire community a period of grace in which to seek its salvation.

Fire by Prescription

When he coined the term *prescribed burning* in 1942, Raymond Conarro was seeking a middle ground between two opposing opinions almost Manichaean in their intensity, that had plagued the southern scene for nearly two decades. One pole argued for fire's exclusion. This was the position of foresters, and it gained political handholds thanks to the emergence of forestry bureaus among the southeastern states. It strengthened its grip with the coming of the New Deal, the search for sites to operate CCC camps, and the acquisition of abandoned lands for public or industrial forestry. The CCC created a government presence and what the Southeast had never before had: a force capable of fighting fires. The other pole demanded unrestricted access to fire as a landowner's right and a customary practice. Each assembled evidence in support of its own position, and each could highlight failures by the other side. In classic debate style it was easier to find flaws in the opposing case than to promote one's own. As the shouting increased, the public, confused and annoyed, threatened to tune out altogether.[4]

Conarro proposed "prescribed fire" as a compromise between no fire and all fire. The concept recognized that fire was useful, economical, and culturally understood while offering guidance from science and bureaucracies to elevate it beyond folk remedies and casual woods burning. It offered control by regulating how and when and where fire could be kindled. Within a year the Forest Service accepted the practice for the Florida National Forest and subsequently throughout the Southeast; a decade later the National Park Service adopted it for Everglades. In 1962 the Tall Timbers fire ecology conferences became a megaphone to broadcast the concept nationally.

The southern controversy was unique only in that it persisted. Elsewhere, as with the light-burning brouhaha, fire protection had smothered open fire, and as more of the population moved off the land and into cities, a larger fraction of the American populace became further estranged from any personal experience with burning. They knew it only as annoying smoke and blackened scenes that interfered with picnicking and fishing or as headline-grabbing disasters. Urban standards slowly squelched even such rituals as lawn and leaf burning. Free-burning fire belonged in the backwoods and the distant public domain. For much of the populace the issue was not whether fire should be wild or prescribed but why it mattered at all.

The fire revolution changed that perception, but in reintroducing fire as a topic and a presence it created another set of polarities. The Southeast wanted burning with a human hand firmly on the torch. The West, ideally, wanted nature's fire, with humans removed to the margins. Both sides dismissed fire exclusion as mistaken and unworkable and so found common cause in seeking to contain it. But how to meld their interests and keep the cause from splintering into sectarian factions was less obvious. Whatever resolution the alliance found, it would have to be acceptable to government bureaucracies that were not inclined (and could ill afford) to cede all control over fire.

The solution was a more expansive definition of prescribed burning that could embrace all forms of introduced and tolerated fire. A new definition could absorb burning for site preparation in loblolly pine plantations, burning for hazard reduction in longleaf, burning to improve pasture, and burning to enhance wildlife habitat in coastal marshlands. It could include burning to halt encroachment by woody shrubs and forests and burning to inoculate against brown spot disease. It included fire to reduce logging slash, fire to transform chaparral to grassy watersheds, fire left to burn in the Mogollon Mountains, and fire as a feature of the pristine Wild. It embraced fire that converted landscapes, fire that maintained them, fire that restored them. It could unite industrial pulp plantations with mountain wilderness. It bonded open-range ranchers in Florida with NPS rangers in the High Sierra and included categories (and places) of burning that had never before been linked. It was, in a sense, a multiuse fire. And it proposed a single policy, acceptable to all parties, both those who wanted fire liberated and those who wanted only to grant it a longer leash. It offered a regimen for scientific and bureaucratic guidance in place of folk woods burning and nature-driven let-burning. As the revolution propagated, regional guidebooks for prescribed fire appeared for the Southeast (1965), the Intermountain West (1966), Northern California (1967), the Southwest (1968), and the Great Plains (1980), along with the scores of examples published by Tall Timbers.[5]

By 1971 the fire revolution was moving from manifestos to field manuals. The National Park Service had written its restorative agenda into the Green Book in 1968. In 1974 the Department of the Interior rewrote policy to allow for both the BLM's goal to suppress and the NPS's goal to restore. In 1978, after a series of stutter steps, the Forest Service enacted wholesale reforms. The organizing principle of the new era was a philosophy of fire by prescription.

* * *

In 1971 two national symposia, one in the Southeast and the other in the Southwest, addressed the new era. The gathering in Charleston, South Carolina, was organized by the Southeastern Forest Experiment Station and was simply titled Prescribed Burning Symposium. The conclave in Phoenix, Arizona, was assembled by the Southwestern Interagency Fire

Council and officially addressed Planning for Fire Management, but of course that meant finding ways to incorporate the explosion of knowledge and enthusiasms overtopping the fire revolution.

For two days, 450 attendees at Charleston spoke and listened to the state of knowledge as understood by the "most knowledgeable scientists and laymen of broad and varied training and experience from industry, universities, and state and federal agencies." It was a mini–Tall Timbers conference focused primarily on the region and on the status of prescribed fire, by now a universal term. The rapporteur noted that the group "did not meet as a prejudiced group to defend or to preserve the use of prescribed fire." But that disclaimer was boilerplate; the issue was decided. The last institutional holdout in Florida, the Florida Parks Service, had taken the pledge in 1969. "Consensus was essentially unanimous that prescribed fire, when properly used in the South, is an almost indispensable management device," generally beneficial with few "sustained deleterious" consequences.[6]

There was no credible opposition to prescribed fire in principle. The issues were practical. When? How frequently? At what cost? And what about air quality and liability for escapes? How should fire restored (or maintained) interact with other practices? What did it mean for careers? What made the gathering significant is that the U.S. Forest Service sponsored it and published its proceedings, including the fascinating Q&A sessions. The torch lit by Tall Timbers was being handed to the region's main landowners and policy institutions. In 1977 the Florida legislature passed the Hawkins Bill, which allowed the state Division of Forestry to prescribe-burn hazardous fuels on private lands. Prescribed fire became as much an emblem of the New South as NASCAR.[7]

The gathering in Arizona included a good sampling of academics and agencies, but it was largely a Forest Service affair. The Southwest Interagency Fire Council was chaired by Lynn Biddison, Region 3 director for fire control; Deputy Chief Edward Schultz delivered the keynote. At the core was an update on Project Focus, a complex computer simulation model under development at the Riverside lab to test various fire plans against anticipated conditions. If values at risk again mattered, if fire contributed benefits as well as damages, and if fire was to be revived as well as removed, then fire planning became far trickier than just finding ways to deliver more force to firelines as fast as possible.

To many fire researchers and to some fire officers, prescribed fire seemed the treatment of choice. Any kind of controlled or semicontrolled burning now went under its rubric. Instantly, prescribed fire seemed everywhere. But much of that tally came from renaming programs such as slash disposal and type conversion for watershed. Outside the Southeast, the number of acres actually burned was tiny and the number of fire-restoring sites small, located in places such as Yosemite or Wind Cave national parks. Much of the West's prescribed burns were trying to catch up (unsuccessfully) with the slash left by accelerated logging or to increase useful runoff and forage, not to ecologically renew landscapes degraded by overcutting, overgrazing, and overly thorough fire exclusion.

In truth, lumping fires set to improve habitat for elk with fires to knock down red slash pejorated the grand project. It tried to consolidate into a single central concept a widening gyre of practices. It allowed critics to dismiss prescribed fire as a smoke screen, a cynical enabler for commodity production. Something else had to relocate prescribed fire from the coastal plains to the western mountains. The means, the intellectual fulcrum, was wilderness. The trick was to fuse the prescribed with the untrammeled.

* * *

In 1971, Les Gunzel, chief ranger of Saguaro National Park and the critic in absentia of bulldozers on the Glacier Wall fire, waited for two inches of summer rains to fall on the Rincons and then permitted 10 of 11 lightning strikes to run "their natural course." He called the outcome a "natural prescribed fire." By the time the concept was written formally into the park's fire plan in 1974, Saguaro had allowed 24 of 46 lightning fires to range freely. Eager for an alternative term to replace *let-burn*, other parks and some forests seized on the Saguaro experiment, renaming the outcome *prescribed natural fire* (PNF).[8]

The prescribed natural fire sought to liberate fire without ceding at least nominal administrative control. It released fire on parole, and so long as it remained on good behavior, carefully monitored, it was allowed its liberty. Like prescribed fires set with driptorches on loblolly plantations, the prescribed natural fire had as its essence a "prescription," a detailed set of conditions that specified what was to be done and how. If lightning kindled a fire in a suitable zone—predesignated and part of a plan—and burned under a set of fire-behavior conditions, it was deemed "controlled" or at least not threatening. It was a prescribed fire set by nature. If the burn left its designated zone or if the fire environment changed in ways that altered the fire's behavior, if in official language it "exceeded its prescription," then it became a wildfire and would be suppressed.

To partisans of the revolution, the prescribed natural fire was a deft solution that permitted fire to return in ways that avoided the overtly manipulative burning done in slash piles and pine plantations. It was intrusive, at least intellectually, but no worse than radio-collaring wolves or trapping rogue bears. Tightly monitoring a fire was as close to not managing it as possible. The concept combined the scientific rigor of the prescribed burn with the free-burning naturalness of wilderness. To cynics, however, it seemed an oxymoron, or a sleight of hand in which managers hid what was really going on by shuffling it among variously named bureaucratic shells. The essence of a prescription was that it was humanly constructed, not natural, and the essence of the natural was that it was untrammeled by human meddling. To this charge sympathizers might reply that the prescribed natural fire was another example of modernism's paradoxes of self-reflexivity, like Russell's paradox or Gödel's proof. Ardent fire officers shrugged, ready to accept whatever intellectual contortions were necessary to get fire back as fully as possible.

In the West the prescribed natural fire became the ideal, and the prescribed fire set by people was valued primarily as a surrogate or enabler to it. The Park Service led the restoration projects, first at Sequoia-Kings Canyon, then at Saguaro and Yosemite, and then, at least on paper, at Yellowstone. Larger parks in particular sought out landscapes that might fit. By 1971 the Forest Service was planning experiments in the Selway-Bitterroot Wilderness, scene of the 1934 fires that had led to the 10 a.m. policy. Later it added the Gila Wilderness. Leopoldian green fire turned red as flames spread over scrub- and pine-clothed mountains.

There were two fascinating outliers. One was Alaska. Until a final settlement of land claims, which did not occur until 1980, it practiced a sort of PNF lite in which control meant loose-herding fires it had no hope of stopping. Suppression could mean a scale of options, not an on/off switch. Policy was analog, not digital. The other outlier was South Carolina, where the Francis Marion National Forest, a place long plagued by incendiarism, devised an eastern alternative to the PNF, what it termed Designated Control Burn System. DESCON removed the requirement that a fire be "natural." The forest could accept any fire, of any origin, even arson, so long as it advanced the goals for the land. Chief Forester McGuire signed off on the plan in 1973 as another exception to the 10 a.m. policy. When it went operational the next year, the forest accepted 4 of 62 wildfires. Between prescribed burns and DESCON-managed fires, the incidence of incendiary wildfires dropped dramatically. The program effectively co-opted the arsonist who was now doing what the agency needed and would do on its own.[9]

* * *

There were setbacks, or what partisans feared might be setbacks, as fires smoked in Jackson, Wyoming, blew out of the Selways, or drew scorn on op-ed pages from blowhards who declaimed that the only proper response to a fire was a bigger hose. But the wilderness ideology shielded the general notion of restoring fire to the backcountry; the concept of a prescription spared the PNF from sneering charges that the agencies were abdicating their duties; and rapid developments in fire science granted administrative cover with the prospect of forecasting the life cycle of PNFs and identifying suitable locales. There was no going back. There were only the questions of how fast to move into the future and whether agencies would drive themselves or be driven by circumstances outside their control.

To partisans, the revolution seemed mostly a case of having the right ideas and transfiguring them into the right policy. Fire, however, is about context: it synthesizes its surroundings. So it is, too, with fire practices, which take their power from their setting. Prescribed fire requires a landscape shaped by a suitable culture of burning. Only the Southeast possessed such a culture on a regional scale; elsewhere it existed in patches such as the Flint Hills and a handful of western tribal reservations. Among the federal land agencies it had vanished. A culture of prescribed burning was not something that could be created by text in a manual. It had to grow out of experience in the field and be handed across generations. Instead, an implicit belief prevailed that science could invent a modern culture. It could not.

The science was far less competent than the gleaming labs suggested. Even as they enthused over administrative interest, researchers warned that their models were being torqued well beyond their stress limits. The fact was, a landscape of restored fires, especially big fires, was an unknown. Sixty years of prevention and suppression had beaten down the fire scene. A new generation thought of big fires as 1,000–5,000 acres and a monster as 50,000. No one knew any longer how fires burned over long weeks on a landscape; no one had witnessed multiple blowups from a single fire; and no one really understood how one burn could check or boost another. Agencies demanded, or insisted that the public demanded, a degree of control they could never muster.

It all happened at a time when weather and landscape fuels were benign, when it was possible to tinker with notions in the field as in policy, and when there seemed time enough to piece experimental evidence with action. But nature did not distinguish among modernism's nuances nor care about the niceties of word craft. It became impossible to finesse actual fires as agencies could ideas about fire in a staff meeting or a congressional hearing. Fire had its own logic. When fires returned in swarms and big fires began to soar to the 100,000-acre range, when climate turned rabid and fuels metastasized, a considered compromise seemed less sensible. The PNF looked less clever. The set-piece prescribed fire looked feeble.

But while the revolution raged, fire by prescription seemed an ingenious way to merge causes and reinstate fire on the land. Whatever their personal qualms, objectors mostly kept silent, like patrons of the Latin mass after Vatican II. They sensed the old ways could not return, yet they could see no future to the new, and so they continued to do what they had done throughout their whole careers.

Total Mobility

Once again, wildfire was the catalyst. Whether or not the Forest Service wanted to act unilaterally, and regardless of those who thought it ought to, the 1970 fires showed it could not. The issue was how the USFS and everyone else might cooperate: by what principles, under what kind of institutions, and with what sort of governance charter. In short order, an interdepartmental memorandum of agreement established a National Wildfire Coordinating Group as a forum and facilitator, the Boise Interagency Fire Center had its status boosted, and the Riverside lab began development of Firescope as a mechanism to coordinate on the fireline.

Some say the immediate prompt was a 1971 rafting trip down the Colorado River by Secretary of Agriculture Clifford Hardin and Secretary of the Interior Rogers Morton. Around evening campfires, they discussed ways their agencies might cooperate. Others point to the damages and costs of the 1970 season and the logic of interdepartmental coordination, which had long precedents, including a 1943 memorandum of understanding. Regardless, Secretary Morton posted a letter to Secretary Hardin in November recommending they seek ways to improve joint fire activities. A favorable reply came from Hardin's replacement, Earl Butz, in early December. There matters rested until the presidential election cycle had run its course.[10]

On January 24, 1973, an interagency task force assembled in the auditorium on the eighth floor of the Interior Building. The Forest Service sent representatives from the national forest system, state and private forestry, its director at BIFC, and forest fire and atmospheric science research. Interior fielded representatives from BIFC, BLM, NPS, the Bureau of Indian Affairs (BIA), and the Fish and Wildlife Service (FWS). There was a clear inequality in agency firepower and aspirations. Thanks to presidential directives in response to the oil embargo, the room was cold and dark; so was the bureaucratic atmosphere. There was no one designated as chair.

Eventually Henry DeBruin of the USFS assumed the role of moderator, and the group began to identify points of potential cooperation and to consolidate its musings into four "options." The scope of future cooperation could (1) form a National Fire Coordination Committee, (2) provide for initial attack on a local basis and extended attack through an interagency zone fire center, (3) establish interagency zones of responsibility, and (4) establish a national wildland fire service. The gathering debated what to call itself, finally arriving at National Wildfire Coordinating Group (NWCG; by not labeling itself a "committee" it was exempt from the Committee Management Act). But its inchoate nature also meant it had no operating instructions. Eventually, it concluded that a charter signed at the secretarial level would work best.[11]

Meanwhile, two other political developments threatened to destabilize the existing terms of understanding. One was a serious proposal to reorganize all the federal land agencies into a Department of Natural Resources that, if adopted, would fundamentally alter the administrative landscape. The other was a report, *America Burning*, by the National Commission on Fire Prevention and Control, that proposed to consolidate all of the country's fire problems under a single federal bureau. Congressional hearings followed and then legislation that created a National Fire Prevention and Control Agency (NFPCA). The NFPCA claimed wildland fire as part of its charter. A Department of Natural Resources remained a pipe dream. While both challenges were faced down, the fire community recognized that it would either reorganize voluntarily or have a new order imposed by decree. More big fires lurked in the margins.

In March, Interior Undersecretary John Whitaker wrote Secretary Butz to urge implementation; Butz agreed in principle, and the USFS and BLM assumed responsibility on an interim basis with DeBruin as chair for the first year. The gathering established five standing working groups for training, prevention, communications, research needs, and safety—and five ad hoc working groups for qualifications, fire-danger rating, air operations, agreements, and retardants. It agreed on a calendar for meetings with one prebudget session in Washington, DC, one just before fire season, a field meeting oriented around a specific problem area, and a review session at the end of the year. The formal cycle commenced in March 1974 at Rosslyn, Virginia, at which point the Forest Service insisted on including a state representative to the group (Ralph Winkworth of North Carolina assumed that role in 1975). Not until March 16, 1976, however, was a formal memorandum of understanding signed by the two secretaries.

The internal pressures mostly pushed in the right direction. Even as the memorandum made its tortuous way through the maze, both the USFS and BLM were having their organic acts rewritten. The key was the Forest

Service: it could stall the revolution by passive resistance as well as by brute rejection. But it seemed caught in the rapids and ready to ride them through. Its director of Fire and Aviation, Hank DeBruin, considered himself a "mover and shaker" as opposed to a "housekeeper," and he prided himself as a clean-desk man rather than a cluttered-office putterer. It was no surprise that he assumed the chair at the embryonic NWCG gathering. In 1974 he addressed the Tall Timbers conference, held that year in Missoula, Montana, to explain how the Forest Service was undergoing an internal reformation. It was changing its division title from Fire Control to Fire Management. It was shifting from a policy of universal fire suppression to one of selective fire by prescription, from fire as a semiautonomous function to fire as an adjunct to land management. It was moving from firefighting autarky to interagency (and international) cooperation, from fire as a singular topic for study to fire as part of interdisciplinary teams.[12]

With regard to interagency coordination, the operative term was *total mobility*: the complete interchangeability of equipment, crews, and procedures among the entire national fire establishment. The concept was simplicity itself. It meant that help could come from any source: the BIA could dispatch pumpers to the BLM, the BLM to the Forest Service, the Forest Service to the Fish and Wildlife Service. Their hoses would connect, their crews could function as identical cogs in a common machine, and overhead teams could travel throughout the country as they did between Forest Service regions. It meant the closest force could attack a fire regardless of jurisdiction; a BLM engine might be nearer a snag fire on a national forest and so be dispatched, or a Forest Service air tanker might drop retardant to hold a fire on a national park until local crews could arrive. It was impossible to object in principle.

Yet in practice, little by way of equipment or procedures was standardized. FWS crews did not fight fire as the USFS did, nor did they want to. The NPS did not automatically dispatch bulldozers; the Forest Service was not inclined to flush the stumps of felled snags. The agencies' fire programs had been created to serve their unique missions, and their fire operations could not be interchangeable unless their charters were. Depending on perspective, total mobility was either an exercise in the implausible or an expression of American e pluribus unum. Still, it was preferable to proposals to organize a national fire service, which could impose the desired uniformity but at the cost of splitting fire management from land management and thus violating a basic tenet of the fire revolution. Instead, in the American system, the federal agencies had to be simultaneously standard in practice and distinct in purpose.

The NWCG was the mechanism to make that concept a reality. Yet the NWCG was a slow, tedious, and in its fledgling years unwieldy institution that operated through consensus. It was, in its inception, strictly a talk shop, not a mechanism for governance: it had no authority to impose decisions. It achieved some political equilibrium by granting the Forest Service three representatives to the group (one for the national forest system, one for research, and one for state and private forestry), thus recognizing its vaster bulk in the national scene and ensuring it could not be overwhelmed by a clique of small fire agencies from Interior. Yet even the practical problems were formidable. Agencies had evolved different standards for training and certification, different operating procedures for aircraft, different protocols for dispatching, and different funding streams. And of course they had distinctive charters. The issues of institutional culture and status were no less awkward: the establishment Forest Service would have to cede some standing to the nouveau riche strivers from Interior.

They could all, however, rally for a firefight, and that was the mission the NWCG wisely claimed as its point of departure. Inevitably perhaps, the group would come to be housed at BIFC.

* * *

When BIFC officially opened, it was for the BLM a national fire dispatching center and for the Forest Service a regional air center co-located with the Boise National Forest fire operations, which also happened to house the USFS national radio cache. They shared buildings but not operations. The Forest Service had a national fire coordinating center in Arlington to orchestrate interregional transfers. But if doctrines including total mobility, closest forces, and joint dispatching were to succeed, they needed genuine integration. The fire revolution demanded a symbol at the highest level of operations.

In 1973 the Forest Service decided to upgrade its commitment at BIFC from a regional to a national facility. It yielded to the BLM's insistence that the BLM would furnish the director. Jack Wilson became BLM-NIFC director, and Bob Bjornsen was named Forest Service director. But the alliance also meant that the BLM had to upgrade its aerial operations into something closer to parity with the USFS; the upshot was that Interior established an Office of Aircraft Services and stationed it at BIFC. That initiated a long process of reconciliation between standards, some of them fundamental and some decorative (e.g., the Forest Service required that helicopters have stripes painted on their rotors so they could be visible from above; USFS fire crews could not ride in helos not so certified). The NWCG became the forum for thrashing out such differences.

The same consideration, however, applied to crews. What standards of training and experience would allow them to be equivalent? After the 1956 tragedies, the Forest Service began investing seriously in training. The other agencies had little more than brief "fire schools" for their summer seasonals. Such training was a matter of safety as much as efficiency and pride. Fire bosses and planning chiefs had to know what specific crews could or could not do. Again, the matter went to the NWCG, which painstakingly crafted a National Interagency Fire-Qualifications System beginning in 1976. NIFQS replaced the Forest Service practice of "red carding," which specified what fire jobs someone could do based on prior experience and apprenticing. A sector boss from the Florida Division of Forestry had to have the same basic training and field experience as one from the Siskiyou National Forest in Oregon; a sawyer from New Mexico must have the same basic skills as one from Minnesota. NIFQS had to identify all possible fire tasks, create formal courses for them (including instructional materials), decide what on-the-ground experience was required, and then certify that a person had met those standards. Among the criteria were basic metrics of physical fitness, which originally took the form of an aerobic step test. The database was, in effect, a master roster of the national fire militia. Those who completed each rank were issued an interagency fire job qualification card, the modern red card (inexplicably, the Park Service's were pink).[13]

While the process allowed for interchangeability, it also broke the old folkways by which fire expertise was handed down generation to generation through work on the line. Old hands from the Angeles National Forest, for example, recalled how they learned fire from tough, plain-talking graduates of the CCC years. Tex Strange would take a prospect to a ridgeline above a fire and ask three questions about the fire, which revealingly took the form of commands: "Tell me what it's gonna do. Tell me what you're gonna do. Tell me when it's gonna happen." A few minutes later he would return, and the novitiate would recite. If he passed he could advance to the next level up the fire hierarchy. If not he returned to his brush hook. Because fire control was not a professional position (i.e., it did not require a degree in forestry), fire officers tended to stay where they grew up; they knew fire intimately on their turf. It was an effective system that produced some of the best wildland fire officers in the country. But it was not a system that could underwrite a national program of total mobility, nor could it cope with that other expression of total mobility, the wholesale movement of minorities, and particularly women, into the ranks. By substituting the classroom for the field, NIFQS could.

So it went, with topic after topic—prevention, the design of pumps and engines, aircraft, radios, mutual-aid agreements. Those were the easy tasks. Gradually, because it was the only forum available, the NWCG became the mechanism for reconciling procedures for prescribed burning, for managing smoke, for monitoring PNFs, and so on; working groups were added for each new theme. Regional guidelines had to be generalized to accommodate national standards. The tasks spawned new positions such as burn boss, jobs that demanded their own standard training. Whether the outcome was a vicious circle or a virtuous one, more and more of fire management was brought into the bureaucratic maw of the NWCG process. Its sticky, ungainly search for consensus, while frustrating to partisans eager for reform, also prevented it from becoming oppressive.

Besides, BIFC and the NWCG were the only games in town. The other Interior agencies piled in. The National Park Service signed on to BIFC in 1974, the BIA in 1976, and the FWS in 1979. The fire center became the tent pole of the national suppression infrastructure—a unique institution that other nations eventually sought to copy. The Park Service located its national fire office on site. Any institution who wanted to (or had to) participate in big-fire mobilizations needed a presence. The BIFC story, moreover, serves as a cautionary reminder that the fire revolution was not only about restoring fire but also about fighting fire better. Perversely, BIFC glamorized big wildfires, capturing national media with its updated version of the firefight as battlefield and leaving gritty prescribed burns and remote PNFs to the deep-buried appendixes of public attention.

* * *

Total mobility might provide an intellectual rationale, and BIFC could do the mobilizing, but the pieces still had to function together on site. They had to be able to communicate, they had to follow common procedures, and they had to understand the evolving command structure. They could not rely on a shared if unwritten culture because they came from different regions and institutions. But the situation was even worse in Southern California because cooperators went beyond land agencies to include urban fire and other emergency services. Unless you could orchestrate operations on the ground amid the pandemonium of a Santa Ana–driven fire bust, written agreements were just so much kindling. The solution was Firefighting Resources of Southern California Organized for Potential Emergencies (Firescope).

Like all major innovations, Firescope pooled many resources together. One was the software embedded in Focus, although even in-house sympathizers at the Riverside lab regarded it as "a great idea" that "worked fairly well" but whose voracious demand for information meant "there was no way that the field could ever use it." Another was the hardware required for infrared mapping—telemetry that could be used for other purposes. Then there was the Forest Service's experience with its pre-BIFC National Fire Coordination Center, a huge task in an era in which orders were written by hand and organized on hanging wall charts and information was largely qualitative. As its first director recalled, an exchange might go like this:

"How are things goin'?"
"Okay."
"You got anything?"
"Not much."
"How are your crews doin'?"
"They're not too busy."

And there was the rapid downsizing of the California aerospace industry, which liberated engineering companies eager to find new jobs.[14]

But the critical catalyst was the 1970 fire bust and the staggering complexity of its callout. Under the California fire disaster plan, the state Office of Emergency Services (OES) siphoned in suppression resources from throughout California—and through agreements with the Forest Service, throughout the West. They came from the Feds: the USFS, the BLM, and the NPS. They came from the state: the California Department of Forestry, the National Guard, Conservation Corps camps, and OES's own reserves of engines and support. They came from local authorities: cities from San Diego to Los Angeles to Oakland provided resources, as did counties from San Bernardino to Humboldt, and especially the contract counties of the South Coast—Los Angeles, Ventura, Kern, and Santa Barbara. Outside fire crews poured in, from Forest Service hotshots to the Southwest Forest Fire Fighters, Snake River Valley laborers, and local Hispanic field workers. Fire engines by the hundreds funneled south. Some 28 air tankers flew missions, and a fleet of helicopters dumped water, retardant, and burnout flares. All this happened within days and had to be supervised for two weeks.

In the aftermath the state convened a task force, OES revised its Fire and Rescue Mutual Aid Plan, and the local agencies pondered their in-house reviews. But the propellant force came from Aerospace Corporation, which had headquarters outside San Bernardino and had watched the spectacle unfold beyond their windows; they were sure there had to be a better way to fight fire, one that might also lead to government contracts to replace those they had recently lost from a downsizing military. They lobbied vigorously in Congress. Just before the end of the 1971 fiscal year, Congress authorized $900,000 "to strengthen fire command and control systems research in Riverside, California, and Fort Collins, Colorado," and added recommendations for certain kinds of hardware and displays of the sort in which Aerospace specialized.[15]

Everything converged on the Riverside lab. Stan Hirsch, who had developed infrared hardware at Fort Collins, was transferred to the lab, where he met Dick Chase, who had run the NFCS before being sent for a tour to Riverside as a liaison with Focus. Chase then recruited two operations research analysts, Randy van Gelder and Romaine Mees. Between them they anchored the two poles, hardware and software, of Firescope. Meanwhile the ad hoc funding led in 1973 to a five-year program with a formal research charter. The hardware was developed, and it even inspired a massive (if unusable) display console, but its payoff lay in many spin-offs for its telemetry technology. As so often happens, the real wealth was the software. That software had to operate in two contexts. One, it had to work on actual events—this became the Incident Command System (ICS); and two, it had to assist decisions and dispatching for limited resources and often unlimited requests—this became the Multi-Agency Coordination System (MACS).

While they all shared fire, the various agencies had little else in common, and a seemingly unbridgeable cultural chasm divided urban from wildland services. There was skepticism about the value of a fully integrated program, and plenty of chauvinism since the project brought together the five biggest fire departments in the nation. At one point the Los Angeles City fire chief declared he was withdrawing because he "didn't see the need." Dick Millar, fire director for Region 5, told him the USFS was a bigger fire organization and could haul engines from Maine to Santa Monica if it needed to and that through Firescope the Los Angeles Fire Department could tap into that network. Slowly, Millar recalled, "he began learning there was more to this than just his small area of L.A. City." Dick Montague, then on the Angeles National Forest, put it simply. "That was our motto: We agree to agree, but we're going to fight getting here, okay?"[16]

Even their dialects were strained to the point of being mutually unintelligible. One considered a *tanker* a water truck, the other, something that dropped retardant aerially. What one called an *engine*, the other called a *pumper*. Wildland fire called its deciders *bosses*—crew bosses, line bosses, fire bosses. Urban fire scorned the term as appropriate only for flunkies;

real deciders were *chiefs*. They wore different uniforms with different insignias of rank. Wildland and urban fire agencies used different equipment, exploited distinct tactics, refracted their scenes through different institutional prisms. Terminology was only emblematic. The real issue was, Chase recalled, "everything." Everything had to be discussed, negotiated, returned for comments, and then redone, one iteration after another. "Everything."[17]

Gradually, the human factors started to gel. By 1974 the project was ready for training exercises involving the Forest Service, California Department of Forestry, Ventura County, Los Angeles County, and the City of Los Angeles. It was the first time Los Angeles county and city had ever staged joint drills. The experience demonstrated that ICS could work. Then the Forest Service decided to move its South Zone dispatch office into the same Riverside building that the CDF region used; this tested MACS. Everyone quickly understood, however, that the issues were bigger than a single complex event or fire bust. To be effective ICS had to permeate each organization in its routine operations; MACS had to standardize operating procedures if total mobility was to work. The flames were only a part of a fire emergency, which would also involve traffic control, evacuations, medical assistance, and plain old politics. The ICS had to embrace the totality of demands during an incident. Its DNA was evolved to accommodate any kind of emergency.

The package was ready for unofficial implementation in 1977, just in time to catch the next wave of California blowups. The following year Firescope's charter ended. But the nature of the problem meant that Firescope could not remain in Southern California. Both Region 5 and the California fire master plan brought resources from the north to the south (and theoretically, from south to north); that meant the rest of California had to sign on, too. In 1980 California OES agreed, and the Forest Service extended the program throughout Region 5. But because California, through the Forest Service, was part of a national system, the rest of the country might need to follow. Certainly the MACS model was relevant (maybe mandatory) for BIFC. The question of a national adoption of ICS went to the NWCG for consideration, and it commissioned a study. Within the Forest Service there was pressure to quarantine the project to Southern California, much as prescribed fire had long been sequestered in Florida. But when it was pointed out that two systems would mean interregional transfers would not be possible, Chief Forester McGuire ruled for service-wide application. The National Association of State Foresters lobbied for adoption. In 1982 ICS was rewritten into a National Interagency Incident Management System (NIIMS). And in a remarkable turnaround, in 1987 the National Fire Protection Association adapted NIIMS for structural fire.[18]

The interagency idea was no longer a fantasy of wishful do-gooders. It was a reality without which fire management could not work. A unilateral world had become a multipolar one, and it reached far beyond firelines. From California the program eventually relocated to the Federal Emergency Management Agency (FEMA). The Incident Management Teams that FEMA dispatched after the terrorist attacks of 9/11 and in response to Hurricane Katrina were spawned amid the furor of the 1970 fire bust.

* * *

As fissures in the old order opened, they created niches for new institutions. A handful of the newcomers, including BIFC and NWCG, rooted into permanence; mostly they attended to fire suppression, galvanized by a sense of crisis. Other interagency institutions came and went like fireweeds occupying temporarily vacant land. Among them were hybrid entities called fire councils.

The first such gathering had appeared in Missoula in 1955. The Forest Fire Research Council (later Intermountain Fire Research Council) assembled researchers, line officers, and academics. The Western Forest and Conservation Association adopted the concept, initially hoping to place one such organization in each western state. It quickly became obvious that the fire community lacked the mass to support anything like this density, and geographically grouped forums emerged instead. California and Nevada paired up; so did Oregon and Washington, and Arizona and New Mexico. They became, in a sense, regional variants of the Tall Timbers conferences that granted fire research a chance to mix with fire officers and administrators and to draw the attention of national authorities by passing resolutions.[19]

As with policies, so with research: the agencies sought to sponsor their own inquiries, not simply absorb what the Forest Service published. Speaking for the NPS, Bruce Kilgore noted that the agency would use Forest Service research "whenever possible," but he stated firmly that the fire labs had "been weighed heavily toward fire as it relates to economic interests—such as slash burning or fuel break maintenance or other fire-control oriented research, with little experimentation relating prescribed burning to the natural role of fire in forest ecosystems." The Park Service had distinctive needs and would have to sponsor its own investigations. The other Interior agencies followed its lead, though with less vigor.[20]

The councils helped bridge agencies, connect the field with labs, and converge national interest with local lore, or more broadly, historical eras. Councils plugged gaps as the national fire infrastructure fractured into a mosaic of regions and agencies. For a while they published proceedings—a valuable supplement to the literature when space in national journals was limited and fire conferences were still a novelty. For scientists the leap was relatively easy and valued. Like the breakup of AT&T's telephone monopoly, which crippled a national lab but sparked an industry, so the breakup of the Forest Service stranglehold on research could liberate fire science but demand a mechanism by which to transmit across common lines and causes. The councils provided that means.

Fire councils helped hold the pieces within a common force field. They were, for fire's intellectual cohort, the equivalent of the updated mutual-aid agreements and interagency pacts being negotiated. Particularly in the East, where the federal presence was smaller, states were pooling resources into regional alliances for suppression and prevention. The councils and fire-control consortia complemented each other. Canadians began participating; British Columbia joined the Northwest Council, and Alberta, the Intermountain Council. But mostly they were a transitional institution at a time when the national order was both crumbling and reconsolidating.

Over time, science proved too weak a force to hold a diverse community of inquirers and practitioners, and in the end, fire scientists wanted to talk more with one another than with fire officers. Those councils with strong links to field operations persisted as a kind of civil-society complement to what remained overwhelmingly a government-sponsored project. The others withered. But for a while, and in a period when information was power, they helped negotiate between the invisible colleges of specialized researchers and the chartered organs of government.

Propagation

Institutions and professions tend to resist stress imposed from the outside and respond instead, if slowly, to strains from within. Where a bureaucracy has long identified itself with a profession, as the Forest Service did, the inertia can be profound. As the USFS finally accepted fire and ecological thinking generally, it renarrated its history by ignoring the cries of outsiders and finding antecedents in fellow foresters such as H. H. Chapman, Harold Weaver, and of course Aldo Leopold. But that internal reform would not have happened without the outside voices. None was so powerful as Tall Timbers.

The Tall Timbers conferences became major events in the almanac of the American fire community, not only in theme but in venue. Tall Timbers had been established to create a working landscape for science in which management was also research; that was not a model that could be easily exported. But the conferences could. There had thus far been nothing like their big-tent revivalist approach to fire. From time to time the Forest Service had called conferences to talk about major matters (as it did in 1936 after the adoption of the 10 a.m. policy) or new topics (such as aircraft in 1957), and scientists reported their findings at professional gatherings such as the annual meeting of the Society of American Foresters. But there had never been a recurring conclave like that convened at Tall Timbers, nor one that brought so many voices together. By 1967 Tallahassee could no longer hold them.

The conference took to the road. In 1967 it went to Hoberg's Resort in California, site of Harold Biswell's demonstration plots. In 1970 it trekked to Fredericton, New Brunswick, strengthening an international connection. In 1971 it returned to Tallahassee but took Africa as its theme. In 1972 the tour traveled to Lubbock, Texas, home of Henry Wright, an avid proponent of brush burning. In 1973 it was back to Tallahassee, but with a European motif, full of invited speakers. In 1974 the conference met in Missoula, cohosting with the Intermountain Fire Research Council on the theme of fire and land management. It was during this event that the Forest Service publicly took the pledge and announced its conversion. In 1975 the conference went to Portland, Oregon, for a look at the Pacific Northwest. By then Tall Timbers had held 15 conferences, and the published proceedings amounted to what observers considered a "fire Bible," full of revelations from the Book of Nature as well as Chronicles and Prophets. With a supplemental bibliography by Ed Komarek, its proceedings became the world's premier archive on fire.

In 1972 TTRS undertook another job normally done by public agencies. It organized a task force on ponderosa pine management, specifically, an "evaluation of controlled burning in ponderosa pine forests of central Arizona." The similarity between southern yellow pine and western was striking, and prescribed fire had apparently established itself on the Fort Apache Reservation thanks to Harold Weaver and Harry Kallander, who had presented the story to a Tall Timbers fire conference. The rest of the task force consisted of Harold Biswell, ecologist Richard Vogl (a Tall Timbers habitué), and Roy Komarek, Ed's brother and representative of TTRS. The restoration of burning in the late 1940s had spared the reservation the worst of wreckage so typical of the montane West, but it was not enough, and a spurt of logging had added urgency. Big, damaging fires were appearing, fires that had not been seen before, powered by dog-hair thickets, windfall and slash, and a general deterioration in forest health. Here was an area that closely resembled the longleaf hills around Tall Timbers; perhaps a similar regimen of burning could spare it from further decay and conflagration.[21]

From its origins the Tall Timbers project had been viewed as a David and Goliath struggle. With the Forest Service's declaration at the Missoula conference, however, the contest ended. If David's sling had not killed Goliath outright, it had knocked sense into him. By then both Henry Beadle and Herbert Stoddard, Tall Timbers' founding financier and its patriarch, respectively, had passed away. A decade after they began, E. V. Komarek mused about "the future of our Fire Conference. When we started we had no idea of carrying these on forever, and it looks like that by 1975 we should take a hard look and perhaps terminate" them. They had "certainly accomplished our purpose and then some"—that being "not to set up a scientific journal, but to stimulate the work in fire research and see that the proper aspects were presented to the public at large. We have succeeded in this way beyond our farthest hopes." In 1975 it was resolved that the Portland conference would be the last. Three years later Tall Timbers underwent a fundamental reorganization. Interestingly, that same year so did the Forest Service.[22]

* * *

The fire revolution began as a revolution in values, not of science. For its evidence it pointed to the evolutionary presence of fire, humanity's long alliance with the torch, and a naturalist's understanding of fire adaptations. Its experiments were nature's. Its lab was the Tall Timbers landscape and its neighboring plantations. It had to chastise institutional science as it did fire agencies because formal science had mostly ignored fire, and when it studied fire, it did so as an exogenous disturbance rather than an emergent property of the living world. Academics had been as remiss as the agencies. By 1975, when the fire conferences ended, even the Forest Service had accepted fire as part of land management, and ecology was absorbing it as a legitimate topic for inquiry. Revealingly, when the USFS sponsored a conference the next year, as though a sequel to the series, it was titled "Fire by Prescription Symposium" and held in Atlanta.

The conference venue proved particularly apt. It was difficult to get fire into most scientific journals; it was hardly taught in universities, except here and there in forestry and range schools. (Not until 1991 did wildland fire acquire its own association and journal.) Nor was it a routine topic at meetings of professional societies. A fire community did not exist as an autonomous presence. So a conference was an ideal medium to bring the scattered members of the clan together, particularly those who had hybridized with other fields; conferences were the intellectual equivalent of interagency agreements. In 1970 the Society of American Foresters and the American Meteorological Society inaugurated a biennial conference devoted to fire and meteorology.

Much as the Tall Timbers conference went global, seeking alternatives to the American way of fire, so did science. The view of Earth from *Apollo 8* complemented the metaphor of spaceship Earth: the planet was a single ecosystem. In 1969 the International Council for Science established its Scientific Committee on the Problems of the Environment (SCOPE), which subsequently sponsored fire symposia on Australia, the circumpolar boreal north, the Mediterranean realms, and elsewhere. Some symposia overlapped with American themes, many did not. Meanwhile, fire conferences went international, as Tall Timbers had in 1971. The North American Forestry Commission, under the auspices of the UN Food and Agriculture Organization, sponsored a "Fire in the Environment" survey of the continent in 1972. That was the year of the UN's Conference on the Human Environment, which led to the Man and Biosphere Program and a global network of biosphere reserves—a striking alternative to American-style wilderness. In 1973 the National Science Foundation selected the quantitative study of fire ecology for special consideration, spurring academic interest.[23]

Suddenly the subject, so long somnolent amid the ashes, seemed to rage into every combustible nook and corner of scientific inquiry dealing with landscapes. In 1973, with the need to consolidate as urgent as the desire to push onward, Arthur Brown, recently retired as director of Forest Service fire research, rewrote Davis's classic *Forest Fire: Control and Use*, and in 1974 T. T. Kozlowski and C. E. Ahlgren edited a book titled *Fire and Ecosystems* that brought together the elders of fire ecology into a summary volume that included not only E. V. Komarek, Harold Weaver, and Harold Biswell but others such as Robert Humphrey and Silas Little, isolated into subregional ghettos, and had chapters written by Canadians, an Israeli, and a South African. Fire belonged to the world.

In 1977 the interested parties—SCOPE, MAB, the USFS, the NPS, and the NSF, to which Stanford University joined by hosting the event—sponsored a world-spanning conference on fire and Mediterranean ecosystems. Virtually everyone who researched or managed fire in any of the five Mediterranean climate zones on Earth attended—Europeans, Israelis, Australians, Chileans, and Californians. The conference instantly defined a community. E. V. Komarek delivered the banquet address and, like a court poet singing a favorite saga, retold the story of the Tall Timbers fire conferences. The next year SCOPE, MAB, the Forest Service, and the East-West Institute sponsored an even more ambitious symposium at Honolulu on "Fire Regimes and Ecosystem Properties." It argued that fire informed almost all biotas, not just those with Mediterranean climates.[24]

But even as academics enthused over new discoveries and geographies of burning, those responsible for land management believed enough was known to change fire on the ground, and they sought working syntheses. Although official policy had changed to one of fire by prescription, practitioners needed the information to write those prescriptions. In April 1978 the USFS convened a National Fire Effects Workshop in Denver to consolidate state-of-knowledge reports on soil, water, air, flora, fauna, and fuels. When finally published, each summary had a different colored cover, which led to their description as the Rainbow Series. In a complementary exercise, a massive bibliography was compiled, what became known as FIREBASE, and which later evolved into the Fire Effects Information System. That year, too, John Hendee, George Stankey, and Robert Lucas published *Wilderness Management*, which included a chapter on fire, as an Agriculture Handbook.[25]

The heroic age of fire research, sparked by strong personalities, had thrived before World War II. The postwar era sparked a golden age in which research relocated into labs, and it was believed that correct knowledge might do what mechanical power had not: put fire in wildlands to proper use. In this new era fire management would spread by hammering the Bunsen burner into a driptorch.

Model Fire

No one had campaigned harder for a fire lab at Missoula than Jack Barrows. When *Red Skies of Montana* was released, he stood outside the Wilma Theater in Missoula dressed in a smokejumper costume to promote fire protection and research. He had come from the old school of field foresters, a protégé of Harry Gisborne, and he relied on statistics gleaned from fire reports for much of his analysis, supplemented by field observations and devices such as fuel sticks for measuring moisture. But he knew that modern science was rapidly shucking the modes of the participant-observer and the statistician for laboratory experimentation, quantification, and modeling. When he became director of fire research for the Forest Service, he pushed to transfer more research into labs. When the Missoula facility opened, he served as its first director.[26]

Yet there was an unreality about the expectations many had for science. One of Barrows's major prelab programs, Project Skyfire, had sought to understand better the lightning that kindled so many problem fires; behind that study lay the loopy hope that it might be possible, through weather modification such as cloud seeding, to suppress fires by suppressing the lightning that caused them. Had not the war shown what organized science could invent? Research existed to aid operations in the field; the more fundamental the science the more widely useful it might prove. However devoted to basic science, the ultimate goal was not to accumulate data or peer through nature's veil but to underwrite field operations.

The nuclear core of fire science was to understand free-burning fire — what burned and how it burned. If you knew that, you could predict the dangers you faced, you could plan fire-protection systems, and you could forecast how fast and fiercely fires would spread. You could plot where smoke would go, and you could make informed guesses about fire effects, both economically and ecologically. The Rainbow Series and the efflorescence of fire-ecology conferences notwithstanding, the urgent tasks of fire management pointed to the same need: a model of fire behavior. Fire-danger rating, fire suppression, prescribed burning, fire planning, fire dispatching — all based their decisions on how a fire would behave, and they generated data through simulations, not the tedious collation of fire reports. Algorithms and quantitative data counted, not qualitative observations and narratives. The days of Tex Strange standing on a ridge in the San Gabriels or of Jack Barrows counting lightning strikes from a lookout on the Kaniksu had passed.

* * *

The National Fire-Danger Rating System was the master project of the early 1970s. Its design had been set in 1968, Barrows parceled its pieces out among the labs, and because the scheme had evolved out of Gisborne's analogue meter, the Intermountain Experiment Station at Fort Collins was assigned to integrate them with a target date of 1972.

Preliminary versions were tested in 1970 in Arizona and New Mexico, then revised and used operationally in the Southwest in 1971. Some 150 stations from Florida to Alaska provided feedback along with a half dozen federal and state cooperators. Its systems architecture was in place. By the end of the year John Deeming, the project leader, published a "philosophy" that spelled out what the system would and would not try to do or be. The next year a fire-behavior module became available from the Missoula lab along with stylized fuel models to help simplify the required input. The quantitative needs of the model determined the inputs, which is to say how the fire environment was to be characterized. Still, the algorithms were rough, the model was primitive, and the complexities of fuels proved daunting. The project contented itself with state-of-the-art knowledge rather than waiting for that "hypothetical day" when they would know all they needed. In 1974 a revised version was published, and in 1978, as originally planned, a final update completed the program.[27]

Project leaders knew they would have to revise. Users wanted more specificity so that the overall system would work better in their particular locale—this the designers had expected. The system needed greater sensitivity in terrain, fuels, seasonality, and range of outputs. In this final report its authors replaced "philosophy" with "principles." The surprises came from the need to create an index for drought and to incorporate live fuels into the fuel modeling. The stunner revelation was that fire officers cared less about the syncretic systems of the NFDRS than about one subroutine within it. They had plenty of feel-the-duff indexes for fire danger; a national system was for administrators eager for a universal standard. What the field folks hungered for was a general model of fire behavior.

* * *

When the Missoula lab opened, it was charged with devising a model that could link environmental inputs to a prediction of fire behavior and hence of danger. Two physicists, Hal Anderson and William Frandsen, along with aeronautical engineer Richard Rothermel, oversaw the transformation of two wind tunnels and a combustion chamber into a cache of hard numbers. They began with the experiments of Wallace Fons, then at the Macon lab. Anderson contributed a critical concept by noting that fire spread was an expression of the conservation of energy as fuel combusted. Gradually, they simplified what burned into stylized models for fuel (nine originally) and how they combusted into a semi-empirical model for fire behavior. In January 1972 Rothermel published the most famous equation in wildland fire.

The formula, Rothermel said, was "developed for and is now being used as a basis for appraising fire spread and intensity" for the NFDRS. It had plenty of limitations. It was two dimensional, it assumed steady-state spread, it demanded that fuels be uniform and homogeneous, and it required that fuels be dead and vary only by moisture content—all properties that could be studied under lab conditions but had little more than casual correlation to the world outside the lab. Natural fuelbeds were lumpy with grasses and woodlands and broken between surface litter and conifer canopy; dead logs mingled with flush shrubs, and mountains and gorges tilted fuel arrays. Winds blustered and eddied, and giant plumes were the diagnostic signature of blowups. But fire officers were desperate for a model that could help forecast fire behavior. Even better, the Rothermel model would "permit the use of systems analysis techniques to be applied to land management problems." The model was a module that could be plugged into the architecture of NFDRS, Focus, or Firescope's original ambition to provide fire-behavior forecasts to assist dispatching.[28]

It was a dazzling achievement, but one that at first seemed likely to remain buried in the black box of equations that was the engine of the NFDRS. Then in 1976 at the Missoula lab, Frank Albini, a PhD mechanical engineer and graduate of Cal Tech, published a guide to "estimating wildfire behavior and effects," or what he likened to a "short course in fire behavior estimation." While models, he wrote, "span the spectrum of fire-related decision-making," they had to be packaged in forms that users could handle. The NFDRS did that for fire-danger rating; the other uses for the Rothermel model needed a more general device. Albini supplied it by an ingenious series of nomographs that linked fuel models, fuel moisture, and wind speed to flame length, rate of spread, and fire intensity. He expanded the range of fuel models from 9 to 13. Rothermel taught the new tools to fire behavior officers at the National Advanced Resources and Technology Center later that year. In 1979 Bob Burgan wrote the equations into a computer program suitable for the handheld Texas Instruments TI-59 calculator. Observers might date the new era from the time when TI-59 calculators hanging from belt loops replaced Buck knives as status emblems. There was no going back.

The model was an administrative godsend that helped leverage the new era of policy. Prescribed fires, prescribed natural fires, wildfires—all demanded judgments based on what the fires were expected to do, and if those choices could be validated by a mathematical model, then they might withstand legal and public challenges better than the intuition of a veteran who had watched flames run up and down the same canyons for 30 years, much less that of a seasonal smokechaser watching a snag fire skip into the crowns. In short order, the Rothermel model powered virtually every software program that required an estimate of fire spread. Its inventors watched with a mix of delight and dismay as the model was put not to the limited uses for which it had been designed but to the uses the fire community needed. It propagated with the velocity of a Fuel Model 3 fire across a prairie. Before its run was exhausted, it transcended its origins in the NFDRS, leaped continents and oceans, and as much as anything from a lab might, apotheosized. It stood for an era of fire research as Michelangelo's *David* did for the Florentine Renaissance.

Yet it also came with flaws and unexpected consequences. It rarely worked in the field without serious adjustments and fudge factors. Rothermel himself warned that it should supplement, not supplant, observation. To an outsider its use could seem perverse in that its predictions were so adjusted by local ad hoc factors that one might as well rely on persistence forecasting, that in the absence of a change in weather tomorrow's burn would look pretty much like today's. Its dramatic appearance and its promotion as the behavior module for NFDRS (and everything else) shut out rivals from the other labs. In particular, the Focus (and later Firescope, NFMAS, and Firecasting) projects at Riverside devised a promising alternative approach, until the authorities decreed that the Missoula lab owned fire behavior modeling and banned competition. In many respects, the Riverside lab never recovered. In that sense the model helped unbalance the internal dynamics of Forest Service research.

Its early—premature—adoption committed the American fire community to a particular technology pathway that became ever more difficult to leave. What the QWERTY keyboard did for typewriters, the Rothermel model did for fire-behavior research. Repeatedly, American models fared less well for fire-danger rating and behavior predictions than those from other countries, but so early and utterly had the country committed to it that the fire community could not reverse. It would have to tinker endlessly rather than start over. Years after he retired, Dick Rothermel noted simply, "It's had a long life."[29]

* * *

As fire required fuel, so the fire-behavior model demanded fuel models. Wildland fuels, Hal Anderson asserted, were "the heart of the matter." Fuels fed the fire, and the parameters of fuel models fed fire-behavior models. Out in the field, fire took its character from the combustibles it burned. In laboratory models, fuels took their character from the demands of fire-behavior algorithms. Real-world environments had to be stylized to fit the model. The fire model specified what fuel models should be. Fire officers would have to make the world conform to their understanding.

What fire behavior was to fire science, fuels were to fire management. The fire triangle metamorphosed into a fire syllogism. To manage land, you have to manage fire. The best way to control fire is by controlling its fuels, and the best way to modify fuels is to burn them under controlled conditions. Prescribed burning became the treatment of choice not only because it restored fire as a natural process but also because it offered an economical means to mitigate fuels. A survey of Forest Service research concluded that the highest priority, both in "potential payoff and probability for success," lay in "hazard reduction through fuel management." The way to contain wildfire was to restrain the combustibles that fed it.[30]

In this way fire by prescription addressed the yin and yang of fire management. By restoring or emulating a natural process, it could enhance ecological well-being: it put the wild back into wildland. Elsewhere, by controlling the abundance of combustibles, it improved fire protection; it made suppression easier. Tamed fire could substitute for wildfire. In fact, fire-behavior models were better suited to prescribed fire than to wildfire because some of the variables could be specified and selected in advance. Fire-behavior research, fuels projects, and prescribed fire advanced in a kind of intellectual three-legged race.

When the Forest Service first adopted fire management, it often meant a more integrated program of fire protection, one that included presuppression, which mostly meant treating fuels. In 1972 the agency adopted a 10-acre policy for planning to complement the 10 a.m. policy for suppression. It created a National Fuel Inventory System and began hosting conferences on fuels. In 1974, as the NFDRS was released with its catalog of fuel types, Anderson published a concept paper on "appraising fuels"; James Brown wrote a *Handbook for Inventorying Downed Woody Material*; and Owen Cramer assembled for the Forest Service a state-of-knowledge compendium, *Environmental Effects of Forest Residues Management in the Pacific Northwest*, later adapted for NARTC. To cynics, slash was a problem in solid-waste disposal. To those with more elastic conceptions of prescribed fire, the volume offered hope for expanding controlled burning. Quickly, the process of inventorying fuels expanded into conifer crowns, sagebrush shrublands, southern rough, and other fuel arrays; even giant sequoias fell under the spell. By now manuals for assessing fuels and for burning them by prescription dappled the institutional landscape like slash piles.[31]

The dark side of this project was that it intellectually reduced landscapes to caches of combustibles and subtly allowed fuel appraisal to replace fire ecology. It committed fire management to a technological pathway much as the Rothermel model did. It shifted attention from fire's restoration as an ecological rehabilitation to fire's use to assist suppression. The modeling of fuels suited a discipline such as forestry, which had long prided itself on its mensurational skills: fuels treatment looked like silviculture by other means. Fuels could be weighed, counted, quantified. Fuel reduction could serve as a numerical index of prescribed burning's success in ways that amorphous "naturalness" could not—a hugely attractive proposition to an agency overturning a long-held policy. The fuels-fire nexus tended to support the old fire-control agenda instead of a fire-restoration one. Besides, because so much of the fuels problem was the outcome of accelerated logging, better fire control was a means to support that harvesting.

Each region had its own characteristic fuel issue and hence style of prescribed burning. The heartland of prescribed fire, the Southeast, had long argued that only routine burning, preferably with strip fires, could beat down the raucous rough that made fires wild. The West gradually discovered it had a comparable issue, a slow-motion train wreck in which a conifer understory had invaded montane woodlands and transformed surface fires to crown fires. The giant sequoias were almost invisible (and placed at risk) because of the white fir and incense cedar that had sprung up with fire's exclusion. The solution was selective cutting, piling, and burning. Meanwhile, the explosion of logging in the postwar West lathered landscapes with slash—probably the most hazardous fuel complex on the continent. As natural wildfires receded under the blows of mechanized suppression, artificial wildfires stoked by slash were replacing them. Burning those debris-choked clear-cuts not only dampened their fire hazard but helped prepare them for restocking (modern forestry was effectively industrializing slash-and-burn cultivation). Where the cutting was heavy, smoke could smother valleys in palls, but smoke was a physical-chemical by-product of burning and so could be factored into the modeling.

Fire control, fire use, logging—all seemed to spiral into a grand convergence. Just as each region had its own fuel dynamic, so each evolved its own research program, its characteristic style of prescribed burning, and its champion fuel complex. Special research projects targeted the southern rough, postlogging "forest residues" in the Pacific Northwest and Northern Rockies, invasive conifers in the Sierras, and perhaps most famously, chaparral in California. Chaparral became for fuels what the Rothermel model was for fire behavior. The southern rough was too intertwined with a problem region, and logging slash was too tainted with commodity production to serve as national paradigms. Instead, the agency looked to California, which seemed poised to do for the fire revolution what it had done early in the century when Coert duBois published a universal model for suppression, *Systematic Fire Protection in the California Forests*. California was the future—everyone said so.

* * *

As Californians dominated the roster of Forest Service fire directors, as California's conflagrations drained the national fire budget and underwrote the national agenda, so the story of California chaparral became the national narrative for fire management. Chaparral accounted for the peculiar dynamics of the blowups that seemed the biotic counterpart to its endless tremors. If unchecked they would cascade from one tragedy to another, block the South Coast's growth by development, and consume the national fire budget. But the reverse was equally true. If the fire-fuel dynamic of chaparral could be parsed and points of intervention identified, then proactive fire management could replace an interminable chain of one big smoke after another.

Chamise (*Adenostoma fasciculatum*) acquired celebrity, if not charisma, when it was adopted by the Riverside fire lab as a model species. It became shorthand for the complex of combustibles subsumed under the chaparral label. As the lab identified fuel models suitable for the NFDRS, and as it investigated under Project Flambeau the properties of mass fire, it had to simplify the tangled slope of mountain brush into properties relevant to fire behavior. The lead scientists were Clive Countryman, who headed the fire-behavior research project, and Charles Philpot, a young forester who later became national director for fire research for the Forest Service. The model species inspired a model study, published in 1970 as "The Physical Characteristics of Chamise as a Wildland Fuel."[32]

What rendered chamise so extraordinary was the way its biology expressed itself in traits that accented its flammability. It underwent seasonal changes in moisture and volatiles and secular changes in its physical properties. Eerily, as chamise passed through its annual phenological cycle, the tendency to burn grew in sync with drought and winds, reaching a maximum during late fall Santa Ana conditions. As it aged it transformed more of its biomass into available fuel such that the longer it remained unburned the more susceptible to burning it became. Before 5 years it hardly burned at all; after 25 years it burned with increasing intensity. All this could be plotted on graphs and compared with the properties of other fuels (only red slash approached aged chamise for flammability). Once burned, it reseeded prolifically in the ash or resprouted from its buried ligno-tuber. It seemed, in fact, a perfect phoenix species with a life cycle primed for rejuvenation by immolation.

Chamise did not grow despite fire but seemingly because of it. The plant sprouted fuel, the fuel burned, the fire renewed the plant so it could refuel the landscape. The system was almost mechanical in its regularity, like the piston in an engine that sucked in combustibles and oxygen and then compressed them until the spark of ignition exploded the mix. Since it was impossible in Southern California to rebuild the mountains, calm the winds, or prevent arson and accidental ignitions, the only way to shatter the cycle of conflagrations was to target the fuels that powered it. The fire revolution suggested the ideal means was to use fire against itself. The life cycle of chamise showed how and when to do that and why the strategy would work.

Projects took two forms. One was to carve fuelbreaks across the terrain. The other sought to replace wildfire with tame fire, to prescribe-burn chunks of chaparral landscapes just as the flora entered its most volatile phase. The patches would break the continuity of the fuels that allowed flames to pour over the mountains. It would renew the biota while reducing the savagery and scale of conflagrations. The Southern California scene could, in principle, resemble the southern strategy of burning off the rough. But chaparral grew on mountains, not coastal plains; Santa Anas were not the steady northers that followed cold fronts. The adjacent urban fire services worried that burns were too risky and the results too unpredictable. In the Southeast, so the saying went, "every day is a burn day." In Southern California fire officers instead operated under a Rule of Three: three days of Santa Ana winds, three weeks to adjust fuel moisture content, three months to accommodate the end of rain. It was harder to find the right moment, and the penalty for failure was unthinkable.[33]

Experienced fire officers knew—had personally seen over their own careers—fires recurring on the same slopes and in the same ravines; they knew the where, the what, and the when. They designed operational projects to burn preemptively and break up the swaths of fuels that made vast fires possible. They designed fuelbreaks as levees to channel wind-driven flame rather than as dams in a futile attempt to block it altogether. The scheme was operational by 1977; four years later the region was burning 11,000 acres a year, targeting a 25- to 30-year return interval—exactly the cycle of chamise flammability.

Still, the mountains held more than chaparral, and research tailored its prescriptions accordingly. Natural fires were impossible to use: the longer and more widely fire lingered on the land, the more likely it was to blow up. Most underburning occurred along forested mountain summits. To improve the brush there for burning, researchers experimented with goats and sheep to chop it up and elsewhere with mechanical surrogates such as bulldozers, anchor chains, and masticators from San Dimas Center blueprints. In 1981 Lisle Green of the Riverside lab consolidated that know-how into a guidebook for "burning by prescription."[34]

It appeared that California had fashioned an integrated fire-management organization that fused fuels mitigation, prescribed fire, suppression, and research into a virtuous circle. Chamise was a model fuel ideal for a model of fire behavior that in turn could inform a model national program of fire management.

* * *

The most public expression of this fuel-informed program—so visible it could be seen by commuters—was a network of mountain-gridding fuelbreaks. That strategy had a long history in the region. Local authorities were advocating (and financing) fuelbreaks as dual-purpose firelines and trails as soon as the mountains were gazetted as forest reserves. There were fuelbreaks in the San Gabriels before the Forest Service assumed their administration. They provided, in principle, a means of access, a method to break up continuous fuels, a rudimentary fireline ready to activate, and a visible display of administrative resolve. Unsurprisingly, the nation's grandest experiment also emerged from California; the Ponderosa Way was a 650-mile fuelbreak built during the 1930s that ran from Bakersfield to Redding. It was California's dark double to the Prairie Shelterbelt scheme.[35]

Over the years fuelbreaks have displayed a cycle of senescence and regeneration much like that of the flora they strip away. After each disastrous fire season, existing fuelbreaks are scraped clean and widened, and the system expands. Then they decay. They are expensive to maintain; other needs clamor for the money, and critics scorn the ridgeline scratchings, which they regard as ugly and useless. The secondary system overgrows. Only the primary roads and those deemed most essential receive maintenance. Then the flames rush over the landscape, the public demands protection, and the fuelbreaks return. The life cycle of fuelbreaks shows an uncanny similarity to the rhythms of the chaparral in which they are embedded.

During the 1950s California hosted a massive revival under the doctrine of conflagration control. One variant looked to the forested Sierra Nevada and almost greenbelt-like landscaping along the summits. The other built on the Transverse Range and pumped up the classic chaparral design that loped from ridgetop to ridgetop like an inverted Great Wall. In 1958 the Pacific Southwest Experiment Station launched the Fuel-Break Program, a cooperative undertaking with the California Department of Forestry and Los Angeles County, and bolstered the science behind it (an offshoot of Operation FireStop). By 1965 the region boasted more than 384 miles of fuelbreaks, all of it over 200 feet wide (and much over 300 feet); it had 188 miles of fuelbreaks over 110 feet in width and another 120 fuelbreaks under construction—and there were 20 outright brush conversion projects. Within five years Countryman and Philpot published their seminal conclusions on the fire-fuel cycle in chamise.[36]

After the 1970 fire bust, the political pressures for more protection prompted yet another, larger outburst of fuelbreaks, this time blended with prescribed burning. The fuelbreak became one of the most distinctive features on the landscape: it was to the region's wildlands what interstate freeways were to its cities. This time the clearing was assisted by herbicides (Agent Orange emerged from the program). And this time, too, officials were challenged by environmental concerns. In 1972 Region 5 submitted an environmental impact statement for its proposed "brushland management." The fuelbreaks were still spreading like the tendrils of catclaw acacia when the 1977 fires broke out. There were plenty of critics who detested the mechanical scrawlings as political graffiti. But science, fire agencies, and even the fire revolution all seemed to point to fuelbreaks as vital for the control of going fires and essential to the prescribed burning that would move the landscape away from conflagration and into a quasi-controlled state. When alloyed with projects to improve wildlife habitat, range, and watershed, fuel modification became the treatment of choice throughout the region. It reduced a lot of quirky biology to codes suitable for the prevailing physical models of fire spread. It seemed to align the region's relentless drive toward suppression with the national interest in prescribed fire.

The reactions came later, when it became irrefutable that Santa Anas simply blew embers by the millions over fuelbreaks and when it became obvious that prescribed fire befouled the air, encouraged conversion to grasses (often invasives), and was too volatile for routine use. The reclassification of lands into wilderness or other protected uses limited the machines, roads, and chemicals that a fuelbreak system required. The demand to convert adjacent land to housing was implacable. The fuelbreaks could stop neither the big fires nor suburbanization, and they only further damaged a traumatized biota. Still, like conflagrations and exurbs, the fuelbreak cycle continued through the 1970s.

Perhaps the deeper issue, also not apparent in the decade, was the limitations of modeling. Pinning a fire program on chamise was a lot to ask of one shrub and one study. Perhaps inevitably, what happened with the Countryman-Philpot fuel model was similar to what had occurred with the Rothermel fire-behavior model: the research was pushed beyond the range of its meticulous measures and boundary conditions. The simplifications that made it useful also made it suspect. No matter; fire officers used whatever was at hand, and if it came with a scientific imprimatur, so much the better.

Chaparral was more than chamise, chamise was more than fuel, and prescribed fire was more than hazard reduction. Fire management meant more than expanding fire suppression to include fuels projects. The often dismissed brush was a complex biota, full of nooks and niche species, not reducible to mechanical analogies. Its landscapes were subject to invasive species, and recurring big fires might simply convert large patches to pyrophytic grasses. In this regard, prescribed fire might be no different in its outcomes than arson. Moreover, demographic changes in fire's workforce were stripping out the people whose intimate knowledge made prescribed burning work. Models might substitute for expert opinion in a lawsuit, but not on the ground. While fire science did brilliantly what fire science could do, the fire revolution demanded more than predicting wildfire and reducing fuel. The type conversion that mattered was from indigenous species to suburbs.

* * *

A paradox of modeling is that it must simplify the world, but once codified, that simulacrum becomes the basis for action. Agencies and officers are expected to ground their decisions on what the model describes or on the simulations the model derives. With the emergence of computers, modeling moved from data processing to data creating as simulations replaced experiments. Once Congress mandated that the Forest Service conduct certain kinds of planning and budgeting exercises, the agency turned to models, and the models at hand became the medium for understanding. Landscapes of oak and elk had to conform to that virtual world.

But like fuel models that reduced layered, mixed-conifer forests, redolent with cultural connotations, to bulk density, mean packing ratio and loading, management models and decision-support software had to ignore much of the political and social landscape. Wilderness celebrated cultural and spiritual values; it was not a module to plug into a scenario. The Endangered Species Act of 1973 counted species and measured habitat, but not through the parameters required by the Rothermel algorithm. The outburst of outdoor recreation—by many indexes the largest economic activity on the public lands—sought blue skies and green woods; these were not inputs recognized by the models that guided planning. As critics and advocates argued over fire, they typically spoke in incommensurable languages. One saw flame burning amid cultural concepts and ecological goods and services. The other saw it converting fuel complexes to ash, soot, and gases. The fuels-fire nexus tended to support the old fire-control agenda rather than an environmental one.

When the Forest Service celebrated fire's management, it often meant a more integrated program of fire protection, one that included fuels. For a decade and more, research had promised a way to counter the diminishing returns from throwing bulkier machines and more crews at fires. Like the Rothermel model, which assumed a steady-state fire in which what happened in one part of a burn had little influence on other parts, fire research reached a plateau. For a while each new revelation could send shockwaves of hope through the fire community. Then more data, or with the rapid evolution of computers the ability to store and move untold mountains of information, reached a point of diminishing returns, a new normal in which fresh discoveries had less influence over fire management outside of research itself. What happened in Macon or Missoula no longer shaped the flaming front of the fire revolution because the really troubling issues were not, in the end, scientific questions at all.

Parks Are for Burning

The two federal agencies that figured most prominently in the revolution were the Forest Service and the Park Service. The USFS, custodian of Smokey Bear, loomed large because it controlled so much of the establishment, stood for what critics thought was wrong, and thus took most of the hits. The NPS, the first federal agency to adopt the revolution, claimed most of the kudos. But of course the story was more complicated than this. The Forest Service moved on the ground almost as soon as the Park Service did, and the Park Service's advances, through striking in their symbolism and rhetoric, were in practice deeply compromised. If the parks were the hares of the movement, dashing boldly about the landscapes, the Forest Service was the tortoise whose steady pace caught up.

Restoring fire was a daunting undertaking for a small, widely dispersed, and often fractious agency. The NPS lacked the institutional mass of the USFS or BLM, lacked foresters' claims to professional standing, lacked a strong synergy of research and administration or of disparate parts that could cooperatively contribute to the whole, lacked a profound commitment to fire as something bred into its genome. It distrusted both the wilderness movement and the drive to found more management practices on science as challenges to the long-held prerogatives of its managerial class. Its loose administrative order struggled to find a place for fire as something actively managed. In most parks fire control remained in the Protection Division, while prescribed fire resided in a newly organized Division of Resource Management.

What the agency did have was a nimbleness that allowed it to move quickly. Its feudal structure permitted reform-minded barons to experiment widely: what happened (or did not) occurred through personalities and at particular parks. Mostly, though, it had the blessings by and large of the environmental movement and a tailwind of public support for pristine nature. Errors of the sort that landed the Forest Service in court or the BLM in the stocks of public opinion were skimmed over when committed by the National Park Service. The Park Service had political space to maneuver. During the 1970s it stood for—or seemed to the public to stand near—the vanguard of the fire revolution.

In reality it often stumbled. The Leopold Report was one challenge, selectively exploited to ward off the more restrictive wilderness movement. Similarly, the Robbins Report was absorbed in ways that either fettered science as an auditor or co-opted it into resource management. An agency accustomed to long-serving directors became politicized and underwent a fast churn of leadership. Ronald Walker lasted less than three years, Gary Everhardt just two, and William Whalen three. The belief that the NPS should emphasize its crown-jewel natural parks was challenged by a drive for more recreational units ("parks are for people"), for urban parks that could bond the agency with the majority of voters, and with increasing pressure from a more bumptious public that led the agency to emphasize such visitor services as law enforcement and emergency medicine. Alaskan parks and monuments, established between 1977 and 1980, effectively doubled the geographic dominion of NPS management. The agency was unsure how much of its land and attention to commit to "vignettes of primitive America" and, within those places, what the right mix of fire might be.

For some superintendents this was an old dilemma updated. The parks had been established as "pleasuring grounds" for the public and were expected to remain "unimpaired" for future generations. The Forest Service had its multiple publics and multiple uses, but the Park Service had two directives that might either complement or compete. Having to both use and protect was not a bad metaphor for the problem of fire management in which fires had to be both restored and controlled. For the Park Service the "use" half of the equation was itself split into two competing factors, one of which wanted natural ignition and the other of which insisted that prescribed fire and probably some landscaping in the form of fuels projects were needed, too. If that formulation tended to oversimplify the choices, it also sharpened them. Either fire returned, or it did not.

* * *

The Sierra parks remained the hearth and nursery. Everglades had broken the agency barrier in 1954, but like its landscape, it was an outlier, with controlled burning another exotic feature along with sawgrass and alligators. The Sierra parks were, together with Yellowstone, the cornerstones of the system. What happened in Everglades would tend to stay in Everglades. What happened in Sequoia-Kings Canyon and Yosemite could disperse throughout the country.[37]

Here amid the Big Trees and High Sierras, Starker Leopold had advertised the problems, and Harold Biswell and his students had found solutions. The University of California, Berkeley, Whitaker's Forest furnished the demo plots. Sequoia-Kings Canyon pioneered both let-burns in the backcountry and prescribed burns in the frontcountry, even amid the redwood groves. Bruce Kilgore, a Leopold student, became the public voice of restored fire. Even the Forest Service assisted in the person of a Pacific Southwest Station researcher, Harry Schimke, who helped transfer fire-behavior expertise. But in any park the critical character was its superintendent. Sequoia-Kings Canyon's superintendent John McLaughlin stood by the program and insisted it move ahead despite qualms from the public, uneasiness from the neighboring national forests, internal friction among his staff, and operational glitches including escapes.[38]

Through Biswell and his students (notably Jan van Wagtendonk) and Bob Barbee, newly appointed to head resource management, the program propagated into Yosemite. Another Biswell protégé, James Agee, argued that by the early 1970s Yosemite had come into parity "with that of Sequoia." The Sierras became a breeding ground for a generation of Park Service fire reformers. Much as California had pioneered western mining—there was always an "Old Californian," a survivor of forty-niner days, among the prospectors to tell the novices how to do things—so the New Californian taught the greenhorns of the modern fire rush what to do. The exemplar spread beyond the NPS. In 1971 Bob Mutch and Dave Aldrich of the Forest Service, then charged with designing a natural fire program on the Selway-Bitterroot Wilderness, visited Sequoia-Kings to study the emerging national paradigm.[39]

The 1972 Tall Timbers conference contained a symposium on Fire in the National Parks, a centennial celebration of the park idea. It was a coming-out party for the fire program, and it laid down the lines of future development. Bruce Kilgore described a meticulous experiment in which five acres were burned to determine how fire might cleanse the understory of encroaching sugar pine and white fir. An at times impulsive doer, Peter ("Pyro Pete") Schuft, chief ranger, spoke of operations. And Superintendent John McLaughlin addressed how the program might align with public sentiment and politics. The order in which the three men spoke was significant: it implied that science would inform, operations apply, and administration coordinate with the outside world.[40]

In reality, the project was an omelet, as it had to be. What drove the program was not experimental science but the undeniable fact that fire had been a part of the Big Trees and their larger environs for millennia and was not going away. By any metric it was "natural." The choice, as Kilgore explained, "is not whether to burn or not to burn; the choice is merely when, how, and under what conditions." Schuft described an impressive escalation in projects that had put 69 percent of the park into let-burn, had prescribe-burned 13,000 acres, and had constructed 27 miles of fuelbreaks. He judged that "we now have enough acreage burned under varying conditions to evaluate what has been done and determine methods to complete the job." With adequate funding—he thought $20,000 a year would suffice—the park could complete in five years the needed burning in critical areas and would be into "reburn cycle." McLaughlin was more circumspect. He believed the public would "wait and see," that its current attitude was a "definite plus," that the case for natural fire was clearer than for prescribed fire, and that he was "quite certain" it would be "woe to anyone who makes a mistake." The ultimate trial would come when the burning migrated into the vicinities of the General Sherman and General Grant trees. Kilgore and McLaughlin were proved right, and Schuft, wildly overoptimistic.[41]

Other parks signed on. In 1971 Saguaro had introduced the concept of a natural prescribed fire. Yellowstone, too big to ignore and too big (so everyone assumed) to fail, released a fire plan in 1972 that committed 340,784 acres (15 percent of the park) to natural fire. Horace Albright, at 82, still a power behind the throne for the agency, thundered against the program. "If you do not stop this fire policy," he warned director George Hertzog, "I'll have to enter the defense of Yellowstone." The park defied its aging patriarch, and in 1975 virtually all of Yellowstone except its developed areas moved into a natural-fire regimen. In 1973 Wind Cave National Park on the flanks of the Black Hills inaugurated a prescribed fire program on its grasslands and pine steppes. Grand Canyon sponsored a research program intended to whet prescriptions before reintroducing fire. Grand Teton released a full-service plan for fire by prescription. By 1974 Rocky Mountain, Carlsbad, Guadalupe Mountains, Shenandoah, and North Cascades all had published restoration plans; Glacier, Isle Royale, Redwood, Lava Beds, and Point Reyes were in the pipeline. The agency sponsored Wilderness Fire Management Workshops (an influential second event coincided with the 1974 Tall Timbers conference at which the Forest Service proclaimed its commitment to fire in land management). On paper the Park Service had over three million acres of land zoned for natural fire. All it took, it seemed, was a recognition that fire belonged, some science to refine prescriptions, and political will. The way to restore fire was to stop fighting it.[42]

Then the grand project hit a snag. In 1972 Grand Teton had proposed a bold plan—too aggressive for the regional office, which told it to scale back and proceed more cautiously. The next year the park attempted two prescribed burns, both of which fizzled but still emboldened fire officers. In July 1974 the park circulated its revised plan through an environmental assessment. Critics including Adolph Murie objected to deliberate burning in favor of natural processes, while others such as Richard Baldwin, a USFS retiree, argued for better fire-behavior knowledge. Then on July 17 nature stepped into the arena with a lightning-kindled fire in Waterfalls Canyon. For six weeks the fire crept and puffed in a patch burned the previous year, while ranger-naturalists interpreted the smoke to visitors and spokespersons declaimed that "at no time was the fire's behavior unpredictable or was the fire uncontrollable." The fire smoldered through the summer. On September 10 it moved upslope and more than doubled in size. Within a week it was 1,500 acres, and by September 19 it had grown to 1,900 acres. The size of the burn mattered less, however, than the size of its emitted pall. Jackson Hole was smoked in. A curiosity had become a nuisance, and a natural marvel, a biotic geyser, had morphed into a public health menace.[43]

As so often, the program rose or fell with its superintendent, and the success of its public reception changed with the receptivity of the national media. Superintendent Gary Everhardt upheld the program. CBS and NBC news reported the event sympathetically, and *Time* explained the episode and supported the policy behind it. The Waterfalls Canyon fire became a cause célèbre: it seemed to demonstrate that public skepticism and political sniping could be overcome by firm will and vigorous public relations. Still, the regional office at Denver convened a workshop in December to review what it called "natural fire management." The 35 participants came from the NPS, the USFS, and fire research. In a detailed "critique" the group noted the special features that distinguished prescribed natural fire from prescribed burning, discussed means to promote PNFs to the public ("an ad hoc inter-departmental committee to approach the National Advertising Council"), urged system-wide guidelines for planning, and emphasized the inadequacy of monitoring. They worried over smoke and fretted about clumsy terminology ("we can only imagine the visitor's feeling when he is confronted with our jargon"). Participants spelled out recommendations for both the region and the agency as a whole.[44]

The next year, Everhardt became director of the National Park Service, and within a month he issued a memorandum that confirmed his "personal interest and involvement in the fire management program"—a first, as historian Hal Rothman observed. But Everhardt understood that a successful program demanded more than enthusiasm. The agency undertook a series of measures to bring some order to its sprawling fire dominion. It clarified what to do and what to call it. It strengthened integration with the national fire community. It cleared a lot of bureaucratic scrub at one boring but brilliant stroke by releasing an environmental impact statement for its entire fire enterprise. It also began to assemble formal guidelines to provide the rules of engagement that its administrative guidelines for natural areas, the Green Book, glossed over. Each measure deftly repositioned the NPS's national fire project.[45]

The first response, staff directive 76-12, standardized procedures and terminology and served as an interim service-wide manual. "Fire management" officially replaced "fire control." Fire management was placed within resource management, and fire suppression, prescribed burning, and research were to fuse into "a cohesive program to perpetuate the resources entrusted to park management." The exercise aligned the NPS with the NWCG and NIFQS and argued that the special tasks of the Park Service required it to create new positions for monitoring and prescribed burning. The agency had its own mission, and it needed its own enablers. Fire officers at Sequoia-Kings Canyon led the effort to identify jobs, qualifications, and training. Beyond defining common terms, the directive extended standardization into fire reports, into mechanisms for funding, into research, and into fire planning. The memo noted approvingly that the NPS had led the campaign to reintroduce fire by means "of natural ignitions."[46]

The second maneuver, publishing a single environmental impact statement for the service-wide fire program, instantly granted it blanket approval, unlike the Forest Service, which had to negotiate every prescribed fire on every forest, allowing for an unbounded number of objections, whether substantive or procedural. The proposal inverted the usual discourse by insisting that it was fire's suppression, not its resurrection, that needed an impact statement to continue. The Park Service was seeking to restore what had been unwisely removed. Early on, then, the NPS had a richer administrative mix of gasoline to diesel in its driptorches. While it had no carte blanche to burn, it did have far more freedom to maneuver without having to constantly legitimize every experiment.

With these preparations the agency had the rudiments of a collective system ready to crystallize. For the Park Service the tricky question was how to plan for fire restoration park by unique park, superintendent by jealous superintendent. Without its elaborate preparations, a plan in place, and a committed superintendent, Waterfalls Canyon might have been a disaster. And what happened in one park affected every park—the public and Congress were unlikely to distinguish among them. Some collective security was essential. The process that had led to an approved plan for Waterfalls Canyon had to go national.

On November 1, 1976, Everhardt signed a new directive that announced a system-wide review to just that end. A task force would design a national structure and devise operational procedures to make it happen. The Washington Office put up supplemental monies. The task force members included many of the rising leaders in NPS fire, including David Butts, John Bowdler, Robert Sellers, William Colony, and Larry Bancroft—representatives from the Sierra parks, Everglades, Glacier, Yellowstone, and Rocky Mountain. A few months later another staff directive, 77-1, outlined the evolving structure. Within the NPS, fire received, for the first time, recognition as a line function separate from rangering. And the Park Service elected to affiliate with the interagency movement. It signed on to BIFC and the NWCG and worked to bend national developments to support its special ambitions. No more than any other agency could it fight big fires by itself, and no less than the others, it could be harmed by funky lexicon and procedures useful to the USFS or BLM but at odds with its peculiar needs.

Like Britain in the European Union, it yielded its autonomy reluctantly, but it could not afford to stay out. While it had supported a national fire program since 1928 and had used the CCC to erect an infrastructure, its moves had been defensive: they sought to protect the park against a threat to its fundamental charter. Those threats remained. In 1972 the 2,680-acre Moccasin fire in Mesa Verde highlighted the quandary of fire within a protected archaeological site as flames exposed new ruins and damaged others. But the assumption endured that fire in many parks, as the Mesa Verde episode seemed to demonstrate, was something extraneous to the NPS's fundamental charge, or as part of the agency's mandate to protect, it was of a piece with vandalizing visitors. With good administration in time it might become vanishingly small. The reforms of the 1970s marked the first time the Park Service seriously sought to integrate fire into its larger charter as a permanent task that would not, even in principle, disappear.[47]

The project climaxed in 1977 with the publication of a national fire policy directive, NPS-18. The manual superseded all predecessors. It specified how exactly the NPS would speak about fire, how it would manage it, and where fire resided within the agency. The objective of fire management was "to achieve the resource objectives of the area through prevention of human-caused wildfire, to minimize the negative impacts on the resources from all wildfires that occur, and carefully guide the prescribed use of fire as an integral part of the resources management program." While quoting the Department of the Interior manual, that wildland fire "usually causes destruction or deterioration . . . and degradation," the new policy identified the growing menagerie of labeled fires—human-caused, natural, prescribed, prescribed natural, wildfire—while herding them into two administrative corrals. They were either wild, which were to be suppressed as economically and with as little environmental damage as possible, or prescribed, which promoted agency goals and might be started by nature or by fire officers.[48]

The directive moved fire matters from the voluntary to the compulsory by requiring every park to assess its fire issues and write a plan to address them. Plans might call for "simple suppression" in places where fire was neither wanted nor needed, or they might invoke "complex fire management" where fire had to be incorporated into routine operations. Such plans mandated research adequate to understanding the history and ecology of fire on every site. Critically, NPS-18 created the template of a model fire plan. While it recognized that each plan "should be unique to the park," the template allowed access to the latest thinking and established a standard of best practice that could spread some uniformity among the 278 units of the system and would also be useful in court or amid public controversies. To implement it, the agency created a Branch of Fire Management, a field office of the Washington office's Natural Resources Management Division, and located it at BIFC. What had emerged almost spontaneously from an archipelago of first-mover parks had become a system-wide standard.[49]

* * *

Under Gary Everhardt, with the Sierra parks as models, the National Park Service committed to the complexity of fire's management. As often as not for the American public it was the poster child for the revolution. But future directors had other priorities. Not all parks and superintendents agreed with the directive, and few rangers were willing to risk a career on something as chaotic as free-ranging fire. Fire suppression was not easily dislodged from its privileges, particularly when smoke appeared on the horizon and fighting it had its own funding. Parks were for people, not for burning. The 1960s and 1970s were by and large benign times nationally for wildfire; big fires stayed in the Northern Rockies or Southern California. When prescribed burns escaped, they scorched tens or maybe hundreds of acres before being wrestled back into confinement. (The notorious Waterfalls Canyon fire barely reached 1,900 acres.) Enthusiasm for wilderness, and the environmental movement generally, allowed room for experiment. For the early decades of the revolution, the Park Service was a free rider on the national infrastructure. Interagency agreements granted it access to abundant resources while it contributed little in return beyond rhetoric.

In 1977, with NPS-18 barely distributed to in-boxes, lightning kindled a fire in Bandelier National Monument in northern New Mexico. The La Mesa fire blazed over 23,000 acres, forced an evacuation of park headquarters, burned a landscape dense with relics and ruins (more vulnerable to line-building bulldozers than to flames), and threatened the Los Alamos National Laboratory with a nightmare scenario that had wildfire ripping through stockpiles of plutonium. Restoring fire, it seemed, might enhance natural resources, but it could also destroy cultural ones. Whatever their ambitions for naturalness, places like Bandelier needed adequate suppression forces and a plan to dampen the likelihood of another blowup.[50]

The La Mesa fire was fought from the onset. The Ouzel fire of 1978 tested the concept of a prescribed natural fire in one of the nation's crown jewels, Rocky Mountain National Park. Rocky Mountain was not known as a fire park: most of the landscape had burned before the park was established. (One of the founders of ecosystem ecology, Frederic Clements, published in 1910 a study of its "lodgepole burn forests.") The park tested the belief that ideas fledged in the Sierra Nevada might take flight elsewhere. The first conclusion was that they could not, not easily.[51]

Rocky Mountain became an *experimentum crucis* because its pioneering plan was written by David Butts, one of the architects of NPS-18, and subsequently the director of the NPS Branch of Fire Management at BIFC. The park boldly decided in favor of natural rather than prescribed fire, reasoning that deliberate burning "results in an artificial system." In any event, lightning fires were rare, so when Rocky Mountain got one on July 24, 1973, the Junction fire became "a test of the system." The smoke was sighted first by visitors who rushed to put it out. As Butts explained, "The Rangers got here just in time to stop the suppression activity, carefully placed the logs back where they had been, took a deep breath, and started to explain why we were letting fires burn." The park stationed an interpreter at the scene and staffed a fire guard "around the clock to prevent the fire from being extinguished." After five days, rain washed the burn away.[52]

So on August 9, 1978, as lightning at last kindled another snag within the designated PNF zone in Wild Basin, the park elected to monitor it. In fact, rangers again rushed to stop some backpackers from extinguishing the burn. For weeks nothing happened. Then the fire ran up the slope into the krummholz. Again the park watched, confident that the fire had done what it would and would soon expire in winter snows. Instead of a snow shower, however, the park got chinook winds on September 15. The fire rose up, then poured down Wild Basin like an avalanche. Only a topographic fluke, a small ridge that caused the winds locally to skip upward, spared the town of Allenspark from incineration. The park ordered an all-out suppression effort, which lasted another two weeks. Since it had nominally been a "controlled" burn, Boulder County cited the park for violation of air-quality regulations. On October 3, superintendent Chester Brooks ordered a board of review.[53]

The board, consisting of representatives of the NPS, USFS, Colorado State Forest Service, and a fire ecologist from nearby Colorado State University, wanted to spare the policy—too much of the fire revolution was at risk. But it was unsparing about the policy's implementation on the ground. The park, it concluded, was not organized for fire management (it was not historically a fire park), it had inadequate understanding of fire's local history and ecology, and the plans had not considered factors outside the park proper. The park had not even followed its own plan and in truth did not have the resources to do so. In the delicate words of the board, it found "that certain deficiencies in the plan may have conspired to prevent users of the plan from making proper decisions." As Hal Rothman summarized the issues, there were "too few resources, too little scientific information, and a public [and maybe a park staff] that did not understand."[54]

What Rocky Mountain had was what many parks outside the Sierra nest had: a statement of philosophy, not an operational plan. It had adapted the Leopold Report when it needed to adapt the NFDRS and BIFC mobilization guides. It had only casually monitored the fire through lookouts and occasional recon flights. When strong winds were forecast for the 15th, it was too late to initiate suppression. The resource management program noticed the blowup when the convective column was sighted from the windows of the Frank Lloyd Wright–designed headquarters at Estes Park. Given fire's rare occurrence and explosive character in the Front Range, it might never be possible to install a program of natural fire. The Park Service again revised NPS-18 and set about rebuilding a new fire plan for the park. Until then, suppression would remain the order of the day.

And that was really the issue before the national parks. Big fires are always emergencies, whether a place is prepared or not. The national fire program would succeed or fail according to the thousands of choices made by hundreds of parks. The fire revolution was not about bold visions and standing aside to let fire reclaim its rightful inheritance. It was about slogging research, scarce money, operational tenacity, administrative patience, and sheer luck. Big ideas are everywhere. The challenge is to turn them into text, image, or place—to take a family saga and write it into an *Absalom, Absalom!* or take a picturesque mountain and render it into *Twilight in the Wilderness* or a notion of rewilded fire and loose it in the High Sierra. Nature did not conform to bureaucratic specifications. Fire would not obey prescriptions.

The real upshot of a collapse like the Ouzel fire was the silencing effect it had on other places, causing some officials to think twice about restoring fire and others to ensure, or at least declare, that no such disaster would happen during their tenure. It was easier to defer a decision until all the research was in, which of course it never is. Or to delay ignition a year or two until conditions were more ideal, which they never are. Or to send in smokechasers to knock down a fire that no one would otherwise know about. There were always good reasons to suppress fire or not to light up. There were few rewards for restoring fire, and only opprobrium for a failure. The agency did not want experiments. It wanted demonstrations of proven fact.

It was a monumental task for a middling agency, and as the Park Service nationalized the program, it became harder to compartmentalize failure. A breakdown in Rocky Mountain would ripple out to Isle Royale and Shenandoah. A dozen bold parks had become points of positive infection. If those big parks held, the little ones would follow. So it might also prove with reforms overall.[55] Big fires tended to be those that started amid surface litter and moved into the canopy. The fire revolution began at the top and had to find its way to ground.

The BLM Builds Out

Reform in the Department of the Interior came as part of a revolution of rising expectations. In the Bureau of Land Management it had the largest federal landholder, an agency intent on evolving into a multiuse counterpart to the USFS. In the Fish and Wildlife Service it had the keeper of the Endangered Species Act, possibly the most radical of environmental statutes, and in the growing archipelago of wildlife refuges, sites for loosely governed experiments in fire policy. In the Bureau of Indian Affairs it had a few enclaves of prescribed burning that might become vanguards of innovation or might sink into unmindfulness. In Alaska it had a battleground—the last frontier—of public land dispensations. In the National Park Service the Department of the Interior (DOI) boasted the most visible of the revolution's firebrands. Throughout, each agency had undergone fundamental changes in its mission.

But while the Interior agencies had in the Forest Service a common rival, they did not have a common champion to challenge it. Most Interior agencies were consumed by the problem of orchestrating their scattered holdings into something like coherence. The NPS struggled to find common ground among an archipelago that included Glen Canyon National Recreation Area, Gettysburg National Battlefield, and Mount Rainier National Park, each of which came with a unique congressional charter. The FWS had to bring system to a scatter diagram of wildlife refuges. The BIA wrestled with reservations among historically distinct tribes; Cherokees had little in common with Hopis. Each agency struggled to find its own coherence quite apart from efforts to join its DOI siblings.

Still, the BLM was big enough that it might bring the others into a loose Interior orbit, so Interior established a Fire Coordination Committee, chaired by Jack Wilson of BLM-BIFC, to help collate and mass their voices in ways that could counterbalance the Forest Service. In reality, interagency institutions such as BIFC were the preferred medium of exchange. Like Australian colonies before federation, each more inclined to look to London than to one another, Interior's agencies related to one other through NWCG rather than a departmental body. Precisely because they were not harmonized into a common organ, and because with the exception of the BLM they had evolved by accreting disaggregated reserves of one kind or another, the Interior agencies were more sensitive to local conditions, more agile regarding novel ideas, and harder to wrangle into a common enterprise.

Yet they all faced fire, and they all faced a fire revolution at exactly the time they confronted a reformation in their missions. Those three strands braided together. Wildfire was always there, both a remorseless chore and an occasion to cut line through bureaucratic hesitation. But throughout the 1970s the Interior covey was spared the wrenching burns that could overturn institutions as they did biotas. Particularly for agencies that lacked entrenched fire programs, the decade was a superb time in which to move from fire control to fire management. Not all made the move. In retrospect, it was the country's last such opportunity.

* * *

Through the 1960s the BLM had campaigned for parity with the Forest Service. From 1963 to 1971 its directors had come from the USFS; Boyd Rasmussen had been a deputy chief when Udall recruited him. During the 1970s that vision was realized, at least in statute.

The BLM began the decade searching for confirmation that the public lands it oversaw it would retain and should manage for multiuse goals. It attained that goal in 1976 with an organic act, the Federal Land Policy and Management Act (FLPMA). Meanwhile, the 1970 National Environmental Policy Act placed on the BLM the same legal requirements as those for other federal agencies; the 1971 Wild and Free Roaming Horse and Burro Act and the 1978 Public Rangelands Improvement Act restated, in more modern terms, the agency's origins in grazing and range. The 1973 energy crisis renewed the BLM's links with mining. It had to conserve National Historic Sites and comply with the Archaeological Resources Protection Act. After FLPMA, it had to review for wilderness. And throughout, it had to cope with a rising controversy over Alaskan lands, sparked by the 1971 Alaskan Native Claims Settlement Act. It ended the decade with its Alaskan empire dismembered and with the so-called Sagebrush Rebellion, in which its old clientele of ranchers clamored to have vast chunks of its estate ceded to the states or even privatized (to them).

Amid such fundamental turmoil, fire was a recurring but minor anxiety. Through firefighting, BIFC had underscored the BLM claim for recognition and became a mechanism for unifying the agency's disparate holdings. Total mobility and the NWCG dampened its need to overbuild its own capacity because it could draft from others. NARTC relieved it of having to create an autonomous fire academy (it had a general training facility in Phoenix). It became the big brother of Interior fire agencies, further enhancing its national stature. It ceded research needs largely to the Forest Service, which had old concerns of its own with grazing and so could transfer studies. It sponsored the occasional conference on a regional topic (such as Alaska), but mostly it participated in those hosted by others. It sought to emulate the Forest Service's equipment centers by sponsoring hardware to suit its particular needs. A few experiments went national, including the automated lightning detection system, which was originally developed for Alaska and the Great Basin in 1975 but by 1978 had spread to 11 states. Others, like the behemoth eight-wheel-drive articulated Dragon Wagon, proved too expensive for routine use in sagebrush playas and too cumbersome elsewhere (shipped to a fire on Michigan's Upper Peninsula, it promptly sank in muck). And like its sister agencies, it began incorporating computers into dispatching.[56]

Unlike the NPS, the BLM was not an advocate of the fire revolution. Instead, it had desperately sought to make itself into a facsimile of the Forest Service—the USFS of the 1960s. When he described the future direction of the BLM at a 1970 Missoula conference, R. R. Robinson, then director of BIFC, declared that "Fire Control Officers are not land managers; they are usually staff specialists to such managers." Management decided and FCOs executed. While he accepted the need to modify a total suppression policy, Robinson defined fire management as the art of manipulating fire "to attain desired objectives" through the knowledge of fire behavior. Prescribed fires helped reduce slash hazard, improve range, create fuelbreaks, and decrease the threat of insect attacks—all goals of multiuse economics, not of general ecology. Fire in wilderness was a special consideration that could be handled as remote fires in Alaska were. In an explicit rebuttal, Bud Moore of the Forest Service argued that fire management and land management were inextricable, that each had to inform the other. The "anchor point" of the future, Moore insisted, rested "in responsible men and women who know their land intimately, understand ecological processes and have reverence for the earth." The BLM was using wildfire to build overall capacity. The Forest Service was exploring ways to redirect its hard-won capacity toward bolder ambitions.[57]

The formal adoption by Interior and the BLM of a modified fire by prescription policy in 1974 had little effect on the ground. Not stressed by another catalytic fire on the order of the Elko complex, the BLM concentrated on upgrading its capabilities to dispatch crews, helicopters, air tankers, and lightning detection networks. As it came to fight fire on a Forest Service model, it began to take comparable casualties; in 1976 the USFS Mormon Lake hotshot crew was burned over at the Battlement Creek fire in Colorado and suffered four fatalities. Like the other federal agencies, the BLM was caught in the vortex of the 1977 season but exited unscathed. The mass callout appeared to confirm the validity and power of BIFC as a national institution, which indirectly redounded to a BLM that nominally oversaw it. The main costs and damages of that explosive season fell on the Forest Service.

What mattered most to Interior was Alaska. There it fought the big fires aggressively. The agency acted, as Robinson asserted, as a fire service, and it battled blazes as it never had before. The real issue, however, was not the ability to amass firepower but the capacity to decide the purpose behind it. The real story, as Moore had predicted, was Alaska's land.

Under the Midnight Smoke

Alaska was special. Like India for the British Empire, Alaska had granted heft to the BLM fire program. Less encumbered by history, the BLM could here adopt the latest technology and move to the front ranks. Significantly, the agency began the decade by renaming its Division of Forestry into a Division of Fire Control, while Congress allocated $500,000 to boost its suppression strength. It put the money into aircraft and organizing Native Alaskan crews.[58]

The decade was framed politically by the contest over land ownership. In 1971 the Alaska Native Claims Settlement Act (ANCSA) ended one squabble by appropriating $1 billion and granting 40 million acres of land to 12 native corporations. But it opened another squabble by authorizing the Secretary of the Interior to consider reclassifying up to 80 million acres for transfer from the BLM to other federal agencies. FLPMA was still five years in the future, so unlike the Forest Service, the BLM was unable to officially manage its holdings under multiuse doctrines; instead, like the USFS, it saw its hegemony spalled apart. What it could do, though, was fight fire.

The complement to ANCSA was a 1971 symposium convened by the Alaska Forest Fire Council and Alaska Section of the Society of American Foresters, sponsored by the Forest Service and the Army's Cold Regions Research and Engineering Laboratory, on Fire in the Northern Environment. In many ways it was a regional clone of a Tall Timbers fire conference; Ed Komarek gave the keynote and admitted that even he had once believed that "fire had no natural place in the boreal environment." Now he knew better. Outside the cold rainforests of the coast, every environment had its fire regime, and virtually all of the boreal forest had regenerated from burns. Of 27 speakers only 1 was from the BLM, and he spoke on the "values protected" by the agency's Alaska Fire Attack Policy and Fire Control Action Plan. That year over a million acres burned, and the BLM fielded its first female fire crew. The next year almost a million acres burned.[59]

National politics provided the big frame. In 1972 the Alaskan pipeline was approved to bring crude oil from Prudhoe Bay to Valdez, and the Rural Development Act (RDA) allowed towns and villages to apply for funds for fire protection. Despite having no holdings outside the coast, the Forest Service furnished basic infrastructure by administering the RDA, by conducting research through its Institute of Northern Forestry in Fairbanks, and by adapting the NFDRS to Alaska's regions. The State of Alaska began assuming responsibilities for itself, using the statehood act to claim its allotted lands and the RDA to bring protection to places outside federal jurisdiction. In 1973 it announced that southeast Alaska lay within its bailiwick (although that meant little since most of the land was in national forests). The next year a joint federal and state land use planning commission identified a suite of likely future fire problems. In 1976 state fire protection expanded its role into the Kenai Peninsula and Anchorage region and then over roads and into the private and state lands south of the Alaska Range. Anticipating a still-greater burden, it sponsored a study of California as a possible model for a state-run protection system with both wildland and urban jurisdictions. Meanwhile, the BLM concentrated on adapting advanced techniques of firefighting to the vast interior, with little ecological sensitivity other than adjusting suppression to local oddities—withholding bulldozers on permafrost, backfiring from rivers, and pre-positioning forces in such places as Bettles, Eagle, and Nome according to weather forecasts. Fire officers experimented with aerial backfiring. In 1977 two million acres burned.[60]

And so it went. Year by year, a brash and steady uptick of technology transfers brought more firepower. In 1973 its aircraft fleet was upgraded as Interior created the Office of Aircraft Services. Reconnaissance aircraft increased to a dozen. The smokejumper program swelled from 44 to 95, the nation's largest corps. In 1975 the automated lightning detection system installed its first stations; remote automated weather stations soon followed. The BLM experimented with explosive cord for fireline construction. It reorganized field units to facilitate rapid detection and attack. It continued to tinker with weather modification. In 1977 it fielded its first hotshot crew, and smokejumpers attacked a record 291 fires. That year 2.2 million acres burned.

Such was the pattern, as two dialectics, one political and one climatic, discoursed across Alaska. The political dialectic concerned land ownership and fire responsibilities, increasingly between the State of Alaska and the BLM, and then among the Interior agencies themselves, no longer bound by the claim markers of the early-prospecting BLM. The environmental dialectic flickered between wet years in which suppression seemingly knocked fires to a pittance and dry years in which fires rebounded with a vengeance. When Interior adopted fire by prescription as a departmental policy, BLM Alaska restated its title to the Division of Fire Management, upgraded evaluation procedures for fires that escaped initial attack, and crafted a flow chart to guide decisions in dispatching. But little on the ground really changed.

At a 1976 fire symposium in Anchorage, George Turcott, an associate director of BLM, addressed the "enlightened attitude that now constitutes the state of art of fire management." The agency would accommodate prescribed fire, but its basic goal was "to prevent the disaster fire!" It would not tolerate let-burns. It considered wildfire an emergency that demanded immediate action. It reconfirmed that "the highest priority is given to controlling disaster fires by aggressive prevention and suppression efforts." In practice, the best way to stop conflagrations was to extinguish fires while they were small, and that was what the bulked-up BLM was doing. What might seem to outsiders as a tolerance for free-ranging fires was, in fact, an admission—a temporary one—that the Division of Fire Management lacked the crews and matériel to hit all fires hard. What was true for Alaska's human economy, a cycle notorious for its booms and busts, was equally true for its natural economy. In 1980 a literature review conducted by the Institute of Northern Forestry on request from the BLM summarized the state of knowledge.[61]

The deep driver was partition. Paradoxically, by allotting eight years for Interior to decide on its land allocation, ANCSA returned the BLM to its historic role as a dispenser of public land just as the agency received an organic act to stop that practice and begin continuous management. Caught in a crossfire between aggressive development and preservation, Congress did what it did best: it balked. President Jimmy Carter broke the stalemate by invoking the Antiquities Act and proclaiming 56 million acres as national monuments and 40 million acres as wildlife refuges. The loser was multiple use, whether by the Forest Service or the newly enchartered BLM. Finally a compromise was brokered in 1980 during the last light of the Carter administration as Congress enacted the Alaska National Interest Lands Conservation Act (ANILCA).[62]

ANILCA designated 104.1 million acres as "conservation" lands, of which 57 million acres went into legal wilderness. The remainder the BLM would manage until a final disposition by 2000. From 290 million acres in 1960, the BLM would administer 61.4 million under multiple use and 2.2 million as National Conservation and Recreation Areas. The NPS would expand its holdings from 7.5 million to 52 million and the FWS from 18.7 million to 76.5 million. The Forest Service would gain 1.3 million acres. In the end, the federal estate would encompass 65 percent of Alaska land, the State of Alaska would have 24 percent, Native Corporations 10 percent, and private landowners 1 percent. The character of Alaskan fire management would evolve as this grand partitioning unfolded. To lubricate the transfers and anticipated coordination, ANILCA established an Alaska Land Use Planning Council.

The new federal landowners did not share BLM's enthusiasm for suppression. They viewed BLM Alaska as they viewed the USFS in the Lower 48. The State of Alaska, like all states, wanted protection but wanted it under its own jurisdiction. The other Interior agencies transferred perceptions and practices hammered out and honed in the Sierras and St. Mark's to their new domains. They wanted true fire management, not "fire suppression lite." In 1978, anticipating a final dispensation, an Interagency Land Managers Task Force convened to invent procedures to coordinate their affairs, which led to a pilot fire-management plan for the Fortymile region. In this way, allowing for adaptations to the peculiarities of history and geography, Alaska's was the national story in miniature. As the federal estate rechartered, it shifted from simple fire control to more complex fire management, though with a disconnect between announced principles and actual practice. Fire's last frontier passed.[63]

It had all happened quickly. Within 20 years the fire program had moved from uniformity to fragmentation to an interagency consortium. The BLM reluctantly released its grip over Alaskan fire—this was, after all, its proudest program. Like the British raj fearing that the loss of India would implode the empire and doom Britain to minor status, the BLM pursued ways to reclaim as a fire organization what it had lost as a landowner. In 1982 it proposed, in the guise of the Alaska Fire Service, to do for the new consortium of Alaska landowners what it had once done for itself.

Fin, Fur, and Fire

It was a turbocharged time for America's gaggle of wildlife refuges and the agency that administered them. The environmental movement had helped orchestrate its miscellany of refuges into a system and had boosted the agency's responsibility for treaties governing migratory birds. It began the decade as the Bureau of Sport Fisheries and Wildlife (known to its sister agencies in Interior as "Fin and Fur") and as overseer of the 1969 Endangered Species Act, which then upgraded in 1973 into the most potent piece of environmental legislation from the era. That same year the Convention on the International Traffic in Endangered Species, which the agency would also oversee, opened for signature. In 1974 it was renamed the Fish and Wildlife Service and ended the decade as one of the two big winners of ANILCA. But it was for fire a time of breakdown and even shame. The agency lacked for fire management the institutional contexts it had acquired for endangered species and wildlife refuges. What it did right it could not promulgate throughout the system. What went wrong affected everyone.

It should have been the FWS's time to shine. If the BLM's instincts were to stop fires, the FWS's were to start them. The refuges had by far the greatest experience in prescribed burning of any federal agency. The

FWS had officially tolerated controlled fire since the early 1930s. Refuge workers, if not their managers, tended to come from the adjacent populace; they understood local mores and, where burning persisted, local practices. But if it had the hands-on, one-foot-in-the-black knowledge that most of the national agencies lacked, it lacked the institutional bulk that bolstered (and sometimes burdened) the others. It had no organic act that defined the place of fire. It had no national office to supervise fire operations—to assist technology transfer, to disperse the latest science, to inform practitioners about policy reforms, or to tinker with adaptations to the NFDRS. It lacked the ability to train to NIFQS standards or to represent the agency in the NWCG. It had no corps of fire officers. The FWS turned to Tall Timbers for a forum and to the resurgent Department of the Interior for backup.

What it knew was intensely personal and specific to its refuges. What it did not know was everything else. As long as issues remained on local sites, the arrangement was good enough. But crises resulted when local breakdowns went national. The good, the bad, and the ugly converged in the summer of 1976 on the Seney National Wildlife Refuge on the upper peninsula of Michigan, where Ernest Hemingway had set his short story "The Big Two-Hearted River." During the heyday of the Great Barbecue, as V. L. Parrington called it, the land had been logged off and burned clean. The town of Seney had vanished in the flames. "Even the surface had been burned off the ground," Hemingway wrote. Then it had been protected, recovered as mire and thick spruce forest, and as the woods returned, so did fire. There had been flare-ups in 1954, 1964, and 1967. By periodically burning away peat, the fires seemed to have invigorated habitat. But it remained a traumatized land, for which fire might enhance or inhibit restoration, or perhaps both. The refuge had not experienced a major drought in 40 years or a significant fire for 70 years.[64]

On July 7, as that deferred drought settled in, the six-person refuge staff ignited the Pine Creek prescribed fire. Original plans had called for a 40-acre site, but the Michigan Department of Natural Resources, with which the refuge had a mutual-aid agreement, talked them down to a one-acre test burn. It smoldered in deep, and now dry, muck. On July 30 lightning kindled what became known as the Walsh Ditch fire in a recently designated wilderness area. A reconnaissance plane from the DNR reported the smoke the next day. When refuge manager Jack Frye tried to inspect it on the ground, he was unable to penetrate the dense conifers and hip-deep muck. On August 3 the Pine Creek prescribed fire revived and pushed outward. What happened next is known with surety only in its outcome, not in its inner mechanics, thanks to conflicting accounts. But for a while Seney National Wildlife Refuge became a cause célèbre over the new order of fire.

The refuge staff strained to contain the Pine Creek burn, eventually stalling it at 200 acres with the assistance of the DNR. Meanwhile the Walsh Ditch fire, monitored from the air, had swollen to 1,200 acres. When it reached 1,800 acres and threatened to bolt out of the refuge onto state lands, the refuge requested the mechanized DNR to help halt it while BIFC arranged a national mobilization that brought in crews and equipment from 29 states and dispatched an Interior overhead team to direct operations. The fire team elected to burn out the entire wilderness area. The Pine Creek fire was finally drowned after diverting a creek with bulldozers. By September 7 the Walsh Ditch fire stood at 64,000 acres, and demobilization commenced. With drought continuing and new organic soils exposed by the burns, however, the fire began reburning areas and blasted outward, highlighted by a seven-mile run overnight that carried it onto state forest land. BIFC sent everyone back, this time with helicopters and air tankers. With the flush money stimulating the local economy, arson fires sprang up. When it all ended, Walsh Ditch had a fireline perimeter of 88 miles around 72,500 burned acres. Suppression costs exceeded $9 million. It was not only the largest fire of the year but the most expensive for Interior up to that time. (The FWS had held the previous record with the 1969 fires at Kenai.)

Accusations swirled like embers in a stiff wind. The refuge staff had not been tested by a fire of any kind in five years, it did not have experience in prescribed burning, and it lacked fire-danger rating capabilities. The refuge's outdated fire plan did not include the mapped wilderness, and it had no national guidance. It simply could not cope, certainly not with two problem burns. The refuge was uncertain what it could do with a wilderness fire. It had no preapproved prescriptions and did not wish to unleash DNR's mechanized arsenal. The federal government that year shifted its fiscal year from July to October, leaving a vacuum during which it was unclear whether the refuge even had money to pay for suppression. It was unsure where to turn for help until a phone call to the national office suggested BIFC, which in the absence of fire activity elsewhere, funneled the national fire cache into Seney. Without lines of authority and clear rules of engagement, everything got exaggerated: the power of the burns, their damages, the muddled response, and the political blustering. Paradoxically, subsequent surveys of the burned area suggested the fire had, overall, enriched habitat. A board of review exonerated refuge manager Frye as having acted prudently "under the circumstances."

Clearly those circumstances had to change. The problem was systemic, or more precisely, the issue was the absence of any serious integration within the FWS or between it and the rest of the American fire community. The FWS began easing into the national infrastructure. In 1977, at the request of the Forest Service and BLM, the FWS cycled staffers, six in all for one month each, to BIFC to start learning the national system. A June 1977 directive for fire management in the National Wildlife Refuge System, "effective immediately," recognized that the agency would have to "implement certain provisions gradually," particularly until training could make the agency current in best practices. It nevertheless required that every unit of the refuge system would have a fire-management plan by December 1, 1978.[65]

* * *

Rapid transitions, however, are exactly the times when accidents are more likely to happen. Fire breakdowns had always occurred, would always occur, but they no longer stayed within the metes and bounds of local refuges and local lore. At least the agency could comfort itself that its southeastern refuges knew how to conduct prescribed burns. But on February 28, 1979, shifting winds carried a controlled burn on Okefenokee NWR outside its plots. Tractor-plow and tractor-harrow operators attacked it and then found themselves caught in flames. The tractor-plow operator died from his burns.[66]

The episode was rare in the Southeast, but it was instantly understood by fire officers as a classic illustration of the difficulty of transitioning on a fireline. The usual scenario involves the confusion of an organization moving from initial to extended attack. On the Pocket fire at Okefenokee it applied to a phase change from prescribed to wildfire. During a fast changeover, an organization may be as confused as the flames. To some critics the lesson portrayed an institution amid the gusts of fast-changing times, as it struggled to orient itself among unexpected rushes of flame.

The FWS problem fires attracted enough attention to require some bureaucratic attention. For most agencies the fire revolution meant a reformation in its existing organization; for the FWS it meant creating an organization in the first place. The agency advertised for a Fire Management Coordinator, which was filled by Art Belcher, a long-serving fire officer with the BLM. It moved him to the Washington office for a year to write a manual and to "indoctrinate" him into the FWS. Then it transferred him to BIFC. As Belcher recalled, "I was the full-time fire person. And they thought if they hired me all their worries were over and they would not really have to get into the fire business." The FWS fire program, he admitted, "was me." He shared a part-time secretary with the BLM.[67]

His FWS bosses and colleagues told him many times, he recalled, "We don't really want to be in fire." It lay outside their core tasks—it was not what they had signed on to do. The 1977 directorate went into agency files under the 13th (and last) category of policies, in the back forty of filing cabinets beyond the directorates on fencing, off-road vehicles, and grazing and haying. After Belcher sent out position descriptions and performance standards to the regions, he said they "threw them out. And what they did was go out and just pick somebody who was within the region, without fire experience, and move them into the position, because they needed somebody in the position." Clearly, they did not want anyone who was not fully initiated into the agency's culture.[68]

To outsiders, the FWS was beginning to seem a problem stepchild, though granted patience and guidance from its sister agencies, it could be brought into the national fold. It did not get that, because in 1981 it had a botched burn at Pungo Lake in North Carolina, for which the North Carolina Forest Service sued. Then smoke from a prescribed burn at Felsenthal NWR in Arkansas drifted across Highway 82 and caused a fatality crash. The coup de grâce arrived when the agency had a double fatality at a fire at Merritt Island NWR, Florida, which drew unwelcome congressional attention. One of the victims was the son of a well-placed federal judge who personally knew Sidney Yates, chair of the House Interior Committee. During hearings the FWS was raked over its own coals. It could ignore fire no longer, and its modern history of fire management grew out of those ashes.[69]

Fire from an Enemy Sky

Of all the federal land agencies, the Bureau of Indian Affairs was the most curiously situated. When the revolution started, thanks to Harold Weaver, it far exceeded anyone else in the West in prescribed fire, and it did it without special appropriations, simply as routine operations; it considered controlled fire the cheapest way to reduce fuel in forests and promote forage in range. Despite official unease, a reasonable expectation was that the agency would lead fire restoration throughout the western public domain. Besides, the agency and its nominal wards were undergoing a fundamental redefinition of purpose for which "restoration" is an apt description.[70]

But the restoration at issue concerned sovereignty, not natural processes. For the BIA the era of the fire revolution was less about what to do and how to do it than it was about *who* would do it. The agency was consumed by political turmoil of a sort that did not translate readily into changes in fire practices. The tribal lands remained under an awkwardly split jurisdiction between tribes pursuing self-determination and a BIA managing their lands "in trust." Their fire programs seemed to occupy a parallel universe. The agency was unwilling to add a fire revolution to its political one.

In the 1950s the long trend to assimilate Native Americans had culminated in a policy of outright termination—the cultural equivalent of fire suppression. The resulting pushback proceeded in near lockstep with fire reforms. In 1961 a pan-Indian conference convened at the University of Chicago and then led in 1962 to a "Declaration of Indian Purpose" presented to President Kennedy. At the same time a National Indian Youth Council organized as an advocacy group for a new generation. A series of reform legislation and court decisions began addressing education, health, tribal courts, and access to natural resources (notably fishing and water rights). These culminated as protest in the American Indian Movement and as a legal regime in the American Indian Civil Rights Act (1968) and the Indian Self-Determination and Education Act (1975). In 1971 the Alaskan Native Claims Settlement Act resolved issues by disbursing money and restoring lands. In 1972 protesters seized the BIA building in Washington, DC, and demanded a reordering of federal-Indian relations. An American Indian Policy Review Commission, the counterpart to the PLLRC, oversaw a panoramic survey of topics and issued a report in 1977. All this set in motion a series of restorations—of tribal identity, of religious sites, of historic places, of control over natural resources, and even the return of some ceded lands. What happened in fire, as a national program moved from a hegemonic agency with a single policy to a pluralism of lands and practices, had a thematic and chronological counterpart in American Indian history.[71]

* * *

The two reformations would seem ideally suited to reinforce one another. In the 1960s perhaps none of the federal agencies was better positioned to redirect the fire revolution than the Bureau of Indian Affairs. It had managed tribal lands since 1910 through a forestry branch, which meant in theory that it applied best practices. In reality, this had evolved into a doctrine of separate but equal. Of course monies had never matched those invested in the USFS or BLM, and few tribes had anything like comparable fire programs. Nor had they the bureaucratic rigor of forestry-driven agencies, which left many tribal lands less upset by fire exclusion. Nor did the reservations make a "system" in the way national parks or wildlife refuges did; for good or ill, it was tricky to transfer what happened on one site to others. But by presenting fire as something from American Indians' cultural heritage, it would seem easy to promote fire's revival as a visible emblem of reclaiming the land.

Circumstances certainly appeared favorable. In some places, sheer lack of capacity had meant that fires were not hammered heavily, and in others it meant an institutional edifice full of cracks and nooks had tolerated some deliberate burning. When the Tall Timbers task force traveled west in 1972 to Fort Apache, San Carlos, and Hualapai lands, it noted that "more controlled burning has been done on these Reservations and over a longer period of time than in any other forested area of the western United States." What had been dismissed as "Paiute forestry" was reincarnated as the highest avatar of fire management. The doyen of BIA burning, Harold Weaver, had even been asked for his imprimatur as Yosemite geared up to reintroduce fire.[72]

The political circumstances, however, did not match the environmental. Even as the Tall Timbers Task Force toured the Arizona mountains, it noted that prescribed fire was not keeping up with need, that wildfire was seizing the woods unburned by controlled means, that the quickening of logging—the major source of tribal income at Fort Apache—would exacerbate the fuels crisis. But this was the showcase site in the BIA. The agency had no coherent fire program. It applied general DOI policy, submitted to the department's Fire Management Coordinator, and joined BIFC and the NWCG, though not actively. It participated in a smattering of interagency projects, such as a joint helitack crew operated out of Mesa Verde National Park. It evolved no distinctive identity with regard to fire: it was too dispersed, too rent by political conflicts, too unsure about the dynamics between tribes and agency. While other federal institutions fretted over the proper balance between fire exclusion and fire restoration, the BIA juggled the fast-morphing imbalances between it and the tribes, some of which began creating their own forestry and fire programs. There was no comprehensive Native American model fire program. There was little means to leverage a good program to other reservations.[73]

Besides, the upshot of a generation that had been better educated into modernity was that many Native American foresters argued for suppression. They saw fire as they were trained to think about it in universities or other agencies, and they perceived it as a source of jobs, money, power, and prestige. They wanted to control those programs, not reform them. The fire revolution was an opportunity to swap hands over the apparatus, not to trade a pulaski for a driptorch. Arguments to pull back from a hugely expensive suppression policy seemed like another excuse to shortchange the natives.

Some reservations, such as the Flathead in Montana, embedded within an active fire culture, absorbed and redefined the revolution on their own terms and created alternative categories for wilderness and PNFs. Fort Apache and others shifted from broadcast to slash burning, withdrew from public visibility, and withered as potential exemplars. Too many others simply lacked capacity to mount a fire program on their own, and the BIA was too compromised and confused to orchestrate a cooperative arrangement among reservations with little in common other than their history of being compelled to sign treaties with a land-grasping United States. Most, that is, simply let the revolution pass them by like a high wind—perhaps the one invoked by D'Arcy McNickle in his 1978 novel, *Wind from an Enemy Sky*. In their inability to translate idea into practice, they did not differ so much from the majority of national forests or BLM districts.

What compelled most federal agencies to respond to the revolution was the disaster fire. For the Forest Service they came in cascades. For the FWS they came spasmodically, but with enough power to catalyze reform. The bad fires of the 1970s, however, blew by the BIA. The 1971 Carrizo burn at Fort Apache could have been an all-hands alarm for the agency—the Tall Timbers task force certainly thought so. "Had there been an adequate controlled burning program," it argued, "the wildfires of 1971 would have been easily controlled with little expense and damage. In fact, the fires might have been permitted to burn without any suppression effort since they would have caused no appreciable damage and would have gone out shortly, or would have remained small or light." Instead, the program began a slow fade. In that recession it could stand for the agency that ambiguously held its lands in trust. A BIA that might have been first among equals sadly slumped toward last.[74]

The Empire Strikes Back

Other agencies might wish fire away. Many saw the fire revolution as another challenge to business as usual rather than as an opportunity to reform what they did on the land. Their political instinct was to burrow deep and let the flames of an environmentalist prairie fire pass over. They saw fire as an emergency, not a relationship. If fire vanished they would cheer and go back to the quotidian world of hosting visitors, banding migratory birds, and fencing out cattle. But no such option was available to the Forest Service. Fire was too broadly distributed across its lands; the agency was too integral to the national infrastructure; and fire was too close to the core of its identity, a strong nuclear force that held its far-strewn particles together. Most of the revolutionaries' ardor was directed at the Forest Service. The revolution would fly or fold according to its response.

The USFS had to cope with wildland fire as one of a volley of assaults on its mission. A decade before, it had believed it was ahead of uncertainties by pushing for the Multiple Use-Sustained Yield Act. At the 1970 Intermountain Fire Council symposium, Joseph Pachenec, director of the Intermountain Experiment Station, asserted that "multiple use provides a mechanism for ecosystem planning and management—the only such mechanism we now have." Whether one viewed the public domain as commodities, ecosystems, or sacred groves, the traditional thrust of the Forest Service had the means to cope. No other strategy could.[75]

Now that synthesis had come unglued. In December 1966 Merle Lowden, director of Fire Control, had established a Fire Policy and Procedure Review Committee to review pending issues. The two pressing concerns were integrating aerial operations and handling fire in wilderness. Because so many Forest Service aircraft were committed to fire, the eventual solution to the first was to merge fire and aviation into a single division. And because the committee thought that "wilderness should be maintained in as natural a state as possible," it worried about the potential removal of essential fire-control technologies such as the helicopter, the need to modify the 10 a.m. control standards, and the prospect of introducing prescribed burning to "maintain the natural environment of wilderness." If a modified suppression strategy worked in wilderness, then it might be extended to low-value, remote, and hazardous lands as well.[76]

As partisans within the Forest Service rallied to the revolution, they recognized that internal inertia, the force of culture, was a primary drag. Charles Philpot concluded that the "public" was less an issue than "our own set of built-in prejudices." An overemphasis on concerns regarding fire's role and fire use, he argued, were "preventing a vigorous attack" on the fundamentals. They created impediments that did not otherwise exist. And Thurman Trosper, now an assistant to the director of the NPS for Environmental Affairs, recalled an episode from his years as a fire-control officer with the USFS in Idaho. A snowfall on the eve of hunting season led to a mix of unburnable patches and plenty of warming fires, and he had elected to let those hunter fires forage through heavy "jackpots" of windfall and overgrowth. For three weeks the fires burned through nature's slash piles, gnawing big holes and de facto fuelbreaks in the clotted woods and "doing nothing but good." Then the regional forester flew over, saw 30 to 40 fires smoking unattended, and demanded they be extinguished. "So we rounded up the crews and as incredible as it may seem we did in fact put the fires out."[77]

Countless fire officers could recall variants of that story. Most did not need a regional forester to remind them: they had internalized the edicts. So it would be no casual exercise to let a new fire policy propagate. It would not simply happen any more than fires would be left to roam freely everywhere. Which makes it all the more surprising how rapidly the Forest Service pushed the agenda of the revolution—at least formally within its upper hierarchy—amid a chronic crisis of agency identity and at a time when it was pounded in courts and Congress and subject to three administrative plans to be subsumed into a mega Department of Natural Resources. Overnight, resistance in principle collapsed. What it meant on the ground would test the trickle-down theory of policy promulgation.

* * *

Throughout the 1970s, the USFS translated the new ideas into policies. The Interior agencies and the newcomers, such as the U.S. Fire Administration and Department of Defense, moved in lurches and epiphanies. That was not Forest Service style. It moved steadily, pausing to keep its complex parts in sync, constantly adapting to Congress and critics. Year by year it advanced a new fire agenda, stride by bureaucratic stride, probing, testing, declaring, and shuffling toward the only end point possible, a thorough reconstruction of the fire program. During the tenure of Chief Forester John McGuire from 1972 to 1979, it completed that chrysalis.

The two national movements, the environmental movement and the fire revolution, ran side by side. In 1971 the terms of engagement were effectively announced. The Public Land Law Review Commission issued its report, Congress passed the Alaska Native Claims Settlement Act, and the Forest Service released the findings of its Roadless Area Review and Evaluation (RARE I). It accepted the Church guidelines for clear-cutting and a "Framework for the Future" that called for a better balance in its programs (translated as "add more wilderness"), and it created Woodsy Owl. It then hosted its first symposium on prescribed fire, announced the initial planning for a natural-fire program in the Selway-Bitterroot Wilderness, and funded the first phase of Firescope. Following a fire policy meeting held in Denver, the Division of Fire Control was renamed the Division of Fire Management and granted an expanded staff role.

The Denver gathering, prompted by the 1970 fires, was the first such conclave since 1954 and the last mustering of the old guard. It confirmed that "the general fire control objectives as presently stated by the Forest Service are sound and adequate." Yet it noted changes in public mood and expectations, and in order "to meet the requirements of the public," the committee reaffirmed the 10 a.m. policy and required "strict compliance in all Regions." Still, there were geographic nuances and wilderness considerations that urged a "modified" suppression program such that the "time will come when a more flexible approach should be considered." The enabler of such a transition was research. Craig Chandler, director of fire research for the USFS, noted that an experiment in limited suppression "involves a calculated risk, and a big one. We are gambling that our knowledge of fire behavior, our ability to predict weather changes, and our ability to rapidly suppress fires when necessary are all good enough that we can anticipate and prevent conflagration fires from breeding in and escaping from the test areas."[78]

Then the ancien régime passed. It helped that the new chief forester, John McGuire, appointed in April 1972, believed that public wishes trumped professional desires. "These national forests are public property," he asserted. "The public can decide to manage them any way at all." The task of the agency was to determine what the public wanted and "then do it." The sticking point was that "the public" was actually many publics, that there were many venues by which their intentions were made known, and that a multiuse matrix could not accommodate them all, particularly when some uses explicitly excluded others. McGuire accepted the tenets of the fire revolution; before his reign ended in 1979, its principles were encoded into internal reorganizations and the Forest Service *Manual*.[79]

The gears began turning, lubricated by the agency's investment in fire science. A National Fire Planning exercise got under way to upgrade, integrate, and standardize programs throughout the national forest system. It had begun with the intention of better orchestrating the various parts of fire's administration, many of which had become specialties in their own right, particularly presuppression. Complementing that project was the release of a National Fuel Inventory System, which sought to bring some order to fuels issues, now partly rationalized by the NFDRS and its fuel models. It was easy to identify logging slash, and common to reduce it by burning—a practice now brought under the aegis of prescribed fire. It was trickier with natural fuels, which was another term for nature, and trickier still where they occurred in wilderness, which led to the most dramatic break with the past, the White Cap Project.

Behind it was Bud Moore, back to Region 1 from his Washington Office tour and work on the NFCS, reawakened to the wondrous landscape of the Bitterroots he had encountered as a youth, and perhaps penitential over the damages he had helped instigate while in the Powell District. He was resolved to preserve its wild state by restoring fire. He found allies in Bill Worff and Orville Daniels, and he recruited Bob Mutch and Dave Aldrich to write plans for the White Cap area of the Selway-Bitterroot Wilderness. He managed to hold off the hard-core pulaski-and-shovel men.

On August 17, 1972 Chief Forester John McGuire signed off on the scheme as an exception to the 10 a.m. policy. The next day lightning started a snag fire on the slopes above Bad Luck Creek. The Bad Luck fire was the first PNF by the Forest Service. It burned a 648-square-foot patch. The challenge came the next year when the Fitz Creek fire blew up, escaped its designated bounds, and had to be attacked when it moved out of prescription; presciently, the order was to suppress only the fugitive portion and not the entire fire, which grew to 1,200 acres. A bleary-eyed crew boss called in to corral the escape exclaimed that he'd like to "meet the SOB who told us to let this one go." But most of the environmental community cheered when *Audubon* published an account for its national audience.[80]

Meanwhile, the Gila National Forest in New Mexico began to accept select lightning fires. The setting had several advantages. It could use the Gila Wilderness, the nation's oldest. It was relatively self-contained, a cluster of the Southwest's fabled sky islands. Much of the biota was ponderosa pine, accustomed to millennia of routine surface burning. It was remote from communities that would object to long-duration smoke. In 1975 under the instructions of the Gila's fire officer Don Webb, Lawrence Garcia, then 22 years old, wrote a fire plan to restore fire. The project began cautiously, much like that at Saguaro; it built out slowly, learning incrementally, starting amid monsoon rains before probing out toward early-season burns. In 1980 the forest commissioned a fire-history study from Thomas Swetnam, later head of the Laboratory of Tree-Ring Research at the University of Arizona. When, after 1978, policy permitted "confinement" as a suppression option, the forest expanded its experiments, and then decided it needed to supplement natural ignition with prescribed burning. Suppression remained the rule for most starts, but the number of burned acres began to rise as suppression loosened its grasp. Almost uniquely, the forest learned to fight fires and light them at the same time. Quietly, the Gila became the premier site for restored fire in the Southwest.[81]

These were dramatic events: their symbolism rang brazenly throughout the American fire community. Yet their numbers were few, their acres piddling, and their ecological impact negligible. They were proof-of-concept experiments, significant as the tip of a wedge, and their impact would depend on continued blows to drive that wedge and pry apart the old ways in an agency suddenly convulsed with change imposed from the outside and whose instincts were to circle its engines and do what it did better than anyone else—fight fire. As a 1977 review of the fire revolution's early impacts noted, "The complexity, the associated risks, and the present rigidity of fire policies have caused many line officers to exercise extreme caution.... It had been the bold Forest Supervisor who would intentionally allow a fire to continue to burn into the next burning period."[82]

Even suppression, however, had become dauntingly complex. A Fire Chiefs National Meeting held at Alexandria, Virginia, in late fall detailed in hundreds of pages both the fine- and coarse-grained complications of running a fire program apart from introducing natural and prescribed burning. Those concerns went international with an FAO-sponsored, Forest Service–hosted conference on Fire in the Environment that assembled fire specialists from the United States, Canada, and Mexico under the aegis of the North American Forestry Commission. The USFS representatives summarizing its lessons at a post-symposium retreat—Robert Martin, Bud Moore, Merle Lowden, and Art Brackebusch—all located the core issues for fire within land management. The clarity of fire's management depended on the clarity of land management.[83]

With enough exceptions, the 10 a.m. policy might itself become the exception. Many of the early reforms, however, tended to emphasize presuppression. The idea was that a better job of fuels management and a selective need-based buildup of initial attack forces would improve suppression and reduce costs. In effect, some of the burden of fighting fires would be shifted to prefighting activities, and shrewd transfers would lessen the program expenses. But such a scheme required planning in its own right. That was the task of the National Fire Planning project; its organizing concept was a 10-acre policy.

The 10-acre policy was "to presuppression planning what the 10 a.m. policy is to fire suppression planning." It established as a planning goal the holding of all fires ("100 percent") to less than 10 acres, thus complementing the 10 a.m. policy's ambition to control all fires by 10 a.m. the morning after their report. The 10 a.m. policy had sought to hold to a minimum not only damages but also the costs of suppression, which were, for practical purposes, not limited by budgets. The 10-acre policy sought to shift that emergency money up front—the ounce of prevention expense that would yield pounds of suppression savings. Focus, the NFDRS and the Rothermel model, published at this time, promised to quantify the calculations, and the National Fuels Inventory System would rationalize decisions about the magnitude of the hazard requiring treatment. The 10-acre policy would save the 10 a.m. policy from its excesses.[84]

As the Gale Report subsequently observed, "Some Regions and Forests saw this as an impossible goal and did not attempt to reach it; others took a more literal interpretation." All, however, escalated presuppression activities, which had their own emergency funding. The project received further encouragement from the 1974 Forest and Rangeland Renewable Resources Planning Act, whose supporting document announced as a planning level for presuppression that the agency should reduce the number of fires exceeding 10 acres by 2 percent. The upshot was a 134 percent increase in presuppression costs, without any change in suppression costs, which bloated by 78 percent.[85]

Add to that frothy brew the passage of the Rural Development Act, which expanded Forest Service involvement for rural fire protection, and it was, all in all, quite a year.[86]

* * *

Throughout 1973 the smoke reports of new administrative brush fires poured in. Under the purview of the International Biological Program, the National Science Foundation sponsored a Fire Ecology Project that granted university-based ecologists a chance to challenge Forest Service fire science. Gene Bernardi, a sociologist who studied fire-prevention messages at the Pacific Southwest Research Station at Berkeley, had won a sex-discrimination suit the year before and filed a class-action suit on behalf of all women in the Department of Agriculture, thus challenging fire's inherited workforce.

The fire organization itself changed titles and tasks. The venerable *Fire Control Notes*, originally established in 1936 to support the 10 a.m. policy, amended its title to *Fire Management Notes*. The Division of Fire Management adopted the doctrine of total mobility, and accordingly realigned its role at BIFC. Though by far the largest force on the field, it committed to interagency cooperation and joint operations where possible. Through the International Union of Forest Research Organizations, it expanded its cross-border interests beyond North America by including fire among exchanges between the United States and the Union of Soviet Socialist

Republics. The United States, led by McGuire, went in August to September 1973. The next year the Soviets sent a delegation to the United States. (The Americans concluded they had little to learn from Soviet firefighting, but that the Soviets were "at par" or better in the mathematical modeling of fire behavior.) And then it faced a very different challenge from alternative models of fire administration.[87]

The most unexpected came from National Commission on Fire Prevention and Control, convened in 1971, which issued its final report, *America Burning*, in May 1973. Its vision was urban, although it spot-welded with wildlands at the rural scene. The 1972 publication of a best-selling memoir about fire in the South Bronx, Dennis Smith's *Report from Engine Co. 82*, gave afterburners to its publicity. While the commission recognized fire protection as a local responsibility, it saw a federal role for reporting, research, and assistance with costs when protection exceeded the resources of a local community. These sentiments found expression in a Fire Prevention and Control Act, which led to a U.S. Fire Prevention and Control Agency (later, U.S. Fire Administration) and a national training academy. The loose wording of the act, however, granted the agency nominal responsibility over all fires, including wildland. Incredibly, the Forest Service had to expend energy brushing that potential competitor aside. The act also established 20 years for federal firefighter retirement: it, too, applied across the board, from military bases to backcountry ranger stations.[88]

Still, all this was only a premonition of what was to come. Urban values and retirement patterns would help fracture the workforce. And within another 15 years urban America would so crowd against wildlands that it would mangle the pyrogeography of the country.

* * *

In 1974 attention turned to that other polarity, wilderness. When the Intermountain Fire Council and Tall Timbers held a joint conference in Missoula, the Forest Service used the occasion to announce its conversion. Its director of Fire and Aviation Management, Hank DeBruin, delivered the first presentation; Bud Moore, recently retired as Region 1 fire director, gave the conference summary; and appearing in the middle as featured speaker was Chief Forester John McGuire.

DeBruin formally announced the Forest Service's reformation, explaining how the Forest Service was moving from "fire control to fire management," and from a unilateral program of suppression to a commitment to total mobility, interagency cooperation, and modern fire management. Moore spoke on "Fire, Land, and People," noting that "wherever fire is a factor . . . and that's nearly everywhere," fire by prescription found those prescriptions in the purposes of the land. That was the theme that everyone at the symposium shared. "Without these, little foundation exists for fire management activities." Others at the conference described the spectrum of ongoing Forest Service fire operations—suppression, research, Descon, the White Cap.[89]

McGuire addressed "Fire as a Force in Land Use Planning." Fire, he affirmed, was one component of "the total forest ecology," and so fire's management could not be "separated from total forest management." While fire was now a "sometime-enemy" and a "sometime-friend," he observed that the important question was not just "what, among many alternatives, we will do with our land," but "who will decide." He put his hopes into planning of a sort Americans were not accustomed to but that the Forest and Rangeland Renewable Resources Planning Act, passed earlier that year by Congress, sought to promote. Afterward he presented a National Fire Management Award to Dr. Bruce Kilgore for his pioneering work at Sequoia-Kings Canyon.[90]

The chief forester rightly saw that the purposing of land was the essence of the fundamental controversies slamming the agency—in particular, that portion of land that was to partisans not fungible, that could not be moved in and out of multiuse categories. For most wilderness advocates RARE I had not gone far enough. In 1975 the Eastern Wilderness Act was enacted, and a Forest Management Review System was activated to help implement it. Two years later the USFS commenced RARE II. The issue could not be finessed: by its nature, compromise regarding wilderness seemed to mean loss. In 1978 the Endangered American Wilderness Act allocated 1.3 million acres in 10 western states. The Alaska National Interest Conservation Lands Act added 57.5 million more.

The other implacable crisis was clearcutting, and because it had forced the agency to defend itself in court, its potential capacity to log at all was questionable. In 1975 the Fourth Circuit Court of Appeals upheld the Monongahela decision, setting in motion the need for a legislative remedy beyond the MUSY Act. In 1976, at Forest Service request and all but mandated by court rulings, Congress passed the National Forest Management Act, which replaced the 1897 Organic Act. That same year the BLM received its first congressional charter. Despite political scandals such as Watergate, the traumas of Nixon's resignation (and that of Vice-President Spiro Agnew), and stagflation, the environmental movement had pushed on. If fire management followed from land management, then the changes in one had to yield changes in the other. There was plenty of planning ahead.

Unfortunately for the contestants, fire was not at the negotiating table or in the courtroom. It obeyed its own statutes. The next year a wave train of big fires struck the Forest Service throughout the West. They rumbled with special ferocity through California and the Southwest, blasting everything from the Klamath to the San Bernardino, Vandenberg Air Force Base and Bandelier National Monument. The agencies spent more money than ever before, and more in presuppression than had seemed possible. Yet the fires still burned.

* * *

In his Missoula address Chief Forester McGuire noted that 1974 marked the 50th anniversary of the Clarke-McNary Act. That statute, building on the 1911 Weeks Act, had completed its historic task: all the American states had a forestry bureau through which federal cooperation could be funneled. Increasingly too, state and federal agencies had to fight fires cooperatively, which meant extending total mobility to those state bureaus. The fire revolution had to deal with suppressing malign fires as well restoring benign ones. Two years before McGuire spoke, the Forest Service shuffled its cooperative fire program. A year later it was reorganized into a Cooperative Fire Protection Staff.

By then the suppression part of the fire revolution needed attention. It was clear that, as a national doctrine, the 10 a.m. policy had exhausted itself. It worked well as a guide for particular fires; it failed as a universal standard for dealing with all fires. Like Clarke-McNary it had played out its historic role. Exceptions chipped away at it. Attempts like the 10-acre policy to revive it by shifting attention to presuppression such as fuels projects were also in disgrace. Spending more emergency monies early did not reduce emergency spending later. They both rose.

The sums involved were not trivial. In 1965 the agency spent $6 million in Fire Fighting Funds (FFF) on presuppression. In 1970 this rose to $11 million, then to $25 million in 1973 and $85 million in 1976. Inflation accounted for only a part of this increase. Similar escalations occurred with the other federal fire agencies as they sought to build up their capacity in the face of otherwise stagnant budgets. Over the same decade suppression costs rose 78 percent. To satisfy the planning level for presuppression laid out in the Recommended Renewable Resource Program in accordance with the FRRRPA would require a 90 percent increase in funds for a 2 percent reduction in fires greater than 10 acres. Such numbers caught the attention of the Office of Management and Budget, which demanded an analysis and an explanation. McGuire added his own request. In July 1976 a study team was assembled from fire, planning, and policy analysis. Its findings led to a summary by Robert Gale.[91]

The Gale Report is one of the fire community's minor masterpieces. Couched in the language of economics, the report explains how the situation had evolved and what in recent years had jolted such dramatic cost escalation. The immediate cause was the "unregulated implementation of fire plans at the local level." Behind that, however, was the 10-acre policy, and behind that strategy the 10 a.m. policy, and behind that project the availability of special funds to implement it. Behind everything was the sense of fire as a hostile threat. As the opening paragraph on fire in the Forest Service *Manual* stated, "Forest fires cause serious damage . . . Effective fire control is essential." And as the Gale Report noted, this has been interpreted as "a policy of fire control and exclusion to the fullest possible extent." All fires are bad, all resources need fire protection. "It fails to recognize any benefits from fire, or the existence of those resources which do not need total (or any) fire protection." While later *Manual* elaborations acknowledged the need for economic considerations, the preamble (the "Zero Code") set the tone. Fire planning took its orders from those stern proclamations.[92]

The Gale Report said little about reintroducing fire to enhance ecological goods and services, or the restoration of natural processes to wilderness, or reducing the heavy hand of mechanized suppression on the land, or other concerns of the fire revolution. The values it promoted were fiscal. Fire suppression as a driving policy cost too much and achieved too little. Until it was possible to relate fire control to impacts on resources, the "Renewable Resource Program document is of little value as a fire planning tool." (For 90 years foresters had attempted to establish just such quantitative relationships and failed; the only explicit market values at risk were timber berths that had been sold; the rest was speculation.) The upshot was that there was little likelihood that the present system would self-correct. The entire Forest Service apparatus had to undergo a radical overhaul. The two halves of the fire-revolution equation—to restore fire and to suppress it more sensibly—converged. Not for the first time, nor the last, a formal review urged the creation of a "National Fire Plan."[93]

That was too much to expect. Instead McGuire reported to Rupert Cutler, assistant secretary of agriculture, that the agency planned a series of internal adjustments. It would rewrite its policy manual to reflect the shift from fire control to "management by resource objectives," and it would introduce a Fire Management Fund that would abolish the emergency presuppression monies and replace them with an accountable budget. Not least it would revise fire planning to accommodate this shift in ends and means. In January 1977 it convened a National Fire Research Planning Conference at the Macon lab to report on the current state of knowledge and recommend future initiatives. Chief Forester McGuire met with the regional foresters to review the fire program. In December a task force headed by Mic Amicarella redrafted a replacement policy to become effective in February 1978.[94]

* * *

More than a policy change, the 1978 reforms reconstituted the entire Forest Service fire program—the most far-reaching overhaul since the 1930s, and perhaps the most comprehensive in the agency's history. No other federal agency had anything like it. The National Park Service had encoded the philosophy of the fire revolution into its 1968 Administrative Guidelines for natural areas, but had made fire's management optional rather than mandatory and had left implementation with individual parks. In 1977 it issued service-wide standards with NPS-18, though its promulgation was neither a policy nor a reorganization. The Forest Service's reforms were both.

The Forest Service reformation crossed all branches and was spelled out in its *Manual*'s Zero Code. For the National Forest System, the new policy affected how USFS fire officers should handle fire. In place of a universal standard, control by 10 a.m., it stipulated that they should take "appropriate suppression action." Suppression became an elastic concept that could range from confining a fire to a specified place and set of conditions to outright and immediate extinction. To assist in original decisions about what was "appropriate" a round of fire planning was readied for 1978, and to aid fireline decisions for fires that escaped initial attack the agency created a "situation-analysis process." For State and Private Forestry, Congress passed a Cooperative Forestry Assistance Act, which repealed and replaced all earlier separate pieces of legislation across a broad range of activities, fire among them. So powerful was the drive toward cooperative, if not coordinated, programs that the Division of Fire Management itself relocated to S&PF. For fire research, a revised charter arrived, the first since 1928. And Congress made the agency the conduit for America's investment in international forestry, including fire-disaster aid.

Not least, the reforms dissolved the old emergency funding mechanisms in favor of budgeted accounts that, in theory, could impose some fiscal discipline on spending, which is to say, would encourage the agency to reallocate its attention and might grant incentives to do what the fire revolution urged. As one skeptic observed, "Congress tried to appropriate fire suppression funds like any other line item." The assumption was that the new order of fire would be cheaper. That, however, was the argument the fire community had made, and remade under changing circumstances, for most of the century without much evidence to support it. To underscore its intentions, Congress in 1979 required a cost-benefit analysis of future fire budget requests. The embryonic software to simulate fire-management scenarios went on growth enhancers. The Office of Management and Budget decided to play with fire.[95]

The Forest Service remained the mainspring in the American fire system. Directly or indirectly, it turned the gears, and had in many cases designed the cogs. Now it was set to power the fire revolution. By 1979 it had more designated natural-fire areas than the NPS, and in the Northern Rockies, with some three million acres in the program, it allowed 29 PNFs, two of them outside legal wilderness. It remained only to continue applying principles to practice and tweaking the system to scale up. The new edict specified full agency-wide compliance by 1983. But as any smokechaser felling a lightning-split snag or a burn boss watching winds turn squirrelly will tell you, that "only" was not something that a mind could just will and a hand just do.

* * *

Pondering over his career as an early BLM director from the perspective of his post as a researcher at Resources for the Future, Marion Clawson frequently observed that "one reason both the Forest Service and BLM were effective in carrying out agreed upon policies and decisions was that each had been forced to fight forest and range fires. One does not fight a fire by a committee or by public participation." Yet now committees and publics ruled, and fire's management meant more than answering an alarm. Both the USFS and its fire mission lost their siren-song clarity. Its fire program resembled post-Apollo NASA as that agency floundered to shift from Moon shots to a complex and overbudget space shuttle with no commanding objective other than to continue to fly. Its research program began a slow unraveling, not unlike that of AT&T's industrial labs after its mandated breakup. Just as the fire program, after a long decade in the garage, hit the road, the wheels came off.[96]

The revolution stalled. The revolutionaries may have seized the commanding heights, but they seemed incapable of taking the fields and villages below. For the next dozen or so years the agency seemed unable to translate ideas into action, and because it remained the prime mover for American fire, the rest of the system acted sluggishly as well. The reasons are many, beginning with the 1978 congressional elections that began the shift toward a Republican ascendancy (both Dick Cheney and Newt Gingrich entered the House). Stagflation still ravaged the economy. In July 1979 President Jimmy Carter addressed the country's "malaise." The Iranian revolution soon restored the energy crisis and shifted attention from domestic to international affairs. The 1978 reforms were, for the fire revolution, its highest scorch marks.

Any policy is only as good as the workforce available to apply it. After 1978 the workforce for fire management in the Forest Service began to decline in numbers and change in composition. When the fire revolution began, the morale of the USFS was among the highest in the federal government— Kaufman had thought it exemplary. Fifty years later agency morale scraped bottom. Its workforce had passed through a looking glass as early retirement obliterated cumulative centuries of field experience, an influx of "-ologists" arrived with little interest in serving on a fire militia, affirmative action forcibly refilled the ranks, and programs and practices, particularly with regard to logging, so offended some employees that they formed dissenting sects. An agency once famous for the collegiality of its workforce underwent repeated purges and schisms. Chief McGuire had been right: fire management was not only a question of what to do but who would do it.

In 1979 the case of *Bernardi v. Butz*, the class-action suit filed in 1973, was resolved when the Forest Service, on advice from the USDA Office of General Counsel, agreed to a consent decree that stipulated that its California workforce should have representation comparable to the general civilian workforce. In practice this meant that, over a five-year period, women should occupy 43 percent of jobs in all series and pay grades and that the number of women in supervisory roles (GS-11 through GS-13) should escalate dramatically. In 1981 the district court approved. The civil rights revolution met the fire revolution.[97]

It had become a truism in the fire community that what happened in California would disseminate nationally. Now the cadre of California fire officers that had colonized so many other regions would be replaced by an influx of newcomers, novitiates not only to the agency but also to fire. How they would amplify or disrupt the fire revolution remained for the future to determine. It was not just that women had replaced men, or that ecologists and wildlife biologists would stand with foresters, or that college kids punching numbers into TI-59s had stepped into the boots of old firedogs, but that the court-imposed method of the change, its magnitude, and the suddenness of the demographic and disciplinary upheaval shocked the system. It would take a bureaucratic generation or two to absorb the impact.

What gradually emerged was a softer, gentler Forest Service, more sensitive to the aura of the times. It was a penitential Forest Service, rising from its knees after the beatings it had taken over wilderness and clear-cutting, and a more pluralistic Forest Service, inching toward a demographic profile that better resembled the country's at large. The overhaul of its workforce continued, though there were plenty of dissenters and footdraggers and revanchists for whom restoration meant not reinstating fire so much as the old order. By the end of the century, looking back, one veteran researcher summed up the consensus of the old guard when he mused that the Forest Service had become a better organization to work for but not, he thought, a better organization.

The Rhythm of Revolution

Revolutions have their rhythms. As with many, the fire revolution started with an idea, a wind sheer in values, a "desertion of the intellectuals." It sprang from hope and rising expectations rather than despair and grinding failure. For the federal lands, it commenced at the top with what might be called coups of policy and administrative reordering. An initial euphoria prevailed—the reforms all seemed so obvious and redemptive. Then this revolt, like so many before it, faced its counterrevolution and Thermidorian reaction before reaching an awkward equilibrium that left it as far removed from the end it had sought as it was from its point of origin.

Over a 50-year span the American fire revolution had followed the narrative arc of a three-act drama. From 1962 to 1978 it flourished amid a rising action of enthusiasms. Then came Act II, pushback, and a counterrevolution in politics, policy, land use, and even climate. The country began polarizing—red states and blue, wilderness and exurban enclaves, deluge and drought—while sliding into various states of indebtedness. The stresses tore at the Forest Service and left it enfeebled, unable to lend its mass to the cause. By the time the revolution rekindled in the mid-1990s, the circumstances that had stoked the revolution now turned against it. Act III watched the American fire community try to build up faster than the accelerating rate of deterioration. The era climaxed with a series of blowups.

There are many other ways to parse this chronicle. Many milestones, for example, came at 10-year intervals. In 1968 the NPS announced the new fire policy. In 1977–78 it issued NPS-18, while the USFS reorganized its spectrum of fire programs and encoded the changes in its *Manual*. In 1988 wildfire overran Yellowstone National Park, advertising the new thinking to the public along with its potential costs. In 1998 a revived revolution underwent a quiet reorganization. The 2008 election brought a political shift that attempted to reorder the institutional landscape even as the natural landscape was in outright insurrection.

Yet as the 1970s wound down, however dismally for the country, the fire community could look back on a decade of accomplishment. The legislative and intellectual firefights had hammered out a template for modern fire management, much as the 1910 Big Blowup had forged a persistent template for big-fire suppression. The reformation was about cultural choices and the politics necessary to chisel those values into the stone of legislation and agency organization charts. That goal it achieved. Policy, principles, tools, and knowledge were all now adequate to move fire from conference room to the field.

Over the coming 30 years the policy would be emended, restated, elaborated, tweaked, and harmonized, but its fundamentals have endured. So have the terms of discourse about it. The arguments for restoring fire, for attaching fire to land management, for shunning simple suppression, for understanding fire as both friend and foe and occasionally both at the same time—all read today as they did in 1970. Controlled burning had once thrived with wooden matches and kerosene-soaked ropes dragged behind horses; prescribed fire did not need propane flamethrowers and aerially dispersed delayed-action ignition devices to put fire on the ground, though it had them. Nor did those monitors observing prescribed natural fires require Apollo-level sophistication in computing to forecast which places could easily accept fires and which could not. After tens of thousands of published pages of research, high-tech equipment that old smokechasers could never have imagined, and organizational studies that could fill the libraries of MBA-granting universities, the necessary knowledge was both equal to and unequal to the task of restoring fire. It was inadequate in that it could not forecast what would happen over long times and down chains of causes; it was more than adequate to put fire on the ground.

By 1978, everything the fire revolution required was present—motive, means, opportunity. Yet it did not happen.

* * *

All revolutions falter and most fail. That the fire revolution fell short of plunging the American landscape into a full-immersion baptism of benign burning should surprise no one. What is interesting are the reasons for its stumbles. They are many, as long as the checklists that seemed to expand preparatory tasks before a prescribed burn could be ignited. Mostly, though, they fall into a few categories that could be classified under workforce, the limitations of prescribed fire as a doctrine, the liabilities of using fuel as a metric, and outright politics.

The workforce issue was one of size, character, ardor, and know-how. Year by year, there were fewer fire personnel. Affirmative action and early retirement forced a sudden, massive turnover. The old ways of knowing fire were plucked out before new methods could root. Many, though far from all, of the old guard resisted, and many, probably most, of its replacements did not see fire as a vocation or a rite of passage, and they watched instead as errors with fire gutted promising careers or deflected money away from more compelling tasks. A bad fire could end a career; a good fire would not advance it. Men who had risen to rank and power and for whom firefighting was the signature of their identity were as reluctant to surrender that life as big-tree loggers, cattle ranchers, or dam-building engineers. As with staff, so with agencies: none actively sought out fire as a defining or informing mission. They responded to the fire revolution as they would to a wildfire: it was a problem to deal with. If it proved surprisingly easy to encode fire management into policies, it proved equally difficult to instill it into staff. Especially if the Forest Service became dysfunctional, that malady would propagate throughout the American fire community.

Oddly, prescribed fire proved flawed as an informing doctrine. Initially, because it allowed both fire's use and its control, it seemed the perfect compromise. The Southeast showed how simple its application could be: you assumed you could burn unless something stopped you. In the West, however, you could burn only after all the somethings on an ever-growing checklist had been satisfied. In the West the culture of fire was wrong. Fire officers and infrastructure had come of age with suppression, and they required reeducation and retooling to convert to new purposes. Restoration through prescribed fire demanded patience and time, neither of which were abundant resources, and a tolerance for risk and error, which was even scarcer. Sites that seemed to cry out for fuel-reducing burns (think chaparral) had partisans to argue against it. Not all natural fires were benign smoking snags deep in the mountains; some were landscape-upheaving conflagrations. The aggregated effects of smoke, community opinions, heavy fuel loading, agency skills, and geographic margins for error meant that prescribed fire was not an undertaking that could immediately substitute for suppression.

As one frustrating year after another passed, it became apparent that the Southeast, the hearth for prescribed fire, was less a paragon than a fire nation within the nation. Unlike the western lands, burning was broadly based both on land and in society. Here, prescribed fire could build on generations of burning, adapting old traditions to new forms; the West had extinguished its heritage of anthropogenic fire and would have to build one artificially, like anthropologists teaching indigenes how to make their traditional crafts. In the Southeast, sites for burning were typically surrounded by other similar sites, which allowed for some slopover and resilience. In rural settings there was an expectation for fire and a tolerance for burning that had vanished from the West.

Prescribed fire was not simply a universal tool in an agency shed—something that science could hone and institutions train novitiates to perform—but a deeply rooted cultural practice that could be modernized but not readily invented. Places that did successfully adopt it did not just allow restored fire: they made it mandatory. Policy specified targets and incentives. (In 1979, for example, Parks Canada rewrote its policy in such a way that parks were under law to burn at least 50 percent of their historic average area burned.) The sad fact was that prescribed fire did not travel easily outside the Southeast or the Great Plains prairies. It became a boutique practice, not an informing one. A general policy of fire by prescription would struggle. Even Parks Canada required another 25 years to move from fire cache to landscape burning.[98]

The list of obstacles, large and small, could go on. The reality was that few fires and scant acres burned. The dazzling gestures remained more symbolic than on-the-ground reality. During the 1970s, the National Park Service set 126 fires that burned 1,656 acres. Over that same decade it experienced 4,159 wildfires that burned 833,017 acres. (Although the numbers grew significantly in the following years, roughly 90 percent of NPS prescribed and natural fires came from four parks.) The Forest Service kept poorer numbers, but the outcome was much the same, or worse. Fire remained on the land mostly due to wild, not prescribed, fire. If the revolution were to succeed, it would have to boost its burning by several orders of magnitude.[99]

Yet the niggling numbers were far from the whole story. The agencies were making an initial attack on a startling concept—they were literally playing with fire. It was a breathless decade of hotlining, hotspotting, and smokechasing an idea with the power to reform whole landscapes and the institutions that administered them. To partisans the dismal numbers mattered less than the momentum behind them. The new regime would take time and patience and a tolerance for risk, but as the 1980 fire season approached, it seemed possible to tear down the remaining barriers and sweep the bureaucratic caltrops off the road.

They could not know that the world the fire revolution had set out to seize was not the world the revolution would have to hold. However raucous the revolution, nature had held its peace, and national politics had stayed its partisan blustering. However daunting the fire scene of the 1960s and 1970s seemed to those living it, it was trifling compared with what would end the 20th century and announce the 21st with long drought, extended fire seasons, insect-plagued forests, landscapes piled with combustibles like cordwood, a rural buffer buried beneath exurbs, a polarized polity, and everywhere fragmentation—of once-uniform public lands, of a workforce, of fire institutions, and of social consensus. The future the revolutionaries got turned out to be a future they feared.

New Normals, 1977 and 1980

Two fire seasons. The first confirmed the failure of the old regime. The second affirmed the difficulties of the new one.

If doubters remained after 1970, the next giant bust in 1977 swept them away. The year was an almanac of fire. Month by month, the red-letter days piled up. The flames burned coast to coast, and beyond. The year began with the usual run of early-spring fires in the Southeast, though in record or near-record numbers, until May rains quelled them. Then the action moved to the Southwest, where in June every national forest in Arizona had a campaign fire, and a lightning spark on the Pajarito Plateau in northern New Mexico blew up to 23,000 acres. The La Mesa fire burned within half a mile of Los Alamos and forced archaeologists to become night-shift line scouts equipped with flashlights to direct bulldozers around unexcavated ruins. (In one notable instance a senior staffer simply turned off the dozer engine. The fire would pass; the ruins would be destroyed.)

More fires broke out in a wide swath over the West and took hold in Alaska. They continued through July. This time they began claiming casualties: three firefighters were killed in Utah, four in New Jersey, and one in Washington, while in California two helicopters collided, killing one pilot and critically injuring the other. Riverside and San Diego counties began to burn. Outside Santa Barbara the Sycamore fire incinerated 200 houses. Gradually, the fires began concentrating along the West Coast and especially Alaska, which was adding new fire busts daily. On August 1, while an international symposium convened at Stanford University to discuss the "Environmental Consequences of Fire and Fuel Management in Mediterranean Ecosystems," nature delivered the keynote by unleashing a barrage of dry lightning to spark fires that eventually burned 410,000 acres in California. Two fugitive fires in the Ventana Wilderness of the Los Padres National Forest merged to become the Marble-Cone fire, the second-largest wildfire in California's recorded history. To prevent the flames from entering the Carmel Valley watershed, bulldozers were permitted to hack lines across the northeastern ridges. The Alaskan fires multiplied to 680, of which 42 attained project size, and burned 2.29 million acres. Mercifully, come autumn, the outbreaks faded almost as fast as they had fanned. California avoided Santa Ana–driven conflagrations. Or almost did. On December 20 a fire started under a howling Santa Ana in Honda Canyon at Vandenberg Air Force Base. It promptly blew up and burned over a party scouting it that included the base commander, fire chief, assistant chief, and a dozer operator—four more fatalities. The final tally recorded 33,276 fires and 3.3 million acres burned.[1]

The season field-tested Firescope and threatened to sink BIFC, which in August for the first time activated the Multi-Agency Coordination System to establish priorities and synchronize dispatches as it shipped 1.56 million pounds of supplies for 8,600 firefighters. It dispatched air tankers across North America and mobilized fire crews from New Hampshire and Arkansas for duty on the Marble-Cone fire in California. The year seemed, to those who lived through it, like a specter returned from the past. In fact, it bore a close resemblance to a long pedigree of ancestral fire years. It had some quirks—fire officers had to request permission to loose D6 Cats in the Ventana Wilderness outside Big Sur, maneuver around space launch complexes at Vandenberg, and navigate dozers around Anasazi ruins at Bandelier, but it was a bad year much like earlier bad years. It offered a difference in size, not in kind. There was one response—control—adapted to peculiar circumstances. The year had more area burned than any year since 1970 and perhaps since the 1950s, but the total was far less than the great fire years of the past. Two-thirds of the burning occurred in the Yukon Valley.

* * *

Then, as though keyed to the sea change wrought by the 1978 and 1980 elections, 1980 marked the change from fire control to fire management. This was perhaps the first fire bust of the new era because it was not just a question of too many big fires overwhelming suppression forces but of too many kinds of fires and too much uncertainty about how to deal with them. In 1977 the agencies had to handle one kind of fire widely distributed in exotic settings. In 1980 they had to cope with different kinds of fires, each of which demanded distinctive treatments. Three problem fires, in particular, triangulate the new era and its awkward relationship with the Forest Service.

Wildfire: Panorama Fire. Suppression might still look back to the iconic Big Blowup for inspiration, and presuppression planners might scrutinize logging slash as the fuel behind the conflagrations of contemporary times, but the fuels of the future were the housing developments spreading like tumbleweeds across the South Coast. On November 24, an old-model Southern California fire met new-model fuels with calamitous results.[2]

The fire began at the hands of an arsonist at Panorama Point on Highway 18 along the south flanks of the San Bernardino Mountains. Fed by dense chaparral and smog-killed trees and shrubs and steered by steep slopes, the flames shot up to the crestline, where they met Santa Ana winds. The flaming front joined the winds pouring down the mountains. It rushed downslope and into the outer suburbs of San Bernardino. Within an hour 280 homes were destroyed, another 49 homes were damaged, and 64 other structures were turned to ash. The fire front moved along the mountain flank, outracing engines, taxing Firescope, and frustrating fire-behavior models as embers spread in leaps and bounds from house to house. It ended when the winds withdrew, leaving 24,000 blackened acres, 4 deaths, and 77 injured civilians.

This was not the fire front the Missoula lab modeled, nor the fuels the Riverside lab meticulously measured, nor the suppression tactics NIFQS qualified for. The Panorama conflagration was a hybrid, a monstrous mutation, but one of the sort that was establishing a new norm for Southern California. It was not enough to build up traditional fire-suppression resources, which could not tackle fires careening through subdivisions, and not enough to set up the multiagency coordinating groups at BIFC, which could not hope to respond in time. It was not enough to invest in presuppression projects, even fuels projects, because the combustibles at risk were houses.

Concerned parties organized into a Foothill Communities Protective "Greenbelt" Program to explore the prospects for segregating what they were already referring to as an "interface" between wildlands and cities. The really uncontrolled behavior, however, lay on the urban side. The megalopolis had behind it economic pressures as powerful as—and more continuous than—Santa Anas. It would take more than the occasional catastrophe to remove shake-shingle roofing from houses and surrender swaths zoned for commercial tracts to greenbelts.[3]

Prescribed Fire: Mack Lake. On May 5, to remove logging slash, prepare sites for replanting, and improve habitat for the endangered Kirtland's warbler that thrived in jack pine of a particular age, fire officers on the Huron National Forest set what they characterized as a prescribed fire. Ignition coincided with an approaching dry cold front. Within 90 minutes the fire had spotted outside the designated zone. As the south flank broke into multiple fire heads, the resulting wildfire blitzed over seven and a half miles in the next three and a half hours before the front passed. The escape ripped over 24,000 acres, killing a tractor-plow operator and blasting the little Michigan town of Mack Lake. It was the largest fire on the Huron since records began in 1911.[4]

A detailed reconstruction of the fire's behavior by researchers at the North Central Forest Experiment Station yielded one of the classic case studies in fire science. It brought to bear algorithms of energy release, fireline intensity, and rate of spread, along with tests on the NFDRS. A major discovery was the presence of horizontal roll vortices, later replicated by experiment under laboratory conditions (and published as a color cover for the journal *Nature*). But the story of the fire was not about the yet-to-be-discovered phenomenology of large-scale combustion and how this might improve fire suppression. It was the lethal collusion of old problems and new circumstances: logging slash mingled with critical habitats, a fire organization transitioning to prescribed burning, the Kirtland's warbler and the fire revolution. The Mack Lake fire was the lower Michigan counterpart to the 1976 Seney fires in the upper peninsula.

In retrospect it was the wrong fire at the wrong time, lit for good reasons and under the right circumstances of policy. The collision should have been as predictable as the passage of the cold front. Instead, it yielded public furor, agency caution, and a regional recession from the national fire scene. This was the second stumble in the Lake States, which had once dominated America's pyrogeography and now commanded attention by their failure to manage both kinds of prescribed burning. The exemplary study of the fire's behavior had the ironic effect of advertising the price of failure.

Fire by prescription might be the national policy. It was a bold fire officer, however, who willingly pushed the prescription and risked becoming the next subject of an intensive inquiry into the nation's next infamous escape. Part of a fire's written prescription was the record and résumé stored in a personnel file.

Prescribed Natural Fire: Gallagher Peak. To complete the hat trick, a natural fire slipped its leash in Idaho. In early July lightning kindled two snag fires on the Targhee National Forest. Both smokes rose within designated zones for PNFs, believed secure because they were within high country basins, a natural burning block. One, over the course of the next six weeks, self-extinguished at under half an acre. The other, what became the Gallagher Peak fire, burned briskly. For a while it doubled in size nearly daily.[5]

Because fuels were generally light and sparse and the setting at over 8,000 feet, it sparked no great concern. The fire program conducted aerial surveillance, wrote an analysis in case of escape, and allowed the fire to do what the new policy encouraged. It burned. On July 21–22 a solid rain fell over the region, and smoke vanished. Four days later three small smokes were sighted, all within the existing perimeter, as the burn smoldered in scattered logs and pockets of windfall. Then typical August weather asserted itself—warm, dry, and windy from repeated frontal passages. A fire on the nearby Challis National Forest blew up. The Gallagher Peak fire gathered heat, swelling from 300 to 500 acres in an afternoon, and then it, too, blew up, blasted out of its prescribed zone, and prompted a major suppression effort. The governor of Idaho denounced the Forest Service in the media. Throughout the Northern Rockies the order went out to suppress all new starts, whether in designated fire-management areas or not.

That was the immediate aftershock. The intermediate tremor was a not unexpected revision in the forest's fire-management plan to better accommodate drought effects on large-diameter fuels, such as those that had held the fire during the rains; conveniently, a revision of the NFDRS in 1978 did just that. Thousand-hour fuels with fuel moisture content under 10 percent put any fire out of prescription. The longer-term shockwave was more subtle. Every escape led to revisions, each innocuous and specific in itself but which collectively added up to greater restrictions that in turn became nationalized. The PNF might become more principle than practice. Different agencies, moreover, made such calculations in ways that rendered their shared borders impermeable to managed fire.

Escapes are inevitable: all varieties of fire escape. The Gallagher Peak fire escaped, but so did the Mack Lake, and for that matter so did the Panorama. Probably each kind of fire activity had a comparable ratio of successes to breakdowns. But suppression failures were accepted more readily; they led to calls for more investment in suppression and more aggressive attack. Prescribed fire failures led to greater restraints and a pulling back from the borders. Natural-fire escapes inspired linguistic sleights of hand, as if a new name for the practice would banish its failures. Over time, suppression would grow stronger and fire by prescription weaker. It was the nature of the dynamic. Especially when escapes attracted political attention, as they did that summer in Idaho, they were an unwelcome distraction. The fire revolution had to justify itself at every episode. Fire suppression could just put its head down and order more crews and engines.

The 1980 season was not one for the record books. It just showed from where those records would be coming.

CHAPTER THREE

HOLDING

He [Reagan] does not have a commitment to the environment. We don't have an environmental policy in this Administration.
—ANNE M. BURFORD, EPA ADMINISTRATOR (1985)[1]

This is the first time I've seen one of my plots burn. Burn, baby, burn.
—DON DESPAIN, YELLOWSTONE PARK ECOLOGIST (1988)[2]

The Cold and the Dark

The counterrevolution of the 1980s was not at first obvious. It was diffuse, indirect, and selective. It was an aspirational decade for private operators, global institutions, and small federal agencies with big problems. It was a dismal decade for large federal agencies that had to absorb national stresses and were surrogates for nation-states. The counterrevolution came through competing, incommensurate land uses such as wilderness preservation and urban sprawl; through fumbling efforts to roll back environmental legislation; through ideological commitments to downsize the presence of the federal government in American life. While the counterrevolutionaries failed to wipe away the fire revolution, they so damaged its institutions that it could not advance. What had been a broadly common cause pulled apart and polarized. Even nature joined in. The 1980s were the last opportunity for serious reform.

Instead, American fire entered a long, lost decade. Long, because it lasted most of a dozen years. Lost because it squandered the inheritance bequeathed by two decades of dramatic reform. That loss did not seem obvious at first. Like a broken wave washing on shore, the momentum of reform carried the action onward for a few years, which made the early 1980s a time of consolidation, a kind of climax and coda to the fire revolution. Yet the originating storm had dissipated. The decade disabled the Forest Service preferentially, but since the USFS was still the axle on which the American fire community turned, the system became dysfunctional. In its place the fire community went through a cycle of stalling, stopping, and suppressing.

The stall came from the Reagan administration. As the Republican Party moved steadily to the right, as it rallied and redirected southern fury about forced desegregation against the place of government generally, it sought to shift power from the public to the private sector, and within the public sector to favor military over civilian purposes and to develop the public lands for commodities. What it could not dismantle it could disable. The Reaganites stalled environmental legislation, which appealed to loggers, ranchers, and oil companies; loosened regulations over finance, which appealed to Wall Street; and retreated from civil rights enforcements, which appealed to the racial caste system of the South. Even ardent Reaganites such as Anne Burford, briefly head of the EPA, were forced to conclude that the president had no interest in the environment.

It was not simply what the administration did but how they did it, and the belligerence they displayed, that mattered because they provoked a reaction that hardened into a political polarization that still persists. Wildland fire became partisan in ways it had not been before. A progressively toxic political environment reduced the federal government's capacity for further reform. The old edifices of fire protection endured the political equivalent of an ember attack. In particular, the Forest Service, still fire's indispensable institution, was left wounded and bleeding.[3]

Of course, fire protection by federal agencies had always been political. It affected matters of public safety and public land—how could it not be political? Its major developments had followed the conservation agendas of the two Roosevelts—the Republican Teddy during the Progressive Era and the Democratic Franklin of the New Deal. But like foreign policy, fire protection had transcended particular parties. The fire revolution had proceeded amid the Kennedy, Johnson, Nixon, Ford, and Carter presidencies without undue interference. The Civil Service Reform Act of 1978 created a senior executive service to promote leadership, but it also permitted political appointments for up to 10 percent of an agency's management positions, a prospect of creeping politicization. After Stewart Udall's tenure as secretary of interior, none of those administrations invested much interest in public land. But fire policy had floated on the rising tide of government-sponsored conservation. The 1980 election changed that narrative.

Even nature joined the revanchism. In its ancient rhythms of drought and deluge, the American West segued into deluge. The Great Salt Lake overflowed its levees. Climatologists, brooding over Milankovich cycles, predicted a coming ice age. Rains pummeled duff into sludge. In many landscapes it proved difficult even to burn slash piles. In 1982, when *Fire in America*, a fire history of the United States, was published, the fires that most interested the intelligentsia were thermonuclear; a review in *Scientific American* aligned the book with simultaneously published studies of a nuclear winter, in which conflagrations kindled by a nuclear war would blot out the skies with soot and plunge the earth into ice. Even hypothetically fire transmuted into the cold and the dark. Incredibly, in 1984 the Forest History Society released a fire-history documentary, *Up in Flames*, in which it concluded that monster fires were a thing of the past.[4]

In retrospect the revolution had flourished amid a benign climate in which bad fire years erupted perhaps twice a decade. While no fire officer from 1962 to 1992 thought his times unchallenging, those years appear quaintly benign after the cycle switched to chronic drought, and the bad fire years came five to six times a decade. The revolutionary years were nature's grace period. They made it relatively easy to restore fire—it was a matter of fighting a climate of opinion, of creating a resolve to restore fire, not of fighting both nature and society at the same time. Those years offered an opportunity to swap tame fire for wildfire. But with a few exceptions the inheritors of the revolution failed to scale up their operations to significant numbers. When the climate again favored fire of all kinds, wildfire reclaimed the field and made it easier to reinstate suppression as a default setting.

Studies of civil disturbances have observed that while fires amid riots are common, they are largely symbolic. They create a rallying point, raise emotions, and coordinate the action. When those fires die down, so does the riot. That is largely the scenario of the fire revolution. When it had fires, it flourished. When nature set fewer, and people failed to fill that void with their own burns, the fervor passed. While conviction remained that fire's restoration was right and necessary, it had few facts on the ground and clustered in a handful of favored sites. When mass fires returned in 1988, they shattered the rhythm of reform. A fire community had to transition from hotline to holding. The era inflected into a night shift.[5]

The gap in actual fires matters because fires paid for fire management, and much of fire management was learned on the job. The labor force was overwhelmingly seasonal, with firefighters rotating through for two to three years. Fire officers could retire after 20 years of active service. Between

1980 and 1986, two generations of seasonals and a third of the career of a fire officer would have passed without major experience—this on top of mandated changes to promote workforce diversity. Ideally, those years should have had prescribed burning fill the void, but they did not. And while the wet cycle lasted only through Reagan's first term, it imprinted on his second term a sense that fire was a trivial matter and a task from which government could unburden itself.

The suppression phase appeared without conscious intent, spreading like chestnut blight. Americans began to recolonize their countryside through an out-migration of urbanites. They were filling the once-rural landscape with houses, thus removing a buffer zone that had mediated between wild and urban. Sprawl did to America's landscape what deregulation did to its financial infrastructure. In fact, *sprawl* may be too benign a term when applied to the West, where development more resembled a splash. But by whatever name, such developments were interbreeding with the miscellany of natural hazards they touched, be they floods, hurricanes, tornadoes, or fire. Hamlets or enclaves or exurbs lay outside the structures of government other than counties—that had been part of their appeal. Of course they expected fire protection regardless.

The three trends braided together. The Reagan administration diminished the attention granted to fire management. It was something best done by the private sector or not needed at all. High-volume logging, particularly in old-growth woods, once more put the Forest Service at the center of controversies, undoing the achievements and any goodwill wrested during the previous decade. The agency was caught by the backlash against the Reagan initiatives and even internalized that dissent. Fire research nearly expired. Like its governing bureaus, fire management found itself without a galvanizing mission, lacking both clear ends and adequate means. Suppression reclaimed its lost ground, doing within agencies what Defense did to the federal government overall.

During the early 1980s, the weather was favorable, and the political climate had worked against a pluralistic program of fire management. But then the weather turned sour, and political will was redirected toward suppression. Wet and dry, red state and blue, wild and urban, green fire and red—fire management was the atonal outcome of the intersecting dialectics. Not until fires and a Democratic administration returned did the revolution revive. By then it was transformed, and the legacy of the lost decade weighed heavily on both unburned, fuel-sodden woods and ever more cumbersome protocols. Until that time, fire management struggled to regroup and to hold as best it could what it had so boldly grasped.

Words on Fire

In the pause of the early 1980s, the revolution consolidated. If flames sputtered, words spewed out. Major writers revisited themes and updated formative books on the environmental movement. In 1980 Sally Fairfax revised Samuel Dana's 1956 classic, *Forest and Range Policy*. In 1982 Roderick Nash wrote a fourth edition of *Wilderness and the American Mind*. Then it was Marion Clawson's turn with a 1983 reconsideration, *The Federal Lands Revisited*. In 1984 Michael Frome released his latest inquiry into the Forest Service. After *Whose Woods These Are*, he had written the Forest Service entry for the Praeger series on government agencies and then agreed to write a column for *American Forests*, only to have it canceled because of his escalating criticisms of Forest Service policy toward wilderness. Now he revised and updated *The Forest Service*, absorbing the controversies of the past decade. Between 1979 and 1983 the founding works of environmental history appeared, announcing a new subdiscipline. There was something in the air, if not the water, beyond pollution.

Prescribed fire had come of age. It left its southern-exceptionalism home and took to the road, soon as widely distributed as rockabilly music. Along the way it absorbed local variants that had survived: pasture burning in the Flint Hills, type conversion burning for watershed in the Mazatzal Mountains, habitat burning in Dakota pothole prairie, slash burning on the Coast Range, and free-ranging fires in the Mimbres—all lumped under a common rubric. To advertise that fact and to better regulate it, the fire community sought to synthesize those accomplishments in books, guides, and manuals.

A veritable library flooded shelves. In 1980 Tall Timbers published a detailed index to its conference proceedings and organized the first workshop on fire history—history as preserved in age-stand classes, fire-scarred trunks, and soil charcoal. In 1982 *Fire in America: A Cultural History of Wildland and Rural Fire* and Henry Wright and Arthur Bailey's *Fire Ecology: United States and Southern Canada* were published. Two SCOPE volumes conducted similar reconnaissances for Australia and the circumpolar boreal. The next year, in a massive gathering of the clans in Missoula, it was the turn of wilderness fire to claim its own dedicated workshop and symposium. Russell Dickenson, director of the NPS, gave the keynote address, and Rod Nash delivered the banquet speech. Shortly thereafter *Introduction to Wildland Fire* appeared, a totally reconceived alternative to Brown and Davis's *Forest Fire*, which expanded the scope of fire beyond forestry and its management beyond the Forest Service.

That satisfied the academics. In quick succession fire officers got an updated guidebook from NWCG for *Prescribed Fire Monitoring and Evaluation* (1982); a handbook on the Rothermel model, *How to Predict the Spread and Intensity of Wildland Fires* (1983); and a planning guide for *Wilderness Fire Management* (1984), followed by a state-of-knowledge review of wilderness fire by Bruce Kilgore. A shelf full of NWCG and agency-specific guides for planning and smoke management tumbled off the presses, complete with burning prescriptions for those environments that promised most of the action. They were, in a sense, the operational equivalent of the Rainbow Series on fire ecology that had succeeded the Tall Timbers conferences. By the mid-1980s the fire revolution had moved from proclamations to policy to practical manuals. Together, they codified the revolution.

Meanwhile, another kind of code pointed to the future. In 1980, prodded by the National Forest Management Act, the Forest Service released its computer-run Forest Planning Model (Forplan). The next year the Intermountain Fire Council sponsored the first conference on computer modeling for fire management, and the information technology revolution got personal as IBM brought its PC to market (*Time* magazine named the machine its "Person of the Year" for 1982). Enthusiasts hoped mechanization would "assist managers in coping with complexity at reasonable cost." In 1982 Firefamily consolidated most of the existing software into a digital fire cache.

Everyone wanted more information. Yet information by itself was a drag on operations, and fire officers and administrators looked expectantly to information technology to bury the overload and complications within a black box of software, even as the new devices prompted still more data and introduced other levels of complication (or distraction). Those aspects of fire management that had ample quantification, such as fuels and fire behavior, synched readily with the machines, which further ratcheted research and policy in like directions. The digital revolution and the fire revolution had found their interface.[6]

Suppression Storms Back

The process began slowly, first noticed (where else?) in Southern California, but it soon became a national condition, like a naturalized pest. Though few communities accepted an urban identity—they were fleeing unlivable cities, they were living closer to nature—the reality was that they were exurban enclaves that expected certain urban services, particularly fire protection. The movement was the flip side to wilderness: it carved single-purpose preserves out of previously generic countryside. But where wilderness sought to restore fire, the wildland-urban interface (or *I-zone*, as it came to be called in California) wanted complete protection. As wilderness partisans campaigned for surrounding buffer zones, so the exurban migration sought protection against fire and smoke. Fire suppression found a new cause.

At first it seemed remote from the themes that had riveted the American fire community for 20 years. It had a dumb name—wildland-urban interface (WUI)—for what most regarded as a dumb problem. Some even referred to it as "living in the stupid zone," given the history and likelihood of wildland fire in such areas, and the not-so-bright addition of more fuels in areas without established fire protection. In the 1970s an average of 405 structures burned annually in rural and exurban settings. By the 1990s that more than doubled to 932 annually, and a decade later, that number had tripled. Over the course of the decade wilderness fire swooned, then self-immolated at Yellowstone in 1988, while the geeky-named WUI grew into what became not merely a counterimage but a full-blown counterreformation. Wilderness grew through increasingly difficult acts of Congress. The interface responded to unleashed market forces.

The problem achieved its definition in California. Here fire officers had long spoken of a rural-urban fire problem. By the 1960s the rural landscape was hardening into specific categories of wildland or converting to exurbs and suburbs. While California's amorphous, once-ranchland scene had allowed for large fires, it had also absorbed them. There was slack in the landscape and some decision space for fire officers. That once-open range was now fenced and increasingly crowded from both sides. As Craig Chandler sardonically remarked, "Fires which 50 years ago would have disturbed only deer, bears, and an occasional sheepherder now burn out three campgrounds full of tourists from Iowa, the retirement ranchettes of two retired vice-presidents of General Motors, one old peoples' home, a hippie commune of two. They also make page 2 in the New York *Times*."[7]

The 1970 fires and Firescope forced the urban and wildland fire communities to find common ground, at least for operations (the WUI was first mentioned by name in the season's postmortems). After the 1980 Panorama fire, the Foothill Community Protective "Greenbelt" Program referred to a wildland-urban boundary. During the 1983 wilderness conference, the concluding speaker startled the audience by speculating that the era of wilderness fire would end by 1990 and would be replaced by some other informing problem, most probably the exurban or intermix fire. In 1986

James Davis of the Riverside lab popularized the wildland-urban interface as a concept (the problem was visible outside the lab's windows). While to most researchers the issue seemed intellectually marginal, and to most line officers, geographically on the fringe, by mid-decade the Forest Service fire research program faced a threatened extinction from budget cuts, and Forest Fire and Atmospheric Science Research (FFASR) promoted the wildland-urban interface as a politically relevant source of funding.[8]

Spark and sprawl, the wild and the urban, when stirred together made a metastable alloy of institutions as much as landscapes. It went national when the Forest Service, the U.S. Fire Administration, and the National Fire Protection Association (NFPA) decided to jointly sponsor a National Commission on the Wildland/Urban Interface Problem. *Fire Command* magazine ran a special feature issue in 1987. A national symposium and workshop, "Protecting People and Homes from Wildfire in the Interior West," was held in Missoula in September 1988, while flames gushed over the Old Faithful Inn. James Davis took the topic to the *Journal of Forestry* in January 1990. The Oakland Hills fire of 1991 concentrated minds, particularly those of urban fire strategists. The next year another conference in Missoula advanced the initiative. The National Association of State Foresters, FEMA, the National Commission on Wildfire Disasters, and the NFPA all weighed in with reports.[9]

If wilderness fire bonded fire management to the environmental movement, the wildland-urban interface connected fire's management to an exurbanizing America. Over the coming decade the issue went from an annoyance to a black hole. It compelled fire officers to shift from managing large blocks of land to a fractal frontier that unbalanced the geography around it. Fire protection was about edges, and the interface multiplied edges with the fecundity of bark beetles. It pushed fire management toward suppression as a first response: it was impossible to refuse aid, and decades of elaborating mutual-aid agreements often demanded it. And as it first did in California, the wildland-urban interface drove fire suppression toward an urban fire-services model that had little to do with the fire-restoring inspiration behind the fire revolution.

Instead, it threatened to redirect suppression away from wildlands altogether, much as the Reagan administration had redirected funding from civilian to military projects. To many fire officers, dispatching engines and air tankers to houses along the border had as little to do with land management as sending the World War II battleship, the USS *Iowa*, to Lebanon to lob a few shells at a civil war. It seemed a massive misallocation, or at least a distraction. In fact, they were witnessing the massing forces of a counterreformation.

Interior: Building Up from Breaking Down

It was harder to reorganize institutions than ideas. It was made more so by the absence, with a few exceptions, of catalytic fires and by an administration emboldened to roll back the frontiers of the environmental movement and to reduce, not reconstitute, civilian agencies. Rather than ecological goods and services, the Reagan administration wanted traditional commodities, and rather than wilderness, it sought to move the timber industry more intrusively into untouched public lands. Where earlier clashes had largely stimulated the fire revolution, these mostly stymied it. The Forest Service bore the heaviest strain. The Interior agencies—each unhappy in its own way—either survived or thrived.

But fire, agencies, and federal administration had their own logic, and no one in 1980 could predict how they might align. The BLM had timber from the O&C Lands and lots of mineral rights, including offshore oil, to keep it visible. For all, the one rule of thumb was that fire was more likely to get funding as a problem than as a promise, and if you are a problem, be a big one. In the early years of the decade, this bureaucratic formula blessed the FWS; in the latter years, fortune favored the NPS. The era selected for the small, the nimble, and the dispersed that could emphasize a single theme or species over the larger and more cumbersome institution that had to integrate contradictions within itself.

* * *

The Fish and Wildlife Service thought it had fenced off fire management by appointing a national coordinator and letting him draft a manual. That smug assumption changed dramatically in 1981 after the lethal Ransom Road fire at Merritt Island NWR. Congressional hearings subsequently grilled the Fish and Wildlife Service and forced the agency to scale up to national standards and integrate with sister agencies, and it was granted the funding to do it. As FWS fire director Phillip Street recalled, "We realized the need to have standards, and have protocols. And we needed to have more training. We needed to have better equipment. We just needed to really professionalize our fire program." FWS had to truly integrate with the national fire community. The task was vastly simplified by a decade of NWCG coordination: the FWS did not have to invent a program, only adapt the national standards, training packages, and guidelines, and often hire fire officers out of other agencies (usually in DOI). It held its first fire management seminar at BIFC in September 1983.[10]

The program ramped up. At the start of 1981 it had one full-time fire director and a part-time secretary at BIFC. By 1983 it had 64 full-time fire specialists and a budget of $4.5 million. After the next crisis year, 1988, when another 1.5 million acres burned on refuges in Alaska, staffing went to 214 dedicated fire specialists with a budget of $16.4 million. The program was filled with fire officers who had learned their craft in other agencies, further quickening the FWS's national integration.[11]

But there were trade-offs. Refuges lost some autonomy, and fire could easily become a square peg amid the round holes of a refuge program and the FWS's disciplinary biases. In 1988 the Northern Prairie Wildlife Research Center published the first bibliography on fire, wetlands, and wildlife. "Our close association with this project," its authors summarized reluctantly, "leads us to conclude, however, with some dismay, that a predictive science for this field is a distant goal." Fire was not analogous to captive breeding, artificial nests, trapping, or other traditional wildlife management practices. Moreover, fire staffers might have closer identification with fire peers across agencies than with coworkers at the refuge.[12]

Still, during the 1980s the Fish and Wildlife Service fire program was born and then flourished; it hired where most agencies were downsizing. The agency also found itself punching above its weight on national affairs as environmentalists turned to the Endangered Species Act to do what the Wilderness Act could no longer do. As the agency rose, so did its fire mission. The FWS became a fire agency even if fire management threatened to assume an existence of its own.

* * *

For Alaska, too, the decade witnessed progress. Almost overnight the fire revolution, adapted to the oddities of place, became the basis for active management. The reason is that the Alaska National Interest Conservation Lands Act resolved the core politics of land use that continued to tear at the public lands of the Lower 48. This allowed it to create collective institutions. And it helped that the two big fire years, 1988 and 1990, came after plans had been written.[13]

It helped, too, that the winners in the partition of Alaska were the FWS and NPS. When the BLM Division of Fire Management reconstituted itself as the Alaska Fire Service in 1982, the FWS was just beginning its escalation into the national fire scene, and its Alaskan refuges could coevolve fire programs without having to reconcile with a preexisting agency program. The NPS, too, was riding the crest of enthusiasm for natural fire—it would soon centerpiece its achievements at the Missoula wilderness fire symposium—and Alaska offered boundless arenas for PNFs. The Forest Service had representation on the committee only through the Institute of Northern Forestry. The State of Alaska ratcheted up its program, piece by piece, in step with planning elsewhere in the state. On federal lands modern fire management was, from the onset, the organizing principle.

The critical experiment in interagency fire planning commenced at Fortymile. On an Alaskan-sized spread of 12 million acres, a planning team assembled by the NPS, FWS, BLM, USFS, BIA, State of Alaska (Forestry, and Fish and Game), and a Native corporation (Doyon) worked to reconcile competing purposes within boundaries that made sense according to standards of fire behavior. In its basics, the choice was whether to modify a suppression program to accommodate some fire use or instead to establish a full-spectrum fire-management program. The complication was that only the BLM had on-the-ground capacity to do much with fire; it resisted the loss of its empire, which was founded on aerial suppression; and because it had the means, it was in a position to determine the ends. Still, when completed in October 1979, the Fortymile fire plan demonstrated the ability of a diverse group to reach consensus on a process if not on a product.[14]

The BLM wanted to do on others' lands what it had previously done on its own. This was unacceptable to the NPS, especially. Once ANICLA converted the monuments to parks in 1980, the Park Service declared it would handle all aspects of fire on its lands itself if the BLM refused to perform under its directions. The FWS and Alaska Fish and Game also wanted as much fire restored as possible. The BLM, however, sought to retain as many lands as it could under fire protection or, in the language of the discussion, to restrict "limited suppression" to those areas that rarely burned anyway. Elsewhere, some form of suppression—labeled variously as critical, full, or modified—would be the operational norm; suppression would remain the assumed response, and only its degree would vary. The NPS wanted recognition of PNFs and their monitoring as a legitimate basis for fire management, and in Alaska, maybe a defining basis. It was willing to call on the BLM for suppression where needed, but it sought to complement that capacity to achieve a "full spectrum fire management" program. When the agencies were unable to find mutually acceptable language, the NPS prepared to act on its own. Eventually, NPS director Russell Dickenson went to his counterparts in Interior and elicited recognition that each agency had its own mission and needed to manage fire accordingly. Alaska thus recapitulated the national fire revolution, with the BLM assuming the role of the Forest Service.[15]

The 1982 result was the BLM Alaska Fire Service (AFS), a round of interagency agreements, and the creation of a forum, the Alaska Interagency Fire Management Council (a boreal NWCG). Fire institutions had undergone a phase change. The BLM was no longer in a position to decide policy unilaterally, nor could fire protection substitute for land management. The land agencies in AIFMC had voting rights; the fire agencies did not. As Robinson had declared earlier, the fire program executed or made recommendations; it did not decide. Still, the BLM sought to maximize suppression. In 1983 the AIFMC rejected the AFS proposal, arguing instead for limited suppression where "the environmental impacts of suppression exceed the effects of fire or where the exclusion of fire may be detrimental to the fire-dependent ecosystem." In effect, a fire was innocent until proven guilty.[16]

These principles were worked out through a master fire plan for Tanana-Minchumina, a sprawling domain of federal, state, Native corporation, and private lands, which in turn became a template for other regions. An updated plan for Fortymile followed, then for Kuskowwim-Illiam and Copper River Basin in 1983; for the Yukon-Togiak, Kenai, Upper Yukon-Tanana, Seward Koyukuk, and Kobuk regions in 1984; and for the North Slope in 1985. By 1988 all of Alaska had a modern fire-management plan. Each expansion, every evolved iteration, increased the proportion of land subject to limited suppression. By 1993 that figure reached 47 percent. Meanwhile the Alaska Division of Forestry (ADOF) steadily expanded its responsibilities from south of the Alaska Range to include towns north of it. By 1985 the ADOF reached to McGrath and Tok as well as Delta and Fairbanks. By 1988 its zone of protection extended over 134 million acres.[17]

Far from vanishing, fire suppression remained the governing principle for half the federal lands and all of the state jurisdictions. The AFS added two hotshot crews (Chena and Midnight Suns). It upgraded its system of remote automated weather stations and lightning-detection network. Instead of adapting the NFDRS, it adopted the Canadian Forest Fire Danger Rating System, which more precisely forecast indexes within its boreal setting. It commissioned relevant studies, such as a regionally based fire history, to better appreciate the pyrogeography of the state (and revealingly, about options for reclamation on firelines, particularly in permafrost). Strikingly, it looked eastward as well as southward for extra resources once Canadians established a counterpart to BIFC and negotiated a mutual-aid agreement with the United States for support across the international border. Alaska followed suit with an Alaska Interagency Coordination Center at Fairbanks. All this was in place when the next fire busts rumbled through in 1988 (2.1 million acres), 1990 (3 million acres), and 1991 (1.7 million acres).[18]

That Alaska was ideally positioned, both geographically and historically, to absorb the fire revolution was no guarantee that it would do so. Other places seemed well suited but failed to kindle, or faltered after a flare-up. Alaska had a scant legacy of suppression, which had never been able to impose its will as ruthlessly as in the Lower 48; it had oil money from the completed pipeline; and it had a cadence of big fires within the scope of agency capacities, and helpfully, those fires had come after plans were in place. But only after hard negotiations, visionary leaders, and just plain luck did Alaska manage to pass through the flames of the revolution with remarkably little scorching. Alaska became to free-burning fire what it was to brown bears.

* * *

As the Alaska saga revealed, of all the Interior agencies, the BLM was closest to the Forest Service in size, mission, and capacity to attract controversy. The FLPMA had stabilized its charter, but because the nominal organic act did not repeal earlier statutes, it also "codified the ambiguities" and thus sharpened the politics of minerals, grazing, and protest. To detractors it remained the Bureau of Livestock and Mining. Something similar happened with fire, which made suppression the path of least resistance.[19]

Even after FLPMA, the BLM was an agency of maverick responsibilities that would take years to corral. The BLM soon had 24.6 million acres in wilderness study areas, grazing allotments shrank, and it confronted protected species issues with desert pupfish and bighorn sheep. It tried to cope with invasive species such as cheatgrass, spreading through the Great Basin like necrotic bacteria. It had its own eccentric analogue of the Endangered Species Act in the Wild Free-Roaming Horse and Burro Act. Consolidation of programs led to the imposition of national standards, most vexingly with regard to environmental regulations. Then the Carter administration, in response to the energy crisis and the Cold War after the Soviet invasion of Afghanistan, began speaking of "national sacrifice areas," all of which seemed to be on public lands in the West.

The BLM, like the Forest Service, found itself in the middle. It was doing business differently, and its traditional clienteles reacted. In 1979, beginning with Nevada, five state legislatures called for the reversion of BLM lands to state control. Presidential candidate Ronald Reagan counted himself among the Sagebrush Rebels, and as president he appointed James Watt as secretary of interior to boost the interests of the extraction industries. Environmental groups rallied in protest. The upshot was a standoff.

Decentralization, a "good neighbor" policy that involved local communities in decisionmaking, and the mending of the general economy quelled the Sagebrush Rebellion. Internal reorganizations split off responsibility for offshore minerals. But the decade was one of truncated budgets and staffing, in keeping with Watt's proclamation that "We will use the budget system to be the excuse to make major policy decisions." As the Reagan administration strengthened its grip, one observer noted that the BLM had "four times the land to manage with one seventh the personnel and one third the money."[20]

Its fire program continued more or less as before, anchored at BIFC and the AFS. Fire seemed a neutral safe-harbor mission, attractive to all its constituencies, and one helpfully financed by emergency funds. Two paths might diverge in the yellowed sagebrush, but both, it seemed, wanted fire protection. The agency contracted for some research and cultivated an interest in equipment, even launching a fire equipment journal and newsletter (a rival to *Fire Control Notes*). Its lightning-detection network granted it national visibility. It complemented hardware with software by installing an Initial Attack Management System, modeled on Firescope and Focus programs. In the thickening of interagency agreements it gained more than it gave. It tenaciously held to the word-shopped phrasing that while BIFC was a true interagency operation, it was a BLM-directed facility. And while the ANILCA saga had left the agency diminished in Alaska, the Alaska Fire Service was a genuine presence in the American fire community and a vital mover in BLM fire.

Unlike the Forest Service in 1978, the BLM did not undergo a reorganization in its fire financing, and like other Interior agencies, it exploited the emergency presuppression fund to bulk up its otherwise stagnant or declining budget. A GAO report noted this underfunding, which had led to understaffing (from 1981 to 1990, BLM range staff had decreased 25 percent) and meant the agency was unable to enforce laws, much less advance initiatives. The same stresses worked on budgeted fire programs, but the peculiarities of off-budget fire funding spared suppression, unbalancing a fire-management mix. There was little money for planned prescription burning, but there was always money to fight fire.[21]

Yet even as the fire program propagated on the ground, its national presence remained muted. The thrust of interagency operations meant the BLM was always automatically invited. It jointly staffed dispatch and coordination centers. It participated in national conferences but rarely originated one; and because most conferences dealt with research, or research applied to practice, it had no special presence since it lacked labs or experimental ranges. Apart from its lightning-detection network, it never sparked an innovation that spread throughout the community. Its primary ecological obsession was not with fire-thirsty pyrophytes such as wiregrass or big bluestem but with cheatgrass, whose containment demanded fire exclusion. Its rancher clientele was eager to burn off brush and juniper to promote more pasture but reluctant to burn routinely to improve rangeland, which could mean reduced stocking. With flat funding, it relied on emergencies to pick up the slack and build out, and this reinforced fire suppression as an informing project. It had no fire-management poster child akin to what the longleaf pine did for Tall Timbers, the red-cockaded woodpecker for the FWS, or the giant sequoia for the NPS. Within the fire community, it was perhaps best known for its Alaskan smokejumper corps. The 1987 and 1988 seasons put BIFC, and the BLM's role there, into national consciousness. Then the administration of Bush the elder quieted many of the controversies that the Reagan years had fanned.

It survived, it seasoned, it bulked out, and it continued to be the big brother for fire suppression among the Interior siblings. What the Forest Service had to hold together, or tolerate as regional divergences, Interior could parse out among its various agencies. The FWS could claim prescribed fire, the NPS natural fire, and the BLM suppression. By the early 1990s the BLM no longer required its Alaskan dominion in order to maintain a critical mass for its national fire program. Compared with its old green-shirted rival, it might lack resources and a heritage of grand firefights. But the parity of recognition it had long demanded it was getting.

Carrying the Flames

No agency had ridden through the fire revolution so high in the saddle as the National Park Service. For the informed public, the NPS was the face of reform. Alone, it had managed to express the new doctrines without having a new charter imposed from the outside—it absorbed the Leopold Report from within. It had field tested both prescribed fire and prescribed natural fire not only on western landscapes but in the iconic setting of the Big Trees. Its creation story from the Sierra Nevada was told and retold at every national conference. With ANILCA it gained twice: once in land, and again, to its credit, by insisting that those lands begin with the new order of fire and not by imposing the old one. Under total mobility it had received from the national establishment far more than it had given (nor had it much that the others could use). It had ample heapings of public tolerance that allowed it to weather early what might have been withering skepticism about prescribed and natural fires.[22]

Yet in some respects the national project was a Potemkin village. A few genuine triumphs in the Sierra parks and south Florida anchored a street of false-front plans that had little substance behind them. NPS-18 might require every potentially combustible unit to submit a plan; it could not ensure that those parks had either the will or the resources to write or implement them. Despite the faltering pace of an in-house research program, its larger fire project did underwrite a commitment to gather basic intelligence about fire history and ecology that over time granted the agency a shelf of reports and a substantive knowledge unrivaled by any other institution. Besides, the agency saw itself as distinct. Its mission as a national park service came first, not just in fire but in all matters.

Still, the pressures grew to integrate with the reassembling American fire community. The Alaskan parks doubled the size of the NPS's holdings and more than doubled the magnitude of lands under fire management. It had to assert its own interests, but within the context of the Alaskan Interagency Fire Council and the AFS. At nearly the same time it added Big Cypress Preserve to its domain, which soon doubled its prescribed burning acreage (with Everglades, the two units accounted for nearly 80 percent of the agency's total). Meanwhile, Congress and the Office of Management and Budget wanted evidence of planning similar to that adopted by the Forest Service to support fire appropriations. In 1981 the NPS released Firepro, a computer-based program to identify workforce needs for a normal fire year. (Never mind that normal years were rarely the problem. The issue was the blowout year, but no agency understood how to plan for those, which is why interagency alliances were necessary.)

Firepro shifted fire funding from Park Service operating funds to a special account in Interior administered through the BLM. Without outside support the agency could never have moved its ideas into action, but granting fire a budget separate from the park mission was a recipe to split fire from land management. That year, too, it fielded three hotshot crews, one each at Grand Canyon, Yellowstone, and Sequoia-Kings Canyon, as its contribution to the national suppression arsenal. It began experimenting with adapting a light-hand-on-the-land backcountry technique to firefighting, what evolved into minimum-impact suppression tactics. (A fire's scars might be visible for a handful of years, but suppression's might last decades.) The agency began publishing *Park Science*, which surveyed sponsored research in the parks, a good portion of which concerned the fire-history studies required by NPS-18. Research, planning, prescribed fire, suppression—by the early

1980s the Park Service had reorganized the full gamut of its involvement and its engagement with the national community.[23]

As always, the reforms proved awkward to implement. The Reagan administration reduced the NPS budget, which by 1983 left Firepro an unfunded exercise. The Alaska integration took hard-nosed negotiations to keep the AFS from reinstalling suppression. Three hotshot crews proved too much to sustain, and they soon dropped to two, then one. Even policy was up for reconsideration. In 1983 a delegation from Sequoia-Kings Canyon, led by superintendent Boyd Evison, visited the oracle himself, Starker Leopold, at the University of California, Berkeley, faculty club for a consultation and 20-year review of his seminal report. Leopold was surprised at how fire management had become "a bit more complicated than I ever had envisioned."[24]

Over the years his eponymous report had been split and refracted through two prisms. One, the agency's, interpreted it through the NPS organic act, which at its core argued for unimpaired naturalness. The other, the public's and environmentalists', saw it through the wilderness movement, emphasizing the parks as a collateral variant of the legally wild. Neither wanted much intervention: the ideal was to let natural processes find their own way. Yet Starker followed his father's vision. There were five tools of land transformation, Aldo had argued—the plow, the ax, the gun, the cow, and the torch. Through them people had remade landscapes, often trashing them. Those same tools could be used to restore ecosystems; one need not wait for natural processes alone. Restoration required "judgment, followed by action."[25]

It is "not resolved simply by 'allowing natural ecosystem processes to operate.'" In running a park, Starker wrote, "you make management decisions every day, many of which are of necessity arbitrary. But they need not be capricious if you have in mind a firm goal toward which management is directed." If a site is ready to burn, "it makes little difference to me whether the fire is set by lightning, by an Indian, or by Dave Parsons, so long as the result approximates the goal of perpetuating a natural community." So with wildlife. If predators could not do the job, then trapping and shooting should. "I am much less afraid of this kind of decision-making than I am of adopting inflexible rules of conduct—such as 'we must wait for lightning to start the fire; we must wait for a mountain lion to work over those deer.'" It was okay to manage the backcountry through natural processes. The frontcountry, open to visitors, should not be left to chance. In some places, he said, "a chain saw would do wonders."[26]

There was no easy solution. Leopold's coda was more than the Park Service was ready to accept, but there were critics primed for both more and less intervention. Shifting policy from re-creating historic scenes to preserving ecological processes had allowed flexibility in doctrine and practice. It was impossible to restore the past, and nature could make the needed adjustments without people's understanding every scintilla of ecological cause and effect. The sinkhole in the road was how to incorporate people—those in the past who had helped shape the scene and those charged with managing it today. In the sequoia groves, for example, the natural regimen of burning proved inadequate to reproduce the kind of fire history recorded in scarred trunks. American Indians had clearly burned, and the recession of fire had coincided with their removal, not with the onset of organized fire suppression. In the hands of foresters such as Thomas Bonnicksen, these issues became a critical stake to drive through the heart of the vision of naturalness or the musings of deep ecology. The original scenes were never natural: they were the outcome of nature and people interacting. Vague notions of naturalness were impossible to implement, a laissez-faire management was dangerous, and a program that could not identify measurable means and ends was sloppy and counterproductive.[27]

But serious intervention could rouse opposition as well. Appropriately, when just that controversy went public, it returned to the Big Trees where the Park Service had birthed its fire program. The casus belli was the 1985 Broken Arrow burn at Giant Forest in Sequoia-Kings Canyon. Eric Barnes had grown up at Three Rivers outside the park entrance and had worked seasonally for many years on its fire crews. When he revisited those treasured sites during the fall burns, he was horrified. The Giant Forest was deeply marred by scorching and appeared at risk. He thought of Edith Wharton's definition of the perfect horror story as "about something that cannot happen, must not happen, and does happen." He began a private campaign to launch an inquiry, writing to the park superintendent, regional and national directors, Harold Biswell, Save the Redwoods League, Senator Alan Cranston, and Representative Morris Udall, chairman of the Committee on Interior and Insular Affairs. In later years he likened the burning to the explosion of the space shuttle *Challenger*. The park assigned Bruce Kilgore to lead an in-house review. Then it accepted the need for an external audit under a panel of academics chaired by Norman L. Christensen. The panel met at the park on June 30 for a public forum and then held two days of tours and talks. It dispersed for research on assigned topics before reconvening in October to craft a final report.[28]

Barnes viewed the scene with the eyes of a landscape architect as well as his remembered youth. He understood that "Certainly, the burning issue today is not FIRE versus NO FIRE. Rather, the issues now are complex, trickier, less tractable perhaps." In its drive to restore processes, the park was ignoring the unique objects the Big Trees represented. They could not be replaced for several thousand years; they, not naturalness, were the focus of proper preservation. The core concern was not just restoring fire but preserving sequoias. The "showcase" sites needed more delicate treatment beyond crude slashing and burning, which only "profaned" what John Muir had christened the "temple grove." Simply restoring fire did not by itself guarantee the survival of the sequoias, and if done poorly it might damage them.[29]

In truth, not everyone was happy with how the NPS was reintroducing fire. Discounting Horace Albright, a throwback who disliked any kind of fire but who was still a power behind the throne, most critics supported the broad goals but fretted over how they were implemented. The burning was too little to be effective and too intense to avoid disfigurement and irremediable damage to irreplaceable objects. Even Biswell—who agreed with the urgent need to restore fire, applauded the park for its efforts, and did not want field operations slowed while waiting for more research—noted that the program was not only failing to burn enough but was also not reburning. In the 20 years since the program had begun, it should have burned the original sites at least once more and perhaps several times.[30]

In what was becoming an NPS tradition, the Christensen Committee managed to agree with all sides. It defended the park's goals, agreed with its fire plan in principle, with some reservations, and accepted the criticisms of how and where the burns were executed. It distinguished between "restoration burns" and "maintenance burns" ("simulated natural fires"). It insisted more research was needed. It conceded that aesthetics and scenery were valid concerns. It urged the park to reconsider its dismissal of "Indian fires" as unnatural and hence irrelevant. The committee noted that the existing groves had grown up amid routine human burning and that it was likely people would need to continue burning in the future.[31]

A weird historical symmetry emerged. When fire suppression began, most proponents believed its major investment would come only at the beginning, that immature nations passed through an era of abusive burning, as people passed through childhood fevers, and needed suppression to assist that passage. Likewise, the fire-restoration project had assumed that once a landscape had been reset to something like presettlement conditions, active management would no longer be necessary. History supported neither belief. To thoughtful observers, it quickly became apparent that practice could not easily be segregated from philosophy.

The report concluded with financing. To conduct the intensive landscaping preparations that Giant Forest needed to prepare for burning required far more money than the park budget allowed. Unlike suppression or presuppression, prescribed fire could not tap into emergency accounts. A first-order treatment was expensive, and because maintenance burns would be required in perpetuity, those expenses might diminish over time but would never expire. As with highways, there was always more enthusiasm and money for original construction than for the endless labor of repaving and patching. Prescribed fire in the West was likely to be costly. In any event, the argument for a pause in the burning program coincided with a diminution of program funds. The Reagan administration's strategy to use the budget to make policy was beginning to bite. In 1985 and 1987 passage of the Gramm-Rudman acts sought to tame the awful deficits by bringing down spending. There was more to do and less to do it with.

* * *

Where the NPS made its reputation was with natural fire. Natural fire was prescribed fire's pristine double. The relationship between the two practices was reciprocal: prescribed fire gave natural fire a rigor that allowed it to be absorbed within institutions, while natural fire granted to prescribed fire an identification with wild nature that lifted it above a mere tool of choice. Such burns were distributing the goodness of the wild to landscapes in need of rehabilitation or redemption. Even deliberately kindled prescribed fires were a means of transitioning toward a more fully natural regime. They were part of a strategy and a system even if they embodied a paradox that fire might be equally wild and managed like a wolf with a radio-tracking collar.

The premise behind restoration could not be simpler: quit suppressing, stand aside, let nature work out its own destiny. As a cadre of NPS fire officers explained to the 1983 Wilderness Fire Symposium, granted the uncertainties about our knowledge and future conditions, "minimum intervention is the wisest and the most conservative management strategy." It seemed possible to substitute intellectual certainty regarding the idea for uncertainties about its manifestation. People did not need to know or control all the outcomes, as with an agronomic or silvicultural system. Nature would absorb the ambiguities and accommodate the range of outcomes. The prescribed natural fire was not an oxymoron but a paradox, like Bohr's principle of complementarity. For fire officers a prescribed natural fire could be both controlled and freeburning as, for physicists, an electron could be both wave and particle. Which dominated depended on what task was needed (or what question was asked). What mattered was how the concept actually played out in the field.[32]

The NPS was the most aggressive agency in distributing prescribed fire to far-flung sites. But then it had proposed the most startling expansion westward by making the Big Trees the poster child for fire's restoration. By 1982 burns were under way from Crater Lake and North Cascades to Devil's Tower and Pinnacles. More daringly, it promoted prescribed natural fire in such places as Hawaii Volcanoes, Big Bend, Isle Royale, and Bandelier. With deliberate prescribed fires to enable more natural fires, the only limits to the expansion seemed to be smoke, adjacent communities, and superintendents' tolerance for risk. Whatever the actual acres burned, the symbolism of natural fire made it a force multiplier for the fire revolution. It was, at one and the same time, a useful practice, an ecological corrective, and a penance for past misdeeds.

The Park Service continued to trumpet its successes. If it was a lightning rod for controversy, that was because it stood taller than the other agencies. It attracted public attention because it operated in America's most revered landscapes. The Giant Forest's prescribed burns were contentious because they were symbolically significant not only for the culture of fire but for American culture at large. In the words of the Christensen Report, it was "an important paradigm for our general understanding of the role of natural disturbance in wilderness ecosystems." In the wake of the controversy the agency again revised NPS-18.[33]

Still, if its prescribed fire program had to recede for a while, its wilderness fire was thriving—at least on paper. And what salvaged the finances of its fire programs overall was a return of drought and big fire seasons in the latter half of the 1980s. In 1988 those two themes, wildfire and wilderness fire, converged with what seemed like apocalyptic fury. The National Park Service found itself in the eye of a coniferous and political firestorm.

A New Old Federalism

Institutionally, the fire revolution meant breaking the old order into parts, rechartering and strengthening those parts, and then attempting to reassemble them into some kind of coherent whole. The national push toward interagency arrangements was only its most visible expression. Quite apart from the Reagan administration's New Federalism, the fire community was rearranging its constituent pieces.

What happened with federal agencies happened throughout the system as a great realignment spread through the American states, nongovernmental organizations, and the fire institutions of other countries. Without intending to, fire stumbled and lurched toward what was a new alignment of powers that needed a new constitution to govern them. As the politics of fire went deeper, wider, and farther, and as the economy of fire became both smaller and global, the old focus of firepower, the nation-state (or its surrogate, the national agency), struggled to hold them together.

* * *

The American states had long been on the frontlines of fire protection. In the 19th century, New York, after reserving the Adirondacks, had created a model of rural fire control that C. S. Sargent carried to wider deliberations through the National Academy of Science's Commission on Forests and from which other Lake States, in particular, took inspiration. Some states had parks and forests—miniatures of the country's public domain; these expanded during the Great Depression as cutover lands reverted to states in tax delinquency. Other states mandated fire taxes to support fire control by industry (e.g., the Idaho idea of fire associations that spread throughout the Northwest). Some required counties to provide protection and in some cases contracted with the counties to provide the service. Most, in deeply rural landscapes, deferred to volunteer brigades and local landowners.

What brought some shared order were programs that joined the states to the federal government. The binding began with the Weeks and Clarke-McNary acts but expanded into a broader Cooperative Fire Protection Program by which the federal excess property program could funnel war-surplus hardware through the Forest Service to state allies and a fire-prevention agenda that included the Smokey Bear campaign. In 1972 the Rural Community Development Act expanded the scope of federal interest in exurban fire. The 1978 Cooperative Forestry Act repurposed the whole project. In 1986, reflecting the tidal tug of state interests, the Forest Service relocated Fire and Aviation Management from the National Forest System to State and Private Forestry.[34]

In a manner that would have delighted Alexis de Tocqueville, the states began organizing among themselves. In 1920 the National Association of State Foresters was founded, creating a parallel entity to the federal program. What moved the politics beyond talking and lobbying was emergencies: fire disasters had a way of concentrating the mind. The 1947 Maine crisis sparked the Northeastern States Forest Fire Protection Compact. It did for fire in the East what the Colorado River Compact did for water in the West. The idea subsequently spread—the South Central States Forest Fire Compact and the Southeastern States Forest Fire Compact, both in

1954, then the Middle Atlantic Interstate Forest Fire Protection Compact (1956), Lake States Forest Fire Compact (1989), and Big Rivers Forest Fire Management Compact, which joined Illinois, Indiana, Iowa, and Missouri. The compacts provided a political alliance that mattered especially where the federal presence was light.

More scares, more activism. After the 1967 fires, the Forest Service and a group of concerned state foresters drafted a proposal for a "National Fire Emergency Program." The idea merged with the Office of Civil Defense–sponsored National Fire Coordination Study, which had recently completed its survey. The proposal argued that "any fire, in any state, that became an emergency situation would be attacked with forces from adjacent states by federal, state and private fire fighting organizations under inter-state and inter-agency agreements." As Eliot Zimmerman, Cooperative Fire Program director, put it, "The proposal was an expansion of the interstate compact philosophy to nation-wide application." Governors Conferences, the National Association of State Foresters, and the American Forestry Association agreed and called for legislation. Though Congress failed to act, the need persisted, and what evolved instead was a series of ad hoc measures and crisis responses that provided coverage but lacked logical coherence or much fiscal control. A cobbled-together consortium better suited the American political temperament, with minimal investments up front and emergency expenditures to cover the bad years and glue the mosaic together. A national fire-protection system thus evolved much like national health care.[35]

Measured by area protected, by numbers of fires, or by spending, the states far exceeded the federal government. Open flame had disappeared from America's vernacular landscapes largely because the states had provided fire control in the name of safeguarding the public and protecting property. In 1970 the country budgeted $113 million for fire protection, of which the states contributed 84 percent (and private landowners 2 percent). Of 121,736 fires reported, state and private lands had 106,768 (88 percent). Of the 3,279,000 acres burned—a blowout year for the feds—2,560,000 (78 percent) scorched state and private lands. Except in areas with a dramatic break in land use, such as apple orchards replacing conifer forests or shopping malls supplanting chaparral, fires burned across cadastral borders and were fought by whatever crews and machines could be rallied against them. By mass and numbers, the states overwhelmed the federal establishment.[36]

Still, few states or local communities had the capacity for fire protection on their own. They needed supplements, and somewhere in the custodial chain that need led to the federal government. While it was one thing for the states to do their job with stimulus monies from Clarke-McNary funds and federal excess property, it was another matter when they began to rack up big suppression expenses that they could not cover or began to contribute to national mobilizations or when their interests and those of the Forest Service began to diverge. Yet all this happened as the fire revolution matured. Long cultivated as allies, the states became, in some respects, also rivals.

* * *

In their original alliance, the interests of the Forest Service and the state foresters converged with laser sharpness: they both sought to reduce the number and impact of fires. That their bureaucracies had common origins in forestry furthered that bond. To outsiders they were the writhing limbs of a single leviathan.

The fire revolution challenged that condominium by rejigging the partnership in two ways. One, it displaced the Forest Service as a hegemon. The states could seek pacts among themselves and with other federal agencies, and they could integrate more fully into the evolving reconfiguration of the national fire establishment. And two, their interests and those of the feds split with the shift from fire control to fire management. The federal agencies required many kinds of fires and responses to wildfire; the states wanted only one. They were charged with fire protection, and they had no interest (or capacity or authorization) to engage with fire pluralism. But even as the states and feds diverged in the wilderness, their paths recrossed on that hybrid landscape where the wild and the urban faced each other over a line in the duff. Here they again shared a common purpose. They did not, however, share a common sense of rights, roles, and responsibilities.

The USFS and states had long cooperated within state borders. With the advent of total mobility, the Forest Service could call on interagency resources from around the nation to assist, all of which increased the massed firepower of suppression. At Forest Service insistence the National Association of State Foresters had contributed a representative to the National Wildfire Coordinating Group. In 1975 the NASF joined BIFC. At the same time, the state fire brigades, like the National Guard, became a kind of reserve militia that could be activated during emergencies. The size of the workforce potentially available for fire control expanded.

As always there were trade-offs, because total mobility required common standards for training and fireline qualifications. The states would have to meet those standards, which few could afford, especially if they relied on volunteer fire departments as first responders. If the states wanted federal assistance, they would have to satisfy national criteria just as the various federal agencies did. No longer could all parties just show up at a smoke and work, neighbor to neighbor, plows and engines, over their shared fenceline. They had to be integrated into the incident command system. Their radios had to communicate with each other. Safety and liability issues meant they had to wear common protective gear. Over the long run, the process raised the caliber of state fire organizations but only after protest and appeals for federal aid to assist with the upgrades.

The deeper issue concerned what this militia should do. With few exceptions, where state bureaus had responsibility for managing substantial state forests and parks, state agencies sought to put fires out as quickly as possible. It was not just that the pluralism of the fire revolution had no appeal; it had no legislative grounding. But as more of the national fire load shifted toward the wildland-urban interface on private lands that had previously furnished their own fire protection or had it provided through state agencies, the power of suppression ramped up. The practical weight of protecting the metastasizing WUI, with its scattershot enclaves and long, pyrophytic perimeters, placed a heavy thumb on the scales that favored suppression. As the interface grew in size and importance, so did the influence of the states, which increasingly saw themselves not just as recipients of federal grants but also as deserving partners who demanded standing.

The tortuous narrative of American federalism that entered a new phase with the civil rights movement had found a parallel in fire management. In simple terms, the states wanted federal funds without federal strings attached, while the OMB was reluctant to finance matters that properly belonged with the states. The solution was to keep budgeted monies low and federal influence slight until forced to respond to crises. So long as emergency monies gushed out of the Treasury, issues remained rhetorical and academic. When the spigot tightened, the crises pointed to the need for a new fire constitution.

* * *

The states were nothing if not varied. But it is surely not accidental that the two geographic poles of the fire revolution, Florida and California, boasted exemplary state fire organizations. Each pioneered one of the apexes of the modern fire triangle—Florida with prescribed fire and California for suppression. Both had to wrestle with what had become by the end of the

20th century the defining issue of fire management—the scrambling of the feral with the urban (neither state had its state agency do anything with natural fire). By the mid-1980s each had found ways to fuse its traditional emphases with the emerging problem fire of their place and time. They published manuals. They promoted model fire codes. They became powerful presences in the American fire community.

In its origins Florida conformed to the standard formula for southern forestry. It enrolled in Clarke-McNary in 1928 amid a landscape that was, as the saying went, cut over, grazed over, and burned over. In 1931 the County Forest Fire Control law authorized matching funds by which the state and counties might pay for fire protection—in effect, an internal Clarke-McNary program for the state. A decade later the Florida Division of Forestry (FDOF) converted to prescribed fire as an acceptable strategy for fuel reduction because without it the state could not hold back wildfires. The FDOF loosely regulated private burning through permits. The program acquired some early rigor in 1949.[37]

The Florida Division of Forestry, primarily a fire organization, boomed with the rest of the state during the postwar era. Most of the explosive growth occurred along the coast; the interior remained the preserve of citrus orchards and ranchers. To balance its rapid conversion from rural to urban geography, the state established a suite of programs for nature protection. Public preserves multiplied, though not at the same pace as private suburbs. By 1972 the FDOF had enrolled every county under its aegis. State fire stations and lookout towers were as distinctive a feature of interior Florida as fire stations were to cities, and it was burning nearly 3.9 million acres through fires wild and prescribed. Smoke towering over the land was as much a feature of the Florida horizon in the dry season as thunderstorms were during the rainy season.

Then the squeeze came. The fragmentation of the landscape made free-burning fire harder to tolerate, newcomers ceased the burning that old ranchers had done forever, and a significant portion of the land was held by absentees. Burning became both trickier and more essential. Prescribed burning could not keep up with needs, and wildfires made up the difference. Fires became tougher on both public and private lands. The FDOF campaigned for legislative relief, which arrived in 1977 in the form of the Hawkins Bill, which reduced some of the uncertainties surrounding burning and, more astonishingly, allowed the FDOF to do the burning that landowners were unable or unwilling to do on their own. Overgrown plats were treated as a public nuisance, not unlike vacant lots in a city. No other state had anything like such authority.

Still, the burning lagged, even as more land was set aside and more houses sprang up on the lands not fenced. In 1985 what the FDOF had dreaded might happen did. Fires roared out of the rough of Flagler and Volusia counties and into Palm Coast, where they forced evacuations and incinerated 250 houses. That got political attention. The fires were both a useful warning and a distraction because, while the forces of protection were not keeping pace with development and the Florida rough, they also pointed to emergency services, not land management, as a likely response. The FDOF and Florida Parks were falling behind on their own public lands. The only way to keep the larger landscape in hand was to encourage burning by private landowners.

The old habits, however, were being fenced in by a new society stringing houses and shopping malls through the countryside and by an encroaching legal environment built on the barbed wire of liability. In 1987 the FDOF commenced a program to certify prescribed burners. When the Florida Supreme Court ruled that landowners and contractors doing the burning for them could be sued for damages that resulted from smoke or escaped fires, the difficulties threatened to overwhelm the prospects, like kudzu taking over a woods. The state rallied, fielded a blue-ribbon committee, assembled the various concerns into a single piece of legislation, and passed the Prescribed Burning Act of 1990.

It was a unique piece of legislation, perhaps the most remarkable fire statute ever authorized in the United States. For one thing, it created a disposition to burn: it did not just tolerate fire, it encouraged fire. It identified burning as a property right, considered it in the public interest if conducted under appropriate rules, and limited liability to "gross negligence." A second innovation was to leave to FDOF the determination of what the guidelines and suitable rules might be. The Division of Forestry became the keystone agency for responsible burning. Its reach applied even to federal lands because smoke management was a state task under the Clean Air Act. If a landowner burned on his own, he had no legal protection. If he submitted to FDOF guidelines for training and authorization, he did.

Anchored, institutionally, at Tallahassee and, in terms of flammability, by fire-prone landscapes that could burn annually from Eglin Air Force Base to Everglades National Park, Florida became the epicenter of prescribed fire. State parks and nature conservancy sites were burned for ecological reasons, which had the side benefit of reducing fuel and hence the ferocity of wildfires, but the Division of Forestry burned to help advance its mission to protect life and property. It did plenty of hard-core firefighting—it had to. But it also became the hub for the many institutional spokes that managed fire in Florida. It did the job that, nationally, the U.S. Forest Service had long done.

* * *

If FDOF had begun as a typical southern forestry bureau before going on steroids, California's Department of Forestry (CDF) began as a normal western one before a relentless series of crises put it on growth enhancers. The state was always and everywhere prone to fire. When foehn winds blew, California was notorious for the savagery of its fires west of the San Andreas Fault; when lightning saturated its forests, lands east of the fault were perforated with burns. Like postwar Florida it faced daunting challenges from exurban development that rolled over its landscapes like an earthquake wave. But unlike Florida, each tick of its institutional chronometer moved it in the direction of suppression. California entered a chronic state of emergency.[38]

In 1905, the year the Forest Service acquired the national forests, a state Forest Protection Act created an effective Board of Forestry with authority to field fire patrols in times of "particular fire danger" with the costs borne by counties. Other acts followed in 1919, allowing California to join the Weeks and Clarke-McNary programs, and a Compulsory Patrol Act in 1923 required counties to provide fire protection or contract with the state to do it. There was never enough money, and the agency became dependent on a whopping state emergency fund to cover costs. During the Depression, CDF relied on the CCC and Works Progress Administration to supplement lost revenues. It dreamed of a statewide "master fire plan."

As with Florida, the inflection point came during World War II. But where the FDOF repositioned itself around prescribed burning, CDF opted for full-scale suppression on a wartime footing pumped up by massive emergency expenditures under a Wartime Powers Act. It became in effect California's department of defense. In 1945 the California Disaster Act transferred that authority to civilian agencies and what evolved into a state Fire Disaster Plan. Once again federal and state interests aligned when the federal government established a grant program for civil defense akin to that of the Weeks Act. An equipment windfall resulted, under the control of the Fire Advisory Board. The state entered its own cold war on fire. By then California was also devising a replacement for the CCC in the form of inmate honor camps as a source of labor for forestry and fire projects.

The postwar housing boom thrust suburbs not only amid formerly rural lands, as in Florida, but against mountain wildlands. Wildfires arced across that boundary like sparks jumping oppositely charged plates. Thirty years before flames rushed through Palm Coast, they were slamming against Malibu and all the Malibu wannabes that had sprung up in the South Coast. CDF found itself in a chronic firefight, edging more and more toward fire protection as a dominant mission, assuming more and more the posture of an urban fire service out in the countryside. During big busts, an Office of Emergency Services coordinated among the various fire-suppression agencies—federal, state, and local. Then the 1970 fires galvanized Firescope and a new California Emergency Plan, which included a Fire and Rescue Mutual Aid Plan. The 1977 fires tested both Firescope and the plan, encouraging upgrades. The tax-protesting Proposition 13 passed in 1978 only shifted the funding away from regular to more emergency outlays.

Through the 1980s the triangular pattern of reinforcement continued. Big fires prompted more comprehensive plans and extra funding; OES and CDF bulked up and strengthened interagency exchanges, and the system became dependent on emergencies and supplemental revenue or revenue-in-kind such as the inmate camps and crews. In 1987 CDF bowed to the obvious and changed its name to the California Department of Forestry and Fire Protection (and 20 years later to CalFire). It was second only to the Forest Service as a firepower in California, and for that matter, in the nation. Along the way it had attempted prescribed burning for fuel reduction but faltered in the face of smoke issues, complications along the interface, and budgets. It was pulled by the gravitational force of urban fire services, especially the Los Angeles County Fire Department. Its presuppression projects involved building codes to eliminate shake-shingle roofing, clearances around structures, and pre-positioning engines and aircraft.

What FDOF was to prescribed fire, CDF was to suppression in the new landscapes of late 20th-century America. It pioneered the fundamental tactics of fighting fire where the wild met the urban. Both states testified to a capacity to innovate and adapt to local circumstances without making every incident a federal case. They could become major players—hosting conferences, promoting blue-ribbon commissions, publishing handbooks. Yet their strength was also their weakness. They tended to be captured by their local surroundings, which made it difficult to transfer what they learned elsewhere. They needed some additional mechanism, some other agency, to promulgate, integrate, and leverage their innovations to the national stage.

The diaspora of California fire officers through the Forest Service had spread a California model of wildland fire. By the end of the 1980s that model was reviving thanks to the I-zone, and it was not always welcome. The Ugly Californian was not yet a fixture in the federal fire community, but the trend to replace a central tenet of the fire revolution, that fire and land management must bond, with an all-hazard emergency service model was smoldering in punk and waiting for favorable conditions to spread. While most of the country hoped that what happened in Florida could relocate elsewhere, they hoped that what happened in California would stay in California. History, however, suggested that little of what began in California remained there for long. Fifty years after the fire revolution started, the fire directors of the Forest Service and the National Park Service had both begun their careers in California.

Privatizing Fire

When the fire revolution began, most firefighters on federal lands were federal employees who relied on government-sponsored science conducted in government labs and who used equipment designed at government equipment centers and furnished by the Federal Supply Service. When fire protection needed more resources, it reached to cooperators, most of whom were also government based. But some were private companies that rented aircraft, logged under contract, and supplied hand crews and bulldozers on request or catered meals to fire camps. A civil society for fire barely existed.

The deliberate and de facto privatization movement of the 1980s unsettled that arrangement. Private vendors probed the market for supplying equipment; consultants, often retired fire officers, set up business; NGOs appeared as think tanks and then as active cooperators. What critics would denounce as a fire-industrial complex was rapidly forming and would likely siphon off any prospective cost savings by creating political lobbies. Whether done by the public or private sector, fire protection was funded by public monies.

These developments coincided with an increasing globalization of fire: the interagency model became international. Even environmental groups, so critical for decisions over land use, went global. For most of the 20th century, the economy of fire management had progressed from local to national scales. Now, pulled by multinational organizations and international treaties, the nation-state reluctantly ceded its monopoly, though it continued to pay the bills.

The easiest moves toward a privatization of fire were those that dealt with ideas. An Ecosystem Research Group of university and agency scientists at the Andrews Experimental Forest in Oregon evolved the New Forestry concept. Cascade Holistic Economic Consultants wrote citizens' guides to Forest Service planning. Island Press reorganized in 1984 to publish books on natural resource topics, helping break the Government Printing Office chokehold on fire literature. The Society of American Foresters, both nationally and through regional chapters, issued a stream of position papers and resolutions. After the 1988 fires, Tall Timbers revived its fire ecology conferences, staging them biennially. The more remarkable developments, however, were those that engaged ideas with practice, and they did not involve internally reforming the federal agencies or replacing them with market forces. Curiously, both were catalyzed by events in Florida.

* * *

The 1985 Palm Coast fire wrought the worst single-day's damage by fire in the state's history. There had been larger and more intense burns, but none, until now, had blown through the fast-sprouting suburbs that were the lifeblood of the Florida economy. The disaster demonstrated that such fires were not just a California eccentricity. The FDOF had already converted to prescribed fire, if only as an essential auxiliary to fire protection. But landscapes that scrambled woods and wooden houses into an ecological omelet defied traditional methods and were in some respects not even a problem of wildland agencies, and hence they concerned government only as a matter of public safety.

What made the wildland-urban interface fire so problematic was that it did not fall tidily into any single bailiwick or even into an interagency consortium of fire agencies. The wildland fire community would have to reach well beyond itself. In 1986 the Forest Service did what it had so often done in the past and no other entity could: it provided the sinews to join otherwise disconnected parts of the American fire community. This time it joined with the National Fire Protection Association and the U.S. Fire Administration to host a conference on the theme of "Wildfire Strikes Home!" That experience led to an initiative called Firewise.

The NFPA was as old as the Forest Service, established as a voluntary society in 1896, a year before the 1897 Organic Act for governing the national forests. It sponsored research, training, and outreach, but it was most notable for its massive exercise in creating model codes and standards. From time to time it boosted initiatives on such concerns as smoke detectors and sprinklers. Critically, it had long ties with urban fire services and civil engineers. It was ideally situated to formulate codes for the interface, as it had earlier done for rural fire matters and volunteer departments. It could broker between the private sector and the fire community. It was accustomed to both fire service truculence and public indifference. It complemented the wildland perspective by understanding the urban half of the equation. It could do what a land-management agency such as the Forest Service could not. For the WUI it would spearhead the promotional campaign while the Forest Service underwrote research.[39]

Interface wildfires propagated like rats. By the end of the decade, the wildland-urban interface fire was replacing wilderness fire as the informing issue of the wildland fire community. That meant that newcomers to wildland fire such as the U.S. Fire Administration and the NFPA had to be in fire camp, at least politically. By the 21st century the Fire Administration had joined BIFC, and the Firewise program was close to the center of a National Fire Plan.

* * *

More striking perhaps was the emergence of The Nature Conservancy as a fire agency. What drove it into fire was its acquisition of prairie in the upper Midwest. The only way to sustain prairie was to burn it. In 1962 it conducted its first burn, done like the rest of its land stewardship by local volunteers. In 1970 Katharine Ordway began donating money to buy tallgrass prairie on what seemed an industrial scale, first in Minnesota and then throughout the Great Plains. As it burned more, the Conservancy, which had no internal fire culture, came into contact with those who did. In Wisconsin it met an academic clique interested in restoration and accustomed to work on small plots of oak prairie savanna. As it moved into the Flint Hills of the central Plains, it encountered a stubborn culture of burning that dated back to settlement, when farmers found the hills too rocky for plows and abandoned it to ranchers, who burned to boost forage and protect against wildfires. The Nature Conservancy became, by default, a fire agency. By 1985 its involvement had advanced sufficiently for Mark Heitlinger to write a draft manual and, along with Al Steuter, to conduct a training session at Niobrara Preserve in Nebraska.[40]

What made TNC different was its emphasis on ecological burning. What put the program on steroids, however, was land acquisition in the Southeast, particularly in Florida. No landscape burning is simple or inherently safe (campfires escape all the time). But tallgrass prairie is relatively homogeneous; burning the same sites over and over builds experience that leads to competence and confidence; and most of the larger prairie sites abut other lands of similar composition that can accommodate some spillover flames. None of that applied to Florida. The Conservancy had to burn on a scale and under conditions that exceeded what it could expect from local volunteers, and Florida marinated the organization in a regional culture of high-octane prescribed fire very different from that in the Midwest.

The Southeast office decided it needed a dedicated fire staffer. In 1986 it hired Ron Myers, a University of Florida PhD with fire experience then working at the Archbold Biological Station. In 1988 he became its national director of Fire Management and Research. He anchored the program in Florida at Tall Timbers Research Station. He hired Paula Seamon, and together—in Seamon's words—they "grew" a national program. Because TNC did not have a resident fire ecologist, they had to evaluate all its sites, educate local stewards in how to write fire plans, and then approve the results. They created a teaching cadre of experienced burners, mostly with extensive mentoring chapter by chapter for all those sites known to need fire and for those that needed fire but did not yet know it. The program was to encompass not only TNC holdings but all those lands and partners with whom the Conservancy had agreements, had arranged easements, or undertook mutual operations.[41]

These became a serious commitment as TNC rapidly expanded. It had completed a Natural Heritage Network for all 50 states in 1974, launched an International Conservation Program in 1980, and, in the same year Myers assumed the directorship for fire management, signed an agreement to help manage the 25 million acres held by the Department of Defense. A year later TNC commenced its Parks in Peril program, targeting Central and South America and the Caribbean, and acquired the Barnard Ranch in Oklahoma that became Tallgrass Prairie Preserve. Fire management figured everywhere—on prairies, scrub pine, barrens, and wetlands, on land held until it could be transferred to public stewardship, on land under conservation easement with neighbors, and on land ripped and burned by Abrams tanks on maneuvers. In 1991 the Conservancy inaugurated its Last Great Places initiative. That year Myers wrote its first full fire-management manual.

TNC had its own version of the two cultures that faced one another across a wildland-urban divide. One fire culture was a tradition of prairie burning maintained by ranchers and prairie enthusiasts. While those chapters burned with the earnestness that seemed endemic to TNC enterprises, the practice had a quality of studied relaxation, not leisurely but measured, like the rolling hills and homogeneous grasses in which it occurred. The

Conservancy absorbed the second tradition from its forced immersion into Florida's fire scene. Burning there was concentrated, was more and more bounded, and was unforgiving. It was a domain that favored toughened cadres for whom burning was something done year-round, not just part of an annual ritual of spring cleaning. Once the program matured, Florida's TNC crews could join any fire brigade anywhere.

The Conservancy was unique: it was a private landowner with a public purpose. It showed private landowners why they needed to burn and demonstrated how to burn with care and sensitivity to ecological goods and services. And it showed public agencies how nimbleness and skill could make up for hordes of personnel and big-number budgets. By owning its land it could avoid the controversies over land use that paralyzed the federal agencies. Its Florida rite of passage put it on the national fire scene. In a fire world rapidly splintering into hostile tribes, it was regarded as an honest broker and became an ideal collaborator.

Reforming a Fire Planet

In the early 20th century, Coert duBois, regional forester for California and inventor of systematic fire protection, declared that backcountry firefighting was an American innovation. It was not, but after two world wars and a depression, the global network for forestry and fire, as for trade generally, had broken. The postwar era witnessed further collapse as state-sponsored forestry imploded along with its sponsor, the European imperium. What happened in America happened across the world; new institutions struggled to replace the old order. What emerged found the International Union of Forest Research Organizations (IUFRO; originally established in 1892) as a medium for researchers and the United Nations, through its Food and Agriculture Organization (FAO), a medium for application and research, a more benign commonwealth of influence.

The United States became the colossus among the world's firepowers. In 1954 S. B. Show, duBois's successor in California, wrote a general manual for the FAO on fire control. Then fire became caught in the general vortex of the Cold War, first as research and then as formal alliances, foreign aid, and disaster assistance. As part of the 1978 reform package, Congress added international forestry to the Forest Service's charter. By the time the Cold War was shutting down, the American fire community found itself eddying toward globalization. But then Americans had always considered their national experiment, however exceptional, as universal.

* * *

Globalization was easiest for scientists, who had a long tradition of transnational collaboration, from the Transits of Venus to the International Geophysical Year, and for whom international conferences were often a way of life.

With funding after the Cuban missile crisis, a Technical Cooperation Program was established among the United Kingdom, Canada, the United States, and later Australia, to research the thermal effects of nuclear weapons. It did internationally what the NAS-NRC Committee on Fire Research had earlier achieved. More awkward was the attempt to meld American and Canadian fire-danger rating systems. The Canadians had evolved a very different tradition and were skeptical about the assumptions underwriting the American approach, and they were wary lest they might become mere hewers and drawers of data. The doyen of Canadian fire science, C. E. Van Wagner, alluded to the perhaps unbridgeable distinctions with a title from Hugh MacLennan's novel, *Two Solitudes*. Cooperation there would be, another variant of the exuberant trade across the border, but good fences, the Canadians decided, made good neighbors.[42]

Meanwhile, colleagues exchanged ideas in scientific journals and meetings. Institutions such as IUFRO and FAO furnished a sufficient context for foresters, and SCOPE provided a setting for fire ecologists. Almost from their onset the Tall Timbers fire ecology conferences had taken Earth as their dominion. Successor conferences increasingly saw fire on a planetary scale. In 1988 FAO published *International Fire Management News*; that year giant fires in Yellowstone made headlines around the world. The next year, highlighting a decade in which "wildland fires caused major loss of life, property, and natural resources in Africa, North and South America, China, the Mediterranean, Australia, and parts of Europe," the USFS, the National Fire Protection Association, Forestry Canada, the Mexican Secretariade Agricultura y Recursos Hydraulicos, the National Association of State Foresters, and the U.S. Agency for International Development (USAID) organized the first of a series of international conferences on wildland fire.[43]

To exchange field knowledge and technology across borders, the fire community turned to FAO, which midwifed study tours and regional forestry commissions. In 1951 FAO had sponsored two Australians to visit the United States, and in 1964 FAO and USAID sent five. The United States did not reciprocate until 1970, but then the two countries began formal study tours on a roughly five-year cycle. This took Americans to Australia in 1983 and brought a party of Australians to America in 1987. Like most globalization of the era, another 10 to 15 years passed before the full consequences would be realized.[44]

The more consistent FAO contribution was the establishment, in 1960, of a North American Forestry Commission (NAFC) for Canada, Mexico, and the United States. From the onset—it met first in 1962—the NAFC included an active fire working group. The group produced a newsletter, translated specialty terms into a glossary for Mexico, and met routinely to exchange information. It sponsored tours beginning in 1968 and followed in 1975 and 1980 (which focused on prescribed fire). During annual meetings the host country would include a short field trip. Mostly, the association trafficked in ideas and built relationships, all of which contributed to building capacity, particularly in Mexico. Still, the group was notorious for meeting rather than doing. The payoff came outside the NAFC proper and was delayed until the late 1990s.[45]

The doing came through direct aid in response to requests. In September 1963 Brazil requested assistance with a five-million-acre fire bust in Parana. As part of the USAID response, the Forest Service dispatched four fire advisers and training aids including films. The experience was deemed a success. When drought and incendiarism threatened the Dominican Republic in April 1965, the United States was again asked to help and sent Merle Lowden, director of Fire Control. Within days he went from the firefight he anticipated to a "full-blown revolution" in which he would "witness strafing, bombing, and shooting incidents, be evacuated from the beach by helicopter, spend five days on five ships, and see American Marine and Navy personnel in action" before returning to his Washington desk. Such was the fog of a fire war.[46]

In the early 1980s the USFS had dispatched fire advisers to the Galapagos Islands and Costa Rica, organized a Spanish-language fire-suppression course for Latin America, and sent teams to the Mexico City earthquake in 1985. Recognizing the need to formalize such requests, in August 1985 USAID and the Department of Agriculture signed an agreement that established a Disaster Assistance Support Program (DASP) within the International Forestry staff group of the Forest Service. Over the next two years a small team from the Foreign Disaster Assistance Office and Fire and Aviation Management (FAM) crafted and tested an "international emergency assistance response process." This looked a lot like traditional mutual-aid pacts, and it developed out of the same templates. More than just fire crises, disasters often needed help with aircraft, communications, and logistics, all of which FAM had deep experience with.

What made the organizational leap possible was the maturation of the National Interagency Incident Management System. ICS could be adapted to any crisis, and with the 1985 agreement it went global. Disaster Assistance teams traveled to Chile, Venezuela, Ghana, and Mexico for fire training; they went to Argentina, Guatemala, and China for fire control. They went to West Africa to help control locusts.[47]

* * *

Closer to home the national borders began to blur. Well before the preliminary negotiations in 1986 that led eventually to the North American Free Trade Agreement, a continental bloc for fire was congealing. The NAFC furnished a useful venue for meetings, but expressions of comity during banquet toasts and exchanges of technical information among panelists meant little until people, hardware, and money could flow freely across borders.

In 1968 the United States and Mexico had signed a mutual-assistance Wildfire Protection Agreement. Updated annually, it appealed for "coordinated action" on wildfires that occurred along the border. A separate agreement to promote scientific and technical cooperation was signed in 1972. In 1983 efforts to enhance capacity began with schooling for Mexicans (and other Latin Americans) at the National Advanced Resources Training Center, which led to Spanish-language translations and subsequent transfers of training programs. A general memorandum of agreement the next year between the USFS and its Mexican counterpart, Subsecretaria Forestal, enlarged exchanges beyond fire (and was promptly amended in 1985). Other agreements expanded with USAID assistance. A major recommitment followed the horrific 1998 fires that swept through the Valley of Mexico, along with much of the rest of the country, and flooded Texas with smoke.[48]

The Canadian border was both easier and trickier. It was easier because exchanges had a long history, and the two countries were both firepowers. It was trickier because each nation organized its fire system on different principles. The process began with the Northeastern States Fire Protection Compact that emerged from the ashes of the 1947 fires and allowed the regional states (which had little federal land or presence) to form a consortium among themselves. A provision permitted New Brunswick and Quebec, with the approval of both national governments, to join as well. New Brunswick did in 1952, Quebec in 1970. Over time, other regions organized similar compacts; Ontario and Manitoba signed on to the Great Lakes Forest Fire Protection Compact.

As with most matters Canadian, such an agreement hinged on federal-provincial relations. While several provinces—British Columbia and Alberta, most notably—had joined American fire research councils, this mattered little. The sticking point concerned mutual-assistance agreements that obligated governments to certain actions. When provinces began joining the Lake States Fire Protection Compact, the U.S. State Department in 1975 "perceived a need" for a more robust and comprehensive agreement between national governments, particularly as Canadian agencies began to tap into resources at BIFC. The U.S. proposal, however, seemed too "centralist" for Canada because it rendered the Government of Canada responsible for what the Government of Canada considered the responsibility of the provinces. The U.S. State Department wanted an agreement between sovereign states; the Canadian Department of External Affairs wanted the provinces and territories to request aid directly as, in effect, delegated agents. Ottawa could not speak for Canada as Washington, DC, could for the United States, and the provinces did not want any national agreement that might compromise their ability to cut deals with the states across the border. In the end, the United States accepted in principle that requests for aid could come from any Canadian institution directly, but that left Canada deciding what such institutions should be.[49]

What finally allowed agreement was the creation of the Canadian Interagency Forest Fire Centre (CIFFC), which provided a cognate institution to BIFC, but it was equally true that the prospect of the U.S. agreement was the prime catalyst for CIFFC. Requests for assistance from one country to the other could now go through equivalent national institutions by formal channels and with mechanisms for compensation. While provinces might initiate a request, CIFFC would coordinate. What drove the interprovincial alliance was the realization that under the treaty, BIFC could decide which Canadian requests to honor in what priority. BIFC could "be the collector and disseminator of information and intelligence on the status of the Canadian forest fire situation"—it would become for fire protection what NORAD was for air defense. The obvious solution was to fashion a Canadian version of BIFC, but as a corporation and not an institution of government. This arrangement satisfied all parties. The Canada/United States Forest Fire Fighting Assistance Agreement was ratified on May 7, 1982, through an exchange of diplomatic notes; CIFFC went operational two months later. During the 1988 fire season, Canadian fire agencies sent pumps, crews, and air tankers south.

By then an international fire community had taken root, complete with formal association, journal, and conferences. What had long been the core and integrating focus of fire programs, the nation-state, was being challenged both within and without. Within, the United States found its national organ, the Forest Service, having to renegotiate relationships among other federal agencies, between it and the states, and with the private sector and NGOs, even as all of them spoke directly with counterparts elsewhere in the world. From without, the politics of fire had joined a multipolar globe; the world no longer looked unquestioningly to the United States for guidance. Australia had its own national strategy, grounded (in principle) in prescribed fire; Canada had its own fire-danger rating systems and fire-behavior models; and emerging countries could look for advice from FAO, various consortia, or consultancies. For the United States the USFS was no longer the sole source supplier for an American presence. The Interior Department, the states, forestry societies, even TNC established offshore operations. In negotiating an agreement with Canada, the State Department had sought a simple instrument between two states or state-sponsored organs. By the 1990s it was harder to find such an entity.

The nuclear winter alarm of 1982 had proposed that fires anywhere could, through the atmosphere, affect lands everywhere. The instantaneous crisis imagined by its scenarios of thermonuclear war did not happen, and research suggested it could not happen. But like their models in which one climatic state suddenly switches to another, or the Gulf Stream abruptly changes course, the fire and climate community flipped from nuclear winter to greenhouse summer and made global warming the new threat. Instead of turning the earth into an icebox, it would become a Crock-Pot.

The specter was doubly unnerving because fires would not only contribute to climate change but blossom because of it. Whether or not fire ignored borders, its effluent did not. With humanity as its unreliable—maybe rogue—broker, nature's economy of fire globalized.

Lost Decade

For the Forest Service and fire, the 1980s were not merely lost but angry. The political stresses that had pushed and pulled the agency intensified rather than eased; what was slowly healing had its scabs torn off and bled anew. Interior Secretary James Watt set the tone when he denounced environmentalists as "extremists," declared that the Reagan administration would "reverse twenty-five years of bad resource management," opened the floodgates for logging and mining, and drove opponents into the courts if they wanted to compel the administration to enforce existing laws. In the end not only were the Reaganites unable to repeal a single major piece of environmental legislation, but Reagan left office, in the words of Stewart Udall, as "the only president this century who served his term without proposing any major initiative to further the cause of conservation." Even ardent acolytes such as Anne Burford had to concede their conservative hero was a conservationist zero.[50]

Unable to pass the legislation it wanted or to repeal what it disliked, the Reagan administration turned instead to the budget. It slashed taxes, always popular with voters, but was unable to reduce spending, which it shifted away from civilian agencies to the military and seniors' entitlements. In 1985 and 1987 the Gramm-Rudman acts tried unsuccessfully to mandate congressional cost controls, although budgets did decline. Instead, the outcome replaced tax and spend with borrow and spend. OMB Director David Stockman foresaw "red ink as far as the eye can see."

For all the fire agencies, the outcomes were mixed. Where fires led to disasters, agencies did well, paradoxically granted monies to correct politically embarrassing deficiencies. When they tried to methodically translate revised manuals into the field, they struggled. The Forest Service once again found its fire program hostage to other environmental and social issues. To counter those public lands that wilderness partitioned off, the administration sought to deepen harvesting elsewhere, particularly in the Pacific Northwest. It intended to boost timber from 13 to 20 billion board feet a year. That ambition (never satisfied) led to the logging of old-growth and formerly remote sites, many of which failed to earn back the money invested in infrastructure such as roads (though those roads usefully precluded a future wilderness designation). Instead, the timber that appeared to make the Forest Service a net revenue producer actually led to losses, and when the housing market collapsed, the initiative ended with a giant timber-sale bailout of the industry. All this aroused fevered opposition among environmental groups. The Sagebrush Rebellion morphed into the Wise Use movement. Deep ecology clashed with free-market ideology. NGOs and critics sprang up to question whether the agency was even viable any longer.[51]

By the end of the decade the Forest Service itself was internalizing that Manichaean conflict and faced upheaval, if not outright rebellion, within its ranks. Total staffing shrank 25 percent over the decade. While the administration at first resisted the consent decree that compelled a rapid, radical shift in new hiring toward women and minorities, it then capitulated. A turn toward privatization further whittled away staff, while in yet another paradox, it spawned a fire-industrial complex that competed for resources. Even chief foresters no longer came from forestry. Chief Max

Peterson (1979–87) began as an engineer. The fire workforce lost what had always been an adamantine strength: field knowledge learned through years on the line and an appreciation for fire that was part of the atmosphere the agency breathed.

The Reagan recession applied to fires as much as to the economy. An administration that wanted to be convinced that wildland fire was no longer a problem found evidence in the quarterly statements of area (not) burned. Its indifference, however, led to irony. In the end, shrinking appropriations and a dwindling workforce only destabilized the fire program, nearly wrecked the fire research that underwrote policy, and turned over government spending to its most uncontrollable elements, notably emergency suppression. There was not enough money and personnel to do prescribed fire right, yet bottomless funds were there to fight wildfires. In fire, as in other matters, deficit spending became the new normal.

The Forest Service lost control over its narrative. It no longer wrote the larger story. Like fire it instead absorbed the tensions around it. The land, the agency, the politics, even the fires, all polarized. Nothing burned or everything burned. The land was left alone or logged off. The Forest Service slid from an informing presence to a seemingly inept one. It lost the capacity to get ahead of events and became mired in merely reacting.

* * *

The Forest Service did what bureaucracies instinctively do. It organized committees, reorganized programs, and sought to resolve ideological conflicts with planning, in this case as mandated by the 1974 Renewable Resources Planning Act and 1976 National Forest Management Act. Quixotically, it sought to mediate without politics what was at heart a political quarrel. The Pacific Northwest forest management plan of 1984, for example, heightened rather than eased the political stresses over old-growth forests. As Daniel Sarewitz has observed, science could actually make environmental controversies worse because political controversies "with technical underpinnings are not resolved by technical means." Since fire management had bonded to land management, the continued delays in forest plans meant the deferral of fire plans and hence of fire practice.[52]

The surge of fire reforms carried through the early 1980s. The 1978 policy had designated 1983 as the time for all forests to have new plans installed. To assist, Focus was updated. In 1980 Richard Barney and David Aldrich published a review and commentary on "Land Management: Fire Management Policies, Directives, and Guides in the National Forest System," a systematics for the bewildering menagerie of legislation and edicts from the National Environmental Policy Act to the Forest Service *Manual*. A National Fire Policy Workbook, issued in March 1982, translated the official language of the *Manual* into the vernacular of forest plans, and recognizing that full-bodied forest plans might not be ready, it explained the use of "fire management area" plans as an interim measure. Already, however, the deadline was moved to 1985. In what was becoming a minor industry, decision-support programs proliferated to take up the slack. Publications explained procedures to integrate fire management with land-management planning and to analyze fires that escaped initial attack. Without the 10 a.m. policy, fire officers had to choose among possible options. They wanted, at a minimum, a protocol to guide the process. When Forplan became the standard medium for forest plans, fire entered the mix.[53]

In reality, few forests had full-spectrum plans, and fire management rarely meshed gears as programmers wished. It was possible to budget roughly for prescribed fire, but not for natural fire, and by definition, certainly not for wildfire. Plans made a "normal" fire year the basis for calculations, but what rendered fire so intractable was that the normal year was not the problem. The abnormal year accounted for most acreage burned, which was also an index of the biological work done. The fallback position was emergency funding. In the early years after the 1978 rechartering, the budget more or less worked as planned. But when the climatic tides turned, fires blew budgets and ideologies away.

The hoped-for compromise was prescribed fire, which, since its conception 40 years before, had sought a middle ground between excluding fire and letting it run at will. Under ideal circumstances it could replace wildfire with tame fire, and a fire agency could claim it was controlling the fires it was charged with. A lot of prescribed fire targeted fuel reduction as its primary goal, which meant it complemented rather than competed with suppression. Even the PNF flourished, after a fashion; by the mid-1980s the Forest Service had more units zoned for natural fire and more such fires than the Park Service. The problem was complexity and cost. Each failure added more checklists, shrank the opportunity aperture for ignition, and escalated overall expense. Just as the decade was confirming prescribed fire as a treatment of choice, agency budgets were shrinking, inspiring further caution, which added to expenses and transactional costs.

Craig Chandler, then director of fire research, noted that prescribed fire, under the rubric of light burning, had been tried and failed in California early in the 20th century. Why, he asked, "when it was so spectacularly successful in the South?" He gave three reasons: a Mediterranean monsoonal climate, the presence of heavy fuels, and steep topography. "In the East, 2 weeks without rain is a drought. In the West, 3 months without rain is commonplace.... In the East, fuels are small in size and burn quickly.... In the West, logs and stumps don't rot away." In the Southeast, he noted, "and in fact everywhere in the world where widespread prescribed burning is an accepted and successful practice, the land is flat," which means that "sun and wind affect every acre alike, and the fuels dry uniformly." By contrast, prescribed burning "in mountain country is a tricky and expensive business and can only be done successfully by biting off little pieces at a time when conditions are right." He should have added that southeastern landscapes had a relatively uniform culture of burning that promoted fire and tolerated slopovers; the West did not. The South had slack in the system. The West had little tolerance for escapes either on the land or within agencies. Each breakdown led to further constraints, which not only complicated burning but also added to expenses, all of which had to be budgeted.[54]

In the past, emergency funds had kept fire operations afloat. But the Forest Service lost those presuppression monies in 1978, and without bad fire years it could not defend against the steady strangulation of programmed monies and workforce. When actually fighting fire, agency salaries, equipment, and even purchases were transferred from budgeted accounts to emergency suppression. In a big year those savings were significant, and in the way of perverse incentives they rewarded agencies that had large, costly fires. So when fires returned in 1985, all the agencies threw everything they had at them. The cycle continued through 1988. Emergencies drove the program, which meant suppression reestablished its primacy. It did for fire management what timber had done for the Forest Service. It paid the bills.

* * *

In 1986 the Fire and Aviation Management staff was transferred to State and Private Forestry. The intention was to smooth dealings with the states within the context of a total mobility program that now swept across the country. The outcome made little difference in the field. The reorganization that mattered hugely was an upheaval in Forest Fire and Atmospheric Science Research.

As the 1980s began, fire research might have echoed a famous anecdote about the physicist Ernest Rutherford. "Lucky fellow Rutherford, always riding the wave." To which Rutherford replied, "Well, I made the wave, didn't I?" So, too, fire research had ridden a crest it had helped set in motion by underwriting policy reform through information, fire-behavior and fire-effects models, and decision-support tools. It could not substitute for politics, but it could erect the knowledge infrastructure for reform. In 1983 Richard Rothermel consolidated advances in his eponymous model and applied them to field conditions with *How to Predict the Spread and Intensity of Forest and Range Fires*. In 1986 Patricia Andrews oversaw an effort to code those graphs into personal computers with Behave. If knowledge was power, Behave gave power to the people. As nothing else could, it represented to fire practitioners the inimitable value of fire science.[55]

But research was vulnerable in ways few might have foreseen. In the early 1970s its budget had spiked both in real terms and as a percentage of USFS research overall. Then it declined, first rapidly as the bubble deflated and steadily thereafter. By 1978 there had been some increase in real dollars beyond the level it had enjoyed when the fire labs were dedicated. In 1981 that, too, reversed, with no letup in sight. There was no private sector rise to compensate, because fire science was funded by the federal government, and a reduction in funding shrank the enterprise, whether it was done by private or public institutions.

Forest Service meteorology and fire science had merged into a single program (Forest Fire and Atmospheric Science), much as aviation and fire management had. Climate research had swollen in response to inquiries into acid deposition, but those funds were drying up. Fire science, too, lost an old pipeline from the Pentagon and the Office of Civil Defense, which was renamed FEMA in 1979 and removed from the Department of Defense. Then Craig Chandler, the primary conduit for Defense funds, retired as director. To the extent that national interest existed in fire, it was the prospect of a nuclear winter, announced in 1982. Unlike the Cuban missile crisis 20 years before, this new outburst of research funding through the Defense Nuclear Agency did not flow into the Forest Service. In fact, instead of DOD money supplementing fire research, funding effectively went the other way. While fire science had always boomed and busted, now the cumulative curve of crests and troughs was sinking. Research received revenue neither from timber nor from fire suppression. It had no other outside contributors; it began a vicious cycle, or what to partisans seemed a death spiral.[56]

Like a body in shock that pools blood away from the extremities to the central organs, FFASR sought to shed outlier projects and consolidate to core programs. In 1984 Charles Philpot convened a User Needs/Research Planning Workshop to explore how to do more with less and how to get research to practitioners through technology transfer. The next year an evaluation of the program at the Regional Foresters and Station Directors meeting sketched options. The East Lansing fire unit (which served in lieu of a lab) was closed; the Macon lab was dissolved, sending some parts to Athens to form an Eastern Fire Management project. Most research was concentrated at Riverside and Missoula, "as the last bastions of fire research with final retrenchment to Missoula if necessary." FFASR's new director Charlie Philpot had little choice. The FFASR staff understandably reacted with disbelief, dismay, and outrage as the particulars were announced. Few researchers were unaffected, because some in Missoula and Riverside would be transferred to tighten projects, and even fewer, especially those with families and long ties, welcomed a forced relocation. A Michigan congressman prevented the closures at East Lansing. Southerners rallied their politicians to protest so that Macon did not close completely. But the cancer was only put in remission, and the lab succumbed a decade later. For those who suffered through the trauma, the scars endured.[57]

The crisis was doubly devious because the Forest Service still exercised a virtual monopoly over wildland fire science. It conducted the only fundamental research, and as one acerbic critic objected, "Cuts in fire management budgets create new problems where none existed previously." Spare as the western program might be, it was richer than that in the South and opulent by the standards of the impoverished northeast. Under the reorganization much of the country would have no fire science at all. In the end fire research would lack a critical mass; not only did the USFS dominate American research, there was little elsewhere outside of Canada. Just as fire went global, as new responsibilities were added to the agencies, notably the USFS, its funding to do them shriveled. The only solution was to increase revenue, and in order to get new funds fire had to be associated with a political problem.[58]

By December 1986, the FFASR had a new director, William Sommers. It had identified 10 initiatives, though none advanced in the face of continued budgetary erosions, and only earmarks by Democratic congressmen kept parts on life support. High-publicity fires in California and Yellowstone set the pot seasonally aboil. Then the Bush administration "ushered in a dramatic change in environmental leadership," in Sommers's words, that allowed fire research to recover. Slowly, more science-friendly administrations pulled the patient out of intensive care.[59]

* * *

The big paradox of the era was the durability and dominance of fire control. While eliminating suppression generally had never been a goal, the expectation was that natural burning would replace the need to fight fire in wild and remote settings, and prescribed burning would do the same in working landscapes. Better prevention programs would knock down the number of fires. Presuppression projects for fuel reduction would dampen the power of wildfire. Suppression would become one of several possible first responses: it would be an emergency backup, not a default setting. Yet when fire busts returned in the mid-1980s, suppression flourished. In this way the internal shift in fire budgets mirrored that larger funding shift from civilian to military accounts.

The reasons are simple, and in retrospect obvious. Wildfires would continue, and for firefighting to succeed against them, agencies needed a core infrastructure and critical mass of personnel, equipment, and know-how. If a natural fire blew out of prescription, suppression had to put it back in its box. So with prescribed burning and its restorative value, suppression helped to justify the process by calling the burns fuel reduction. Suppression still paid the bills. If wildfires did not happen, programmed budgets would be strained because the emergency funding that came from firefighting would dry up. Total mobility added to suppression's capabilities without adding to its up-front costs.

To the new sensibilities, suppression adapted. It explored minimum-impact tactics. It continued, selectively, to redefine suppression along the lines laid down in the 1978 policy. They could control in the traditional way or contain along its perimeter or confine. In practice a confined fire might be indistinguishable from a prescribed natural fire. The differences were conceptual and (especially) budgetary. A fire confined was still technically a suppressed fire and hence could tap emergency accounts for the costs of observation or monitoring. A PNF was, as its name indicated, a prescribed burn, which had to be budgeted like any other project. By becoming more pluralistic in this way, suppression was able to hold its ground against rival practices. It kept an autonomous funding source, which allowed it to evade the shrinking budgets that hammered other fire programs. As money become more scarce, suppression filled the void.

The plot turn is that big money was not just suppression's salvation but also its failure. Big fires in the backcountry cost big dollars; behind the 10 a.m. policy had been not just an instinctive pyrophobia but a resolve to rein in those emergency outlays. The 1978 policy reforms had also sought to transfer emergency monies to budgeted accounts. But as the revolution's replacement practices stumbled or tripped over the budgetary obstacles littered by the Reagan administration, all of fire management depended more and more on those emergency outlays. The return of big fire years revived the whole fire establishment by putting fire before the public and its representatives.

* * *

The fire community found itself between two poles, one of which wanted it to leave the wilderness and the other of which wanted it to leave the wildlands for the city. Yet the point of the fire revolution was to match lands with fire regimes. In reality, the big story was the dog that did not bark—that is, the fire that was not lit. The lost decade was a decade of missing fires.

By the end of the 1980s virtually every index showed that there was significantly less fire on the land than a century before. Some of that historic burning had been abusive and ruinous. But good fires had been lost along with bad ones. Once they vanished altogether, a fire deficit built up, combustibles stockpiled, and biotas deteriorated. By the late 1980s these effects became noticeable, along with the first alarms over global warming. During the 1980s the experiments of the 1970s should have scaled up at least one, or better two, orders of magnitude. Instead, with a handful of exceptions, they stalled.

A subdued discourse emerged about the origins of those missing fires. The prevailing thesis was that they represented fires that nature had set and people unwisely suppressed. If so, the solution was clearly to pull back from suppression and let fire recover its lost ground. The counterthesis was that most of the unburned acreage resulted from fires people had historically set but no longer did. In this analysis the solution was to rekindle anthropogenic burning on a landscape scale. In practice, though, neither happened. The fire revolution failed to return fire to the land in anything like the volume required and was not even approaching the traffic needed to keep pace with existing conditions, much less to begin burning through the amassing backlog. To this dialectic there was no synthesis. The prospect for restoring the missing fires required a middle, increasingly missing, landscape.

The fire void spread. What was sucked into the vacuum was wildfire.

* * *

On May 18, 1980, Mount St. Helens erupted on the Gifford Pinchot National Forest in Washington. In addition to blast and ash, electrical storms kindled fires. The Carter administration may have ended its political life with a whimper, but nature concluded it with a bang. Similarly the Reagan administration would end with a giant blowup in the Northern

Rockies, its epicenter at Yellowstone National Park. One lesson was that nature held the trump cards. Another was that emergencies claimed attention, and any program that did not affect large numbers of voters—and the public lands were by definition uninhabited—needed some crisis to keep it in the political arena. Particularly in a time of enforced austerity, emergencies speak and money talks back.

None of the American fire community—not its crews, not its scientists, not its administrators—was driven solely by curiosity or markets. The community existed because of fires; because of the fires people wanted and those they feared, because of the fires nature kindled and those people set, or needed to. Without fires, the community lost its raison d'être. In the early 1980s caution over failed prescribed burns, loss of funding, and the stubborn stall over resolving land use blunted the thrust of the fire revolution's push to replenish the fires nature needed and to exchange tame fire for feral. Those years saw few disaster fires. Then the mandala turned, and the fire tithe that nature demanded and people withheld was extorted by force. In 1985 wildfires returned, and they continued, localized in regions, until 1988. The Reagan years ended with a systemic fire deficit to match its budget deficits.

The 1985 fire bust sparked a postseason, agencywide review, which Chief Peterson circulated throughout the Forest Service. The statistics seemed at the time breathtaking. Across the country 83,000 wildfires had burned 2,975,000 acres; the national forests accounted for 10,000 fires and 633,756 acres, "the most since 1934." At $165 million, emergency expenditures were the highest on record. The season began with a declaration of fire emergency by the governor of Florida and the Palm Coast conflagration. It climaxed in August with large-fire outbreaks throughout the West and Alaska. It ended, in late October, when rains finally pounded California smokes into mud. The nationalization of fire suppression had created a 10-month fire season, a bestiary of exotic wildfires, and a record mobilization.

The December review highlighted successes in preparedness and the new policies, which yielded "appropriately managed prescribed fire programs" and "significantly reduced fire suppression costs" as compared with probable expenditures under the 10 a.m. policy. The callout confirmed the necessity for interagency operations. But there were serious deficiencies as well. Suppression performed "significantly below expected levels" from lack of experience, a result of workforce changes and lean fire years. The 1978 policy was not yet "fully implemented," which lessened opportunities for savings and led to "some lack of understanding and execution" by line officers and staff. Still, it was the wildest season since 1977, and the fire community felt some exhilaration as it threw everything short of the Seventh Fleet at the flames.[60]

The new normal continued into 1986, with slightly more fires (86,000) and fewer acres burned (2.7 million). What struck observers, however, was an uptick in burning intensity and the presence of crown fires in montane forests that had historically known surface fires. It was as though a strain of flu had mutated from a seasonal nuisance into a virulent plague. In 1987 the national scene calmed slightly, with 71,300 fires and 2.48 million acres burned, but that mild quelling was upstaged by a whopping outbreak in Northern California, a veritable eruption, the first of what became known as fire "sieges."

The Fire Siege of '87 rocked 775,000 acres, socked in the Central Valley with cloying smoke for weeks, and set new records for interagency mobilization and costs—19,000 firefighters from 42 states at $160 million. The usual suspects lay behind the outbreak: drought, record temperatures, a prolonged barrage of dry lightning, an overextended regional fire load thanks to fires in Oregon. From August 29 to September 5, lightning kindled 1,244 fires, overwhelming initial attack forces and merging into 23 fire complexes. Four national forests—Trinity, Klamath, Shasta, and Plumas—bore the brunt. Rehabilitation efforts over the next several years consumed another $150 million. The suddenness and breadth of the fire bust led a breathless CDF to call the siege "unique" even for a state long identified with large fires.[61]

The cycle climaxed in 1988, an American *Götterdämmerung*, as big fires swallowed vast swaths of the West, mesmerized the evening news, and, in burning over Old Faithful, instructed the American public in the radical policies and violent potentials that could govern the coming era. Over five million burning acres blew across television screens, op-ed pages, scientific journals, and the memories of the largest fire mobilization in American history. What might be considered a Second Big Blowup set new records for suppression costs. Even as the manifesto of the fire revolution burned across America's flagship national park, fire management itself was, like the federal government, committing to a remilitarization. What should have been a celebration of fire restored had become a renewal of the Cold War between fire wild and fire suppressed. What was loudly proclaimed as a reduction in government expenses ended like the savings and loan scandal, with costly bailouts.[62]

The Reagan revolution had announced its arrival by firing members of the Professional Air Traffic Controllers Organization who went on strike. In a valedictory, nature appeared to fire the administration with a barrage of lightning strikes.

The Big Blowout

In 1988 three separate waves converged to make the perfect firestorm. One was climate. The West had swung into drought. For three years some part of the West had been drained of moisture and burned, the center shifting like a migrating magnetic pole. Each outburst exceeded the previous one, then the Fire Siege of '87 blasted past the "average worst" season on which fire plans were predicated. In the summer of 1988, the epicenter hovered over the Greater Yellowstone Region. The second wave was fiscal, expressed through the serial Gramm-Rudman acts. The federal budget had a monetary deficit to match the West's moisture deficiency; structural debt threatened to match the structural deficit in wildland fuel reduction. The two droughts mapped neatly one on the other, and both nudged fire management away from prescribed burning in the face of complexities and costs.

The third wave was the fire revolution itself. The fabulous long surge of fervor and experimentation that had carried reformation had finally broken. Ideas had become policies, and enthusiasms had morphed into manuals. As the 1988 fire season approached, 20 years had passed since the NPS had inscribed the fire observations of the Leopold Report into the administrative guidelines of the Green Book. No one would ever know everything, or know all they wanted; but they knew enough, had sufficient bureaucratic sanction, and held adequate tools to promote the right kind of fire and prevent the wrong kind. But that transition would take far longer than anyone expected. Restoring fire was not simply a case of understanding fire's ecology or nature's way. It meant reconciling the uneasy realms of nature and culture, each of which had its own momentum and laws. In the summer of 1988, American nature and American society collided with a nearly tectonic fury not seen since the Great Fires of 1910. Out of the flames emerged a new creature in the menagerie of American fire, the charismatic megafire.

The season recalibrated the record books. Nationally, 77,750 fires burned 5,009,290 acres. Nearly half the burns (2.2 million acres) were in Alaska, which left the West with 2.8 million acres, or what it had experienced during the drought bubble. What made the season spectacular was the concentration of large fires in the Greater Yellowstone Area, which bestowed on the flames a cultural resonance that big burns in the Northern Rockies backcountry could rarely claim. In the GYA 248 fires burned 1.4 million acres; add in the Canyon Creek blowup on the Bob Marshall Wilderness, and you have half the total acres burned in the Lower 48 that year. Of those 248 Yellowstone starts, 31 (32 with Canyon Creek) were declared prescribed natural fires before they either died small or merged into 10 fire complexes that accounted for nearly 80 percent of the Greater Yellowstone Area total.[63]

To the fire community, those PNFs became a test of the fire revolution's most dramatic expression. To the public they were simply the Yellowstone fires. When the North Fork fire flashed through Old Faithful on September 7, the Summer of Fire burned into the national memory.

* * *

Yellowstone National Park had not pioneered the fire revolution for the NPS; that had happened in the Sierra parks and Everglades. But its enduring elk crisis had sparked the Leopold Report, which culminated in the 1968 policy reforms. And it was *Yellowstone*: huge in land and symbolism, rich in fire history, and eager to reclaim its prominence. As soon as the NPS allowed, it released a fire plan in 1972 that committed a serious portion of the park (340,784 acres) to a natural-fire program. It revised the plan and prepared an environmental assessment for public review in 1975; they were approved in 1976 by the Rocky Mountain Region. The update expanded the domain for natural fires. By then 19 PNFs had burned 833 acres.[64]

The plan identified some patches—the main tourist and concession sites—for automatic suppression. It crosshatched in a "boundary cooperative area" shared with the surrounding national forests. The rest of the park was a "management area." The plan called for a Fire Committee—consisting of the research biologist, the chief ranger, and a fire management specialist—to evaluate starts and recommend courses of action to the superintendent. To critics the document was a statement of philosophy, not a plan of operations. The park would allow as much natural fire as possible. The document was not revised after NPS-18 was published in 1978.

In 1983 Robert Barbee became superintendent. Barbee's career had flourished at Yosemite, where as resource management specialist he had overseen the buildup of a stellar fire program along the lines of Sequoia-Kings Canyon. Yellowstone already had its own in-house research program. It had created its own fuels map, based on criteria independent of the NFDRS models. It interpreted the Leopold Report, which Barbee considered a "manifesto," as an invitation to manage through "natural regulation," which meant as little interference as possible. Its plan included no prescriptions. It was Yellowstone. It did things its own way.[65]

Not everyone was happy. Horace Albright thundered ruination if the park did not suppress fire, but his time had passed. Alston Chase, an academic philosopher turned author, blasted Yellowstone's policies on natural regulation (especially elk, again) and biocentrism as a perverse concession to "California cosmologists." Published in 1986, his *Playing God at Yellowstone* probably did little to educate the visiting public into environmental philosophy, but it did serve as a lightning rod for bolts of criticism. Even within Yellowstone there was some uneasiness with a document that relied for its operations on the "intuitive judgment" of a small cadre of staffers, and the assistant superintendent requested a formal review by outsiders. The team consisted of Robert Heyder, superintendent of Mesa Verde National Park, David Butts, chief of the branch of fire management, Robert Mutch, FMO from the Lolo National Forest and coauthor of the White Cap experiment, and Bruce Kilgore, formerly of Sequoia-Kings Canyon and then at the Missoula lab. They recommended that the park write a modern plan.[66]

The park stalled, however, until 1985, when the Denver regional office recruited a planner who had revised the Rocky Mountain National Park fire plan after the Ouzel incident to update Yellowstone's step-up plan, which governed its use of presuppression emergency funds, but with the understanding that he would undertake a revision of the fire plan overall. On August 1 he submitted that new plan, modeled strictly on NPS-18 guidelines, along with recommendations for the fire program overall. Those unsolicited observations centered on an appeal to "recharter" the program. Bluntly, Yellowstone could neither fight big wildfires nor monitor natural ones according to modern standards. He urged the park to "assume the burdens—not merely the status—of leadership."[67]

The draft plan was unacceptable. The park rewrote those portions that limited its discretionary actions, which meant it operated under a prescribed natural-fire program without prescriptions or decision-point triggers, which is to say, it had a let-burn program subject to the judgment of its fire committee. This resolve also meant that the park's plan could not align with those of the surrounding forests or wilderness. In May 1987 the park signed off on a highly emended text, restating the draft's premises to reflect its particular preferences and revising operations such as monitoring to ensure maximum discretion. It dismissed, as unnecessary and distasteful, any thought of prescribed burning.

Yellowstone had not requested help, did not think it needed anything other than what it had, and did not believe that anyone outside Yellowstone could understand its fires in any meaningful way. In 1981 the park had accommodated 21,000 acres of natural fire kindled by "matchsticks from heaven." Big fires were natural at Yellowstone. The park was bigger than the fires nature might throw its way.[68]

* * *

There matters stood as the trough of drought migrated across the West. On June 14, lightning kindled a fire in the Absaroka-Beartooth Wilderness just outside the park. Another 247 starts followed through July, a few caused by people, most by lightning. Some snuffed out in rocks or along lakes, others swelled and collapsed as a wave train of cold fronts rumbled through the region. The survivors began to bunch and merge. Of those fires in the Greater Yellowstone Area, 31 began as PNFs, 28 within the park. Sixteen of the park's were eventually reclassified as wildfires, and only one outside the park. On July 21, well past the time when surrounding forests had determined that any new start would be attacked as a de facto wildfire, the park announced it would accept no new natural fires. Burned area stood at 17,000 acres, which seemed a staggering amount, but it was just 2 percent of the area within the fires' final perimeters. On July 27, Secretary of the Interior Donald Hodel toured the scene and confirmed that suppression was now the order of the day.

Those existing fires continued to burn despite the greatest concentration of firefighting power in history—9,500 firefighters mustered within the park alone, including 4,220 military personnel, and crews and aircraft from Canada. They failed to stop the fires. Over and again they cut lines, burned out, lost fires and lines, and started over (only 20 miles of fireline out of 400 held). The failed backfires themselves multiplied the fires' dimensions. On August 20, the 78th anniversary of the Big Blowup, the fires doubled in size to more than 480,000 acres. So mesmerizing was the spectacle that hardly anyone outside the Northern Rockies noticed that the escaped Canyon Creek fire in the Bob Marshall Wilderness on September 6 blasted through 180,000 acres. On September 7, the North Fork fire stormed over the Old Faithful area, one icon illuminating another. The park prepared to evacuate Mammoth Hot Springs. On September 11, snow and rain arrived, and over the next two weeks the fires quelled.[69]

A media firestorm had raced along with the flames. In keeping with the park's ill luck, it was a national election year, and almost no other news drew public attention. The media swarmed like moths to the flames. Superintendent Barbee later reflected that the fires were a "great lesson in the peculiar challenges facing a free press. Some coverage was very good, some was sensational, and much was inaccurate." While few people outside the fire community understood the premises of the fire revolution, everyone knew Yellowstone, and they saw it burning day after day, night after night, week after week on newspapers, in magazines, and on television, a grisly serial drama. At first the media had done what their training and deadlines taught them to do: they placed the fires into existing templates. The fires were a disaster, destroying a priceless national heirloom; the firefight was a heroic battle against overwhelming flames; the fires showed the failure of government or the fires showed the wisdom of the National Park Service. Most western fires lasted one or two news cycles. Yellowstone's went on and on, and then it went global. Much as its convective columns spewed ash downwind, the fires sent images and information around the earth through television, popular print, and technical journals. As the fires lingered, the lessons and language changed to something more nuanced.[70]

But then every actor in the Yellowstone drama had his prepared script, every analyst his formulas, and every critic and apologist his narrative. And they all failed. Fire behavior specialists saw their algorithms go up in smoke; even Dick Rothermel scrapped his model in favor of historical precedent. Fire planners watched the flames vaporize their designated zones and prescriptions. Fire teams "tried everything we knew of or could think of," as one incident commander, Denny Bungarz, put it, "and that fire kicked our ass from one end of the park to the other." Total mobility could not mobilize enough crews and aircraft, and BIFC had to appeal to the Pentagon and Canada. Fire revolutionaries watched their presumption that natural fires would be small, benign, and self-regulating swept away in towering fire whirls and rampaging firestorms. The estimated cost begins at $120 million and goes up.[71]

Everyone could agree that Yellowstone's Summer of Fire was a major moment in American fire history. The superlatives trip over themselves. But hardly anyone could agree on what it meant. Were these transformative fires or just mammoth fires? Did they matter because of what they were or where they were? The 1910 Big Blowup created an enduring narrative. The 1988 Big Blowout looked increasingly like a celebrity event, something known for being well known. The meaning of the flames would depend on what kind of story grew out of their ashes.

* * *

In the aftermath of the burn, reviews, reports, analyses, and postmortems grew thick as fireweed. The Greater Yellowstone Coordinating Committee assembled a report on *Greater Yellowstone Area Fire Situation, 1988*. The secretaries of Interior and Agriculture appointed an Interagency Fire Management Policy Review Team. Norm Christensen was asked to do for Yellowstone what he had done for the controversy at Sequoia-Kings

Canyon and head a Post-Fire Ecological Assessment Team. The Government Accountability Office weighed in with its first-ever study on fire. *Western Wildlands*, the *Journal of Forestry*, *BioScience*, even *Forum for Applied Research and Public Policy* ran special issues on the fires. Tall Timbers renewed its fire ecology conferences on the theme of "high intensity fire in wildlands." The park made the fires the focus of the Second Biennial Conference on the Greater Yellowstone Ecosystem. The first international Wildland Fire Conference used the fires as a thematic centerpiece. And then came the books, a flaming front of pictures and texts followed by swarm on swarm of articles. Many major magazines from *Time* to *Audubon* ran features. The publication and commentary went on and on, not unlike the fires. Within a handful of years enough existed to support a bibliography of fire publications. Twenty years later Tall Timbers hosted a retrospective.[72]

At its core, the postfire vortex debated whether the park had behaved courageously or foolishly and whether the fires were good or bad, and through them whether the fire revolution was shrewd or wrongheaded. But the magnitude of the fires and sheer celebrity of the place, the wild swirl that mixed the pull of telegenic flame with the pall of obscuring smoke, overwhelmed the usual instruments of meaning. It was not clear through which prism the fires might, and ought to, be viewed. When the fires became big enough to attract attention, the media first reached for familiar formulas about disasters and firefights, and the public saw fires ripping through a national heritage site while the National Park Service seemed unable (or perhaps unwilling) to stop them.

That misreading became, for the park's apologists, a point of attack. The journalists were wrong, the public was uninformed, and the whole brouhaha was, at base, the outcome of a poorly handled media campaign. The fires were natural, even those started by people, once their fires had moved into the park, and they burned so ferociously because nature was so desperate for them. This—sweeping crownfires—was how the predominantly lodgepole pine forests of Yellowstone burned; such eruptions were not aberrations but exactly the kind of flashy fire through the canopy that serotinous cones required to release their seeds. The summer of fire restored the old pattern of Yellowstone's indigenous conflagrations. Throughout, the park had followed a wise and necessary policy. The fires had performed ecological alchemy: they were renewing Yellowstone. What made them problematic was an ill-informed media blitz and the unwarranted suppression efforts that left fireline scars and wasted money. The park had behaved boldly and bravely. By autumn the national media had largely converted to this story. A big feature by Peter Mathiessen in the *New York Times*

Magazine argued the "case for burning" and set the standard for progressive journalism. Interestingly, this reversal happened before the official review and before scientific committees had released their findings.[73]

That was for public consumption. Within the fire community, two sentiments prevailed. One was awe at the attention the fire had granted. An enterprise that had known only the occasional notoriety of a disaster story, that was struggling to find money to pay for its routine operations and keep research on life support, had become an international spectacle. The fire community shared the glamour of the fires. The second sense was the need to circle pulaskis and driptorches. The fire revolution, to which nearly all members were committed, was precarious; beyond that, they had endured eight years of an administration openly committed to obstructing environmental reforms that would gladly welcome an excuse to hammer the NPS and other public-land agencies and savage contemporary ecological thinking. There could be no internal dissent, at least none leaked to the public. The consensus narrative was that the fires were not harmful and were probably helpful and that the firefight was monumental and heroic.

Still, there were critics, and they clustered around three theses for why the blowout had occurred—bad ideas, bad execution, and bad faith. All the critics supported fire's restoration—that had been decided a decade or two ago. As longtime USFS fire officer and critic Bill Buck bluntly said, the "rightful role of prescribed natural fire in western forest ecosystems is not at issue (nor should it be)." The question was why, in their minds, the fires went bad, which led to how they were handled, which, again to their minds, was where the discourse should be centered. Revealingly, the *Journal of Forestry* became the primary venue for venting.[74]

The bad ideas group found a spokesmen in Alston Chase, whose 1986 book *Playing God at Yellowstone* had seemingly found a fiery coda, and Thomas Bonnicksen, a forester and former NPS ranger-naturalist, now viewed as a renegade, deeply suspicious of natural regulation, who proudly noted that he had written "the first documented internal Park Service critique of the fire management program." Fire policy had to reflect human judgments and required active intervention, a kind of ecological silviculture. The scene was not natural, as wilderness enthusiasts proclaimed. Biocentricism was not a manual for management. Laissez-faire fire management would lead, in nature's economy, to damaging cycles of boom and bust. He regarded the Policy Review Report as "a masterful coverup." Willingly or not, such voices became the intellectual front for critics such as the Wise Use movement hostile to public lands and government generally, which made it easier for opponents to dismiss them as hired guns.[75]

The bad execution group focused on the prescribed fire that was not done and the suppression that was. Early in the fire revolution Yellowstone had dismissed prescribed burning as both impractical and improper. It violated the doctrine that nature should work as unimpaired and unimpeded as possible; setting fires was no different from shooting elk. Defending the park, Paul Schullery and Don Despain argued that "human intervention [had] not created that situation, and human intervention in the form of prescribed burning could not have prevented it." Besides, surface burning on the Florida model could not work because the Yellowstone plateau lacked the right understory. The only possible fire was a crown fire, perhaps every century or two. But critics pointed out that other kinds of prescribed burning and fuel treatments were possible. More pointedly, they noted that the firefight had gone in reverse. You attack fires when they start, you back off when they are blowing and going. Yellowstone had accepted fires, except the North Fork, which had started outside its border, and then weeks later (under heavy political pressure), announced it would suppress them.[76]

Asked by the Interagency Fire Management Review Team for comments, Bill Buck snarled, "Yes, I view the Yellowstone fire season of '88 as a farcical imitation of the management practices that were available and could have made a difference in what happened, were they employed." The issue was "the deliberate misinterpretation of policy followed by the cavalier defense of misguided actions." The Yellowstone blowout, he concluded, "exemplifies the right job being attempted by the wrong people." Those running the show simply were not qualified by hard years on the line.[77]

The review team, which submitted its report on December 15, made the same observation in more gracious language. Fire policy was sound; execution was flawed. It noted discrepancies between agencies over terminology, protocol for PNFs, budgeting procedures, and planning fundamentals, but it was particularly appalled to discover that the Yellowstone fire plan had no written prescriptions. As one academic member, Ron Wakimoto, summarized, "The team quickly realized that Yellowstone Park's fire management plan was clearly not a good example of national fire policy. A 1986 plan was utilized by Yellowstone Park officials, although it did not have the required approval by the NPS Regional Director nor had it been signed by any USFS official responsible for managing the surrounding national forests. The only signature on it, the Yellowstone Park Superintendent's, was written in by an assistant." Worse, the review team realized that half (16 of 32) of the NPS plans for PNF programs had no prescriptions despite requirements in NPS-18. "No other agency's plan was found without written prescriptions."[78]

The 1988 blowout was only the down payment on what promised to be a serious debt. In fact, it recommended that no prescribed natural fires be allowed under any federal agency until standards were met. The conclusion that "parks with plans like Yellowstone National Park's, which lacks criteria for prescriptions, among other things, will be hard pressed to re-establish programs quickly." Three days after Black Saturday, NPS director William Penn Mott had declared a moratorium on all prescribed burns in the national park system. The Policy Review Team recommended extending that freeze across the agencies, a doctrine of total immobility. That, in fact, was the real ecological effect of the fires. They shut down PNFs and many fire programs until the whole system could reboot. They took fire away from New Mexico, Minnesota, and Oregon. The reality is, fire ecology cannot be limited to nature preserves: it resides equally within social systems. The crown fires cresting through Yellowstone lodgepole also overturned an institutional ecosystem, shifting nutrient cycles of information and rewiring energy flows of money. Outside auditors such as the GAO would evaluate progress or its lack. The fire revolution would have to start all over again.[79]

The charge of the last critics was bad faith. Yellowstone commissioned the Christensen report, then buried it, and Christensen had to rewrite and publish its findings in *BioScience*. The park considered the summer's breakdown as one of public relations rather than of practice. Over the next year it was part of a consortium that spent $718,000 in a public education campaign. The gist of the bad-faith critique was that Yellowstone had not done what it said it was doing. It had exploited an NPS policy that allowed prescribed natural fires but stripped the prescriptions out of them and did not subject the result to public scrutiny. It did not see its fire plan as a social contract between the park and the public that laid down rules by which both sides would play. Convinced of its righteousness, it did what it wanted. Others paid the price.

* * *

The Yellowstone fires became an exercise in pyromancy as defenders and detractors struggled to see meaning in the flames. For years afterward the Big Blowout dominated the American fire community: it just had to be a defining event in American history, as significant for fire history as Yellowstone's creation had been to environmental history. Yet a final reckoning is more subtle and counterintuitive, with paradoxes from the shear of the rising flames against the ambient winds of history spinning off like roll vortices.

The core is how little the fires affected the trajectory of the fire revolution. For the public they announced, on a colossal scale, what had begun 26 years before and had been inscribed as NPS policy for 20 years. Even with outbreaks such as Ouzel, Waterfalls Canyon, and Redwood Mountain, the reformation had proceeded below the radar of public awareness. Now it was undeniable. After the Yellowstone conflagrations, the idea that fire had an ecological role in wildlands became fixed in the national consciousness. Big fires joined hurricanes, floods, and earthquakes as part of an annual almanac of media expectations. Free-burning fire reengaged with American culture. For the fire community, however, the fires declared nothing new and, on the contrary, helped to further limit progress that was already stalled. The fires had repeated the 1978 Ouzel escape on a vastly larger scale. The reviews, both scientific and political, republished the same conclusions: the policy barrel was sound, its execution flawed by a few rotten apples of practice.

The fires tested suppression, prescribed fire, and the assumptions of the fire revolution, and they found all wanting under extreme conditions, but they introduced nothing that had not been discussed endlessly. They could not resolve such matters as how (or even whether) a legacy of fire's exclusion might have contributed, though the numbers suggested that the park had burned in one season what would have burned in gulps since the cavalry had commenced fire patrols. One grumpy critic even likened the season to a potlatch. The most insidious outcomes of the 1988 season were indirect. Like all crises they tended to move and concentrate power upward, relocating Park Service fire funds from the NPS to Interior. Once again, the Park Service lost control over its own program, or at least the capacity to negotiate fire within a larger agenda and budget. Much as the American fire community saw the exurban fringe seize the commanding heights of political attention, so, within the NPS, visitor services and its emergency arm, law enforcement, shifted the agency's attention from the backcountry with its fires to the frontcountry with its swarming visitors.[80]

Many partisans argued that the fires were the beginning of a new era. How could something so extraordinary not spark or at least symbolize equivalent responses after the smoke blew away? That year gargantuan burning in Amazonia reached public awareness; NASA's James Hansen testified to Congress that fires, among other indexes, were the signatures of global warming; there was no going back. But more and more the fires look instead like the closure of an old era. The '88 fire bust had climaxed a drought bubble, much as those of 1977 had; when the bubble burst, no new fires spilled out. Nature recovered from its seizure. When fires renewed, the fire community was inclined to look to the borders of the wild, not to its fiery heart. It turned to the Oakland Hills and the exurbs of Glenwood Springs. Since the advent of the fire revolution, fire restoration and wilderness had fed each other like a self-reinforcing dynamo. The Big Blowout signaled a shift to another informing theme.

The saddest paradox was that, for all the talk—and it seemed that there must be a word or image published for every lodgepole pine needle burned—the American fire community missed an opportunity to advance the discourse of the fire revolution. It defended the park by emphasizing that fire belonged in western wildlands, even in iconic landscapes like Yellowstone. But the issue was not whether fire belonged—every informed observer agreed that it did. The issue was *how* it belonged; by what means, according to which rules, at what costs. Restoring fire was like restoring wolves: the means would decide the validity of the end. This was a review that the fire community needed to have within itself if not with the public.

It did not happen. And that lost opportunity seems an apt capstone to a lost decade. Instead, by repeating the hoary syllogisms about natural fire and wild nature, they sustained the old narrative rather than writing a new one. As Don Despain was fond of saying, the burned forest did not come back after the fires because it had never left. Nothing had changed. The policy review was interchangeable with that from Ouzel. So, when the fire revolution rebuilt years later, Yellowstone found itself at the margins, and Yellowstone's fire story had become a matter of scale rather than uniqueness. Its fires did not lead to new policies, new sciences, new ideas, or new narratives. No Ed Pulaski walked out of the smoke. No agency had the fires transcribed into its DNA. The fires did not introduce a new era but closed one that began on Kennedy Ridge in Sequoia-Kings Canyon. In the end, North Fork, Fan, Hellroaring, Storm Creek, Clover-Mist, Huck, Mink, and Snake were celebrity fires, not great ones.

Night Shift

The fever broke. After the flames of Yellowstone, all parties were exhausted, nature seemingly no less than the agencies. The institutional equivalent of cold sweat dripped off committees, panels, and fire caches. The action paused. Old crews pulled back and new ones shuffled out as the fire revolution entered a night shift.

The rule of thumb in fire ecology is that most of the scene stabilizes within three years after a burn. That seems to hold for the political ecology of American fire after Yellowstone. The scattered tiles of the fire establishment were gathered up. If the fire revolution failed to rally, it at least regrouped. This time, however, the controversies over the use of the national forests were not simply fought externally between the Forest Service and other agencies or NGOs but were internalized within the agency itself, formalizing dissent. This time, too, the agencies were no longer allowed to govern themselves without oversight of their fire mission. The GAO, and later OSHA and others, pushed into their work space. Throughout, the politics of divided government put one foot on an accelerator and the other on a brake. George H. W. Bush announced that he would be the "environmental president" and so calmed the most flagrant provocations. But no new initiatives resulted.

The fire revolution banked its coals, then rekindled. The process had to begin over, not quite from scratch, but near to it. While significant wildfires occurred annually, there were, blissfully, no successor blowout seasons that would overwhelm the ability to think as well as to act. Year by year, committee by committee, agency by agency, idea by idea, the recovery proceeded.

* * *

In 1989 attention riveted on after-action reviews of Yellowstone. The smoke had barely cleared when on September 28, a Fire Management Policy Review Team was appointed. Charles Philpot of the Forest Service and Brad Leonard of Interior served as cochairs. The USFS and NPS each supplied two other members; the FWS, BIA, BLM, and NASF one each, to which were added two academics, and representatives from the NFPA and Western Governors Conference. The committee held public meetings, gathered statistics, conducted interviews, and submitted a draft report on December 15. A final report was published on May 5, 1989, in time for fire season. Until then the moratorium on prescribed fires declared by NPS director Mott on August 23 remained in effect.[81]

The gist of the findings was that federal fire policy, including prescribed natural fire, was appropriate and should be reaffirmed, but that "many" existing plans did not meet policies or contemporary standards. Specifically, they often did not have written prescriptions, did not include guidelines for decisions (and documentation of those decisions), did not account for regional fire loads that would determine whether suppression (if needed) was feasible, and did not factor in off-site effects on communities. There were problems with the public review of plans as required by NEPA. Adjacent agencies did not have plans that would allow them to co-manage fires along their shared boundaries. There was inadequate funding and differences in funding between agencies that translated into difficulties in how they coped with mutual fires. The Department of the Interior needed better oversight of its agencies. Most plans failed to accommodate drought indexes. Few regarded the tricky transition between monitoring and, once a fire escaped, active suppression. "Current fire policy or guidelines are subject to abuse." Such misuses of the PNF could "eliminate the program itself—and lose the benefits that derive from it." A long roster of recommendations boiled down to one: "USDA and USDI agencies will periodically review fire management plans for parks and wilderness for compliance with current policy, direction, and the additional requirements recommended by this report. No prescribed natural fires are to be allowed until fire management plans meet these standards." On June 1, the secretaries of Agriculture and Interior announced that all fires in parks and wilderness would be suppressed until fire management plans that met national standards were approved.[82]

The Policy Review Team was careful not to expand their comments beyond their specific charge, which was to examine "the appropriate fire policies for national parks and wildernesses which address the concerns expressed by citizens and public officials about the management of fires on these lands as a result of the Yellowstone fire situation." The team was especially told not to explore other questions of management that might relate to the fires or of which the fires might be representative. Theirs would not be a Leopold Report.

That ambition, however, was latent in the Christensen Report, which assembled 13 researchers, 11 of them academics, to review the "Ecological Consequences of the 1988 Fires in the Greater Yellowstone Area." The committee considered the value of proposed remedial actions such as reforesting, seeding, and feeding ungulates as well as articulating a research agenda for the future and, more expansively, "any recommendations about alternatives and/or options the agencies may want to consider in future management programs." The committee commenced with a workshop on November 18–20 under the sponsorship of the Greater Yellowstone Coordinating Committee. It delivered an interim report on December 1. A second meeting in Denver held January 13–14, 1989, identified goals for the concluding product. A final report followed. When the sponsors declined to publish it beyond spiral-bound photocopies, Christensen summarized its findings for a special Yellowstone issue of *BioScience* in November 1989.[83]

It was too much to expect the researchers to stay with their science alone. The interim report ended with recommendations for "initiating and sustaining a program of ecosystem research." The digest in *BioScience* ended with thoughts on "management implications," with what fire might mean in parks and wilderness sites, and with what kind of human presence is possible and desirable. Chance would persist, knowledge would be incomplete, yet the report argued for research both academic and practical by concluding that "many unknowns can be reduced to uncertainties, and uncertainties to probabilities."[84]

* * *

For 1989 those two commissioned reports were the big news items for fire management. But behind them other events recapitulated the reforms of the early fire revolution that would, in the coming years, shape the cultural context for fire and the social setting for the agencies that had to make sense of fire in the field. One cluster of developments looked to the environmental movement and the ideas powering it, one to the workforce of the U.S. Forest Service, and one to the welcomed reentry of two institutions of fire's civil society, Tall Timbers and The Nature Conservancy, in the form of revived fire conferences and participation in the national discourse over policy. Together they reconstituted and reanimated the fire revolution.

American environmentalism had climaxed with ANILCA and the spread of wilderness study areas to BLM lands. When the Reagan administration attacked its larger agenda for the public domain by shutting down further wilderness, seeking to limit NEPA reviews, and accelerating logging on prime sites, environmental organizations responded by appealing to the Endangered Species Act. A month after the Policy Review Team released its final report, the FWS listed the northern spotted owl as "threatened." A political uproar followed, climaxing a national debate about saving old-growth forests. The crisis centered on the Forest Service (in the Northwest and Alaska) and the BLM (in Oregon). Congress authorized an elaborate study, headed by wildlife research biologist Jack Ward Thomas of the USFS. A temporary release was arranged, an Interagency Scientific Committee on the northern spotted owl was authorized, and the FWS was given a year to determine final status for the owl.

That same year Jerry Franklin of the Pacific Northwest Experiment Station and colleagues from the Andrews Ecosystem Research Group proposed a concept they called *New Forestry* as an alternative to the increasingly acrimonious polarization between industrial logging and environmentalism;

it was to forestry what prescribed burning was to fire management. They took the idea to foresters through an article in *American Forests*, to the general science community through *BioScience*, and to the literate public with a book, *Maintaining Long-Term Productivity of Pacific Northwest Forest Ecosystems*. The essence was to shift attention from trees to forests, to think less of maximizing immediate commodity production and instead imagine ecosystems as a long-term nurturer of fiber and other forest goods and services. The spotted owl only highlighted a deeper crisis.[85]

The controversy and concepts affected fire management variously. The resolution of the spotted owl imbroglio would shape how the land was managed, which would govern what kinds of fire practices were appropriate. New Foresters viewed fire not as a competitor for timber but as part of what made forests work. Even severe fires were less devastating to ecosystems than the clear-cutting in old growth that had become common to meet timber harvest targets. New Forestry did not see dead wood as just fiber or fuel. It imagined snags as part of a necessary forest structure, not simply as potential ignition sources. It considered woody debris as habitats for enhanced biodiversity, not just caches of combustibles. All were "biological legacies" that helped explain how nature could re-create complex ecosystems after nominal catastrophes. By contrast, the kind of industrial strip-mining of woods characteristic of the contemporary Northwest essentially nuked a landscape. Even the eruption of Mount St. Helens had produced less lasting damage.

Other regional surveys for salmon in the Columbia River Basin and old growth in the Sierra Nevada followed in the slipstream of the owl controversy. Behind them was the imperative to "get out the cut," to lubricate more logging particularly among larger trees (and so help large private landowners who had cut out their own stocks and needed decades to rebuild them), and to reward the Forest Service for meeting those targets and penalize it for falling short. The sharpening of pushes and pulls through the 1980s translated into a hardening of positions between contestants, a battle of attrition in a self-described "timber war," often decided in the courts, between environmental groups and the Forest Service. By 1989 the agency had begun to internalize that fissioning. The cipher was Jeff DeBonis.

DeBonis, a self-described "timber beast," had graduated from Colorado State University, labored as a forest planner and facilitator of large clear-cuts in old growth, and seemed the rational face of the modern alliance of industrial forestry and the Forest Service even as he aggressively pushed a big-tree logging agenda. Gradually, however, he began to question what he had done until "a flash of enlightenment went off in my head," as he put it, and he experienced a road-to-Damascus moment in which he went from persecutor to proselytizer. He wrote an electronic letter about the need for the agency to return to its origins and "re-take the moral 'high ground.'" It was a pre-Internet era, but the Forest Service had an Intranet through its Data General system. DeBonis's missive went viral. He was threatened with punishments until others, both inside and outside the agency, rallied to his cause.[86]

Thanks to funding from the Ford and Rockefeller foundations, he was able to organize the Association of Forest Service Employees for Environmental Ethics. AFSEEE (later FSEEE) became a refuge for USFS staffers who disliked the way the agency had bent to political and industry pressures—those denied free speech rights, those told to break the law as they understood it, those inclined to whistleblowing. FSEEE soon numbered 10,000 members, though not all of those were Forest Service employees. A newsletter, Inner Voice, spread the word outside the Data General network to 100,000 readers. Already hit with affirmative action programs, the consent decree, and the infusion of the "-ologists" (as the growing numbers of scientific advisers were called), the Forest Service workforce was now splintering along ideological grounds, emulating the fractioning of its lands. In 1988 the first Women in Fire Management conference was held in Region 6. The agency had come to better mirror American society, but not in the productive ways proponents had expected.[87]

Nor was it all about the feds. In May 1989, the same month as the release of the Policy Review Team's report, Tall Timbers Research Station and The Nature Conservancy convened the first fire ecology conference since 1974. The intent among the 250 participants was to consider whether it was possible to manage fires in places naturally prone to "high-intensity or catastrophic fires." Not only did the issue matter to places larded with lodgepole or jack pine, public opinion and agency directives based on experiences with such fires would "influence management policies for all fire-dependent habitats in the future." In other words, what happened around Old Faithful could affect what happened around Disney World. If fires were divided between wild and prescribed, then prescribed burns set to improve pasture in central Florida were, by this lopsided reasoning, of a piece with fires left to free-burn across the Yellowstone plateau. The conference hosted the first E. V. Komarek Sr. Fire Ecology Lecture, which Norm Christensen delivered. Civil society was plugging the dike holes left by a disintegrating public service.[88]

A larger contribution was a reengagement with American culture at large. The Yellowstone extravaganza had mesmerized the public; now writers sought to transform that spectacle into something like literature. Stewart Udall used the brouhaha over Yellowstone to reissue *The Quiet Crisis* with a new preface that lambasted the Reagan years. The fire community saw books about the lives of its famous and its unknown, life at the top and life at the bottom. Harold Biswell published an autobiography of his career, *Prescribed Burning in California Wildlands Vegetation Management*, revealingly titled around what he did rather than who he was; and *Fire on the Rim*, a paean to the life of a seasonal smokechaser, was promoted as the autobiography of a crew. But a deep connection to national culture required established authors with ties to the literate elite and had to wait.

While the big minds of fire management reflected back on Yellowstone, fire itself pushed on into new frontiers. On July 9 an accidental fire, probably from a tossed cigarette, started along Highway 119 west of Boulder, Colorado. The flames blew out of Black Tiger Gulch to engulf 44 homes in what quickly became an exemplar for the fire of the future—not flames running mostly free in the wild but flames running wild through the sprawl of exurban America. The NFPA commissioned a detailed report under the National Wildland/Urban Interface Fire Protection Initiative, a companion study to its proposed model code NFPA 299, *Protection of Life and Property from Wildland Fire*. Paradoxically, the real threat to fire's management was less from "prescribed" fires in wilderness that might taint prescribed burning elsewhere than it was from fires prowling the unstable border between the natural and the built. The Black Tiger fire created a prototype analysis that pointed to that future. Yellowstone pointed to the past.[89]

* * *

In 1990 everything was the same, only more so.

The Forest Service was being paralyzed by forestry—exactly what it had long prided itself on as its specialty and what granted it intellectual if not moral authority. In cutting old-growth woods, in subsidizing below-cost timber sales, particularly in remote landscapes, in clear-cutting from West Virginia to Oregon, it was condemned in both courts of law and the court of public opinion. (The BLM was even more stubborn on the O&C Lands, but it lacked the breadth of old-growth lands the USFS held and so escaped most public opprobrium.) In the Northwest the Forest Service was locked in trench warfare, and it was losing by attrition. Serious critics joined free-market ideologues in debating whether the agency should be dissolved and its tasks distributed among the private sector. (Mostly that movement was checked by fears that, as Randall O'Toole put it, a crack-up and free-for-all might lead to something worse.) Reforms in fire management were trapped in that larger matrix of agency controversy.[90]

The Interagency Scientific Committee released its conclusions on the spotted owl, which the FWS listed as threatened, in a report that the Bush administration refused to endorse. The main action lay beyond the country's borders—with the UN International Decade of Hazard Reduction, which included wildfire among threats; with the Global Climate Change Prevention Act, which absorbed combustion of all kinds; and with the International Forestry Cooperation Act, which among other projects sent USFS fire specialists to Brazil. In 1990 the USFS, NPS, Tall Timbers, Society of American Foresters, and several academic institutions hosted an international symposium on "Fire and the Environment: Ecological and Cultural Perspectives." But the United States seemed hostage to its own internal fracturing. For the first time in its history the U.S. Forest Service, in the words of Chief F. Dale Robertson, fell seriously short "in a major way of meeting its financed goals and targets as outlined in the congressionally approved budget and appropriation laws."[91]

In response, and building on New Forestry, the agency advanced an agenda it called New Perspectives to pilot test different strategies of land management; in August it committed the entire Ouachita National Forest in Arkansas to the idea. The decision did little to wipe away the crisis, which was beyond the capacity of the Forest Service to control. It meant little to fire management, which had had its own new perspectives for 20 years. For fire the story was the need not for new ideas but for new ways to put the existing ideas into action. On prescribed fire, on suppression, and on natural fires, the agency seemed helpless.

On prescribed fire, not only the Forest Service but all the federal agencies began to cede primacy to the states and civil society. Prescribed burning had become complex. What had been done on Redwood Mountain with a page of handwritten notes could now require a ream of reports and complicated checklists, all embedded within environmental assessments and impact statements. Worse, calling slash burning, which was intended to clean up after clear-cuts, "prescribed fires" only tainted the brand. It made the term appear a euphemism and sleight of hand, or a devious enabler for felling old-growth forest. The smoke hung over populated landscapes such as the Willamette Valley: it put into public view (and into the public's lungs) what might otherwise have been hidden on the westside Cascades. The smoke provided a point of attack to limit the burns, which would force logging to gag on its own woody effluent. Every fire threatened to become, literally, a federal case. Each prescribed burn was a set piece, any component of which was a potential glitch that could shut down the whole. All of this made a striking contrast with Florida. In Florida burning could proceed on private and state lands and amid a culture that generally accepted it. In Florida even the Endangered Species Act compelled agencies to burn to enhance the habitat of the red-cockaded woodpecker.[92]

What the West had was large tracts of wildland. The prescribed fire that most interested the region was the prescribed natural fire. After the 1988 fire extravaganza, even this stagnated as the third critical report of the post-Yellowstone review—everything in fire seems to come in threes—emanated from the Government Accountability Office, which Congress had requested to audit progress on the proposed recommendations from the Policy Review Team and which issued its conclusions in 1990. In May it testified before Congress that its research showed "that in the summer of 1988 both the fires that raged in Yellowstone and the government's fire program were out of control."[93]

It got worse. The GAO team readily accepted, as every thoughtful observer did, that fire belonged, had been unwisely excluded, and needed to be restored even though the task was a high-risk operation. The issue was how to do it, whether the policy recommendations were sensible, and if correct, how well they had been implemented. In the GAO's measured judgment, the program remained a mess. It questioned "whether the new controls are as sound as they appear, and whether the program has the organizational structure required to coordinate firefighting efforts in time of national emergencies." The reforms lacked the means—the money and manpower—to achieve those ends ("an issue not addressed by the task force"). They were also "resisted" by some fire managers largely because institutional incentives favored fighting fires rather than lighting them. There was money and goodwill to cover escaped wildfires; there was neither for escaped prescribed fires. Accordingly, the GAO doubted "whether the revamped fire program will evolve from a program on paper to one in practice."[94]

Rekindling the fire revolution needed more than good ideas and good feelings. In its December report, the GAO elaborated on the slippage that had followed the Yellowstone bust. The agencies (it concentrated on the USFS and NPS) had not had enough funding, institutional coordination, and practice to do the job properly before the outbreak, and they had no more now. In 1978 the Forest Service had 8,444 personnel in regular fire positions, which amounted to 60 percent of federal firefighting resources. That workforce began to degrade the next year, and it continued to plummet through the Reagan years. In 1988 it was down to 4,859—a loss of 40 percent. Since PNFs required suppression capability in reserve should they escape, they could be declared out of prescription if the country had big wildfires elsewhere that drew down the national suppression force. Meanwhile, the agencies had resolved to review every park and wilderness fire plan to ensure conformity with national standards. The goal was to have them ready by May 1989. Instead, as of August 1990, the Forest Service had approved plans for just 8 of 75 wildernesses (8 percent), and the Park Service 3 of 26 parks (12 percent). There was no firm commitment by either agency to accelerate the process of recovery. In 1990 the NPS had 69 percent of the funds it reckoned it needed for prescribed fire, and the USFS had a mere 1 percent (NPS used emergency suppression money for PNFs). Meanwhile, only two of eight interagency preparedness plans had been approved. The national fire enterprise faced a prolonged rehabilitation, especially if starved of money and attention. Already the agencies were looking elsewhere.[95]

They did not have to look far, because exurban sprawl was carrying a new voter-rich problem to their borders. Suppression had to retool to learn how to keep a light hand in the wilderness and a heavy hand in the I-zone. Fire control might be short staffed and stressed by the ongoing upheaval in its workforce composition, but when fires took off, there was always money to fight them. And when the fires came in 1990, they were conveniently segregated by the emerging polarity between the wild and the urban.

That year the big burns were in Alaska—three million acres in all. The few in the Lower 48 that grabbed attention in 1990 were all, in one way or another, fires along borders. On June 25, lightning kindled a fire below the Mogollon Rim on the Tonto National Forest, near Dude Creek, along which private cabins were strewn as inholdings. Forces were immediately mustered to fight it. A microburst downdraft from a thunderhead stirred the flames and pushed them through the exurb, consuming 60 homes (and even Zane Grey's historic cabin), and also the Perryville Inmate Crew, trapping 11 firefighters and killing 6. The fire scoured the Rim until it was stopped at 30,000 acres. Two days later an ignition, likely arson, along Highway 154 in the Painted Cave area of the Santa Ynez Mountains met 90-year-old chaparral and 50 mph Sundowner winds to power a wild rush of flames through 427 houses, 221 apartments, 15 businesses, 10 public buildings, and 4,900 acres around Santa Barbara; one person died. Most of the damage occurred in the first 90 minutes. On August 7, lightning danced across the Sierra Nevada, kindling fires where its skipping steps met record dry kindling. Three spread along the borders of Yosemite National Park. The A-Rock fire burned into Foresta, incinerating 68 structures and closing the park's entrance for the first time since its founding, forcing the evacuation of Yosemite Valley and El Portal. Across the Merced River, the Steamboat fire burned 4,400 acres. And a third fire, threatening the Merced Grove of sequoias, was halted short of the Big Trees. For a while, it looked like a bad movie, *Yellowstone: The Sequel*, and the national media, primed, swarmed over the spectacle. But there was little basis for controversy. From the onset, the Stanislaus National Forest and Yosemite threw everything they had at the fires.[96]

By now, however, prescribed fire seemed quarantined in the Southeast, and the prescribed natural fire was banished to Alaska. The fire community was in a painful pivot away from the friendly flames in the wilderness to the ugly, feral ones along its pushy, fractal urban fringe.

* * *

The next year, 1991, witnessed a slow repositioning among the feds, a quickening pace from fire's civil society, and two defining fires, one brazenly shouted in headlines and one prominent by its silence.

The NPS aside, Interior had not escaped the post-Yellowstone turmoil. Most of its agencies were too small and scattered to establish a critical mass for good planning and too autonomous to allow for coordination. An ad hoc system to speak with a unified voice evolved during the Yellowstone crisis. The Review Team argued to make that experience permanent. Crises and money, like wars, tend to consolidate and to move power up the hierarchy. The result was an Interior Fire Coordination Committee composed of the agency fire directors, later headed by a fire policy coordinator. The project was eventually merged with emergency hazard programs into an Office of Managing Risk and Public Safety (Secretary Morton then split off law enforcement into its own entity). In 1991 the OMB applied pressure by creating a unified fire account for all of Interior, which began to move emergency funding toward a more predictable budget, much along the lines of the Forest Service. That did not inspire reforms, but it did help position Interior's bevy of agencies for action when the political climate improved.

For the Forest Service a decade deferred became a decade forfeited. It remained shackled to the Northwest forests controversy as the federal courts and Bush administration each sought to impose its own solution. The agency tried to paddle through the political rapids with its New Perspectives initiative. Forest Service chief Robertson admitted that "traditional forestry would no longer fly in the federal government," but alternatives were scarce on the ground. Its budget and staff continued to shrink. Amid such angst, reforming fire management seemed almost an indulgence. The GAO might report that the agency was not meeting its fire targets, but the agency could shrug and reply that it was not meeting any targets, and without clear support from Congress and the administration and the resources to satisfy them, it was disinclined to engage in what all critics acknowledged was a high-risk enterprise for both agencies and individuals.[97]

The gap between needs and deeds was becoming a chasm. What helped bridge it was a reinvigorated civil society. Part was institutional and confined to the fire community. An International Association for Wildland Fire was chartered for all constituencies—researcher, administrator, and practitioner—complete with a journal, the first dedicated wholly to the subject. The privatization of equipment picked up. Specialty tools were invented, manufacturers put their own pulaskis and short-handled shovels on the market, and The Supply Cache opened as a vendor for such implements and personal gear. The Joseph W. Jones Center for Ecological Research was birthed in south Georgia, across the state line from Tall Timbers, creating a genial rival with an equal interest in fire and its place on the landscape. Much as Tall Timbers began to house several federal agencies interested in prescribed fire, including a national office for FWS, so the Jones Center became the catalyst for nationalizing prescribed fire councils. Tall Timbers itself reinstitutionalized its fire conferences, staging them every other year. The second was held in 1991 and returned to the longleaf pine biota that had first inspired its insurgency. The Nature Conservancy published its first national manual for fire management, a daunting bid to fashion its consortium of volunteers into a national presence. Prescribed fires faded from the federal lands and migrated onto private holdings.

That was the story of America's fires. They split between the accidental and the unlit, between the urban and the wild, between Yellowstone and Oakland, between Sequoia-Kings Canyon and the Blue Mountains of Oregon. Somehow the country had to pass between them all.

* * *

By the end of that year, the locked plates had begun to shift, ending with an electoral quake. And a posthumously published book galvanized the American fire community.

Globalization—in this case, of environmental sensibilities—helped jar loose the stalemate over clear-cutting and old growth. At the Earth Summit in Rio de Janeiro, the United States found itself ridiculed by critics over its logging policies, even by American delegates (such as Senator Al Gore).

With President Bush scheduled to speak on the last day, the administration wanted to defuse the criticism and "say great things." Chief Robertson helped draft a statement that the Forest Service would eliminate clearcutting and adopt a more systemic approach. New Perspectives morphed into Ecosystem Management. Bush gave the address. Robertson marveled, "That was really something, to get it through the Republican administration." The impasse did not cease, however, until a new administration arrived the following year.[98]

That still left the agency struggling to implement those ideas with a workforce that was drastically downsized and (to many skeptics) indifferently experienced. Bill Buck gave those critics a public voice in his denunciation over Yellowstone's conduct. Those running the program, he charged, were "way over their heads," and for all the "renegade" qualities of that bunch, peculiar to the dynamics of the park, the problems had become national. NIFQS had been "deliberately diluted" to accommodate abject newcomers and unqualified partners; "quota hiring" for affirmative action and the consent decree had created a "double standard," which placed further burdens on Type 1 crews. The Fire Fighter Retirement Program, he claimed, would "have a monumental impact on the development and retention of the most capable." Fire management demanded nose-in-the-ash, shoulder-to-the-shovel, year-after-year experience with real fires, "an empirical knowledge that can only be acquired through years of hotline fire suppression experience." Out of these trials by fire came a hard and inescapable sifting of the qualified from the unqualified. You learn prescribed natural fire by doing prescribed fire over and over. "If you can't cope with planned ignitions (under exacting conditions), you sure as hell have no business toying with natural fire." The breakdown in competent staff was the "weakest link" in contemporary fire management.[99]

There were plenty of people ready to disagree with Buck's old-guard sensibilities. Yet he was right to note that the Forest Service's workforce, no less than its timber mission, had been tied up in courts and controlled by others. Moreover, that workforce began to shift, along with the general American economy, from those who worked with their hands to those who worked mostly with their heads. In 1991 Robert Reich published *The Work of Nations*, in which he argued for a new class of "symbolic analysts." The National Research Council spoke to the dismal state of such analysts in the Park Service, the failure of the Robbins Report to find handholds as the Leopold Report had. For the Forest Service the symbolic analysts would be the -ologists, who had not entered the ranks with a shovel and had little interest in fireline callouts; even within the fire community, simulations of the future were replacing experiences of the past. Then in 1992, at least part of its legal logjam broke when the Forest Service and courts approved a settlement agreement to end the consent decree. This, too, would have to wait for the new administration to fully realize when Mary Jo Lavin was appointed director of FAM, an appointment that testified to the triumph of the managerial class.

Meanwhile the other agencies had quietly, and with less public notoriety and discord, steadily integrated.[100]

* * *

The real breakthroughs came from writers. Inspired by a photo he had seen in 1989 that showed a fire crew outlined against 300-foot flames, Sebastian Junger sought "their awe, their exhaustion, their sense of purpose," and wrote an essay on the 1992 Foothills fire (later republished in his anthology, *Fire*). If he made hotshotting seem an extreme sport, he also made fire of interest to a general readership outside *Audubon* or the *Sierra Club Bulletin*. But what grabbed the literary world and galvanized the fire community was the posthumous publication of a meditation on the 1949 Mann Gulch fire by Norman Maclean. *Young Men and Fire* did for intellectuals what the Yellowstone blowout had done for the general public: it connected fire to the culture. Maclean, a professor of literature at the University of Chicago, was widely known and honored for his earlier collection, *A River Runs Through It*, the title story of which had also been made into a successful movie.[101]

Young Men and Fire was more than a riveting account of a fire that burned over a group of smokejumpers and killed 13 in a remote canyon of Montana. It was a rich text—a tragedy, rendered as only a Renaissance scholar could; a theological tract, after a fashion, by showing how hand and head might craft a species of grace, in the form of foreman Wag Dodge's escape fire; the redeemed story of a firefight in the wilderness otherwise buried in the files of Region 1 and the Aerial Fire Depot at Missoula. The book caught fire, blew into best-seller ranks, and made people who would otherwise have dismissed Western fire as a freak consider fire as a topic of conversation. Quickly, other fire works followed as art imitated art, in a rush of personal accounts and even retellings of Maclean's version of Mann Gulch. Two years later life imitated art, and helped make the 1994 season and the South Canyon fire the reborn revolution's fire of reference. The Mann Gulch depicted by Norman Maclean provided the prism by which a revived engagement of fire and society would be viewed.

* * *

Wildfire itself remained relatively calm. The most notorious outbreak occurred on August 19, 1992, when five lightning starts ran past BLM crews in the grasslands between Boise and Mountain Home and then burst into the mountains. The flaming front raced 18 miles in 10 hours and became at 257,000 acres the largest fire in Idaho history since the Big Blowup. Instantly, mobilization went national, drawing in 2,886 personnel from 5 federal agencies, 17 states, the U.S. Air Force, the Idaho National Guard, local departments, and private contractors, along with an armada of dozers, engines, helicopters, air tankers, light aircraft, ICS teams, the works—the fire establishment emptied its bench onto the court. Air tankers flew from dawn to dusk. One firefighter died. For years the fire's severity served as a textbook example of extreme behavior on soils and streams. Rehabilitation efforts set new records for spending.[102]

Yet Foothills was neither a wilderness fire nor an interface fire, and it was not even nominally compromised with injunctions about how it was fought. It was an old-model wildland burn. What no one remarked on, officially, was the irony that even here—with a smoke plume within sight of the Boise Interagency Fire Center—the fire could not be contained until the weather changed. There was no going back to simple suppression. The next time fires burned near Boise, as they did in 1996, they burned into the town. As crews were beginning to say openly, they were not putting fires out, they were putting them off.

Prescribed natural fire, prescribed fire, wildland-urban interface fire, just plain wildfire—across the gamut of contemporary fires, there were breakdowns. The fire revolution had shaken apart the old consensus, and the '88 season had scattered the new pieces to the four winds. By 1992 the shards had been regathered or refashioned. How those pieces might recombine, if they could, would depend on the return of industrial-strength fire busts, on finding a cultural catalyst to complement if not replace an exhausted environmentalism, on the right mix of personalities, and on just plain luck. The International Association of Wildland Fire sponsored a "global conference, symposium, and workshop" in Missoula on the theme, "the power of politics," later considered by "many" observers as "the quintessential event that helped shape how the wildland/urban interface fire issue" would be viewed in the coming decades. Then the 1992 presidential election stirred the embers.[103]

Lost Fire, 1991

The fire revolution had bestowed on two landscapes the grace of its attention. One was the wild, for which the national park and wilderness were paragons, with natural fire the ideal means of fire restoration. The other was the working landscape, for which prescribed fire was the treatment of choice. Thirty years after the inaugural Tall Timbers conference had announced the two, the pyrogeography of the country had transfigured in ways no one had anticipated. Patches of cityscapes competed with patches of wilderness, and the working landscape for fire, like much of industrial America, had shrunk. The wildland fire community had, it appeared, lost the capacity to halt the new kind of wildfires it confronted along with the ability to set the old kind that it needed to restore. A lost decade ended, appropriately, with lost fires.

* * *

The interface fire that observers were coming to characterize as the fire of the future, or the new American fire, had plenty of examples but nothing comparable to announce its arrival on a level with Yellowstone's valedictory to the era of wilderness fire. But on October 21, 1991, they got it, when Diablo winds—much like the Santa Anas in Southern California but blowing from the east over the hills around the Bay Area—fanned a smoldering fire in the Berkeley Hills into a conflagration that rampaged through an elite neighborhood of Oakland. America's worst urban fire disaster since an earthquake and fire leveled San Francisco in 1906 was under way within hours.

Oakland was an oddity, but only because it was both a cameo and an exaggeration of the features that made the wildland-urban interface itself an oddity in the evolving national estate. It seemed to invert the usual assumptions. It was like a patch of Southern California in the San Francisco Bay area, with a similar Mediterranean climate, seasonal foehn winds, and an abrupt but porous border between city and wildland in which regional parks replaced national forests. Dense high-end housing, not filigree exurbs, abutted fire-prone open space. Its faulty fire protection system was not the outcome of developments leaping beyond jurisdictions but of a proudly urban fire service that had fallen into decay. Unlike the Los Angeles County Fire Department, Oakland's had not coevolved with responsibilities for both cities and wildlands, it had no next-door wildland fire agency as a partner, and it had not incorporated the NFDRS or other accommodations into its parklike environs. It had the only apparatus in the East Bay capable of attacking a high-rise fire, but it did not have air tankers.

It did, however, have a history of fire in its hills. In 1923 flames had bolted through them and onto the University of California campus. In 1970 the hills had a serious scare, the first outbreak of the 1970 fire siege, and another in 1980. Committees debated how to keep the city from intruding into the parks, and the parks, through fires, into the city. In a nutshell, Oakland had rushed into the fire-prone hills, while the wildlands had in effect infiltrated the city through the widespread planting of flammable eucalypts and shrubs. What had at the beginning of the 20th century been grassy hillsides, easily and often burned, had morphed by the end of the century into a combustible scrub of wooden houses and woody flora with a backcountry open to Diablo winds and random sparks.

On October 20, 1991, the Oakland Fire Department knocked down flames that had started as a warming or cooking campfire amid a stand of pines near Marlborough Terrace. Crews mopped up by soaking the perimeter lines. The next day, while firefighters were on site and rolling up hoses, the still-smoldering hot spots disgorged embers, and the flames got into patches of Monterey pine litter untouched by hose lines. The rekindled fire raced up a grassy slope to the ridgeline. There the Diablo wind caught it, and the fire blew up. With stunning speed it burned out the basin below Grizzly Peak. It burned through the Parkwood Apartments. It burned out the Hiller Highlands. It burned out Grandview Canyon and then burned over Highway 24. In the first hour it consumed 790 structures, each of which scattered new sources of ignition. What became known as the Tunnel fire spotted over Lake Temescal and burned through the Rockridge District. When the Diablo winds finally slackened and northwesterly winds returned, the main front—a swarm of new ignitions, building after building—headed southeastward into Forest Park. A new index of fire spread made its appearance—homes burned per hour. Before the orgy of flame ended, 3,354 houses and 456 apartments were ash, and 25 people had died. Total area burned amounted to 1,600 acres.[1]

A cataclysm this horrific sparks reviews at all levels of government, which the Tunnel fire certainly did, from citizen groups to the National Fire Protection Association, from a mayoral task force to California's Office of Emergency Services to NFPA and FEMA. As with all major disasters, the surveys identified many causes, most of which had to happen together to produce results so far off the scale. Those factors that governed fire behavior fell into two general categories. One pertained to the fire environment, the other to fire-suppression capabilities.

There could be little dissent from the observation that the East Bay Hills were a prime natural setting for fire. A Mediterranean climate, seasonal foehn winds, terrain that could channel fire like a coal chute, and pyrophytic vegetation that encrusted the hillside—such conditions would argue for fire anywhere they appeared. That the fire occurred amid a drought and an epidemic of sudden oak death and after frosts had killed eucalypts and then a record hot spell had worsened the circumstances. Yet the values were "extreme, not exceptional," and even the exotic flora only acted as an accelerant by allowing embers to kindle surface fuels and fling sparks from torching eucalypts. The "wildland" fire, however, had burned upslope toward the summit, not through the structures, and in the end, the canyon's flora survived better than its buildings. The fuels that mattered were the houses and especially their wood-shingle roofs. So close were the buildings that they burned one to another, and so combustible were the roofs that they both received sparks and recast them into the wind. The character of the quasi-natural setting allowed the fire to start. The character of the city allowed it to spread.

The capacity to fight the fire was badly compromised. After decades of boom, Oakland had gone bust in the 1970s. Urban services decayed, among them the capacity to maintain the kind of varied fire protection demanded by the mix of landscapes within the city. Over and again, the urban fire service had failed to integrate with wildland counterparts. It did not know of the red-flag warning posted by CDF for the day of the fire. It did not understand how mopup in wildland fuels differs from overhaul in buildings. It did not appreciate how a city, full of internal firewalls, might be breached from the perimeter and find itself assaulted not from the streets but from the air. It had not reckoned with fire-induced power failures that neutralized pumping stations. It had not adopted ICS, and could not function seamlessly with other assisting agencies. It could not communicate on common radios (CDF officers resorted to telephoning dispatchers). It had three-inch hydrants, while adjacent cities used national standard two-and-a-half-inch hydrants. But even if compatibility had been perfect, the fire would have likely exploded because it was moving faster than any fire department ever could. Something else intervened to break down the response.

That something was the wind. It did for Oakland what the earthquake had done for San Francisco in 1906. It simply overwhelmed the capacity to respond. The OES report noted haplessly that "a fire burning 400 or more homes per hour does not allow for normal firefighting tactics—either urban or wildland." Even mutual aid requires time to muster engines, planes, and personnel. Within just two hours the conflagration had reached perhaps 80 percent of its final size. The narrow streets soon clogged with traffic and fleeing residents. It was not possible to move people and cars out as fast as the fire moved in. Converging fire engines met outgoing civilian autos. There was no Maxwell's demon in the box canyon to sort them out. There was no single flaming perimeter or high-rise to focus the action, only hundreds of individual fires—the firefight as melee.

The subsequent committees, panels, boards, and task forces published hundreds of recommendations, ranked by priorities. Some involved simple changes in protocol, such as getting daily fire weather. Many, however, required costly retrofitting, either by the city or residents in the Hills, such as replacing nonstandard hydrants and burying powerlines. Given the parlous state of Oakland's finances, only a fraction of the recommendations could be enacted. But perhaps the most critical need was simply institutional: the East Bay required the fire equivalent of its municipal utility district, or what Southern California had found with its county-CDF-Forest Service triumvirate. The South Coast, however, had a few big, wealthy entities; the East Bay had many smaller and poorer ones. Even the 100,000 acres of regional parklands distributed its holdings among 65 units.

Still, the reconstruction went forward. The neighborhoods rose from the ash, with better fire protection built in. After several stumbles, the East Bay Regional Park District was voted bonding authority in 2010 to expand. A Hills Emergency Forum gathered the various constituencies into a common conversation. A memorial, shaped like a gutted house with a missing roof, was erected at the intersection of Highways 13 and 24. The scars, both environmental and social, slowly healed. Ten years later the Hills Emergency Forum sponsored a review, and then another 10 years after that.

Most of the national discussions over the emerging I-zone concerned wildland fire agencies that fretted over how to cope with exurban landscapes littered with houses in place of logging slash. But the crush of sprawl would compel urban fire services to meet them along their shared fencelines. As the interface matured, and as Southern California had first illustrated, the urban model would subsume the wildland. In the usual formulas, the mantra was to bring the savvy of wildland fire agencies to urban-style fire departments while imposing urban-model fire codes onto the built environment of the newly recolonized woods, and in places like California, a curious hybrid evolved. By the time Oakland gathered to commemorate the 20th anniversary of the Tunnel fire, another dynamic seemed to dominate as the push for urban-style emergency services jeopardized the fire-management mission in wildlands. The new mantra was not just to keep wildland fires from slamming into cities but to keep urban fire protection from intruding into wildlands.

* * *

The early manifestos of the revolution had predicted catastrophe if fire was excluded from fire-dependent biotas. The consequences were quickly visible in Florida, where growth, unless actively hacked back by blade or flame, would overwhelm landscapes — one reason that prescribed fire there flourished with an exuberance akin to that of its living fuels. The eco-shock took longer in the drier West, but by the early 1990s it was undeniable. The disaster fire was the one that had not happened. It was the fire not lit.

The crisis found its poster child in the Blue Mountains of northeast Oregon. The consensus narrative told how, with American settlement, fire had been excluded and the forest restructured by removing the indigenous peoples, by overgrazing the native grasses that had carried frequent fire, by selectively logging the big trees; the coup de grâce was administered by organized fire suppression. Without the cleansing and catalyzing effects of fire, a tangle of grand fir and Douglas fir grew up that threatened to strangle the ponderosa pine and western larch forests that had previously thrived. Starved of nutrients, desperate for water (worsened by recurring droughts), jammed into thickets, choked by their own teeming reproduction, the landscapes succumbed to beetles, dwarf mistletoe, spruce budworm, root-killing fungi, and other pathologies, and when they kindled under the right circumstances, catastrophic fires burned well beyond the range of adaptations they had acquired through evolution. Flames that had historically swept along the forest floor, clearing grasses and needles and the occasional windfall, now leaped into the crowns with a severity that eliminated their capacity to regenerate. Good fire that had once routinely visited the woods yielded to bad burning — conflagrations or an outright drought of flame.

What the Blue Mountains illustrated was common throughout the montane West, which was also the landscape most amenable and attractive to exurban development. Once-benign fires had vanished, and malign fires were taking their place. Worse, the process was a vicious cycle in which the need to prevent horrific fires kept suppression supreme, while the sick, unburned woods just added more fuels. The issue was not whether fire would return: it would. It was about restoring the right regimen of fire, about inventing a culture of prescribed burning, about creating a habitat in which reinstated fire could behave properly. Simply dumping fire into the Blues was no better than releasing packs of endangered wolves. Unless the setting was right, fires could not do their expected ecological work and might even run feral. And the existing circumstances were all wrong. The Blues had no tradition of burning, had landscapes stockpiled with flashy combustibles, and held little or no commercial value in the dead and dying forest, which left the cost of pretreatments terrifyingly high. It was a time when the only acceptable slashing was to the budget.

All these outcomes had been forecast during the light-burning controversy. It had revived when Ed Komarek, alarmed by "the tremendous losses from wildfires and the increased costs of forest fire protection" in the West, had dispatched a Tall Timbers task force to Arizona in 1972. In September 1991, a month before Oakland burned, an interdisciplinary group of Forest Service fire researchers had toured the Wallowa-Whitman and Umatilla national forests. Those who had not seen the Blue Mountains recently were "shocked by the severity and widespread extent of the forest mortality." They resolved to make the Blues a paradigm of what had gone wrong and what it would take to make matters right. Fire management met forest health.[2]

The group identified a dozen things to do, and two dozen difficulties in doing them. Every remediation hinged at some point on fire, and fire's restoration hinged on the "question of scale." Current burning for logging and silvicultural treatments ranged about 1,500 to 2,000 acres a year. A serious down payment would require 20,000 acres or more, or over the Blues entirely, perhaps 60,000 a year. Such an effort would cost time, intellectual investments, political capital, and just plain money. That is what the announced doctrine of ecosystem management required. It was also what the agency did not have.

The old frontier had destabilized whole biotas that could not be reversed as one might unscrew and replace a broken light bulb. An exurban frontier was fashioning a regimen of fire that defied both simple classification and easy correction. What the fire community had once believed it had more or less banished, like medicine's assumption that tuberculosis and measles were tamed or extinct, was returning in more aggressive forms that threatened to infect large swaths of American land and society. Thirty years after the fire revolution declared its ambitions, it had held the line for prescribed burning in the Southeast and created pockets of natural and prescribed fire in the West, but its triumphs were more symbolic than practical, and its scale was laughably tiny compared with needs. What the environmental president had celebrated as a "thousand points of light" threatened to be overwhelmed by a growing ecological darkness and the arrival of apocalyptic conflagrations.

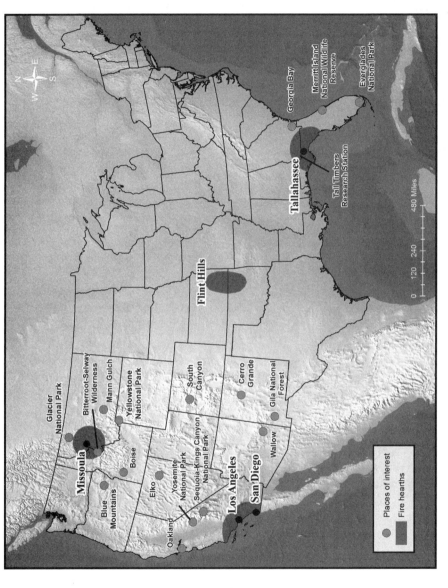

Figure 1. America's cultural fire hearths and major fires (1961–2011).

Figure 2. Sleeping Child fire (1961). At the start of the fire revolution, the towering pyrocumulus remained the iconic image of backcountry fire. Source: U.S. Forest Service.

Figure 3. Yellowstone fires (1988). The kind of image that riveted the media, public, and politicians. Photo by Jim Peaco, courtesy National Park Service.

Figure 4. A talismanic image from the fire revolution, John McColgan's *Elk Bath*, taken on the Sula complex during the 2000 season, also in the Northern Rockies, captured the sense of fire as benign and a pristine part of nature.

Figure 5. The firefighter as Hemingway code hero: the Liebre fire, Angeles National Forest. Photo by Tom Roberts. Source: U.S. Forest Service.

Figure 6. Frank Irving (*left*) and Alvar Peterson (*right*) at The Nature Conservancy's first-ever controlled burn, which was conducted at the Conservancy's Helen Allison Savanna Preserve in Minnesota on April 26, 1962. Photo by Don Lawrence, courtesy of The Nature Conservancy, with permission of Bill Lawrence.

Figure 7. Fire, the Forest Service, and generations: Dale Bosworth and father, Irwin. Courtesy Dale Bosworth.

Figure 8. The patriarchs: E. V. Komarek and Herbert Stoddard Sr. at the 1969 dedication of the Stoddard Lab. Courtesy Tall Timbers Research Station.

Figure 9. The fire revolution comes to Yosemite. Standing (*left*) is Bob Barbee, who made a reputation as head of resource management at the park and later served as superintendent of Yellowstone during the 1988 fires. Next (*standing*) is Harold Biswell. Then Norm Messinger, Wawona District naturalist. Fourth from left is Biswell student and long-serving fire researcher Jan van Wagtendonck. Kneeling (*from left*) are Don Taylor and Bill Jones (chief naturalist). Courtesy Bruce Kilgore.

Figure 10. The prime movers of the White Cap experiment gathering for a 30-year retrospective on the Bad Luck fire. *From left*, Bob Mutch, Bill Worff, Bud Moore, Orville Daniels, and Dave Aldrich. Courtesy Bob Mutch.

Figure 11. Lab meets the field: Richard Rothermel studying the Pack River bridge destroyed by the 1967 Sundance fire. Trained in aeronautical engineering, Rothermel was among the early hires for the Missoula fire lab, and he soon learned that flames in wind tunnels had to make sense in a world of mountains and dense woods. His fire behavior model gave scientific heft to the fire revolution. Photo by Danny On, courtesy U.S. Forest Service.

Figure 12. Norm Christensen, one of the founders of academic fire ecology, thrice called on to chair committees to review how ideas might enter practice (including one of the postseason surveys of the 1988 Yellowstone burns) and later dean of the Nicolas School of the Environment at Duke University. Courtesy Norm Christensen.

Figure 13. Wally Covington amid his beloved ponderosa pines. His Flagstaff plots became the point of ignition for restoring fire and improving forest health by thinning and burning—and served as a lightning rod for critics. Courtesy Wally Covington.

Figure 14. Jack Ward Thomas, where he liked to be. Having led the scientific team investigating the northern spotted owl, Thomas reluctantly became chief forester during a time when the agency converted officially to ecosystem management and workforce diversity and then suffered through the horrific 1994 season. Here he chills out on an elk hunt outside the Eagle Cap Wilderness after the fires ended. Photo by James Applegate, courtesy J. W. Thomas.

Figure 15. Midnight Sun Hotshots in 1996 with Secretary of the Interior Bruce Babbitt among their complement (*second from left, standing*). Courtesy Bruce Babbitt.

Figure 16. California split. If the modern era began in California, it is appropriate that it should conclude with the careers of two Californians who became fire directors for the country's primary federal agencies but who rose through the ranks through different career paths, one through fire restoration and one through fire suppression. *Left,* Tom Nichols began as a fire monitor at Sequoia-Kings Canyon National Parks and concluded as fire director for the National Park Service. Shown here for his retirement photo in January 2014. Photo by Tina Boehle, courtesy National Park Service. *Right,* Tom Harbour began his career as a seasonal firefighter on an adjacent national forest, the Stanislaus, before joining a hotshot crew, serving as fire-management officer for the Angeles National Forest, and finally assuming the directorship for Fire and Aviation Management for the Forest Service in 2004. Courtesy U.S. Forest Service.

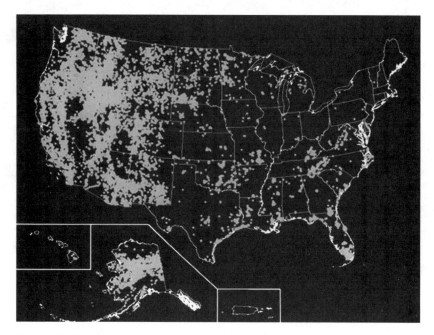

Figure 17. The large fire scene as mapped by the U.S. Geological Survey, 1980–2003. The cartography is basically a record of public lands. Source: U.S. Geological Survey.

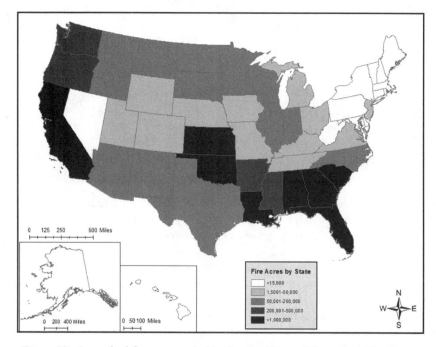

Figure 18. Prescribed fire as recorded by the Coalition of Prescribed Fire Councils for 2011. Source: Coalition of Prescribed Fire Councils, with permission.

Figure 19. America burning as recorded by NASA for 2012. Source: NASA.

Figure 20. Two visions of fire's future: the 2003 Aspen fire in the Santa Catalinas and Biosphere II. Photo by Francisco Medina, with permission.

CHAPTER FOUR

SLOPOVER

The Forest Service simply no longer has a clear mission.
—FOREST SERVICE CHIEF JACK WARD THOMAS (1996)[1]

We are in a national fire crisis.
—SECRETARY OF THE INTERIOR BRUCE BABBITT (1999)[2]

A Rendezvous with the Land

By 1992 the dismantling of the old USFS hegemony was nearly complete. The Forest Service still commanded respect—it oversaw a vast and valued estate, and it held over half the national resources devoted to wildland fire. But no longer could it dominate the others, nor could it even be said that the national narrative was its history or even the histories of the agencies braided together. Interagency standards such as NIFQS prevailed for training, qualifications, equipment, and practice; interagency institutions such as NWCG set guidelines for monitoring and prescribed burning; and interagency fire teams increasingly worked the big burns. Each agency still had a fire policy peculiar to its mission, but they were converging around a common consensus to fight the bad fires and light more good ones. Themes, regions, and personalities replaced agencies as the prime movers and organizing conceits of American fire.

The Forest Service struggled to regain its footing not only in suppression but in all aspects of its former dominion. It no longer controlled fire research (its own fire research establishment was on the ropes). The stresses that had warped its workforce on matters of diversity stabilized, but they were calmed, not resolved. Questions of the agency's mission caused internal dissent and with FSEEE spawned a shadow organization. Even when aligned with the other federal agencies, it could no longer control the discourse. A vigorous civil society ran research plots, bought and burned land, and trained fire crews. Tall Timbers had restored its conferences and The Nature Conservancy expanded its fire capabilities. The International Association for Wildland Fire assumed the global role once claimed by default for the Forest Service. As the I-zone gathered steam, as flames splashed over jurisdictional lines drawn on maps, the states became more active cooperators, and then the counties and even volunteer fire departments. Multiple use as a doctrine for synthesizing many lands under a common administrative banner lay in tatters. Only those agencies with distinctive charters knew what to do and how to judge their actions, and those with all-purpose missions wandered without a political compass.

Still, they all struggled. When the Yellowstone blowout led to a programmatic reboot, it was assumed that recovery would be quick. Parks, forests, districts, and refuges had only to rewrite their fire plans to align with policy and protocol, yet as the GAO soon found, that proved beyond the capacity of most units. And if they did have plans, they were unlikely to have the funding to execute them. Probably the NPS was better positioned than most. Like the FWS earlier, its wheels had squeaked louder than the others, and thus it got the highest ration of budgetary grease. A pass around the compass, however, showed that no one was meeting targets. The issue was less policy—it was adequate for what they had to do—than in-the-field execution. The agencies simply could not muster the funds or the freedom to maneuver or the political will or the managerial incentives to get good fire on the ground. All the policies encouraged and allowed for restored fire; none required it. There was no prescribed fire equivalent to the 10 a.m. mandate.

In the years since 1988, all the parts had gradually gathered together. The Bush administration was at least neutral. Almost a quarter century had passed since the NPS broke ranks and promulgated its own fire policy—as many years had passed between 1968 and 1992 as between 1910 and 1935, when the 10 a.m. policy had been adopted. Thanks to Norman Maclean and Sebastian Junger, wildland fire was finding a national audience among the literati and the culture generally. The Yellowstone challenge had concluded by affirming the principles behind the fire revolution, and it cited flawed execution for the stubborn roster of failures and missed opportunities. Even the critical people were still in the ranks. By 1992 the revolution's founding triumvirate—Ed Komarek, Herbert Stoddard, and Harold Biswell—had passed away. But the people who had translated their vision into policy and plans were still in harness. Bruce Kilgore and Jan van Wagtendonk,

Bob Mutch and Dave Aldrich, Dick Rothermel and Norm Christensen, and those who had come of age during the revolution and would assume leadership later, the pioneer generation—all were ready to begin anew.

That mattered, as Stewart Udall understood. Alarmed at the Reagan administration's legacy, he had revised his *Quiet Crisis* for "the next generation," and with uncanny timing published it as the Yellowstone fires monopolized television screens across the nation. Updating his conservation pantheon, he celebrated the heroes of the new environmental movement, people such as Rachel Carson, David Brower, and Howard Zahniser, and organizations such as the Environmental Defense Fund, the Sierra Club, and even the Club of Rome with its call for restraint on human grasping. He reminded the "oncoming" generation that it, too, had a "rendezvous with the land," and he summoned its members to meet it with courage and humility.[3]

He said nothing directly about fire. Instead, he conjured up the context within which the nation had sought to reform how it managed fire, those environments where fire as a cultural and political phenomenon had to burn. Fire would behave as its setting dictated. Over the past 30 years the American fire community had witnessed a double breakdown—first, through deliberate assault, the fragmentation of the USFS hegemony, and second, at Yellowstone, the fumblings of the interagency apparatus that had been cobbled together to replace it. It would not be enough to simply collect the pieces in a box and run a charge of bureaucratic policy through them and expect them to self-organize. Something else was needed.

As the country headed into national elections, wrestled with the savings and loan scandal, and plunged back into recession, it was as though the institutional chemicals lay in solution, ready to crystallize if the proper precipitant could be found. That cultural jolt came with the changed fire environment created by the election and, as always, with fire itself. By then the land was prepared to respond to the effects of 80 years of attempted fire exclusion, the western climate was spinning into drought, partisan politics sharpened knives, and the fires had mutated into something more vicious and seemingly immune to the suppression strategies that had worked in the past. The narrative slopped over the old lines of containment.

Revolution 2.0

Controversies over the public domain were exhausted into working compromises. Like poet Gary Snyder, the country had moved from the belief that the fire control he knew as a lookout was "Right Occupation" to the recognition that it had been a national mistake, that "the Sierra Nevada is a fire-adapted ecosystem, and that a certain amount of wildfire has historically been necessary to health." In moving to Grass Valley, California, he found that the fire-management role of fire in sustainable forests "has given everyone at least one area within which they can agree." Their shared fire—or shared fire-shaped landscape—was potentially a common cause. So, likewise, new people arrived on America's fire scene and moved new ideas into the fire-governing bureaucracies. The old fire revolution had died; what emerged was a reconstitution, like a classic car remade with modern parts. Call it Revolution 2.0.[4]

* * *

The Northwest timber economy had been buffeted by court injunctions under the Endangered Species Act (most famously involving the northern spotted owl), but the deeper issue was the region's old-growth forests. They harbored many other species at risk but had also become the preferred habitat of the timber industry, and beyond old growth there were the matters of roadless areas and clear-cutting. What was true for the Northwest held for the Sierra Nevada and for much of the country. The sheer turmoil itself had become an issue: the paralysis it induced had infected both the BLM and the USFS and all their other programs, including fire. The Northwest congressional delegation wanted an enduring solution and turned to science for an answer. Unfortunately, the committee charged with investigating options concluded that "it would not be possible to both protect old-growth ecosystems and continue historical timber harvests." On that sour note, "Congress left the problem to the next presidential administration."[5]

The Clinton administration had campaigned partly on an environmentalist agenda, and in the spring of 1993, amid much fanfare, it convened a promised Forest Summit. The resulting Northwest Forest Plan broke the logjam, and it became a model for other resolutions in California and the Columbia River Basin. It is doubtful the higher administration saw fire in similar terms, or in President Clinton's case, even cared much about public lands, but its actions began to pry apart the locked horns that had stymied agencies from doing much beyond responding to emergencies. Ecosystem management promised to do for policy generally what the Northwest Forest Plan did specifically for logging. It quickly became official doctrine for the Forest Service, although what it meant on the ground was ambiguous.

Civil society—this time the Ecological Society of America—helped chisel the concept into a definition. In 1993 the indefatigable Norm Christensen was called on once more to lead an inquiry into "the scientific basis for ecosystem management." As he had in the past, Christensen sought to leverage the particular charge into something like a Leopold Report for a new era in environmental relations. Sustainability, ecosystem management, adaptive management, the contingency and incompleteness of knowledge—all were on the table, a modernized version of the conservation agenda that had inaugurated America's commitment to public lands in the 19th century. It was a replay of the quarrel between parks and forests, wilderness and working landscapes, biocentric and anthropocentric values. Revealingly, the committee's 1996 report argued that "ecosystem management is not a rejection of an anthropocentric for a totally biocentric worldview." It acknowledged the importance of human needs and only wished to place those necessities and yearnings within the long-term capacity of Earth to support them and within concepts of intergenerational equity and stewardship. In acknowledging that ecosystems must serve people, it did not equally acknowledge that people might serve or add value to ecosystems or might be indispensable for keeping them within historic ranges of variability. The chasm between the wild and the working remained; it only shifted to allow more freedom of movement. For its fire example the report chose the sequoia groves of the Sierra Nevada.[6]

But the agencies still had to translate principles into practices. In January 1994 the Forest Service hosted a roundtable at the Rensselaerville Institute. The theme was "Navigating into the Future," in which the agency characterized its status as having "embarked on what could be called a voyage 'beyond the maps.' Ecosystem management, sustainability, biodiversity, forest health—these concepts are taking the agency outside its traditional boundaries." The BLM announced its conversion that same January. The NPS and FWS made the transition easily by adapting their existing guidelines. From an ad hoc announcement, ecosystem management had become a national initiative. By the time the GAO surveyed the institutional panorama, some 18 federal agencies had formally adopted it. "Compared with the federal agencies' current approaches to land management, this new approach will require greater reliance on ecological and socioeconomic data, unparalleled interagency coordination, and increased collaboration and consensus building among federal and nonfederal parties within most ecosystems." Within the next year the Forest Service publicly announced its conversion to a Leopoldean vision with the 1995 publication *The Forest Service Ethics and Course to the Future,* while an Interagency Ecosystem

Management Task Force responded to the GAO's call for coordination with a three-volume summary on the "ecosystem approach."⁷

* * *

Behind those notions lay ideas and disciplines and professional societies that also had to adapt. The reformation most affected forestry. Two concepts proved particularly influential, one theoretical, and the other transitioning theory into practice. The first was the New Forestry, pioneered by Jerry Franklin and his colleagues. It affected ecosystem management generally, especially forests, and it aimed primarily at conserving existing landscapes. It built on the controversies over old growth in the Northwest. The second focused on forest restoration. It targeted ponderosa pine, a widely dispersed ecosystem that was notably upset by fire exclusion and that was also among the prime habitats for exurban development. Together they addressed the two landscapes, the wild and the urban, the pristine and the built, that were transforming the national estate into a land apartheid.

Yet those notions could be hard to reconcile in action and awkward to interpret for fire management. Both pivoted on a concept of forest health, broadly understood, but as researchers readily admitted, forest health was not a measurable scientific concept. So, instead, emphasis shifted from ecosystem health to fire hazard, which meant fuels, which was a metric that silviculture could handle. The process began in 1987 when Congress held some hearings after which the Forest Service prepared a strategic plan in 1988; it was not implemented. The idea revived in 1992, joining forest health with fire. Here the experiments begun by Wally Covington on the Pearson Natural Area around Flagstaff, Arizona, proved critical, not only for congressional attention but for their subsequent effect on Secretary of the Interior Bruce Babbitt. Together, New Forestry and what became known (rightly or wrongly) as the Flagstaff model identified the two biotas that would dominate attention: the forest of the Northwest, aptly symbolized by its majestic old growth, and the montane woodlands ubiquitous throughout the West, best known for their ponderosa pine, with old-growth yellow pine also at risk to chainsaws and fire and to a ravenous exurban sprawl.⁸

The Forest Service updated and retitled its former plan into the Western Forest Health Initiative; this, too, went unfunded and unimplemented. In 1995, while the federal common policy on fire was evolving, the Forest Service redefined the problem in terms of fuels. Fuel treatment was something that the agency could do and measure, and if done correctly it could also make landscapes more resilient against a seemingly unbreakable drought.

Fuels were an easy concept to sell—more fuel meant more fire, big caches of fuels meant big fires. Remove those fuels and you removed, or at least dampened, big fires. When Congress asked the GAO to review the fire program, what it meant was the fuels program, and what the GAO wanted was what it always wanted, better cohesion and accountability, preferably through quantitative measures.[9]

The fire dynamics of the two model forests were very different. When coastal Douglas fir or interior lodgepole burned, they blew up in giant batches. The Coast Range resembled a fault line, like a biotic San Andreas, broken by ruptures on a long-wave rhythm. The classic ponderosa forest was a savanna, with clumps of big trees clustered like megaliths on meadows. It burned regularly, sometimes annually, not unlike its longleaf cousin. The old-growth scenario represented all those forests that burned infrequently but necessarily through forest-sweeping conflagrations. The montane scenario could stand for all those forests that hungered for and sustained medleys of surface burns. The issue before the American fire community was not to reinstate fire per se but to restore the appropriate fire regime.

Where the New Forestry sought to define the forest as more than board feet and biomass as more than fuel, what evolved into the Flagstaff model interpreted the degraded ponderosa woodlands as more than a fire-famished fuel array. Part of what had inspired the experiment was the startling discovery that reintroducing fire alone might fail. Where, through decades of fire exclusion, the landscape had blistered into dog-hair thickets, stockpiled windfall, and downed boles, and had paved over the native grasses with pine needles, fire might kill but not burn up poles and scorch but not consume large-diameter woods, though it could girdle even old-growth yellow pines with slow-combustion smoldering at their roots. What restoration required was the right kind of fire, and the right kind of fire depended on the right kind of habitat. The maladapted forest first had to be manipulated—thinned, given a woody weeding—before fire could help and not harm.

Each proposed a place for a reinstated wood-products industry, though it would not be one based on big-tree saw logs or on clear-cutting. Both recognized that logging did not emulate fire. Cutting was a mechanical process; loggers took the big stuff and left the little. Burning was a chemical process; fire took the little and left the big. There were places where prescribed slashing and selective burning could nudge ecosystems into more historically adapted conditions. But neither industrial logging nor industrial fire suppression did that. In the end, the new thinking about forests and fire blended with the move toward the inchoate doctrine of ecosystem management.

There were plenty of reasons to reintroduce fire. New Forestry and the Flagstaff model explained why fire had to reappear as a regime and listed what might be needed to enable the return of what might be called the prodigal flame.

* * *

Though the Clinton administration sought to defuse the most debilitating environmental controversies, it demonstrated little interest in the enduring questions of public lands and the nitty-gritty of their management. It wanted to avoid problems more than it wanted to find solutions. Still, it granted some space to tinker and collect. These changes in direction coincided with a change in directors as well.

America's fire history had always been about people as much as programs and policies—the latter were often shorthand for the former and a way for a narrative to be crafted without cumbersome roll calls of participants. Similarly, institutions were the means to move beyond the insights of individuals. So the fire revolution, while it was announced by strong-personality patriarchs, had focused on agencies and their policies. Now, as the revolution struggled to revive, attention turned again to personalities. The age of the prophets had passed. The era of critical upper-level administrators had arrived.

In December 1993 the Forest Service replaced F. Dale Robertson with Jack Ward Thomas as chief. Thomas had risen through the ranks as a wildlife research biologist and had headed the Interagency Scientific Committee to Address the Conservation of the Northern Spotted Owl and then the Forest Ecosystem Management Assessment Team, the successor project that followed the Northwest Forest Plan. If anyone embodied in his personal career a statement that the agency was committed to ecosystem management beyond wood and forage, if anyone could by becoming chief advertise a new epoch for the Forest Service, Jack Ward Thomas did.[10]

Thomas had never nurtured political ambitions—he had in fact repeatedly refused to do the more or less standard "Washington Office tour." He had not trained through the Senior Executive Service (SES), and at first he turned the appointment down. He was coaxed and practically drafted by Clinton and Gore, and he yielded with the understanding that his tenure as chief would be short. He would oversee the transition—help retool a sluggish bureaucracy from what it had been to what it had to be—and leave. In 1992 he had delivered the Albright lecture at the University of California, Berkeley, in which he noted the contest over the national forests had become increasingly adversarial, presided over by "professional gladiators," an arena in which scientists were uncomfortable but in which they would have to come "off the bench" and mix it up. His subsequent career showed exactly what that could mean.[11]

Because he was not (as other chiefs) in the SES, he would be a political appointee, to which he objected. A compromise was concocted, and he again acceded, though the serial crises that soon engulfed the agency reinforced his sense that he was not a politician by either training or temperament, even in the guise of an agency administrator. Most of the politically driven frenzy he characterized as "lurch management." He had had some experience with fire—anyone of his generation in the agency would have. And he had memorably found himself at Yellowstone's Old Faithful Inn when the North Fork firestorm had threatened to consume everything in its path, an experience he could not expunge and did not wish to repeat. Yet few chiefs had as tumultuous a tour of fire duty as Thomas. None had visited fire camps as he did, an act many firefighters remembered years later.[12]

The agency's workforce had to change as well as its doctrine, and this too could seem abrupt if not brutal to the old regime. Soon after becoming chief, Thomas was instructed that the agency's "primary mission was civil rights" and meeting the USDA's hiring objectives. What Thomas's appointment said publicly about a commitment to ecosystem management, Thomas's appointment of Mary Jo Lavin as director of Fire and Aviation Management said about the transformation of its personnel. In 1960 it would have been unthinkable to have as a director of Fire Control someone who had not known fire intimately in the field, who had not been educated in forestry or range, and who was not a male. In 1992, with a final settlement of the consent decree, L. A. Amicarella, then FAM director, circulated letters and solicited projects to enhance "workforce diversity." The endeavor quickly segued into a very visible statement the next year when Lavin became acting director, and soon the director in her own right. FAM, Thomas felt, needed someone outside its usual ranks and off the fireline, and especially it needed someone who was primarily a manager.[13]

Lavin had advanced degrees in higher education administration and the Kennedy School of Government at Harvard before moving into natural resource management, first with Washington State, and then with the Forest Service, becoming the agency's first female deputy regional forester— this in Region 6, the scene of the old-growth controversies. (In 1988 the region had sponsored the first Women in Fire Conference.) From there she transferred to deputy chief for State and Private Forestry, and as workforce diversity became an agency priority, to Fire and Aviation Management. She brought a fresh pair of eyes and ears to the USFS fire scene at a time of still-wrenching reforms. She revived a moribund *Fire Management Notes*. She launched a workforce diversity campaign under the title "Faces of Fire" (although one observer muttered that he saw a lot of faces but not much flame). That project merged with the chief's "Workforce 1995: Strength through Diversity" action plan. On her watch, long-standing concerns over the workforce for fire management intertwined with the hiring of women and minorities. Even so, most of the agency's newcomers were the -ologists required to satisfy NEPA reports.[14]

Both appointments symbolized the ways in which the Forest Service and its fire program were pushed and pulled by larger agendas. The program intended to do for workforce diversity, in a sense, what the Northwest Forest Plan had done for biodiversity; at a minimum it should dampen a political crisis. If Jack Ward Thomas was intended to quell the festering crisis about the mission of the USFS, Mary Jo Lavin was intended to confirm and calm the turmoil stirred by the consent decree, firefighter retirement, the decay of the traditional militia, and affirmative action. Neither had sought the assignment. If Thomas had shown himself a gifted scientist and found (as he suspected) that administration was another beast altogether, so Lavin had demonstrated a remarkable talent for administration and was to find that fire management was a program that did not fit tidily into spreadsheets or the lectured categories of the Kennedy School. Managerial gurus could speak loosely of firefights and body counts, but for Mary Jo Lavin those figures of speech became all too literal.

Chief Thomas recalled, somewhat to his surprise, that he more often met with officials from Interior than from Agriculture. As they moved out of commodity production, the mission of the national forests more resembled parks, wildlife refuges, and protected lands than cornfields and feedlots. The Council on Environmental Quality provided a forum akin to that of the NWCG. Thomas developed a good working relationship with Mike Dombeck, formerly of the USFS and now director of the BLM. He believed it was "probably the closest of any chief and a BLM director." He thought the administration favored Interior and "chose to scapegoat the Forest Service," as it had during the spotted owl controversy with the BLM's O&C Lands and again when the notorious Timber Salvage Rider wedged openings for postfire "salvage" logging into forests that the Northwest Forest Plan had effectively shut down. But, creaking and stuttering, the Forest Service was moving to recover from the traumas of the past dozen years and syncopate its mission with that of the other federal land agencies. Often Chief Thomas found himself receiving orders from Interior

Secretary Bruce Babbitt, certain that Babbitt had the necessary authority but unclear whether Agriculture Secretary Dan Glickman had approved. The old ambition to return the national forests to Interior, from which the 1905 Transfer Act had plucked them, was becoming a de facto, though not de jure, reality.[15]

The catalyst was surely Secretary of the Interior Bruce Babbitt. Probably no previous secretary had a résumé so ideally crafted for the post. Babbitt had grown up in a ranching family in northern Arizona before taking advanced degrees in geology, followed by a law degree from Harvard. When he became governor of Arizona, he oversaw landmark reforms in groundwater management. That by itself was not unusual: Interior secretaries were traditionally western governors who spoke for one side or the other of the age's environmental politics. But Babbitt was the only secretary of either Interior or Agriculture who had seen and fought fire as a youth—had like many youngsters of his time and place passed himself off as 18 so he could be hired for occasional fire duty, which he relished as he did outdoor activities generally. Others sought to assist fire management by unburdening the agencies from issues that hampered their effectiveness and hence weakened their fire programs. Uniquely, Babbitt took on fire directly.[16]

He knew fire mattered, and knew it was somehow central to his larger ambitions for the public lands. He read Maclean's *Young Men and Fire*. Then, recognizing that he could not rely on his experiences as a teenager, he decided to learn the contemporary scene by joining a crew and working on real fires. He was told, politely but firmly, that he would need to be red-carded—meet NIFQS standards—like any other rookie. Because he could not afford a month at Boise (as proposed), he arranged for instructors to join him at the Interior Building in Washington, DC, for a string of Saturdays over a six-month period. He recalls with relish his efforts to practice deploying a fire shelter on the roof of 1849 C Street.[17]

In June 1994 he joined the Midnight Sun Hotshots from Alaska on the Bunniger fire outside Grand Junction, Colorado, and later for others. The self-deployments were initially curiosity, maybe a hobby, then a chance to leverage his status as secretary to experience the outdoor world he enjoyed while learning firsthand something about a critical issue. Where other administrators had seen fire as a problem, he also saw in it a promise. Then, during that 1994 season, fires turned lethal, the cost of suppression blew away old budgets, and the fire revolution was poised on a knife-edge ridge to move ahead or fall back. Like many participants, Babbitt wished it to advance; unlike many, he was in a position to toss a few matches into the political tinder. In 1995 he visited Wally Covington's plots outside

Flagstaff, not far from where he had grown up. Impressed with what he saw, and working with Arizona's senators, he arranged for Mt. Trumbull, a forested BLM sky island on the Arizona Strip, to be dedicated to test the theory on an operational scale. Fire became a signature concern of his secretarial tenure.

For 25 years the Interior agencies had noticeably expanded their fire capabilities. After Yellowstone, Interior had begun to pool resources, create a formal fire budget process, and provide oversight with a fire policy coordinator. During the remainder of the 1990s, field programs consolidated, as research did. Interior claimed parity with the USFS and filled in where the Forest Service lagged or stalled. Fire personnel began to transfer from agency to agency, often more loyal to the fire fraternity than to any particular host institution. The interagency vision had become a reality. In 1994 BIFC changed "Boise" to "National" and became NIFC.

* * *

Among its ballyhooed innovations, the Clinton administration, led by Vice-President Al Gore, proposed to "reinvent government." For the land agencies the exercise seemed a "distraction," as Thomas recalled, that had for the Forest Service "almost no good result." What the agencies needed was clarity in their missions and the material support to advance them. For all the "hand grenades," such as reinvention and affirmative action, that the new administration tossed, it did succeed in blowing up some of the worst politically charged logjams on issues that had plagued and prodded the agencies, such as old growth, clear-cutting, and roadless areas, and it had lubricated a pivot toward a genuine operating philosophy of managing for whole ecosystems. Notoriously, it had also left Secretary Babbitt out to dry in his campaign to reform grazing and mining on the public lands.[18]

Then the 1994 elections turned control of the House to Gingrich-led Republicans. Unable to abolish existing legislation it did not like and unable to enact alternatives it did, the House leadership again turned to fiscal showmanship. Reforms slowed. Babbitt recalled that the self-styled insurgents were "ready to dismantle every environmental law in sight" and even submitted a bill to close "excess" national parks. In late 1995 the U.S. government shut down for 28 days. The landmark environmental reforms had come through a Congress that, however fissured by political party, had mutually rallied behind a consensus to halt pollution, preserve wildlands, and protect endangered habitats. The Reagan administration had ended that bipartisanship, and with divided government, dissent became chronic and compromise a badge of dishonor. The old gangs were back in the saddle.[19]

But none of the contested politics—the roads, the logging, the endangered species, civil rights—spoke to fire as a fact on the land. Fire did not vote. It did not caucus. It was not a box on an affirmative action checklist. It could not be summoned for a dressing down at the secretary's office or the White House. It could not be rescinded. It could not be filibustered. It listened to the wind and obeyed the order of woods, brush, and grass on the land—these were the bullet points of its own Contract with America. After several decades of relative quiescence, fire was poised to storm back into national attention. This time it would pursue its peculiar purposes, and this time there was little in the landscape to oppose it.

The Coast Is Toast

That Malibu burns is not news. It seemingly burns as often as Minot, North Dakota, floods, and for the same reason. It lies in a fire floodplain, or more accurately a fire delta where Santa Ana–driven flames spill down Topanga Canyon from the Santa Monica Mountains. If it were not in Hollywood's backyard, if the scene were not populated with celebrities, its fires would be almost banal, and so, it might have appeared, for Southern California overall, which was laced with similar fire corridors. Almost annually something burned. Fires raged around Los Angeles and San Diego as though the Transverse and Peninsular ranges were volcanoes that belched and occasionally overflowed with lava that every so often spread into the settlements that clawed up their slopes. For the country overall, it was mostly spectacle, like a celebrity scandal, and occasionally a disaster. For the American fire community the scene was more. Every decade or so, an outbreak became a disturbance in the Force that warped the contours of the nation's pyrogeography. On big fires, costly fires, deadly fires, California dominated national rankings.

Inevitably, the October 26 to November 4 fire siege in 1993 struck Malibu as it rimmed the perimeter mountains with wildfires, but so vast was the contagion that this time Malibu was upstaged by even worse outbreaks, and the entire complex was telescoped into the Laguna Beach fire in Orange County. With Santa Ana winds pummeling the mountains, with pulses peaking at 92 mph, the fires rolled out day after day, a total of 21 major burns across 6 counties. The North Malibu, Old Topanga, and Laguna Beach fires burned clear to the beaches. The rest burned onto the suburbs that defined the shorelines of the urban South Coast. The outbreak set new records for the mobilization of regional and national firefighting resources, damages, and costs. Images of incinerated homes, weary firefighters, and untrammeled flames spread across the media like an ember storm. The figures ratcheted up, and finally just numbed: 197,225 acres burned, 1,241 structures turned to ash, 162 people injured and 3 killed, damages in excess of $500 million—and this before the winter rains replaced the flames with flood waters and debris flows. National headlines shrieked, "The Coast is toast."[20]

Of the 26 fires reported over the 10-day siege, 19 were known to be started by arsonists, and the others were the outcome of carelessness, whether from a transient's abandoned campfire or power lines arcing in the relentless winds above slovenly kept rights-of-way. That incendiarism made it possible for many observers to deflect attention toward the bad behavior of individuals rather than toward the collective culpability that had transfigured coastal plains and soaring mountains into a crucible of combustion. The real problem was how Americans lived on their land. If arsonists had not set fires, some other ignition source eventually would have. Over the years the leading cause of the South Coast's big burns was power lines that failed or that arced showers of sparks during high winds—it would be difficult to design a better landscape for bad fires. Once a house kindled, it spread to other houses without the need for chaparral or coastal sage to assist the transition. Instead, attention turned toward better protection in the form of more law enforcement to halt arson, more engines and air tankers, more structural solutions to avoid changing the pattern of urban sprawl that made natural fires into cultural disasters. This time, however, the political synapses began to spark in other pathways as well. After the disaster Malibu enacted some of the strictest fire codes in the country.

This time, too, the fires were not so easily brushed aside to ensure that the torrent of development could continue. In his jeremiad on Southern California, *The Ecology of Fear*, Mike Davis transformed the Malibu fire scene into a Hieronymus Bosch–like tableau of everything that was wrong with a metropolis that occupied a special niche in the American imagination as the place that everyone loved to destroy. The 1993 fires joined other natural, social, economic, and political catastrophes—the Rodney King riots, the implosion of the defense industry, chaotic immigration, unsustainable development—into what Davis termed "a theme park of the Apocalypse." That the country was willing to spend endless dollars to defend the indefensible in the Santa Monicas while ignoring urban blight, with its own fatal fires, symbolized for him all that was wrong with a system that privatized profits and socialized losses.[21]

Among its legacies the firestorm of '93 bequeathed an iconic image of what the metastable border between the urban and the wild meant. An aerial photograph of the Laguna Beach hills showed three parallel streets across the ridgetop, with every house but one burned to its concrete pad. The spared house overlooked the steep brush fields that had carried flame to the summit, but it had a tight tile roof and concrete sides and resembled a hybrid between a seaside mansion and an artillery bunker. It served as the image of choice for the I-zone much as the photograph of Old Faithful surrounded by soaring flames did for the conundrum of fire in the wild.

* * *

What, nationally, did the 1993 fires mean? There were several templates on the shelf for interpreting them.

One pointed to the rhythm of big fires in the West. They came with droughts, swelled and burst like climatic bubbles, clustered into two or three years. The firestorm of '93 was the successor to the siege of '87 and the fires of '88, or any number of other batched burns such as those of 1970 and 1977. After Yellowstone the fire scene had calmed, as the old scenario had predicted. There was little reason to believe the 1993 outbreak promised anything different.

Another template characterized the firestorm as the latest of the serial conflagrations that had habitually plagued California. They were a Californian oddity, a freak peculiar to the Left Coast, not likely to propagate widely or to root outside its cultural miasma. Yet their effects could unhinge national institutions. They commanded too many suppression resources and too much of what was, after all, limited political capital. They distorted the entire fire establishment. They threatened to hijack the fire revolution into more firefighting and less fire lighting. The two districts in the entire national forest system with the heaviest fire calls were in the gorges flanking the San Gabriels.

But this time neither of those two prevailing templates held. The 1993 season inaugurated a cadence of conflagrations that defied the rhythms that had formerly governed the marshalling of big fires and big fire seasons. Instead of one or two bad seasons in a decade, the breakouts came almost annually somewhere, and on alternate years nationally. In 1994 the West erupted from Colorado to the coast. The outbreak swept away old records for both suppression costs and fatalities, began rewiring the motherboard of the nation's pyrogeographic circuits, and trashed the old interpretive templates. The American fire community shuddered collectively and admitted that not only the old ways had failed but perhaps the promise of the fire revolution. The South Canyon fire, in particular, shook its foundations in ways Yellowstone had not. Tragedy solidified a resolve to do things differently.

Then came 1996. The season added 17,000 fires and 2 million acres to the legendary 1994 totals: another rampage around the West, another record burn, another helpless struggle to contain what no longer followed the old script. In 1998 the contagion spread to Florida, which seemed to trade its hurricanes for conflagrations. From Memorial Day to the end of July, 2,300 wildfires burned across nearly 500,000 acres, exhausting fire organizations and siphoning in the bulk of the national fire establishment. The fires incinerated over 150 structures (and 86 vehicles), forced more than 50,000 to evacuate, and drew in over 10,000 firefighters and 7 national incident management teams, which granted it a larger muster of firepower from crews, engines, and aircraft than had assembled at Yellowstone 10 years earlier. For a while fires closed I-95 between Jacksonville and Titusville and postponed the Pepsi 400 at Daytona. Some of the early fires were kindled by dry lightning and the later ones by any and every ignition source on the land. For 60 days Florida lived under a state of emergency. Each of Florida's 67 counties was affected, with Flagler, Brevard, and Volusia the worst hit. Costs and damages were reckoned to exceed $1 billion.[22]

What made the fire bust prominent was not the numbers, which were quickly losing their power to impress. What mattered was the shock of the fires. Florida was the fourth most populous state. It was not a rural backwater or an unbounded wilderness, nor was it a polity unfamiliar with fire. It presented an intermix landscape, not an interfaced one, but it was a fire state—always had been. It boasted one of the major fire cultures in the country. It could fight fires as well as anyone and light them better than most. It had passed legislation requiring clearing around housing developments. If Florida could burn like this, so could everywhere else.[23]

The biennial rhythm broke when fire again romped around the West in 1999, then reinstated itself in 2000 as fires swarmed over the West, concentrating with special fury in the Northern Rockies and mocking 90 years of progressively sophisticated suppression since the founding Big Blowup. The breakdown in suppression was matched by an equally horrendous breakdown in prescribed fire, as an escape rushed out of Bandelier National Monument and into Los Alamos. That year some 92,000 wildfires scorched over 7.4 million acres. The 1994 season had doubled the burned area of 1992; the 1996 season had ramped that figure up by 50 percent; and now the 2000 fires boosted the new record by yet another 50 percent. Secretary Babbitt declared the country was experiencing a "fire crisis."[24]

The national fire establishment confessed both to itself and publicly that it was broken. Since the new cycle of conflagrations had begun, wildfire had overwhelmed every apex of the fire establishment's triangle. It had swept over the I-zone in California, defying the five largest five departments in the country. It had barreled across the core of prescribed fire in Florida, indifferent to the state's astonishing successes in policy and legislation. And it had rambled with the insouciance of a grizzly bear over Northern Rockies wilderness. To the charges that it was ineffective, the American fire community pleaded nolo contendere—it had in fact so charged itself.

It was as though the old rules of engagement that had characterized fire management, a working consensus (as it were) between people and nature about how fires would behave and what people would do, had shredded in parallel with the breakdown of a national consensus about the politics of the environment. The old order had plenty of costs and failures, and few people were happy with how the revolution was playing out, either because it had done too little or it had done too much. But the order of things by which fires stormed over the landscape in relatively predictable ways and the responses that one might make to them, however marginal at times, were understood. There had been a working compact that said when and where fires came, how large they grew and how severely they burned, and how people might understand them and protect themselves. That agreement included the possible replacement of bad burns with good ones. Now that implicit compact had been sucked up and lofted away in the convective plumes of a new order of firestorms.

The emerging disorder evolved in eerie syncopation with America's rising political dysfunction. The bad fire years came with election years and served as photo ops. They were not occasions to address the country's fire problems but opportunities to exploit those fires to animate other messages. The 1998 outbreak aligned with the attempted impeachment of a president. The 2000 collapse matched a bitterly disputed election decided on a 5–4 partisan vote by the Supreme Court. The 2003 season in California coincided with a gubernatorial recall. The inability of the American electorate to agree on what they wanted or even on a process by which to reach agreement seemed, in creepy ways, to have somehow channeled onto the land itself. Efforts to "starve the beast" of the federal government only led to a debt burden in unburned wildland fuels as deep and structural as fiscal deficits. A political scene that was prepared to break down and burn up rather than find consensus was oddly mirrored in an American landscape aflame in ways no one wanted and no one could agree on how to change.

Annus Horribilis, Annus Mirabilis

In 1994 Smokey Bear celebrated his 50th anniversary, but few fire officers, hotshots, or engine crews were in a triumphalist mood because the question of which historical template the fire seasons after 1993 would follow was soon answered. None of them. If not exactly unprecedented, the unfolding seasons, beginning in that annus horribilis, could no longer consider the past as simple prologue.

It was not just that 1994 was a bad fire season: it was a traumatic one. Probably no year since 1910 so scarred the American fire community. The Yellowstone blowout had stunned the public with the potential magnitude and savagery of the new era and had set back programs, but it had left the agencies relatively unscathed. It did not challenge fundamental tenets or practices or inspire a major policy revision; instead, it urged the agencies to double down on existing norms, and some, such as the Park Service, saw significant boosts in funding. The fires of 1994, however, struck at the way the fire community saw itself and how it did business. They forced the agencies, from departmental secretaries and chiefs to squad bosses, to question the unthinkable. They rekindled the revolution.

The box score summary is impressive in a stilted way. Some 79,107 fires burned 4,073,579 acres at a cost of $918,335,000—the nation's first almost $1 billion suppression bill. The numbers of fires were average, and the acreage burned was high, but it was not inflated by Alaska. What surprised was the invoice for their suppression, which veered off the scale. There was an undeniable disconnect between fire size and cost. Some fires, such as the Tyee Creek Complex, were big on both counts: 135,170 acres controlled at a cost of $38.2 million. But many broke the logic between area burned and expense. The St. Joe Complex burned 850 acres at $4.7 million. The Chamberlin fire burned 1,214 acres at $3 million. The Hatchery Complex romped across 43,463 acres at $23.3 million. The Ruby fire burned 944 acres at $1.8 million. The reasons were several, but as firefights reoriented themselves from the deep backcountry to the peri-urban fringe, their costs ballooned. More urban and volunteer departments were called to assist. The threat to life and property created political pressure to respond with massive force—a kind of Powell Doctrine applied to wildfire. Since Yellowstone, it had become almost routine to appeal for military assistance, which did not come without high bills of lading.[25]

So, too, size did not reflect significance. The big fires were not larger than many others except in costs. Big fires were no longer a dime a dozen (more like six or seven figures apiece); another big fire was lost in the historical noise. What mattered was the cultural resonance of a fire. The fire that defined the season, South Canyon, burned a scant 2,115 acres. But it overran a mixed crew of smokejumpers, hotshots, and helitackers. Fourteen died. More to the point, they died in a way that seemed to recapitulate, with unsettling fidelity, the tragedy of the 1949 Mann Gulch fire that killed 13 jumpers and was the subject of Norman Maclean's *Young Men and Fire*. It was as though history had passed through a looking glass, transposed 49 into 94, and replayed the horror. Every account of South Canyon found itself paired with a retelling of Mann Gulch; the *NFPA Journal* even ran dual stories, and John Maclean, Norman's son, wrote a book. Life imitated art. Instantly, the entire season collapsed into the saga of the South Canyon fire. [26]

* * *

To the question, "What went wrong?" the answer was, "Practically everything." But wildland fires are inherently messy: they do not occur in office complexes or on shop floors; nothing ever goes according to bureaucratic rules or training precepts. Firefights are, by their nature, improvisations. The hope is that the juggling occurs within known parameters, that adaptations and miscues cancel one another out rather than compounding one atop the other, that crews on a line will not experience an unexpected collusion of contributing causes that can lead to tragedy. Of the almost 80,000 fires that season, including the 40 fires (South Canyon among them) kindled by dry lightning within the range of the Western Slope Fire Coordination Center at Grand Junction on July 2, only South Canyon burned over a crew. That blowup hinged on an exquisite act of timing, of hours, perhaps even minutes, as a cold front blustered through while hand crews were trenching an undercut line. But because the fire ended horrifically, the story of South Canyon unfolds as a string of beaded mishaps, like the Nomexed dead strung out on a spur ridge of Storm King Mountain. Everything that had to go wrong did. As one of those recruited to review the fire, Ted Putnam, put it, "It was just a shitty fire."[27]

The original fire went unreported for a day. Its location on Storm King Mountain was then misidentified as South Canyon. It was left to be monitored, and air tankers were diverted to more active fires. The smoke was visible from I-70 and the outskirts of Glenwood Springs. Smokejumpers were dropped onto a site that hand crews had walked into. By the morning of July 6, 49 firefighters of assorted provenance and experience were assembled around the fire—Prineville Hotshots, smokejumpers, BLM district firefighters, and helitack crew members, an outfit that had never worked together before. Crews were split, there was some uncertainty over command, and a weather forecast predicting the afternoon passage of a cold front went missing. The little screwups, as Norman Maclean had called them in *Young Men and Fire*, multiplied. None would have mattered except that when the winds struck, the crews were strung along an indefensible fireline halfway up the slope. Around 3:20 p.m. the flames quickened; by 4 p.m. the "whole canyon" was becoming a simmering cauldron of heat pocked by smoke and scattered flames and spotting. By 4:18 p.m. spot fires were popping up in the canyon below, and with 40–45 mph winds behind them they shot up the slopes in a handful of minutes. The firefighters found themselves in exactly the wrong place at the wrong time. Nine hotshots, three smokejumpers, and two helitackers died in the flames. It was the worst fire tragedy since 1953.

The response was instantaneous. Thanks to Yellowstone the national media was primed to report fire stories, and thanks to *Young Men and Fire*, commentators had a prism with which to refract this one. But the leaders of the American fire community did not need CNN or the *Denver Post* to prod them into action. Secretary Babbitt had just arrived at NIFC on another matter when he heard, and he immediately arranged a flight to take him to Grand Junction. Chief Thomas learned from a midnight call. He informed BLM director Dombeck, and they flew to the scene that afternoon. In his mind's eye Thomas recalled the day he stood outside Old Faithful Inn as a firestorm raged. He thought he understood something of the awe and terror of a blowup.[28]

They were on site, they spoke with the survivors, they arranged for the inevitable review team. They saw firsthand, up close, and more personally than any would wish, what America's firefight with the wild actually meant and what its costs could be. Babbitt had already gone to firelines as a grunt firefighter himself. Wildland firefighting, he appreciated, had changed; so had the wildlands, whose new circumstances were "making western forests ever more dangerous and explosive." Thomas insisted that "each of the bodies be escorted by a regional forester as it was delivered to the place designated by the family." South Canyon was a BLM fire but had burned over USFS crews. The media did not discriminate, and Chief Thomas accepted responsibility. This fire was personal. No prior fire tragedy had ever attracted such intimate attention. On July 13 Mary Jo Lavin announced a service-wide, hour-long "stand down for safety" that was understood as a tribute to the fallen.[29]

The formal machinery went quickly to work. A fire accident investigation team was being assembled even as the flames spilled down the backside of Storm King. It issued its preliminary report in August. A more in-depth version, with recommendations for improved firefighter safety, was released in October. The *Final Report of the Interagency Management Review Team* appeared in June 1995, without full consensus. This time, however, the agencies would not be allowed to conclude with an in-house postmortem. The year before the Occupational Health and Safety Administration had tentatively encroached into the work world of wildland fire and now mounted a full inquiry of its own. It had three investigators on the scene by July 7. As NEPA had restrained the agencies' judgment on land use, so OSHA now intervened in how they ran field operations.[30]

OSHA's inquiry added little new when it was released on February 8, 1995, though it cited the USFS and BLM for two violations. Its investigation agreed with the "excellent" reports of the agencies. It concluded that the agencies had violated their own standards and protocols, or to put a face on the tragedy, that "no one person or group was responsible for ensuring the safety of the firefighters." It supported "systemic" efforts to reform the larger "workplace" by "suppression preparedness, fuels management, and the wildland/urban interface," for if such fundamental policy issues were not "squarely addressed," firefighters would be unnecessarily at risk. It found that the proposed corrective actions should remedy the problems. Essentially, OSHA served notice that the agencies would no longer be allowed to police themselves, that OSHA would investigate fireline accidents as the National Transportation Safety Board did aircraft crashes.[31]

The agencies already had a face on the tragedy—in fact, 14 of them. Some critics were unhappy that the investigations had made the problem systemic and not named names of those who had failed; in protest, Ted Putnam refused to sign the final report. With unusual unanimity, the American fire community rallied and determined that it would self-reform life on the firelines. South Canyon acted on the fire establishment as the killing of Army rangers in Mogadishu did on the Pentagon and the policy of foreign interventions. The resolve was palpable not to let this happen again. Unsaid but understood was the acknowledgment that there was nothing on Storm King Mountain that justified the kind of aggressive attack that led to dead firefighters. That sentiment ran like an electric current through the community from ground pounders on the line to agency chiefs. When the Forest Service published a new code of conduct for fire suppression, it left no wiggle room regarding the primacy of safety. As Chief Thomas declared, "I gave the order and meant it. Safety first on every fire, every time."[32]

Amid the press inquiries that followed the OSHA report, Jack Ward Thomas noted in his journal one query in particular.

REPORTER: Chief Thomas, is this devastating report a blow to your agency?
THOMAS: Madam, the blow to the Forest Service was losing fourteen of our people to a firestorm. All else is trivial. Good day.

"These were my people," he wrote. "I did not need a report from OSHA to remind me of my responsibility."[33]

He and Mike Dombeck got one anyway, and their successors could expect the same.

* * *

South Canyon was the cipher of the 1994 season, and it served to galvanize the fire community as the Yellowstone fires had the public. But among the 79,107 other fires that year was one in Glacier National Park that, arguably, had as much practical influence on the future of fire policy. The Howling fire resembled a scurrying, overlooked mammal amid an age of monstrous dinosaurs. The future evolved out of it.

The Howling fire began on June 23, 1994, from a lightning strike on the North Fork of the Flathead River near the park's western border. At the time Glacier was one of a handful of parks with the size, funding, and clout to allow prescribed natural fires. A large burn, the Starvation Creek fire by the Canadian border, was under the direction of an Incident Management Team. Suppression resources were strained. To oversee the Howling PNF, the Park Service assembled an ad hoc team to predict the fire's behavior and plan for contingencies. The Howling fire did not blow up, and by late August it was blocked to the east by two other fires, both wildfires, but both controlled through a confinement strategy. The three fires burned together. After the Starvation Creek fire was controlled, the Howling PNF overhead team assumed control for it as well. For 75 days the team stayed with the fire until responsibility was ceded back to the park. All in all, the fire burned for 138 days. By accident, necessity, and daring, the experience with the Howling PNF demonstrated how to cope with fires of long duration by substituting fire behavior knowledge for heavy machinery, by trading land for options to maneuver, and by adapting opportunities presented by nature to the strategic purposes of the park. Fire officers began by making the best of an awkward situation. They ended by inventing a new mode of operation.[34]

Interestingly, the Forest Service had a parallel experience on the Granite fire, started on August 28 in legal wilderness within the Hells Canyon National Recreation Area. A rugged landscape, a failed initial attack, problematic fire behavior, and limited suppression resources were factored into an escaped fire situation analysis that led to some containment but otherwise continued surveillance. On September 11, a regional wilderness fire assessment team validated that decision. "Suppression" had meant "confinement." The fire expired on October 14. A postfire review noted that the strategy had saved money, spared putting firefighters at risk, promoted ecological benefits, and stayed within the rules.[35]

When agency officials convened to translate the 1994 season into a reformulation of policy in 1995, the momentum behind the exercise came from the shock of the South Canyon tragedy and the season's cost in lives, dollars, and burned area. But the outcome led to a steady liberalization in handling wildfires that dissolved the PNF in favor of a reincarnated avatar: "wildland fire use for resource benefit" fires (WFU), along with the fire teams needed to handle them on the ground. To the architects of the new order, the South Canyon fire lent urgency to reform, and the Howling and Granite fires provided the means to enact it. It took another dozen years, more tragedies, more official reviews and OSHA citations, more big science, more escaped burns, and more fumbling, mumbling, and missteps, but the future had its beta test.

Appropriate Response

Among its responses to the 1987 and 1988 seasons, Congress had authorized a National Commission on Wildfire Disasters as part of the Wildfire Disaster Recovery Act of 1989. In its parsimonious wisdom Congress denied the commission any funding, forcing it to rely on donations of kind and money (but none too large). The commission published its 29-page report in May 1994. In addition to the usual suspects, it enlarged the realm of culpability from the agencies and fire programs to "a failure of land management and public policy." Uncontrolled wildfires were not an "unpredictable act of nature." They represented choices Americans made about how to live on their land, both the public lands and the privately owned and rapidly urbanizing peripheries. A solution could not be found by reforming fire programs alone. The land itself had changed, and the only hope of containing fire was to proactively engage the environment within which those fires burned.[36]

The commission forwarded its report to the secretaries of Agriculture and Interior as mandated. Its public venue took the form of a summary article in *American Forests*. To skeptics the report was code for reintroducing a wood-products industry into the wild because it was chaired by Neil Sampson, then vice-president of the American Forestry Association. Was a disguised regimen of slash and burn what the commission meant by "aggressive, proactive" management? Still, the report had the effect of shifting responsibility from fire-suppression policies to the conditions under which the fires burned. Wildland fires were out of control because wildland fuels were. The United States had a national crisis in "forest health." Drought, disease, insects, and fire exclusion—and, the commission should have added, past logging and grazing practices—had rendered the federal estate "an explosive time bomb." Human actions had transformed altered natural conditions into a disaster, and humans would have to begin repairing that damage. If actions were not taken soon, the future would only grow worse.

The commission looked to the recent 1993 California season as a possible prelude. In 1994 the future arrived.

* * *

Within a year the institutional scene was overgrown with reports—virtual bureaucratic fuel ladders of paper ready to erupt into a crown fire of policy. The studies tended to cluster around three themes. Two were obvious: the fires in wild and built landscapes. The third scrutinized the workforce, broadly interpreted, or the capacity of the agencies to put pulaskis and torches on the ground. Fire in the wild, fire in the urban fringe—these could be separately handled. It was trying to combine them into a single agency, or pact of agencies like the federal land bureaus, that threatened to transfer that external tension inside and tear institutions apart.

The shift from timber to ecosystem management as an informing agenda had its counterpart in a redefinition—or parallel definition—of fuels into "forest health." In 1993 a major conference assembled the key researchers for a week's assessment in Idaho, published as *Assessing Forest Ecosystem Health in the Inland West*. With this as a text, the USFS launched its Western Forest Health initiative. Fire's exclusion lay at the core of what had sent the woods into a spiral of decay. A January 1995 conference in Florida, Environmental Regulation and Prescribed Fire, explored the legal, political, and social restraints that still shackled prescribed burning.[37]

Together forest health and prescribed fire furnished a larger context for both fire's restoration and its active management. It moved the debate from fire as a natural process in the abstract to fire as an ecological catalyst, the removal of which could be profoundly disruptive. There could be no neutral stance. It made a case that fire could not be restored successfully without considering its habitat. Fire could not renew itself by itself. It was not ecological pixie dust that magically transfigured the degraded back into the majestic. Messed-up landscapes might only yield messed-up fires.

The urban side was more complex because it encompassed the states, counties, and thousands of local departments. The major players at the table all anted up with special inquiries of their own. The National Wildland-Urban Interface Fire Protection Initiative with the NFPA furnished a relatively apolitical setting, one granted a jolt of urgency by the 1991 Oakland conflagration and the 1993 Southern California fire siege. In 1991 the NFPA issued Code 299, *Standard for Protection of Life and Property from Wildfire*. The Federal Emergency Management Agency, which saw its own involvement (and expenses) rising through grants-in-aid, published its assessment in July 1992, *Report of the Operation Urban Wildfire Task Force*. The National Association of State Foresters contributed a report on *Fire Protection in Rural America: A Challenge for the Future* (1994). Much of "rural America" was in fact a rapidly urbanizing countryside that was pushing out into the woods with the intensity of clear-cutting in an earlier age. The fragmentation that threatened biodiversity also threatened fire protection. Without rural buffer zones, fires were harder to contain, and they spread across institutional landscapes that were maddeningly fragmented, if they existed at all. FEMA grants came only after the fact.[38]

There was not much to unite the two groups: the interface between them often resembled a trench from the Great War. Only the Forest Service was tasked with reconciling two very different kinds of fires, fire cultures, and fire ambitions. Urban and wildland fire organizations had no more in common than cotton farmers practicing flood irrigation had with urbanites filling swimming pools. Only in California had a hybrid organization evolved. The wildland group, mostly federal, wanted more of the right kind of fire. The exurban group, mostly states and local jurisdictions, wanted protection from all fire, and eventually perhaps the extension of urban-style emergency services into the hinterlands. Each saw the other as diverting scarce monies, crews, and political attention from their core mission. The breakup of the old hegemony meant the Forest Service no longer had the power to impose a common solution. Instead, it absorbed that conflict internally, yet another of what seemed endless fissures in American land and life.

For the public the USFS remained the premier fire organization, and rightly and (often) wrongly was charged when breakdowns occurred. South Canyon, for example, was a BLM fire, but it was widely interpreted as a

Forest Service failure. Certainly, the USFS, which absorbed the casualties, responded the most publicly. In 1994 it sponsored a baker's dozen of reviews, trying to meld ecosystem management, fire economics, workforce developments, forest health, exurban sprawl, and safety into a coherent platform. It had to span the spectrum of fires with a shrinking and less-skilled workforce. The outcomes ran across 1995, and in keeping with interagency esprit spread beyond the Forest Service itself.

On January 13, the agency released its "Strategic Assessment of Fire Management in the USDA Forest Service." The responsible committee included researchers as well as line officers and even had representatives from Interior and the National Association of State Foresters. In a dense, thoughtful analysis the report identified three areas of emphasis. One was to fully integrate fire into ecosystem management planning and practice in order to better capture the benefits of burning. A second was to grapple with the wildland-urban interface problem, which was threatening to engulf everything else (in the 1994 season the agency estimated that 75 percent of suppression costs went to interface fires). The third was to address the reorganizations necessary to make it happen. Some barriers, it concluded, were only perceptual. The committee was unable to decide whether FAM should remain in State and Private Forestry, where it could bridge with WUI cooperators, or return to the National Forest System, where it could integrate better with ecosystem management. The report ended with an appeal for "a strong fire research program" and continued international leadership.[39]

There was little new. The report was an internal document, an occasion to consolidate and systematically connect known and expected issues, to rake up into one coherent pile the leaves of insight and events that had fallen since Yellowstone. It was intended to assess "needs and roles" for the coming decade. Its identified goals would underwrite a fire cost study and the Federal Wildland Fire Management Policy and Program Review commissioned after South Canyon. It was submitted by Mary Jo Lavin, director of FAM, William Sommers, director of FFASR, and Steven Satterfield, director of program development and budget.[40]

The formal document, to both the agency and the public, was released in May 1995. The *Course to the Future: Positioning Fire and Aviation Management* was prefaced as a comprehensive strategy that would safely and efficiently "lead Forest Service fire management into the 21st century." While its immediate stimulus was the "challenging" 1994 fire season, which "made us question our basic beliefs about our organization's ability to deal with serious fire situations," it reached back to the "gamut of issues and concerns that have appeared repeatedly since 1985." These distilled into five themes, but they could be boiled down to two hard crystals—firefighter safety and the ballistic cost of large fires.[41]

Throughout, while coded in clear bullet points and graced with quotes from various authorities, the document radiated both determination and frustration. The public and media did not understand. Congress was contradictory. The fire environment was spiraling into a state for which there might be no remediation and from which there might be no retreat. The same problems keep repeating—to be met with the same nonsolutions. It was far from the case that the agency did not appreciate what was happening, or could not see beyond horizons of tree stumps. They understood as only people who saw real fire, year after year, could understand.

They could not act: that was the essence. There were too many competing interests eager to put fire to use to some purpose or another; the pillars of fire rising from Idaho and California were too telegenic and dramatic not to exploit. But that fervor was not put toward reconstituting fire management. The agencies were not bucking just drought and a century of how Americans had chosen to use their public lands, but how they were choosing to live today and how they seemed increasingly unable to govern themselves. The unbridgeable split between the wild and the urban was the landscape expression of the growing polarities of the American experiment. Amid the heady euphoria of the early revolution, it had been enough to change minds and hearts. It proved achingly more difficult to change facts on the ground.

The agencies knew what would not work, and they thought they knew what might work. The all-suppression policy had failed. The fire revolution had stalled. The answer did not lie in policy reforms because in truth existing policy was sufficiently flexible to do what most observers deemed necessary. Even the Howling and Granite fires had been within guidelines. The gist was the same conundrum that had plagued the fire community for over three decades: to move policy into practice on a scale large enough to make a difference. Otherwise the costs would rise in dollars, damaged landscapes, and lives. Chief Thomas added his personal imprimatur: "I expect nothing less than your full participation as we position ourselves for the future."[42]

* * *

Still, official policy was where the symbolism concentrated, like a parabolic dish gathering signals from wide and far and directing them into a single focus. The final report of the Federal Wildland Fire Management Policy and Program Review, issued on December 18, 1995, was a *summa teologica* of the American fire scene.

Secretaries Glickman and Babbitt had commissioned the review after the smoke cleared in the fall of 1994 and charged it with ensuring that "federal policies are uniform and programs are cooperative and cohesive." They sought a single "umbrella" under which to subsume the splintering, mission-driven programs of the federal bureaus: it was, in a sense, total mobility for policy. A steering group chaired by Charles Philpot from the USFS and Claudia Schechter of Interior orchestrated a convocation of research findings, representatives of the federal agencies, their federal "partners" (NOAA, FEMA, DOD, EPA), their state, tribal, and local cooperators, and the interested public. The process began on January 3, 1995, with an announcement in the *Federal Register* and a barrage of letters to potentially interested parties. A draft report was published in the *Register* on June 22. The period for comments was further extended on September 25. The official report was released on December 18.[43]

A massive exercise, the review was the fire equivalent of a constitutional convention—the first of several to come—by which the entire edifice of American fire would presumably agree on rights, roles, and responsibilities. The exercise was less revolutionary, however, than evolutionary; few new ideas were introduced. Instead, the review inaugurated a formal mechanism by which the pieces left by the fire revolution and the fragmentation of lands and fire duties created by the recolonization of once-rural America might find ways to come together.

That was an unstated, and quixotic, ambition. The review did address five major topic areas, presented nine "guiding principles that are fundamental to wildland fire management," and recommended a set of 13 federal wildland fire policies that were not intended to replace existing agency-specific policy but would "compel each agency to review its policies to ensure compatibility." It affirmed the argument that fire simply had to be reintroduced into wildlands, that the WUI was primarily a local matter for which federal assistance of various kinds was possible, and that states and local institutions would have to be integrated into the national infrastructure. It urged legal review of federal obligations in the interface. There were literally scores of tasks, major and minor, that would improve the scene.[44]

Perhaps nothing reveals the document's scope from the exalted to the mundane as its attention to words and numbers. It tweaked the language of policy from taking an "appropriate suppression response to wildfires" to taking "appropriate management response to wildland fires." Those two amendments—from "suppression" to "management" and from "wildfire" to "wildland fire"—wedged open a bit further the options available. They began to blur the hard distinctions that had bifurcated fire and an agency's responses. The goal, as Tom Zimmerman explained, was to guide selection of the "correct action for a given situation." (Boldly, he noted that the new policy would "not automatically translate to a decrease in wildland fire management costs in the future.") The review ended with a demand that the agencies keep fuller and comparable statistics. Astonishingly for an enterprise that had always insisted it was science based, fire statistics were laughable. Data was scandalously "incomplete, difficult to use, and not portrayed consistently," and in critical areas such as the I-zone, the kind of data needed was "only now being identified." More information, however, was not the boulder that blocked the trail.[45]

Throughout, there was a sense of urgency to the text. The time for further study had passed. It was time to act. On that count the review identified barriers to implementation, among them the "widely held view that agency administrators are neither held accountable for failures nor rewarded for accomplishments." In 1960, failure to manage fire was cause for dismissal. By 1995, a far more diverse workforce, a diminished workforce with fewer experienced fire officers, hired for a panoply of tasks, was prepared to push fire to a periphery where emergency funds would not compromise the programmed budget and administrators were neither blamed for wildfires nor penalized for sluggishness in restoring prescribed fire. Even as the fire revolution had insisted that fire and land use were inextricably intertwined in principle, they still diverged in practice. The confused incentives left the field to wildfire.[46]

The interface and the prospect of climate change remained the gravitational disturbances rearranging fire's environment. However much the federal agencies yearned to extricate themselves from the I-zone, the review confessed that elected officials and the public believed the job belonged with the feds and "opposed Federal government withdrawal." There was far more electoral power in the states and local fire departments than in the federal land bureaus. Further, the review recognized that there was "a widespread misconception by elected officials, agency managers, and the public that the wildland/urban interface protection is solely a fire service concern," not an issue driven by land use, zoning, codes, mortgage rates, and misplaced incentives. Fire protection in the interface was treated as a government subsidy, like crop support or highway construction.[47]

The review concluded with the recommendation, almost a cri de coeur, that "all agencies be directed to develop implementation plans that include actions, assignments, and time frames." The secretaries agreed and stipulated that the committee "prepare a joint integrated strategy for implementing the report by no later than March 1, 1996." The management oversight team decided to rely on the Federal Fire and Aviation Leadership Council (FFALC), housed under the NWCG, to furnish staff support, advice, and "implementation support." The FFALC convened the five federal land agencies into a quasi-permanent forum that could continue the work of the oversight team. In January 1996, the FFALC issued a separate report on prescribed fire or how to boost productivity and reduce agency differences. On May 23, the review oversight team issued its proposed action plan. It had 83 items, some for immediate action such as promulgating the "philosophy, principles, and policies" of the review, and (yet again) a reconciliation of fire plans with land-management plans. Other issues would require years and significant investments. Particularly with regard to the WUI, where the political and fiscal costs were high and discord greatest, with almost no drivers of sprawl under the control of the agencies, there was little expectation of immediate outcomes.[48]

The report arrived two months later than the secretaries had requested, yet in time for the 1996 season, in which 96,363 fires burned over 6 million acres. Shouldering his firepack, Bruce Babbitt personally inspected the unprecedented crown fires of the Hochdoerffer fire that had scoured the San Francisco Peaks outside Flagstaff. That fall Tall Timbers hosted its 20th fire conference at Boise on the theme of "Shifting the Paradigm" from fire suppressed to fire prescribed. Like the sour odor of an old fire, the scent of 1994 lingered across the years.[49]

* * *

While fireline fatalities and out-of-whack landscapes overcharged with combustibles were the most public issues of the 1994 season, the bottom-line dollar outlays attracted political notice. The USFS emergency fire expenses ($757 million) were the highest in its history. The federal bill for firefighting nudged the $1 billion mark. This was enough for the USFS to commission two studies on fire economics. One looked at large fires, which seemed the most immediate cause of the unprecedented costs; the other surveyed the entire panorama of fires and how they were financed.

The large-fire study, chaired by Denny Truesdale, recognized that the "dramatic jump in annual cost was partly due to numerous individual fires with total costs in excess of $10 million. The 20 most expensive fires in 1994 cost more than $200 million." The study team scrutinized 6 fires in

4 regions. They identified 7 factors, each of which helped boost expenses, and they made 27 recommendations, each intended to claw back some costs. The outlook was bleak. There were more fuels. Seasons went longer. The wildland-urban interface was a black hole sucking in resources (60–75 percent of costs, the Forest Service estimated), forcing the agency to act "as a surrogate for the Federal Emergency Management Agency." Heavy helicopters were hugely effective in the interface but hideously expensive. The effective workforce continued to shrivel. Traditional fire suppression tactics were pricy and often ineffective.[50]

The report concluded, "Emergency fire suppression expenditures are increasing. The Forest Service is spending more per fire than it has in the past, and more per acre burned." There were few carrots or sticks with which to minimize costs during large fires: all the incentives pointed instead to call out every hotshot crew and air tanker available and be publicly seen to do so. Change was possible but difficult and would have to be part of a wholesale restructuring of fire management. Fire suppression, like prescribed fire and natural fire, was more complicated than in decades past. It would come with more oversight, not only for crew safety but for emergency outlays.[51]

The parallel inquiry, constituted in January 1995 and chaired by Enoch Bell, addressed the tougher issue of how to find the right funding for the wildland fire program overall, not just during big burns. The "Fire Economics Assessment Report," with its ominous acronym, FEAR, produced a systematic primer on the agency's financing of fire operations. While many of its conclusions fit conventional wisdom, some were surprising or even counterintuitive, and the report refused to obscure contemporary problems, which were approaching a conundrum. It found that if 1994 were excluded as an outlier, "real (net of inflation) expenditures for fire suppression have not increased since 1970," and that "real expenditures for presuppression activities and fuel treatments have declined over the past 25 years." Since the 1978 reforms, the money invested in fire had steadily dropped. Worse, the study determined that "current policy and direction will only increase costs, not reduce them." Programmed funds remained, at best, constant: emergency, off-budget funds were plugging the gaps. Like the federal government overall, fire management was sliding into a hardwired pattern of deficit spending.[52]

It did not assume, without more evidence than existed, that a fully implemented prescribed-fire program would be cost effective and expected it was politically "unlikely" anyway. It noted that there was "little relationship between what is written in Forest Plans concerning fire suppression options and what is applied on the ground." Ecosystem management as an informing principle would "require a considerable outlay in research funding that neither the Forest Service nor Congress is currently willing to make." Dazzling new technologies only allowed the agency to spend more. America's unhappy fire scene was the outcome of nearly a century of actions; it would take as long to correct it, assuming the country would rally the needed passion and monies, and the FEAR committee found little evidence of either. Comparing 1970–74 with 1990–94, it found that California had absorbed 33 percent of the total changes in expenditures. Revealingly, it discovered that the USFS Washington Office alone, which oversaw not only the national staff but the fire labs, equipment centers, training centers, and the agency's commitment to NIFC, accounted for 24 percent. Like the national economy, fire's economy was shifting away from production into services. America's fire system resembled its health-care system: it had plenty of money and incentives for emergency medicine, little for prevention and wellness, and scant political interest in serious reform.[53]

FEAR redefined fire management as risk management, but few of those risks were under the control of the Forest Service. Everything that affected fire pushed its costs higher. "Only major changes in fire management policy can change this outcome." The 1995 policy promised such a change, at least on paper; the 1995 Congress argued against it. The elections of 1994 yielded control over Congress to Republicans intent on reducing government and its expenses. The 104th Congress passed into the scene like the downdraft of a thunderstorm, suddenly fanning abstruse concepts into a political blowup. Shortly after the reports were released in September and October of 1995, the federal government began sliding into a manufactured crisis that ended with temporary shutdowns. An incommensurably dividing landscape met an implacably dividing government. FEAR's forecast for more, and more expensive, fire looked prescient.[54]

Trying Fire

Among the "appropriate responses" coaxed out of South Canyon was the question of how to remember the dead. It was, curiously, a topic the fire agencies had never formally addressed. When the Big Blowup killed 78 firefighters, the Forest Service had no ritual to honor them or mechanism to cover burial expenses. Eventually 53 were exhumed from temporary graves and relocated in 1933 to a special plot at Woodlawn Cemetery in

St. Maries, Idaho. The Forest Service rangers and CCC boys in the 1937 Blackwater tragedy got several memorial markers and inspired a medal for forest fire heroism, jointly sponsored by the USFS and American Forestry Association. The Marines who fell in droves at Hauser Creek in 1943 were war casualties. In a sense, so were the smokejumpers overrun at Mann Gulch. Only a few months before, they had ceremonially jumped to the Ellipse outside the White House, paratroopers turned firefighters.

Then matters got confused. The trainee missionaries who died on the 1953 Rattlesnake fire, set by an arsonist hoping to be hired as a camp cook, did not fit paramilitary stereotypes. The inmates who perished on the 1956 Inaja fire might be memorialized by Chief McArdle for dying in the defense of the free world, but only by those without a sense of irony or honor. The El Cariso Hotshots, with their jaunty berets, were caught in the flaming chute of the 1966 Loop fire and viewed through the dark glass of the Vietnam quagmire. The CDF firefighters killed on the 1968 Canyon fire were urban fire crews in the brush and could be honored as such.

Over the years tragedies had led to established rules of engagement to improve fireline safety. The 1957 "Report of Task Force to Recommend Action to Reduce the Chances of Men Being Killed by Burning While Fighting Fire" led to the 10 Standard Firefighting Orders, modeled on the Marine Corps's code of standing orders. After the Loop fire and other disasters and near misses, the 10 Orders were supplemented by 18 Situations That Shout "Watch Out." No one could remember them, not in any way that might be meaningful on a fire that was blowing and going. In 1991 hotshot superintendent Paul Gleason distilled that ungainly catalog into a single system with four parts: LCES. Every fire required Lookouts, Communications, Escape routes, and Safety zones arranged into a mutually dependent system. The South Canyon Investigation Team had concluded that while an understanding of safety issues and training were adequate, application and leadership had failed. OSHA had agreed.[55]

* * *

What, then, did the tragedy at South Canyon mean? There was no obvious social or cultural setting by which to frame the event. The Forest Service had a mournful heritage of tragedy fires dating back to its origin story in the Big Blowup, but none fit exactly. OSHA defined those deaths as an industrial accident, as if Storm King Mountain could have been a construction site or a coal mine. It is easy to overlook the fact that 20 other firefighters had also died in that baleful year. One was a victim of a burnover on the

Oconee National Forest in Georgia. Elsewhere on the national forests, six died in aircraft accidents, two in engine accidents, and one of a heart attack. The other fatalities resulted from crashes, heart attacks, and burnovers involving volunteer firefighters. The VFDs were local people dealing with a local problem, a service to neighbors. Dramatic burnovers accounted for half the 1994 tally; but most years it was industrial combustion, not galloping flames, that dominated the list of causes. Was South Canyon just an act of God and nature? If so, it lacked the moral fervor that can come only from human agency, in which people have some hand in their fate and their deaths have some cultural significance because they were not foreordained and could have turned out differently. Otherwise, the dead firefighters have no more to say than squirrels and deer caught in the flames.[56]

In 1994 there was a void. A mass fatality fire had not occurred for 28 years; the fire revolution in all its manifestations was redefining what a firefighter should do or not do. The chain of traditional lore and stories by which those on the line understood their identity and the potential costs had been broken. Norman Maclean filled that interpretive void; thanks to *Young Men and Fire* a schema of meaning was at hand. Moreover, it was the genius of his book that it spoke in a universal language. The fire was the "It" at the center of things. The story of Mann Gulch was the most basic of stories—of birth, life, and death compressed to a small patch of land and a few hours. The moral narrative was timeless. The smokejumper was Everyman. The elite were the elect, hopeful that they might escape the flames of perdition. The narrative of Maclean's meditation was itself the story of a search for meaning and salvation. The book was published in 1992, shot to best-seller status and National Circle of Book Critics Award, and was reissued in paperback in December 1993, seven months before South Canyon blew up. The South Canyon deaths were the Mann Gulch deaths reincarnated. They would be likewise honored with crosses on the hill and accounts of the young men (and women) who perished. They would receive homage as firefighters in the Macleanesque vision.

But even that proved insufficient. A plaque and memorial trail were subsequently constructed. That pattern of memorialization then reflected back on former tragedies, which had their sites remodeled to emulate the South Canyon complex. (In 2010 the process was even projected back to the community's foundational tragedy, the Big Blowup.) NGOs became active when, after South Canyon, Vicki Minor, a seasonal concession contractor for fire camps, campaigned for what became in 1997 the Wildland Firefighters Foundation to "honor past, present, and future wildland firefighters" with a national monument, financial assistance to families of fallen or injured firefighters, and help with public education and strengthened fireline safety. In 2000 the foundation dedicated at NIFC a Wildland Firefighter Memorial.[57]

Firefighter safety became an obsession and a valued theme that critics tried to wangle toward their particular concerns. Some saw it as a way of further distancing suppression from ill-advised campaigns in places that would be better served by letting fire burn itself out. Some worried that it interfered with the risk-taking that was a legitimate facet of the job, that it might become an all-purpose alibi for caution or lassitude when decisive action was demanded. Others saw careers crushed because of the failure to adhere to the new norms (the 2001 Thirtymile fire resulted in criminal charges of negligent homicide). Like every other aspect of wildland fire, fireline safety had become complicated, scrutinized, political, and institutionally charged.

In 1992 Sebastian Junger had written for a national audience about his experiences on the Foothills fire. The vision that inspired him to journey to the firelines was a scene he had witnessed a few years before and that refracted through his interest in the professional risk-taker. It was as though firefighting had become a government-sponsored extreme sport. Two years later it was impossible in polite circles to invoke such an image. Firefighter safety had taken precedence over all other duties. The cachet that firefighters now claimed was not their traditional appeal to the firefight as the moral equivalent of war nor to bold adventuring, but to status as a tragic victim or as a Macleanesque existential hero. The understanding of what a firefighter was had become a thicket as tangled as its workforce demographics.

Such a state of affairs was predictable. The uncertainty over how to imagine wildland firefighters mirrored the ambiguity over what the nation was doing with fire. If the country could not agree on what fire management meant, it was hard to agree on how culturally to define and valorize those charged with doing it. Especially if the aspirations of the fire revolution succeeded, it would be difficult to justify a death as other than accidental, as a misfortune, as a personal or family tragedy, not a national one. But to hold to the tragic vision so graphically conveyed by Maclean meant a continuation of hardfisted firefighting. The American public was not alone in its confusion. So, in its own way, was the American fire community. Of all the curiosities and coincidences, the most elemental may be that a community that defined itself as science based had its revolution revived not by the latest model from the fire labs but by a posthumous meditation written by a professor of Renaissance literature at the University of Chicago.

* * *

One of the passages in *Young Men and Fire* that resonates most with the fire community is the description of the smokejumpers as they hike, full of daydreams, down Mann Gulch. Maclean imagined them picturing "the mountainside as sides of an amphitheater crowded with admirers, among whom always is your father, who fought fires in his time." Bud Moore echoed that generational theme when he spoke on the occasion of his retirement about what his years in the agency and on the fireline had meant. "To me," he said, "most of all the Forest Service is the eager uncertainty of young men and women as they confront an old pro at their first job in the woods."[58]

Among the curiosities of the Maclean fervor was that it redirected firefighting away from its archetypal narrative, which was a coming-of-age story and often a generational one. Maclean's quest was itself framed around the handing down of lore, from the seasoned Bill Bell to the fledgling Maclean and Moore, and then from an aged Maclean to Laird Robinson, who accompanied him on his journeys. For the country wildland fire was typically a story of disaster. For the fire community it was the story of young men (and young women) finding something about themselves, becoming adults, during a time in their life called fire season. It was a season when you learned new skills, accepted hard work, had adventures, knew the camaraderie of shared hardships and jobs done well, and grew into the person you would become.

The public face of a fire crew was captured by the crew slogan of the North Rim Longshots: "Flame and Fortune," money and action. And for a significant number of them, no doubt that was true. For most seasonals, that first smokechased snag or burnout on a big fire was a rite of passage. For those who stayed in fire, it was not just an adventure or a job but an avocation, a way of life. The old veterans passed along their skills, knowledge, stories, and approbation to the next generation. It was an almost wholly unwritten tradition, a folklore in the fullest sense. Then the fire revolution and, even more, the demographic turnover ruptured that chain.

* * *

The revolution questioned the unambiguous call to arms that had sustained a life on the line. Not a few of the old generation resented attacks on their legacy as making a mockery of their lives. That smokejumpers might have damaged the landscape as much as big-tree loggers or cattle barons undercut the legitimacy of their livelihood, and they were righteously indignant. It was not clear how you converted to prescribed fire, much less to prescribed natural fire, and still held to the heritage that had validated the guild. But the old-timers knew their worth from being tested by fire. So, while they wished the cup of those horror fires could pass from them, they could—both the religious and the secular among them—accept the flames in the spirit of Paul when he wrote that "Every man's work shall be made manifest: for the day shall declare it, because it shall be revealed by fire; and the fire shall try every man's work of what sort it is."[59]

The upheaval in the workforce severed the generational rites that had guided initiations and affirmed the worth of what the initiated did. The clarity of the 10 a.m. policy had brought a clarity of valor as well as an unambiguous standard to operations. That lived heritage was not a tradition for which NWCG-sanctioned courses could substitute. The demographic revolution had dug a chasm between generations, broken the long chain of folklore, and demanded a new story without getting one. The old rituals and tales were lost, leaving another internal cleavage within the fire community, as rites of passage and emblems that were the heritage of one generation failed to speak to the next. Each generation saw its own legacy torn up and dishonored.

Then came the literary blow delivered by Norman Maclean, the first fire-themed book to achieve national and critical acclaim. The old coming-of-age story—comic in the critical sense—had underwritten the better part of a century and had found literary voice here and there in Kiplingesque poems. It was scattered through the occasional western pulp novel by Zane Grey or Stewart White and was found in George Stewart's 1948 fiction, *Fire*. And there was the saga of the Great Fires with their desperate bravura. But those efforts had never propagated much beyond the fire community. They were instantly overwhelmed by Maclean's tragic masterpiece, which rushed over the literary landscape like the blowup it described. The chrysalis by fire from adolescence to adulthood was lost before the dark allure of disaster and death. This was true even for Maclean as *Young Men* overrode his earlier account of his own adventures as a young man, "USFS: 1919." The inherited narrative was a tale for a heroic age, not a working story for the quotidian life of seasonal firefighters. Besides, the coming-of-age story, the generational passage of fire lore in which skills were traditionally passed down, was a narrative and ritual that could link prescribed fire in the Southeast with wildfire in the West. *Young Men and Fire* could not.

Art comes from art. For years afterward almost all writing about wildland fire played off Maclean's formidable text; not a few revisited the Mann Gulch tragedy itself. John Maclean followed his story of South Canyon,

Fire on the Mountain, with a series of studies, a veritable subgenre, on other major burnovers. When the community then began to publish new stories, they typically took the form of online "lessons learned," most of which pertained to near misses, entrapments, and other events that triangulated from those two geodetic markers of modern American fire history.

With the Macleans the American fire community found its voice. But paradoxically it was not the voice of the revolution. It was a voice that harked back to the past and to the firefight as the set piece of plot and the template of meaning. Mann Gulch was a story of suppression, even if it had gone wrong, not a story of restoration. It was a story from the past made new; it was not a new story, the voice and vision of the fire revolution. Its contribution to the new order came indirectly by dramatizing the potential cost of a suppression program and, once leveraged by South Canyon, by helping nudge the rules of fireline engagement. But when *Young Men and Fire* spoke of tragedy, it was the tragedy of lives cut short and of meaning lost in the plume of a blowup, not the tragedy of a land denied its heritage of flame.

The firefight remained the default setting for literature as it was for operations. Until America could decide what it wanted from its lands and how to reconcile that desire with fire, the fire community would struggle to fill the void left by the decay of the old standards and the standard stories. Veterans would default to suppression in the field. Rookies would measure their own pilgrim's progress by the only tale to command consensus.

Red Fire, Green Fire

Like the rest of the fire establishment, research had fragmented during the 1970s, struggled for footing during the 1980s, and stabilized during the early 1990s. But institutional, administrative, and intellectual cracks had become permanent chasms. The breakup of Forest Service research allowed other organizations to appear, though their sum was less than the former whole. How to transfer knowledge from lab to field found new ways to vex overseers. And a conceptual gully threatened to widen into a ravine as traditional red-fire research into fire behavior coped with green-fire concerns about ecosystem health.

The particular issues varied by agency. For the Department of the Interior the problem was to install a basic research program into a bureau like the BLM, to upgrade those that existed in the NPS and FWS, and to create a firewall—a bureaucratic Glass-Seagall divide—between research and management. Nearly everyone wanted science to advise management, but as at Yellowstone, too often this meant injecting researchers into positions of decision as line officers. The nominal auditors and advisers became investors, as it were. For the Forest Service, the issue was bureaucratic triage — to halt the hemorrhaging of research funds away from fire. The existing capacity was underutilized. There was not enough fire science, and what existed was not making a difference on the ground. Though fire science might study facts, it was not changing them. Though fire research could promulgate models as fire administration could with policy, neither seemed to make a difference commensurate with their lofty pronouncements.

And then there was the tension among the sciences themselves as to topic and method. The historic strength of fire research lay with measuring the parameters of fire behavior, now given new vigor as blowup fires threatened crews. But the challenges of ecosystem management required understanding how fire worked in a biological matrix. The one suggested physical remedies, from classic firelines scraped to mineral soil to wholesale fuel rearrangements. The other sought ecological solutions measured by biodiversity, resilience, and a general if unmeasurable condition of "health." Did the biology of fire follow from its physics, or the physics from its biology? The political compromise was a program of fuels modification, though in practice this meant a shift toward the WUI and away from the wild. It was the intellectual equivalent of a renewed program of suppression. The resulting pulls would test the tensile strength of fire research.

* * *

As with so much of the rebuilding, the process gelled when Bruce Babbitt became Secretary of the Interior. His personal inspiration had come years earlier, when growing up in Flagstaff, he had read Wallace Stegner's *Beyond the Hundredth Meridian*. That book recounted, through the career of John Wesley Powell, a 19th-century attempt to organize the process of western settlement through applied science. In the post–Civil War era the country had dispatched four "great surveys" to inventory its western lands before consolidating them into one, the U.S. Geological Survey in 1879. In 1881 Powell became director, recognized that water was vital to the arid West, and tried to rationalize land and water use by creating a master map that would identify critical places. That book, Babbitt recalled, was "the rock that broke through the window." In 1993 he was in a position to attempt for the late 20th century what Powell had tried a century before.[60]

Some things seemed unchanged. Settlement — this time as urban sprawl — was often as chaotic and damaging as ever. But the issues had shuttled from those fundamentally grounded in geology as ores and water to biology in the form of endangered species and protected habitats. The northern spotted owl had been a political and economic nightmare. Perhaps Powell's experience, updated, showed how to avoid future paralyzing controversies. With regard to fire Babbitt sought "a conservation movement that puts prescribed fire back on the landscape and increases the health and the productivity of the land, and reduces the risk and destruction of wildfires that do occur." In his view—he had a master's in geology as well as a JD from Harvard—such a project demanded a firm basis in science. Interior needed to consolidate its research programs as it did its separate agency fire programs. The original creation of the USGS, which brought together the existing surveys, the National Academy of Sciences, and even the Smithsonian, suggested a model.[61]

The National Research Council had weighed in with a report on *Science in the National Parks*, released in 1992, while the Carnegie Institution also urged Interior to consolidate its research program, at that time scattered among seven agencies. In February 1993 Babbitt requested further advice from the National Research Council. The NRC released its conclusions later that year. In March 1993 Babbitt issued Secretarial Order No. 3173 to establish a National Biological Survey. The decision amalgamated existing researchers from the NPS, BLM, Minerals Management Service, Office of Surface Mining Reclamation and Enforcement, Bureau of Reclamation, and the USGS—some 1,800 personnel in all. The FWS supplied the lion's share of personnel (1,300) and of budget. The NPS contributed 170 researchers, and the BLM (which hardly had a program) only 35. Partnerships were created with the states. Ron Pulliam, an ecologist at the University of Georgia, agreed to head the program. Secretary Babbitt made explicit the historical analogy to the USGS and its campaign for a national map. Later that year the NRC released *A Biological Survey for the Nation*, arguing for an inventory of the country's biotic resources—this a year after the Rio Summit and Global Biodiversity Strategy, which had led to the adoption of ecosystem management by the Forest Service. Litigation under the Endangered Species Act promised further "train wrecks" if management did not become more proactive.[62]

Babbitt had sought to act quickly, with the expectation that Congress would follow with an organic act for the National Biological Survey (NBS). Instead, as with his attempted reforms in mining and grazing on public land, he ran into a buzz saw of western opposition and lobbyists, and after 1994 a hostile Republican-dominated Congress and its Contract with America, which advocated rolling back environmental laws. Instead of depoliticizing the issue, the NBS had repoliticized it. The more conspiratorial critics hinted that the NBS was a mechanism by which to find endangered species on private land and then confiscate those properties. Pulliam recalled, "I was told that NBS scientists would soon be cutting fences in the middle of the night and trespassing on private property with the aid of hordes of untrained volunteers, most of whom were probably card-carrying members of radical groups like Greenpeace and Earth First." Babbitt sought, without much effect, to disarm that argument in 1995 by retitling the bureau the National Biological Service.[63]

It was all eerily reminiscent of the opposition Powell had inflamed by seeming to hold settlement hostage while his crews plotted watersheds and dam sites on a national map. Then, Congress had savaged the USGS budget and eventually drove Powell out of office. This time, Congress refused to enact a charter for the NBS or to fund it, leaving it subject to annual renewal orders from the secretary. But there was also hostility from within the reorganized research establishment. The NBS took money and people away from existing institutions—never a happy outcome for organizations. The NPS had struggled for 30 years since the Robbins Report to merge science within the agency, and since 1980 it had published *Park Science* as a system-wide journal to integrate "research and resource management." The FWS relied on staff scientists intimately familiar with particular refuges. The BLM had barely begun. Now that integration was threatened. The scientists no longer reported directly to superintendents or directors—were no longer "one of us" and were instead members of the NBS. Both researchers and administrators fretted over whether the local, site-specific studies that agencies constantly demanded might be satisfied under the new regime. Bizarrely, many researchers continued to work out of the same desks as before.[64]

Eventually the solution was to relocate the NBS into the USGS as a Biological Resources Discipline. The compromise worked, politically and budgetarily. (Babbitt later claimed the outcome was what he had hoped for all along.) But the dilemma was twofold. Interior needed enough critical mass to do serious research on the scale required, and it demanded enough autonomy that it could furnish apolitical advice and serve as an auditor of field programs and policy, that is, as a source of data and neutral criticism. Equally, it had to be grounded sufficiently in particular sites and bureaus that its know-how could be applied and the research was seen to be integral to the bureaus and not the product of narrow-eyed grumblers from the outside. Not even the Forest Service, which had established a Research Branch 80 years earlier, had clear answers, and in any event it was hobbled by defunding.

As the "reinventing government" theme gained traction, however, a collaborative program emerged involving universities and other interested institutions, the Cooperative Ecosystem Studies Unit, which eventually blanketed the country with 17 regional consortia. What remained unclear was how the reorganization boosted fire science. For most Interior agencies research had focused on the fire history of particular units as a basis for fire planning. But such plans also required knowledge of fire behavior and effects, which demanded a more fundamental program that, in the past, only the Forest Service with its dedicated labs had been willing or able to conduct.

* * *

The Forest Service remained enfeebled. Like all civilian bureaus, its research overall had shriveled under the Reagan administration. The antimissile Star Wars fantasy could command billions to protect America from possible nuclear firestorms while research to protect against real-world firestorms shrank. The paradox was palpable: Los Alamos National Lab might develop lasers to down ICBMs but could not protect itself from the free-burning flames of the Jemez Mountains on whose flanks it uneasily resided.

Within the Forest Service, research committed to fire had continued to plummet—between 1979 and 1989 by 47 percent. The Macon fire lab teetered on closing, then shrank to minimal operations; it never recovered from the shock and finally was shuttered a decade later. (The facility had always been a curious hybrid with the state of Georgia, which had provided the building but now pulled back.) The rollback hit the Riverside lab next. Desperately, FFASR had identified some critical issues, such as the wildland-urban interface and climate change, that it was ideally positioned to address. But the I-zone seemed like a California eccentricity, and Republican administrations were hostile to climate change science. The catalyst for recovery, as with so many fire issues, began with the tragedy at South Canyon.

The great question of South Canyon—Why?—would not go away. One member of the investigative team, Ted Putnam, a specialist in entrapments from the Missoula Technology and Development Center (MTDC), had puzzled over the position of bodies and gear, unlike any he had seen before. Existing models of fire behavior seemed to have failed. He asked a colleague from MTDC and the cream of Forest Service fire-behavior researchers at the Missoula lab, five in all, to visit the site with him in August 1995.

They self-organized into an informal team with the intention of reconstructing "the fire behavior in greater detail than the accident investigation report."[65]

They had unrestricted time to ferret out details not available to the original teams. They tested the evidence on the ground against lab data and mathematical models of fire behavior. To clarify their study they did what natural science has always done: they stripped out the moral parameters by shifting attention from the behavior of the firefighters to that of the flames. Revealingly, even this exercise harked back—like so much of the response to South Canyon—to *Young Men and Fire*. To understand why the smokejumpers at Mann Gulch had been overtaken by the blowup, Maclean had contacted Dick Rothermel and Frank Albini at the Missoula lab, who applied their expertise to reconstructing the flames' race up the gulch. Rothermel's conclusion was distilled into a graph that measured the rate of advance of the men against that of the fire. It was, he determined, "a race that couldn't be won." For Maclean the crossing lines were the script of a Greek tragedy.[66]

That scenario, both the fire and its reconstruction, was replayed at South Canyon. In the end, the meticulous details confirmed what even rookie firefighters knew. The fire had synthesized drought-leached Gambel oak, steeply corrugated terrain, and high winds, gusting and unstable, into a blowup. In retrospect, everything favored a rapid burnout: the needle of every compass by which to measure wildfire behavior pointed in the same direction. But such rule-of-thumb understandings had been known forever. The point of fire science was to predict just when those factors might combine with deadly effect. At South Canyon the models had failed. They could no more cope with explosive runs like that on July 6 than could crews scrambling up the fireline on Lunchspot Ridge. The fire behavior team demurely declined to address "the ability of currently available fire models to predict extreme fire behavior" as it nudged "related issues such as human behavior factors." It did not have to. If firefighter safety was to improve, science would have to "stand up for safety" as much as FMOs and agency administrators would.[67]

That an investigation into something so fundamental to the agency had to self-deploy speaks volumes about the unsettled nature of fire research. It had undergone the same workforce downsizing and churn that affected fire management, and to many observers seemed close to the point that it could no longer mount the kind of joint lab and field campaigns that it had once been ideally situated to undertake. The NBS could fill some Interior needs, but neither it nor universities (or for that matter anyone else in the world) was equipped for the fundamental inquiries that could move fire beyond the realm of guild know-how.

The National Academy of Sciences had stepped in with a 1990 report, *Forestry Research: A Mandate for Change*, which addressed ecosystem management but said little about fire. The Forest Service adopted the general vision with a "Strategy for the '90s," pivoting its machinery toward the kind of ecosystem knowledge required to keep the national forests and grasslands functioning. By now fire research claimed 7 percent of the total investment in research and unsurprisingly emphasized its traditional strengths in fire behavior and fuels. Meanwhile, the Macon lab shut down, Riverside went on autopilot, and field experiments persisted at barely survival level.

Even with regard to the I-zone and fuels, research fissioned. One agenda followed a traditional emphasis on wildlands. The way to protect structures was to halt the fires that threatened them, and that meant restructuring the fuels that powered them. Burning exurbs were, in a sense, another expression of unhealthy ecosystems. They could be spared horrific fires by attacking fires while they were in the backcountry, and part of that strategy was to starve them of the fuels that made them into monsters. A second agenda, however, looked at the other fuel caches involved, namely the houses. It seemed that few structures were actually consumed by rolling tsunamis of flame. They burned because they had continuous carpets of light fuels connecting the woods to the house, or because of fuels brushing against combustible structures, or especially because they were hit by swarms of embers that found points of vulnerability. Protection required hardening the structure itself. Under Jack Cohen at Missoula, a Structure Ignition Assessment Model (scrutinizing a "home ignition zone") became the focus of inquiry. Each research approach suggested very different strategies for response.[68]

* * *

The Forest Service also needed research for the same reason that Babbitt insisted Interior needed it: to cope with the demands of ecosystem management as leveraged by the Endangered Species Act. With the accession of the Clinton administration, the USFS acquired a chief whose career had risen on exactly this theme. The premise was that science could furnish an empirical foundation for ecosystem management and test options for policies and field operations. For that, the science itself also had to be good enough—not perfect, but sufficiently coherent to be useful. Ecology was more tentative than either researchers or practitioners wished. At a minimum it required far more data on a broader geographic scope than what existed.

If the agencies' new "guiding paradigm" looked to science for its empirical underpinning, ecological science (and fire science) looked to that management commitment to reinvigorate the scientific enterprise, but unless the science was funded, that was a hollow hope. The 104th Congress not only continued the trend begun under Reagan to shift federal funding from civilian to military science but cleaved away at all research that might feed into policies it did not like, such as endangered species and habitats, global change, social science, and ecology generally. As Jack Ward Thomas noted in 1995, "When Congress funds my wildlife budget at half of what I ask for, and research at one quarter what I ask for, but the timber sale program at two-and-a-half times what I ask for, what do you THINK is going to happen?"[69]

Even had funding been adequate and theory sturdier, critics noted that "the science" was as intellectually fragmented as land ownership. A fire science required the conceptual equivalent of an interagency consortia, an intellectual NWCG. For fire science, in particular, there was a rift between ecology and physics that was not easily sutured together. For traditional fire science, fire's ecology derived from the physics of red fire. Yet the ESA and ecosystem management were about species and habitats—the green fire of biology. Treating landscapes as caches of combustibles said nothing about biodiversity and could actually damage habitats. But no alternative biologically based theory of fire existed. Ecologists had not traditionally fashioned their master models around fire. Fire just did not exist in the founding consciousness or literature of ecology. It was not in general biology texts, which reflected a science that was rapidly evolving along genomic and molecular themes. The internal combustion of the Krebs cycle interested general biology. Free-burning fire did not.

Certainly most observers of the public lands saw fire's influence everywhere. They saw fire exclusion as a leading cause of the damaged biotas that littered the American landscape. They saw revanchist wildfires as a common consequence of degraded ecosystems. And they viewed prescribed fire as a potential catalyst for curing ecological ills in many settings. The "science" as science was not yet adequate. Much as the Rothermel model had been pressed into service too soon and far beyond its carefully drawn boundaries, so ecological models were conscripted to work beyond the scope of their warranties.

Besides, it was not obvious that better fire science could help management outside of set-piece firefights and burnouts because the barriers were not the limitations of knowledge but the capacity to apply what was known. It was not just that one scientific model or another was lame but that science could not resolve the core concerns. The larger, immovable issues were cultural, as synthesized by politics, not scientific. Still, demand for fire science had never been firmer even as support for fire science had rarely been more rickety.

* * *

The fires did not wait. As environmental issues proved ever more intractable, independent science review panels flourished on topics such as water in the Everglades, salmon in the Columbia River, and old growth in the Northwest, all of which were consortia of scientific disciplines as well as of agencies. But no such arrangement existed for fire. The Forest Service had its labs. Interior had the wobbly NBS. NSF sponsored some fire ecology, but not since the early 1970s had it identified the field for special attention.[70]

By 1997 Congress had grown weary of watching houses incinerate along the interface, and they accepted the argument that the uncontrollable fires were the product of uncontained fuels. In its Annual Appropriation for Interior and Related Agencies, it allocated $8 million for a Joint Fire Science Program (JFSP), a partnership with the USFS and five Interior agencies. The intention was "to provide a scientific basis and rationale for implementing fuels management activities, with a focus on activities that will lead to development and application of tools for managers." The program had a governing board of 10 members—5 from the Forest Service and 5 from Interior. It was located at NIFC. For the first time wildland fire had a funding source and a mandate not tied to any single agency—total mobility had finally come to fire science.[71]

The effect was galvanic. By the standards of big science, $8 million was a pittance (it was soon boosted to $16 million). For an enterprise that was accustomed to operating on short rations and which had been on life support a decade earlier, it was a cornucopia that opened fire research to all comers, even universities and private companies. As its charter mandated, the first wave of sponsored projects targeted fuels inventorying and mapping, the evaluation of various treatments, the scheduling of treatments, and their monitoring and evaluation. By emphasizing fuels, the JFSP played to the strengths of the most quantitative sciences, not necessarily the most vital. It adapted policy—in this case, research policy—to what the science could do rather than adapting the science to fit larger policy needs. There was no end of objections possible, and many were made.

So it is easy to overlook what JFSP achieved. The Forest Service hegemony in research had been shattered, much like the breakup of the great industrial labs at AT&T and IBM. Not until the early 1990s did USFS fire research stabilize. The NBS (later USGS) organized Interior fire research into something like a critical mass and counterweight. But it was JFSP that allowed the parts to coalesce around a new institutional order not controlled by any one agency. It helped defibrillate fire science. It provided a forum at which the various players and nominal partners could converge and a fulcrum with which they could leverage projects. It did what the best federal fire programs had always done—stimulate local organizations and give a critical boost to capacity. As the years passed and its funding base expanded, JFSP found ways to nudge its charter into a widening gyre of topics. It became the vehicle for spawning a pact of science consortia that spanned the country—the research equivalent of the interagency coordination centers that handled wildfires and the prescribed-fire councils that were collectively stitching a vast patchwork quilt to cover the nation. A rising plume, it was hoped, could loft all embers.

World Fire

In the 1990s, Earth became in the mind what it had always been in reality: a fire planet. Like business, money, media, and popular culture, fire went fully global. With satellite surveillance, Earth's hotspots could be crudely inventoried. With better instruments, ideas soon followed about how to integrate fire into the atmosphere, into biogeochemical cycles like carbon and nitrogen, and into biogeographic patterns; even historians began to see the past with fire in their eyes, and a global narrative formed out of the smoke. America's "fire crisis" played out amid a scientific consciousness that Earth itself was experiencing a wholesale crisis of combustion. And where ideas went, institutions followed.

* * *

During the 1980s free-burning fire appeared in the general scientific imagination. In the 1990s fire became an object of inquiry, even by big science. Of all the contributing sources, a few stand out. One emerged from the short-lived but fierce debate over the prospects for a nuclear winter as an unintended outcome of landscape burning kindled by thermonuclear war. Its promulgator, Paul Crutzen, had acquired an international reputation from his work with ozone (for which he later shared a Nobel Prize); in 1982 he turned to the atmospheric aftershocks from widespread fires sparked by nuclear weapons. By 1988, as Yellowstone and Amazonia burned to a repeating reel of media attention, James Hansen of NASA, among others, argued that global warming was upon us, fed by the combustion of fossil fuels. Open burning was a factor too, as satellite imagery tracked a planet rent by tropical deforestation and alive with flames. Scientific campaigns were launched to measure the relative inputs of greenhouse gases from open and internal combustion.

Almost annually there was a conference or a multinational field campaign. The United States signed a bilateral agreement with Brazil in 1991, which sparked a gamut of research projects. The American Geophysical Union sponsored two Chapman conferences on global biomass burning in 1990 and 1994. A Dahlem Conference was held in Berlin in 1992. Led by Domingos Xavier Viegas, Portugal began hosting fire-behavior conferences in 1990, renewed every four years. The UN Intergovernmental Panel on Climate Change, chartered in 1988, included open burning among its topics of concern. The International Global Atmospheric Change group oversaw expeditions, several supported by national space agencies such as NASA, which sought ground truth for Earth-orbiting satellites. The most complex was the Southern Africa Fire-Atmosphere Research Initiative (SAFARI) in 1991. In 1993 an international conference on Eurasian fire held at the Sukachev Institute for Forestry in Krasnoyarsk, Russia, was followed by a dramatic experiment in which Bor Island in the Yenisey River was burned off to sample fire-induced effluents and long-term ecological responses. Other campaigns were scheduled for Southeast Asia, the scene of horrific "haze" caused by land clearing and burning in Borneo, for northern Eurasia (Firescan), for Alaska (Frostfire), and for the Mediterranean Basin (Firescheme). The Wildland Fire Conferences begun in Boston in 1989 leaped to Canada in 1997 before continuing to Australia (2003), Europe (2007), and South Africa (2011). Surely, the most striking collaboration occurred in 1994 with the International Crown Fire Modeling Experiment. Canada supplied a site outside Fort Providence, in its Northwest Territories, for field trials. A consortium formed in 1995 brought together over 100 participants from 30 organizations and 14 countries to carry out 18 experimental burns between 1995 and 2001. It became, after a fashion, an Olympics of fire research.[72]

The research on free-burning fire that the Cold War had initiated with military funding had segued into civilian science on open burning. Tests on nuclear weapons, one result of which was fire, had yielded to nuclear winter studies, which in turn were green-washed into biomass burning and global warming. A camel of a topic was pushing its nose, in minor but insistent ways, into the tents of major journals in sciences far removed from traditional natural resources. Still, while no single nation could support more than a periodical for its guild of fire practitioners, the Earth's collective fire interest could sustain a dedicated publication. The International Association for Wildland Fire plugged that gap in 1991 by gathering the entire fire and sponsoring journals for both researchers and practitioners.

What for the United States had begun as interagency cooperation had gone international. Fire science was halted at borders only when it met a fuelbreak called funding. Increasingly, fire's problems, and its climate, management, and science had assumed global dimensions. No one nation could cope with it all, or even hope to understand the context of its own fire scene, without appeal to the brains of colleagues located throughout the world and to the experimental opportunities afforded by other nations. With regard to fire, the United States no longer served as the Forest Service of Earth but one partner in a multipolar world.

Yet much as the Forest Service remained critical to the United States, so the United States was to the emerging global order. No conference or big field campaign lacked an American presence and often significant American funding or contributions in kind. If America's fire programs were dysfunctional, so too often were international programs. The disintegration of state-sponsored forestry was global, and the Earth's fire community scrambled to find institutions appropriate to its needs. Still, for good or ill the American experiment—in this iteration, the belief that humanity could do more with fire than try to suppress it or stuff it into chambers—would spread.

* * *

Ideas were never long segregated from institutions. Someone had to pay for fire science, and someone would need to apply its results or at least reconcile them with existing practices. The earliest international organizations were imperial, further stiffened by the global guild of foresters. After World War II that order disintegrated as Europe decolonized. It reincarnated in the form of multinational bodies like the UN's Food and Agriculture Organization. After the Cold War, it became part of a new world order on fire, most spectacularly symbolized by the attempted integration of Russian fire science.

The United States had treaties with Canada and Mexico and added another to govern mutual aid across the Bering Strait between Russia and

Alaska. USAID, the Peace Corps, cooperative research projects—all continued as requested, with the Forest Service still designated as lead agency for all matters of fire and forestry internationally. Thanks to its buildup in the 1990s, the BLM-sponsored exchanges between smokejumpers in Alaska and Russia.

More striking, perhaps, was the projection of fire's civil society across and beyond America's borders. The Nature Conservancy assumed the lead, first toughening its capacity in Florida and the states, then crossing oceans. It had contributed Ron Myers, its chief fire specialist, to the National Commission on Wildfire Disasters and then to the 1995 federal policy review. Much to the consternation of local chapters, TNC adopted NWCG standards for its own, typically volunteer, fire crews. Meanwhile, in Florida, TNC devised fire teams, an alternative to hotshots, to execute prescribed burns, and during the 1998 bust contributed crews to the suppression effort. Under Myers it expanded from its Florida beachhead into the Caribbean and Mesoamerica. Soon it was burning in Bermuda and Belize, Honduras and Mexico. As its Last Great Places scoured the globe for endangered habitats, conservation scientists estimated that 84 percent of the important sites had badly altered fire regimes, for which active management would be required. In 2002 it partnered with the World Wildlife Fund and others in a Global Fire Initiative.[73]

Yet the hard power remained with governments. The UN became the prime locus for agreements that went beyond mutual-assistance pacts across a single shared boundary. In 1988 it published, through FAO, *International Forest Fire News*, launched what became the Intergovernmental Panel on Climate Change, and moved toward an International Decade for Natural Disaster Reduction (IDNDR) for the 1990s. The FAO expanded its inventories and projects to include a larger share of fire and created a team of fire specialists. It acquired a dedicated fire staffer (1990–91) and with the UN Economic Commission on Europe began hosting seminars in Europe, which in 1996 relocated to Shushenskoe, Russia, for a workshop on fire and the global climate. In 1997 the IDNDR established a Fire Working Group. When the Decade ended, it was succeeded by an International Strategy for Natural Disaster Reduction (ISNDR), which retained the group. In 1998 Germany provided funds through its Office for the Coordination of Humanitarian Assistance for a Global Fire Monitoring Center under the direction of Johann G. Goldammer, long-serving editor of *IFFN*. The GFMC and ISNDR spun out an extraordinary web of regional fire networks incorporating almost all the combustible regions of the earth. With breathtaking speed fire management had propagated around the world.

Like it or not, the United States was part of a global economy of fire, of fire's study, and of fire's management.[74]

Yet while academics might fuss over the finer points of fire ecology or the magnitude of carbon dioxide from biomass burning as a greenhouse gas, and though TNC published its 10 Guiding Principles for maintaining or restoring fire to replace the 10 Standard Firefighting Orders, most governmental interest focused on fire as a destroyer of assets and an agent of disaster. That is what justified government-sponsored outlays for research, for building institutional capacity, for cooperation across borders, and for treaties governing fire as a contributor to climate change. Paradoxically, the combustion that was unhinging the Earth was less towering fire whirls than the hidden burning in engines and coal-fired dynamos.

* * *

The Earth System itself was undergoing a monumental overturning, what was simplified into the phrase "global change," powered by humanity's reconfiguration of itself as a fire creature. The planet's species monopolist had decided to magnify its firepower by combusting fossil biomass. The world fire that could destroy and birth new worlds was now a fire that burned in machines. In 2000 Paul Crutzen proposed that human activity had swollen to such magnitude that it constituted a planetary force sufficient to delimit a geologic era, the Anthropocene. Its trigger point was the conversion to fossil fuels.[75]

The Big Burn of lithic landscapes did more than pump effluent into the atmosphere and unmoor the historic climate. Its effects cascaded through the biosphere and reconfigured how humanity lived on the planet. Internal combustion became the dominant narrative of fire on Earth. The planet divided into two grand realms of combustion: one that burned surface biomass and one that burned subsurface fossil biomass. By 1990 it was estimated that the burning of fossil fuels contributed 60 percent of the greenhouse gases released into the atmosphere. By 2010 that figure was closer to 80 percent.

The point at which internal combustion replaced open combustion was awkward to track because it varied by place and because it did not just add to the sum of burning but partially replaced open fire. ICE differed from FIRE—the burning by Internal Combustion Engines differed from Flame In Real-world Environments—in that it stood outside the traditional constraints of fuel, terrain, weather, and season. It could burn day and night, summer and winter, year after year, through drought and deluge. It burned landscapes exhumed from the geologic past and released their combustion products into a geologic future. It destabilized the realms of air, water, earth, and other fires. It underwrote global warming by saturating the air with heat-trapping gases. It made the wildland-urban interface possible by removing land from agricultural production and putting people into vehicles that fed on oil rather than oats. It made wildland fire exclusion possible by finding surrogates for prescribed flame and by powering the counterforces that made wildfire suppression feasible. More firefighters died, by far, from vehicles and aircraft powered by internal combustion than by flames. Throughout the decade, among all the scientific powers, combustion became an object of legitimate and growing inquiry. The Earth's keystone species for fire had changed how it lived, and little on the planet could escape the consequences.

Still, there was pushback, if not over the science then over its political meaning, as the implications of global warming became better known. American fire science faced special challenges beginning with the hostility of Republican administrations to the notion of anthropogenic climate change and their refusal to fund or sanction studies pertaining to it. But almost as great was the struggle to fit the facts into a usable history. The wilderness ideal that had been so vital to the fire revolution had positioned fire science in the wild, the more pristine the better. ICE, however, put humanity at the center; without people, wildland fire would exist, but industrial combustion would not. The notion that humanity was the driver of landscape fire and that the anthropogenic landscape was the norm came easily to culturally trained fire historians but only haltingly to those from the natural sciences. The whole premise behind the Anthropocene was that humans were the critical agents who had their hands on the levers of planetary change. But the premise behind natural science was that humans had to be removed in order for the models to work. As the decades scrolled past, the prime mover of America's pyrogeography was not climate but people, who were now in fact unmooring climate.

Intellectual skepticisms and political skirmishes did not matter so much to fire officers who saw a new reality taking place before their eyes. ICE was pushing FIRE into new forms on the ground. It powered the WUI that was unsettling American geography, and it fed the greenhouse gases that were reorganizing the atmosphere. They saw their range of maneuvering progressively constrained. The I-zone took away the rural buffers that had allowed them to trade land for control. The changing climate took away the grace period of benign years that had broken up the bad ones and allowed them to regroup. Now the stresses were everywhere all the time.

In 1910 the Big Blowup referred to landscape-scale burning of wild and newly disturbed lands. In 2010 the Big Burn was the remorseless burning of fossil landscapes from the Carboniferous and other lost worlds.

Coming Apart: America's New Pyrogeography

A people's lands cannot be segregated from their society. As America's economy concentrated into a small elite while stressing and thinning out the once-broad middle, as incommensurable classes emerged and sorted themselves into the social equivalent of gated communities, as political parties polarized into sects with incompatible ideologies, as globalization and the growth of regional powers weakened the nation-state, as immigration and social mobility (even through court injunctions and illegal migration) remade the composition of the populace and the workforce, so it was with America's lands. The generic yielded to the specific, and the specific sorted itself into exclusive groupings.

The competing political philosophies of market and commonwealth whipsawed public-land policy with each new administration. The parceling of once-common lands into wilderness, nature preserves, and peri-urban fringes paralleled the emergence of the Super Zips, the neighborhoods where the new elites are concentrated. The two poles, wild and urban, replaced the working middle as defining concerns. Some smaller portions of the public domain received ample attention and relative wealth; most did not. The national fire policy fragmented into regions across which fire practices did not easily travel. What had been a multiuse landed melting pot crystallized into a mosaic of special-use lands, the equivalent of the once-common culture's fracturing into hard crystals of class, race, ethnicity, and gender.

Nature's economy assumed the character of America's society's. Fire simmered or crashed, ever more volatile and more concentrated, into particular settings. More and more fire gathered into a veritable plutocracy. What Alan Greenspan, then chair of the Federal Reserve, descried in 1996 as a market feverish with "irrational exuberance" could as well describe the country's pyrogeography. It was a bull market for burning, and more and more of that burning was collecting into a shrinking number of fires. The American fire scene was headed toward a plutocracy in which 1 percent (actually, closer to 0.1 percent) of its fires did most of the burning and commanded a disproportionate share of resources.

The process had begun in the 1960s, hardened into trends during the 1980s, came to dominate in the 1990s, and became intractable as the new millennium tried to cope with market crashes, foreign wars, unyielding ideologies, and a general breakdown in social cohesion. Dysfunctional government aligned with dysfunctional lands. Design an ideal algorithm for matching a fragmenting land, an immiscible society, and opportunistic fires, and it would look a lot like America at the end of the millennium.

* * *

Legacy fire institutions were not able to counter those trends. The Forest Service, which had long before ceded real control, was left with all the burdens of hegemony and without any of its powers. Instead, civil society was invoked to do what government had traditionally done and was no longer allowed to do. In 1996 at Boise, Tall Timbers hosted a conference on the theme of "shifting the paradigm from suppression to prescription." Yet the real shift was that a private foundation had assumed an unfair and monstrous burden that rightly belonged with the federal government. Tall Timbers accepted it anyway. Beginning in Florida, voluntary prescribed-fire councils sprang up to promote prescribed burning. By the new millennium, so discordant was the relationship between the American public and America's public agencies that The Nature Conservancy was enlisted to act as an honest broker between them.

What appeared to replace forestry as the moral philosopher of public lands was economics. Yet when a national symposium in 1999 sought to synthesize economics, planning, and policy, its texts made for confused bottom lines. Consider, for example, the case of the National Park Service, which had long argued that every fire suppressed was a problem deferred and had boldly broken trail for the fire revolution. Stephen Botti, the NPS national planning programmer, surveyed the agency's state of the union. It oversaw 377 units in 48 states, plus the District of Columbia and overseas possessions. Its landed estate consisted of 82 million acres, of which 21 million were in Alaska. Most units were historical or cultural sites, and every one needed fire protection. Only a handful of them were amenable to fire management.

Botti recast the NPS project in the language of economics. The agency restated its appeal to PNFs and prescribed fires as "capital investments that will return benefits in the future" by reducing the exorbitant costs of catastrophic wildfires and the readiness preparations their prospect demanded. And, in fact, the NPS economy of fire did resemble the nation's. Its fire program was concentrated in its big natural parks. Over the past decade 11 parks accounted for 75 percent of the area burned by wildfire; a mere seven units racked up 87 percent of prescribed natural fires, and another seven (four of which fit both categories) did 80 percent of the prescribed burning. In fact, a whopping 88 percent of prescribed fire came from two units, Big Cypress and Everglades. This looked like trickle-down fire economics. Like America's new service economy, the fire revolution had succeeded spectacularly for a small constituency but had failed to diffuse throughout the system.[76]

A similar scenario of skewed distributions held for the USFS. While most of its visible problems lay in the West, an increasing proportion of prescribed burning took place in the Southeast. Working landscapes and a heritage of burning made it easier to expand in the Southeast, while wildlands, an extinguished folk tradition of fire, and liability law made burning more complicated and expensive in the West. It had solid natural-fire programs in the Northern Rockies, notably in the cluster of wilderness around the Selway-Bitterroot, and another stellar program in the Gila. But even with natural fire the regional biases came through. Bradwell Bay Wilderness in Florida's Appalachicola National Forest was authorized to do prescribed burning, while Lassen National Park in California justified prescribed burning only as a prelude to natural fire. It was a familiar story: culture trumps science. Regional fire traditions determined the choice of treatments.

The country as a whole was both coming apart and reassigning itself into what seemed exclusive categories. Distressingly for those who believed that economics could sort out values and analyze cost-benefits that should underlie political calculations, there was little cause for optimism. The cost of readiness—of preparations, of infrastructure, of fixed expenses—was rising, even as emergency suppression costs boomed and crashed along with fire seasons and trended inexorably upward. The magical thinking that assumed fire management was an accomplished task, which believed not paying would mean there were no costs, went the way of the rest of the federal budget. Beginning with the 1980s fire budgets doubled each decade.

* * *

What drives the American fire scene in any era is always unsettled. At times it is ignition, at times fuel, at other times weather. So it is, too, with characterizing the history of that scene. The prerevolutionary era was dominated by personalities such as E. V. Komarek and Harold Biswell. The revolution was about ideas and institutions and the ways they fragmented, arose, and reassembled. Then outside forces such as climate and politics informed events. Personalities reasserted themselves as the catalysts for a revived revolution. But as the 1990s matured into a long march, America's pyrogeography might best be organized around regions and the kinds of fire each had. The American fire scene, like American society, was sorting itself out into types of fires and the regions that held them. It was hard for a fire in one category to move to another except that prescribed or natural fire could transfer to wildfire subject to suppression. It was hard for fires characteristic of one region to migrate to another.

Take prescribed fire. By the early 1990s there were few critics of prescribed burning among the fire community—in principle. On restored prairie and fire-culture niches such as the Flint Hills, it persisted as a seasonal ritual. It flourished in parts of the FWS that had inherited the practice and that after 1981 imposed a more rational order on it. In the Southeast it was the foundational doctrine of fire management, a benign means to sustain ecosystems, without which even suppression was hopeless. So well established was it that fierce debates raged over what season to burn, not whether to burn, and especially for private landowners, operations hinged on capacity, the ability to apply accepted ideas. Where it did well, it got better, but where it competed with alternate technologies, it struggled. Elsewhere, prescribed fire remained what it had been for 25 years, the establishment's compromise between no-burning and let-burning.

Where it thrived, it expanded. This did not happen as a self-emergent unfolding. It occurred because proponents anticipated criticisms and acted before they became crises. Over and again, Florida was in the vanguard. Then in 1999 it turned previous liability legislation regarding fire on its head. A certified burner could not be found civilly liable unless a plaintiff could prove gross negligence. This did for fire what "stand your ground" legislation did for guns: it shifted the burden of proof from the practitioner to the complainer. Equally, Florida pioneered prescribed-fire councils, and when the federal agencies in 1998 sought a location for a national Prescribed Fire Training Center, it was no surprise it chose Tallahassee. The federal agencies ramped up their national prescribed-fire acreage by burning in the Southeast, which held just 7 percent of the national forest system but accounted for 80 percent of the burning.[77]

But while universally accepted in principle, prescribed fire struggled in the woods. The acreage fired was far less than needed, and it proved difficult to transport the techniques outside the region on an industrial scale. Why? The reasons are the usual ones. It was much harder to restore a practice than to enlarge one. The West had lost its cultural heritage of burning; its fire environment was more complex; agencies even began to record fatalities on prescribed burning crews. The burning did not occur on a landscape scale but as a suite of set pieces—easy to stop, hard to rekindle. In places that most needed fire and should have been easy to burn, the legacy of fire's exclusion had so altered the fuel complex that some preparatory work was required if fire was not to boil over or damage what it was intended to rejuvenate. This was expensive and could be controversial, and made courts as much a part of the scene as fire caches. There was no protocol for prescribing the high-intensity fires that were the natural regime in many high-value landscapes—and this was just to initiate burning. While sold as a prophylactic against wildfire, prescribed fire typically required multiple treatments. If it resembled a landscape vaccine, it was more like a flu shot than an inoculation for polio. It required frequent boosters, repeated every few years, and it could never induce complete immunity. And not least, failures were measured in escape fires and smoke plumes, not in acres unburned. With each passing year there was less slack in the landscape and less political tolerance for escapes. The costs of such failures far exceeded the rewards of successes.

No one could even say with much confidence how much was being burned, much less how much should be burned, only that they were burning far less than what was needed. The federal agencies did not begin to record their prescribed burning on a comparable metric until 1998. What policy tinkering could not fix, perhaps a change in strategy could. In 1995 the NPS introduced a pilot program for prescribed-fire support modules, a mobile crew that could assist local units that otherwise lacked capacity. TNC helped make the concept national, beginning with Florida, and then by using fire-training occasions to conduct burns in targeted sites. Still, the amounts were meager compared with needs. (A sympathetic observer of field trials around Flagstaff noted that Phoenix was paving over more land each day than the Fort Valley Experimental Forest was thinning and burning each year.) With respect to fuels, wildlands had entered a cycle of debt peonage; the agencies were unable to pay the accruing interest, much less the principle. Only in select areas, almost all in the Southeast or in tallgrass prairies, was burning adequate to needs.[78]

And there were always critics to argue against burning for some cause or species. There were protests against burning in tallgrass prairie because a favorite beetle or butterfly might lose its habitat, or in western forests because the spotted owl needed dense woods and crown fires were, by their nature, uncontrollable. In ranching country, escapes threatened winter forage. Around peri-urban settings residents considered black not beautiful and soot a nuisance. Everywhere everyone denounced smoke. The American Lung Association declaimed that it regarded *any* particulates as potentially damaging if not lethal. Always there was some partisan group, full of passion, to protest against prescribed fire, if not in principle, then against this particular fire in this particular place and time. The process of objecting trended only one way: toward less burning. Among fire suppressants, NIMBY has proved more powerful than Phos-Chek or Firetrol.

Every new item on the checklist, every escaped fire at Lowden Ranch or Point Reyes, every lawsuit, every career derailed because of a feral burn, every public outcry and political inquiry, all added to the expense and complexity of successive burns. Fighting fires cost more every decade, but so did lighting them. Outside the Southeast, the dazzling vision that tame fire might replace wildfire faded. In the West prescribed fire would at best supplement, not supplant, the natural or wildland fire that would have to inform the region's regimes. Even archetypal sites such as Sequoia-Kings Canyon struggled. In 1993 Dave Parsons tallied up the record for prescribed burning and concluded that the restored fire-return interval in the groves was 74 years. Before the park was established it had been at least 4.1 years and probably annual. The most renowned prescribed-fire program outside the Southeast had fallen behind and was trailing further with each year.[79]

* * *

Natural fire was the most dramatic and radical of the reforms prompted by the fire revolution and the one most pounded by the Yellowstone blowout. Few serious observers wanted it abolished; even the GAO argued for more. What was needed, they argued, was stronger protocols, surer incentives, and better adherence to the rules of engagement. The aftershocks of 1988 left most of the natural-fire landscape in rubble. A few prime sites rebuilt quickly, but others were ruined and too small, too complicated, or too remote from their purposes to justify rebuilding.

Reconstruction began with the change of administrations. In March 1993 the fire community reconvened in Missoula to review the decade since the last national symposium on wilderness fire. What had changed for the better was the public recognition that fire in the wild might be good and necessary. What had changed for the worse were the practical consequences of having to reboot the system amid tightened scrutiny and a polarizing polity. No one was happy with the status quo. Jim Bradley, a political staffer, declared distress that there was less wilderness and park fire than a few years before. It was, he concluded, "a declining program." Bruce Kilgore noted that NPS-18 had been revised in 1990 but had boosted the number of "objectives" (or factors a plan had to consider) from 3 to 15, and the community had still not resolved "what to do about the natural role of high-intensity, stand-replacing crown fires" typical of many boreal biotas. He noted that the prospect for "severe curtailment" of prescribed fire, announced by the post-Yellowstone review team, was entirely possible. Bob Mutch recounted how the easy areas had been done but that "fully three out of every four wildernesses in many areas of the West" did not even have fire-management plans, much less functional operations. Elmer Hurd Jr., chief of Fire and Aviation Management for the NPS, laid out the political realities in which media, public review, and quarreling political parties exercised "dramatic leverage over the managerial discretion for prescribed fire and suppression strategies." As with prescribed fire, a handful of premier sites, all in the West and Alaska, accounted for the acreage.[80]

Compared to 1983, it was a smaller group that convened this time, more the choir than the congregation. Most of the issues had not changed; most of the participants were the same. In this round, however, enthusiasm for wilderness was not enough to overcome issues of air quality, funding, partisan politics, environmental creeds, and squabbles over when and where to substitute prescribed fire for natural fire. Much of the group's attention pivoted on matters of scale. Small parks were too tiny to allow big fires, but so were big parks, as Yellowstone demonstrated. Mostly, there were not many acres burned for the years invested.

That was the point of departure for another interagency group of "fire managers, wilderness managers, and environmental planners" who met in Missoula in early August under the auspices of the Arthur Carhart National Wilderness Training Center to craft a guidebook. Its invocation declared that in the nearly 10 years since the new directives had been passed down, only "a handful of plans for the use of prescribed fire in wilderness have been approved." In addition to the hindrances recounted at the symposium, the Carhart Center group focused on how fire plans fit with "the overall planning process." The guidebook would provide an interagency template but one adapted to particular wilderness settings; published as a looseleaf notebook, its very design spoke to the inevitability of future amendments. Review, conversations, and training took time. On November 28, the National Advanced Resources Training Center in Marana hosted a week-long course on "National Park and Wilderness Fire Management" that sought to rally the troops and outfit them with the latest administrative concepts, fire-behavior models, and firing strategies. *Wilderness Fire Planning* was finally published in March 1995.[81]

The breakthrough came the summer of 1994 from the Granite and Howling fires. With inspiration borne of desperation, the Glacier National

Park management team had redefined the terms of engagement. It had been mandatory that all fires be classified as either wild or prescribed such that any fire was either all one or all the other. Wildfire remained forever wild; prescribed fires remained prescribed unless they fled their prescriptions, in which case they, too, became wild forever after. But in the summer of 1994, without the resources to suppress—the national system was overwhelmed with fires and fatalities—the fire team treated various portions of the Howling fire differently. The experience was not widely appreciated—it did not immediately become an alternative exemplar—but, through partisans familiar with its story, it entered into the subtext of the 1995 policy review and a shift in wording from "appropriate suppression response" to "appropriate management response." It ruptured the impermeable categories of wildland fire. It promoted the creation of fire use modules—teams to manage wilderness fires similar to the teams created for prescribed fires. It nudged policy toward a transition from wilderness fire to wildland fire. It suggested, and in 1995 it was only the barest of suggestions, that the same strategies that applied to fires in legal wilderness might be applied to fires in wildlands generally. Beginning with the NPS, agencies experimented with wildland fire-use modules, which became for wildland fire use what incident management teams were for suppression.[82]

The management missteps at Yellowstone had tainted the term *prescribed natural fire*. For a few years afterward it lingered, lame and wounded. Then it went into the shop and reemerged as *wildland fire use for resource benefits*. The bestiary of American fire acquired a new species and got stranger—certainly the terms by which it was explained to the public became ever more abstract and opaque. No one dared speak of "controlled burning" because not all those fires remained in control. "Prescribed natural fires" became toxic after 1988. Now the WFU replaced the PNF. Numb terminological nothings replaced the wacky charm of let-burning and loose-herding. Whatever their lexicon, though, such fires remained relatively insignificant on the ground. In 1998, a lean year for fire, 358 WFUs amassed 76,141 acres, or 5 percent of the national total burned. In 1999, a fat year for wildfire, WFUs burned 75,681 acres, or 1 percent.[83]

Outside Alaska, natural fire was a niche enterprise. It had not happened that the wild would spread its goodness over lands in need of rehabilitation; with a few exceptions wilderness fires were not meeting even the needs of wilderness and nature preserves. On the contrary, throughout the West the only significant burning came through wildfires, which were nominally suppressed. Nor was an exalted vision of the wild shaping fire's management overall. That torch had passed to the interface. Sprawl was a more powerful shaper of America's national estate than wilderness was.

* * *

That left wildfire. The largest number of fires, and the greatest number of acres burned by far, came from fires that no one wanted or could manage only by trying to extinguish them. Suppression remained the default setting. In terms of blackening acres, suppression had the advantage that, arson excepted, no one deliberately started the fire, so no one could be held liable for its consequences.

The American fire community understood clearly that bad wildfires were the outcome of its inability to manage fire holistically. But the system's mouth said one thing and its feet did another. Suppression had a similar failure rate to that of prescribed and natural fires (2–3 percent), but no one was hauled into court or threatened with dismissal because an initial attack failed. Instead, a perverse double standard rewarded escapes while fighting fires but punished them when lighting or monitoring them. So long as fatalities did not occur, the economics of fire favored places with lots of large costly burns, and its politics lavished praise and resources when smoke plumes boiled up on the horizon.

Like the rest of American fire, suppression rebuilt after the '88 season. After all, it too had broken down. Bottomless money, whole armies of firefighters, and fleets of air tankers had not halted the flames, and the efforts may well have magnified the burning by repeated backfiring. More did not mean better: NIFC's reach ended at the flaming front. But as the 1990s evolved, two themes, like shears that worked to make a scissors, shaped the new age of suppression. One was the flaming frontier of exurban sprawl. The other was firefighter safety. The first looked to the Oakland Hills Tunnel fire, the second to South Canyon.

The Tunnel fire granted front-page headlines to an obscure topic with a geeky title. It went national in that universal year of decision, 1994, with a symposium in Walnut Creek, California, to honor the recently deceased Harold Biswell. The wildland fire community wanted the I-zone to go away. Wildland-urban interface fires were alien to its culture, they were expensive, and they threatened to divert attention and resources away from what the community thought was its true mission. The interface was, paradoxically, a national problem that could be solved only by local acts. For both wildland and WUI fires, the ultimate solution lay with land management.

For the interface this meant zoning, building codes, and other regulations beyond the bailiwick of the fire community. Fire officers in the Forest Service, NPS, FWS, or BLM had some say in their unit's land planning. They had none in the interface.[84]

The WUI began rewriting the operating codes where the wild met the built. Instead of interagency coordination among land agencies, the interface demanded partnerships with the U.S. Fire Administration and National Fire Protection Association. What joined them, like an electron shared by and joining two distinct molecules, was the concept of fuel, though urban and wildland communities saw that valence from different perspectives. NFPA's Firewise program looked toward building codes and safer materials. Wooden roofing, for example, was critical to conflagration. Urban cityscapes had no interest, as wildlands had, in the use of free-burning fire to rejuvenate senescent sites. The wildland agencies saw a fuel complex created by contemporary urban out-migration as the counterpart to the fuels left from logging, overgrazing, and fire exclusion, and they begged for variants of slashing, cultivating, and burning to reduce the hazard. Ladder fuels that carried fire from the surface to canopies were the equivalent of shake-shingle roofs. And in wildlands there was no surrogate for the ecological alchemy of free-burning fire. It could not, even in principle, be abolished.[85]

Among the big differences was the relative impact on local economies. Urban fire prevention meant substituting materials and designs: it swapped one set of economic practices for another. It meant using tile rather than cedar shakes, but roofs and roofers remained. Wildland fire could mean recreating an economy where one had been removed. Fuels projects could resuscitate industries savaged by spotted owls and anadromous salmon, rural communities depressed by the loss of tax revenue, and agencies stripped of a source of revenue. It meant putting people back on the land, which could renew all the objections that had led to their earlier removal. The politics of fire could not be segregated from the politics of how people used the land.

The temptation was strong to exploit the fire crisis to advance other agendas. In 1995 Congress passed a Timber Salvage Rider Bill, which allowed salvage logging out of burned areas. The volume involved was not enough to revive timber industries, but it was sufficiently inflammatory to spur outrage from environmental groups who saw the sneaky legislation as a backdoor by which to return chainsaws to the woods, like Detroit evading fleet mileage standards by labeling a sports utility vehicle a light truck. To pay, loggers needed to take bigger trees, typically away from towns, while it was the small stuff near the urban fringe that carried fires into the interface.

Other schemes sought to revive logging in the name of fuelbreaks. The fact was, the federal government could not afford to pay for cleanup throughout the I-zone, but private companies demanded more than thinned sticks. Such controversies tainted all fuels projects.[86]

Equally, fuels projects could upset agencies internally. The 1998 Hazardous Fuels Reduction Program built on earlier stewardship programs dating back to 1972, but it directed operations to the WUI. Money went to protect exurban enclaves from wildfire rather than to restore free-burning fire to wilderness. Issues in fire ecology were redefined as fuels problems. The system filled with fuels managers instead of fire ecologists. Perhaps inevitably, agencies such as the NPS took the allocated fuels money and applied it to where the fire problem, as they conceived it, was most acute.[87]

The fact was, the public identified the federal agencies, especially the Forest Service, as fire agencies, and it expected protection. The agencies were convinced that they were not equipped either by temperament or training to be first responders for structure fires. Protecting summer homes and McMansions in the woods was not what had drawn them to the wild or what had roused their passion for restoring free-burning fire. The reality, however, was that the federal land agencies occupied one side of a shared fence, and the fires that threatened settlements, however ill-conceived the premises behind those settlements, often burned out of national forests and parks. The agencies could not disengage even as they watched the exurban fringe propagate like cheatgrass. They would have to join the fight. By the time the decade ended, the WUI was redefining the character of wildland fire suppression; even monster fires had their resources concentrated where flames met houses. After moving from the frontcountry to the backcountry, American fire was moving back again because our frontcountry was moving. The universe of American fire seemed populated only by red giants of ever-larger fires and white dwarfs of settlement.

The other influence was concern over firefighter safety. It argued against hurling fire crews into remote low-value land or wilderness. But it equally argued against pouring wildland crews into housing developments, where few of their skills had much value and the risks were enormous. After South Canyon, fire officers paused before automatically committing people. They matched a "light hand on the land" philosophy with a "light hand on the crew" method. Research on firefighter protective equipment and work environment revived. A 1997 workforce study accented safety as much as diversity. On the 1999 Sadler fire in Nevada, after a reckless incident ended with a fleeing crew deploying its fire shelters, an incident management team was dissolved for its failure to exercise sufficient attention to crew safety.[88]

Suppression still ruled the fire community. It was indispensable, it commanded the greatest resources, it defined the ends and means of fuels projects, and in that respect nothing had changed in the old discourse. But suppression itself had changed. It had to battle flames around houses, and it had to back off from rugged terrains. The character of the firefight as a set piece was blurring. The issue was not simply whether to fight or light but how to do either or how to do both at the same time. The wildland fire community was morphing into more flexible figurations—or maybe a mishap. The danger was that fire's environment was losing its slack and adaptability—was closing down options and fencing off decision spaces.

As the arguments over land use stalemated, fires filled the void. The danger was not that people might choose the wrong fires or execute good ones poorly but that fires of no one's choosing might determine the lands' use altogether.

Transitioning

Transitioning. It happens when a fire leaps from the surface to the canopies, or when a fire crew jumps from the controlled scramble of initial attack to the orchestrated chaos of extended operations, or when a fire program snaps from one state to another, and transitioning is what happened when the shock of 1994 acted on the shards left after 1988. As bad fire seasons repeated themselves, the wildland fire community began self-assembling into a new order. What it needed were upgrades of the old principles or perhaps some new ones, which led to Revolution 2.0. By the end of the 1990s the contours of a replacement regime were in hand.

* * *

In May 1996 Tall Timbers staged its biennial fire conference at Boise. In 1962 fire ecology was practically a neologism; 34 years later "fire in ecosystem management" was becoming a bureaucratic cliché. When it launched that first conference, Tall Timbers had loosely partnered with Florida State University, defying the fire establishment. When it met in Boise for the 20th, it headed a consortium that included The Nature Conservancy, the Forest Service, the National Park Service, the Fish and Wildlife Service, and the Bureau of Land Management. The highest-ranking fire officer to appear in Tallahassee was R. A. Bonninghausen, chief of forest management for the Florida Forest Service. At Boise every agency was represented, and there were even presenters from Australia and Russia. Secretary of the Interior Bruce Babbitt made an appearance.

Two months later Babbitt was on firelines of the Hochdoerffer fire outside Flagstaff and recorded what had changed since he had fought fires in high school. To the good, fire officers had air support (in 1954 there was "nothing in the sky but clouds"); they had the 1995 Federal Wildland Fire Management Policy that allowed for decisions other than immediate suppression; and they had the still-fresh horrors of South Canyon to caution them against risking crews recklessly. To the bad, there was the need for "structure protection," with the I-zone crowding out the options that the liberalized policy had granted. The Hochdoerffer blew with the southwest winds until it ran out of fuel amid lower-elevation pinyon-juniper woodlands. Had there been houses in its path, it would not have been possible to let those flames pass unchecked.[89]

Also to the bad was the overgrown forest, not only an accelerant to wildfire but an obstacle to restored fire. Those woods entangled with combustibles made it nearly impossible to use fire alone to reinstate fire. After the fire Babbitt visited Wally Covington and later went to Mount Trumbull to inspect restoration plots. He toured the nightmare of Lake Tahoe, a potential crucible of flame; he saw San Diego, where the Harmony Grove fire had selectively burned out houses with wooden roofs. He returned to South Canyon to silently renew "my promise to the parents—that we will honor the memory of their sons and daughters by doing everything possible to prevent another tragedy." Within the year he announced a campaign, Fight Fire with Fire, to quicken the transition from wildfire suppression to prescribed burning.[90]

By 1998, a decade after the Yellowstone eruption, it appeared that the fire revolution had recovered. Virtually every fire program had been upgraded, made more robust, and integrated with other parts. The Joint Fire Science Program issued its first call for proposals. A progressively partisan Congress reauthorized the 1978 Forest Stewardship Program to help grapple with fuels, particularly around the WUI, then enacted a Community Protection and Hazardous Fuels Reduction Act the next year. After the first forest fire-management officer conference in 1997, fire-training centers were established in Tucson, Albuquerque, and Tallahassee for general fire management, wildland fire use, and prescribed fire, respectively. An Interagency Airtanker Board was convened, building on national studies on "shared forces" and "tactical support on large fires." In August the NWCG released its *Wildland and Prescribed Fire Management Policy Implementation Procedures Reference Guide*, an attempt to standardize the interpretation of the 1995 policy. In November NPS Director's Order 18 did the same for the fissiparous collection of parks, monuments, recreational lands, and historic sites under its jurisdiction. For FAM in the Forest Service, the year offered, if not a reorganization, at least a revitalization under a concept called FIRE 21 that would position the agency for the 21st century.[91]

That year, too, the concept of "appropriate management response," encoded in the 1995 policy but not yet written into published field manuals, was field-tested in the Northern Rockies as 100 lightning-kindled fires were managed "for resource benefits" on the Flathead, Nez Perce, Payette, Salmon-Challis, and Bitterroot national forests and at Glacier National Park. The season proved, in Tom Zimmerman's words, a "thorough" vetting of the new policy. Equally, the season demonstrated the unbounded reach of the I-zone as the Florida fire establishment was brought to its knees. That it was possible to muster a national response to Flagler and Volusia counties testified to the success of the interagency idea. Crews and engines had become virtually interchangeable, and after Yellowstone, transfers of personnel among agencies became common. The Forest Service lamented that it was becoming a farm system for the other agencies, as it developed apprenticeship programs only to see graduates transfer to the FWS, NPS, or CDF. That pattern extended even to directors. A Forest Service fisheries biologist, Mike Dombeck, was named director of the BLM in 1994 and then became chief of the Forest Service in 1997.[92]

* * *

That was the good news. It was sweetened by a wildfire season that, outside Florida, was one of the mildest on record. Nature had backed off and allowed some respite. The pieces could sort themselves out in the evolving fire mosaic. The bad news came in 1999.

Wildfire returned—the second largest area burned since the fire revolution had begun. Only 400,000 acres separated it from 1996, the worst. The hardest-hit region was the Great Basin; in a single week 1.4 million acres burned. This time nature ganged up with politics, as President Clinton faced an impeachment trial. There was little grace and less forgiveness from either nature or politics. From here on fractious government met a new norm of burning; what before the early 1980s had been big fire years were now small ones. The bull market for burning was headed for a crash.

This time, too, there were breakdowns in the basics of fire management. On the Sadler fire a crew barely escaped the flames after deploying shelters in a panicked rush. On the Lowden Ranch fire a BLM crew ignited a

100-acre prescribed burn that burned over 2,000 acres and 23 residences before a CalFire team corralled it. An interagency review board raked the operation over the politically red-hot coals. Both planning and implementation were "not in compliance with BLM standards and procedures," and the crew "failed to implement the prescribed fire" according to the approved burn plan they did have. The review team concluded, as all such reviews had, that the national fire policy in general and the agency's prescribed-fire standards and procedures specifically were "sound," but the failure lay in execution. Ominously, the two crashes affected both firefighting and fire lighting, and both were BLM operations. Transitions are awkward, often risky events, and by the end of the century the fire community had not yet navigated the passage successfully.[93]

That was the bad news, and it was echoed by the GAO. Since drafted into fire after Yellowstone, the GAO had become a go-to source for Congress. In 1993 the California delegation called on it to discuss aspects of that year's fire bust on the South Coast. In 1994 it had commented at length on the emerging philosophy of ecosystem management. In August 1998 it reported on expenditures for wildfire preparedness and suppression since the advent of the Clinton administration, and in September it turned its steely gaze to the question of forest health and the "catastrophic wildfires" that threatened resources and communities. It concluded, somberly, that the Forest Service might need a decade to complete the required research, that it might need another decade "or longer" to rewrite forest plans, and that "many experts argue that the tinderbox that is now the interior West cannot wait that long." The GAO agreed: inaction would prove the more costly alternative.[94]

In 1999 it published a veritable fire bust of reports, arguing that the nation faced an explosive clash between two accelerating processes, wildland fuels and sprawling communities. The reports came like shift changes: in February, in April, in June, and in August. The threat was a "catastrophic fire" or fire season that would rumble like a tsunami through the new western landscape of overgrown woods and overbuilt houses. Such fires would not only overwhelm suppression efforts but would also compromise the agency's larger mission. The gist of the GAO's conclusion was that the Forest Service—the USFS was the only agency in its sights—understood the problem and had taken measures to correct it, but that it confronted "numerous barriers to effective action" and that these could be overcome only with a "cohesive strategy."[95]

Every strategy had its complications and critics, and the data to decide among them were lacking or inconclusive. There was probably not time to methodically advance from field science to operations, and of course there was the matter of paying for treatments on the 39 million acres then reckoned in need of it. The agency could not rely on its timber program to fund projects, because the commercial timber was in one place and the interface in another and because the loggers wanted big trees and fire sought the little ones. Nor could prescribed fire do the work alone because mechanical methods often collided with other agency mandates. The GAO concluded that the Forest Service lacked the necessary knowledge and had no "performance measures and goals" by which to be held accountable. As much as a "cohesive strategy," the agency needed a hard schedule for implementation. The Forest Service agreed.[96]

In June the GAO reported on the status of the agency's efforts. It recognized institutional barriers as well as those grounded in environmental conditions along with competing publics and conflicting mandates. "A key factor that separates the strategies that are effectively implemented from those that are not is whether the agency treats the issue as an agencywide priority." The GAO did not see evidence of either the strong leadership or the marshalling of funds and resources that would indicate a "sense of urgency" and commitment to reducing the threat as a high priority. In August it expanded its survey beyond fuels to the full gamut of fire management and included the BLM and NIFC as well as the USFS. It elaborated with a September report that examined the shrinking fire workforce, fitness and safety concerns, and radios.[97]

The GAO, like OSHA and the EPA, parsed and restated fire issues into its own vernacular as though prose could be subjected to management by objective. But its purpose was to audit and hold accountable, not to inspire. Moreover, the GAO could say what a federal fire agency could not. When the Forest Service pointed out conflicting mandates or insufficient funds, its message could be dismissed as special pleading. When the GAO noted the same conflicts, its report was heard, if not acted on. As the fire scene intensified amid a deteriorating political landscape, the Government Accountability Office became the broker of default between the agencies and Congress, much as The Nature Conservancy was becoming the broker of choice between the agencies and the public.

The bottom line of its audit, however, was this: 37 years after the first Tall Timbers fire conference, 30 years after the National Park Service had announced its policy reforms, the new-model American fire community was struggling to execute its fast-morphing fire mission, and as the WUI worsened, wildfire was likely to deform and distract the agencies from the larger land-management missions that were the purposes of their existence. The Forest Service, in particular, seemed incapable of writing its fire agenda into its land-management plans, or when it could, of putting projects into the field. While it, like the other agencies, received 85 percent of the funding deemed necessary, it was unable to translate even these sums into effective field projects.

In 1960 Herbert Kaufman had made the Forest Service's ability to endow its rangers with the agency's national ambitions and put them into practice a defining trait, one so pronounced as to potentially become a liability. Almost 40 years later, the Forest Service's incapacity to move ideas into plans and plans into action had become almost legendary. Its workforce no longer absorbed the institution's identity as its own, no longer did what it was supposed to without being told. Nothing synchronized its mission, plans, and execution. Incentives and goals were misaligned. Land-management plans remained unwritten. Fire plans defaulted to suppression, or outside the Southeast, dabbled with experiments in natural fire and prescribed burning. What had so distinguished the U.S. Forest Service among bureaucracies at midcentury had vanished.

Other agencies fared better—the National Park Service especially. But what was true for the Forest Service was true for the wildland fire community overall and for its grand project at fire restoration. They had learned to cooperate and often to coordinate. But in the GAO's understated but often-repeated words, they had not moved from voluntary cooperation to systemic cohesion.

Conflagration and Cohesion

Both the good and the bad were magnified during the 2000 season and then turned ugly.

On January 12 the Forest Service inaugurated an era of metafire with the final report of a National Management Review's examination "of past reviews." A joint product of the USFS, NPS, NFPA, NASF, and a newcomer, the Brookings Institution, *An Agency Strategy for Fire Management* sought to jolt to life the recognitions that everyone seemed to accept but no one appeared able to implement on the scale required. The review bluntly identified four "chronic" problems that "have been identified over and over in many reviews in this decade" yet remained unresolved. Fire and land management were not well integrated. The agency's ability to handle large fires was questioned, which its cooperators took as evidence of a generally moribund institution. The Forest Service appeared headed down the track of the Postal Service, sufficiently hobbled by conflicting politics to be inefficient without being allowed to reorganize in ways to make it competitive.

At the core of the analysis was the proposal to convert to a large incident management organization, which was longhand for a dedicated fire service, but still tethered to agencies and their land missions. It was fire's equivalent to replacing conscription and militias with a professional military.[98]

Big fires remained, as always, the driver of suppression costs and damages, and they seemed as immune as ever to efficiencies. Behind the agency strategy report stood a parallel inquiry by the Strategic Overview of Large Fire Costs Team, which released its findings on January 21. Its organizing theme was a detailed exploration of how management decisions and costs were related on two major fires in California the summer before. The Big Bar and Kirk complexes burned 227,000 acres over three months at a cost of $178 million. The review team's goal was to highlight the need for "higher priority on fire management, in the broadest context," or more strikingly, to "redeem" the agency's fire role for the coming century. Revealingly, it opened by triangulating from the two geodetic markers of the restored revolution: Norman Maclean's *Young Men and Fire* and the South Canyon tragedy, documented by John Maclean's *Fire on the Mountain*.[99]

Everything about Forest Service fire management needed upgrades. Better initial and extended attack capabilities. Better fire planning and fuels reduction. Better postfire rehabilitation. And just plain better ways to "achieve its service and land stewardship mission." In some cases, this meant "fundamental adjustments" in overcoming barriers, and in others, improved implementation of recommendations from past reviews. The reviews were in danger of tripping over one another. On its originating question, the Large Fire Costs Team "did not identify anything that would have significantly reduced the costs" of managing the two complexes, although regional researchers noted that the lack of federal staff meant that positions had been filled by much pricier local, state, and private-contractor alternates at nearly triple the cost. Such fires, and the responses to them, were the contemporary reality of fire management. Perhaps the most remarkable feature of the report was its unremarkable conclusions. It had all been said before. Many times.[100]

The Forest Service said it again on April 13 when it published a densely argued, 89-page reply to the GAO's call for a cohesive strategy. Headed by Lyle Laverty and Jerry Williams, assisted by representatives from the NASF and Interior, *Protecting People and Sustaining Resources in Fire-Adapted Ecosystems: A Cohesive Strategy* argued for resilience through restoration. Get fire back in something like its former regime, and healthy forests would follow, and with healthy forests, ecosystems and the communities interweaving with them would be less likely to face catastrophic fires and would be more resilient to those that did occur. The West needed a prescribed-fire program analogous to that in Florida. But it could not get there without prior strategic treatments for biotas choking on a century of undigested fuels.

The Forest Service's counteroffer to the GAO shifted the discussion from communities at risk to ecosystems at risk. If it sought to mark the place at which ecosystem management and fire management crossed, it noted that the intersection was less a point than a sliding scale. It projected that, "fully implemented," fire-restoring treatments in the West would rise from three-quarters of a million acres to three million acres per year, and that nationally, costs would soar from $75 million to $825 million. The price of inaction, it flatly declared, would be much greater. Still, western fires were seen as a western problem, not a national one, and a matter for the American fire community, not for the American commonwealth. No fire had yet galvanized the country's attention as the 1994 season had the fire community's.[101]

What claimed political attention were fires that burned towns, combusted budgets, and gathered the media. What defibrillated the metafire analyses were megafires. They came with the 2000 season, the worst in memory, not only for its box score of acres burned, dollars spent, and firefighters mobilized (29,000), but because it hit the historic heartland of wildland fire, the Northern Rockies. It smoked Missoula in for weeks on end. It exhausted NIFC. Fire officers from Australia and New Zealand were flown in to assist. On this, the 90th anniversary of the Big Blowup, fire suppression had never been more massive — or more helpless. Then prescribed fire failed at the hands of the federal agency that had spearheaded the fire revolution. The Cerro Grande escape at Bandelier National Monument burned into Los Alamos, New Mexico, amid national hype and horror. The Cerro Grande Fire Assistance Act to repair damages cost $661 million; its final $1 billion bottom line made the fire the most expensive in American history. Cerro Grande singlehandedly exceeded the outlays of the entire 1994 season. Some heads rolled, and some administrators fell on their swords to the outrage of the Park Service, its apologists, and the more fervent partisans of the revolution. Secretary Babbitt accepted blame, paid the losses, and kept the tragedy from festering in law courts and media.[102]

After 1994 the reigning sense within the wildland fire community was that it needed a reboot. After 2000, the sense prevailed that it was simply broken.

* * *

The inevitable policy review got under way, although it would conclude, as they all did, that the policy was right and proper and needed only a few tweaks. The malfunction lay in application and hence accountability.

The GAO insisted again that the Forest Service in particular failed to valence incentives to outcomes; often the areas it treated were not the areas most in need of treatment. Even without Cerro Grande it was apparent that prescribed fire was not adequate by itself, but combining it with mechanical preparations might conflict with other stewardship obligations (and would likely arouse critics and the courts). The problem was worsening—had worsened measurably during the 10 years the GAO had been investigating it. Forty years to treat 40 million acres was a losing hand. The Congressional Research Service issued its review in December, deftly furnishing a tutorial on the fire scene as restated into the language of authorizing legislation, funding, and measurable outputs.[103]

Few observers focused on policy as such. Broadly considered, policy had been adequate since 1978. Each crisis and review had affirmed its basics. Money invested had risen sharply through the current administration, and while suppression costs had spiked and crashed with particular fire seasons, they were undeniably trending upward. Something else was needed. It came on August 8 when President Clinton requested secretaries Babbitt and Glickman to recommend "how best to respond to this year's severe fires, reduce the impacts of these wildland fires on rural communities, and ensure sufficient firefighting resources in the future." The administration also sought "short-term actions" that the federal agencies could take to reduce immediate hazards to communities and firefighters. A month later the two secretaries submitted their report.[104]

The National Fire Plan, as it became known, synthesized the GAO's concern over protecting communities with the USFS's regard for fire-restored healthy ecosystems. The resources needed to support such a program would come from a buildup of fire suppression and a boost in fuel treatments. That meant money, and the National Fire Plan called for a FY2001 budget of $2.8 billion, nearly $1.6 billion above the administration's request. Unlike so many documents before it, the NFP was not a statement of policy but an action plan that, if implemented, would underwrite the most dramatic investment in wildland fire since the New Deal. With smoke from the 2000 season barely cleared from the sky, Congress approved the appropriation in October.

Whether the NFP was the cohesive strategy that the GAO had requested, it was what the fire community could get, and given that the most dramatic event of the fire season had been the self-inflicted blowup at Bandelier, it testified to the rootedness and pervasiveness within the fire community of the notions it proposed. Its funding would allow the agencies to modernize their equipment and training—to make the transition to a professional military model (as it were), though one still bonded to agency missions. That funding permitted the Forest Service, in particular, to replace the revenue lost from diminished logging with another funding stream from fuel treatments. The National Fire Plan marked the culmination of a rebuilding project that had groped about in the ashy aftermath of Yellowstone and then grouped itself into themes and institutions without explaining how they added up to a national project. The two sides of the contemporary fire scene still stared at one another across the interface, but neither had driven out the other.

* * *

In this way, fire's American Century seemed to end where and how it had begun—in the Northern Rockies and California, amid debates about big burns and light burning, amid controversies over the character of American settlement that left towns and homesteaders vulnerable to wildfire. The fire revolution, too, showed remarkable persistence. It was astonishing how little had changed over the course of almost 40 years. The same arguments, much the same evidence, the appeal to the same cultural values, many of the same people, all had aged if not evolved, but they were recognizably derived from the same stock.

What had changed were the fires, the workforce, and the politics. The fires were vastly bigger and meaner. The workforce was more smartly outfitted, trained, and diverse, but so less well adapted to the task at hand that the agencies were opting to hand fire over to a guild of fire specialists. The politics had morphed from bipartisan enthusiasm for environmental issues and a sense of a shared and cumulative stewardship to bitter partisanship in which each side declared the other's ideas null and void, and each new administration sought to undo the record of its predecessor. The various conflagrations—flame, economy, politics—converged with the dotcom crash and, when the epochal 2000 season ended, with a sharply cleaved electorate and a president put into office by a partisan Supreme Court split 5–4.

The worst was yet to come. A slopover was set to boil over. V. L. Parrington had famously characterized America's Gilded Age as the Great Barbecue. America's new Gilded Age seemed headed to a more literal version, a Great Blowup.

Millennial Fire, 2000

The 2000 fire season began early and ran late, and it became a referendum on the American way of fire. For almost 40 years America's wildland fire community had divided the population of fires into two moieties: wild and prescribed. Between May and September both suffered spectacular failures in the West. The season started with a prescribed burn in New Mexico that went feral to become the worst escaped fire on record. The season matured with a fire bust in the Northern Rockies that just would not quit. While the fires broke out in the usual places and burned in the usual ways, they were bigger and tougher and seemed unstoppable. They mocked all efforts to contain them.

What made the season memorable, however, were the fires' political environment and the cultural tinder that fed them. The real firefight was over meaning. Something had snapped. Both within the fire community and outside it, the nation was eager to know what the season meant. Competing groups argued over going direct or indirect on the narrative of wildland fire. They debated whether to accent risky behavior that might lead to tragedy or to point to the deeper risk of inaction. The fire in the mind had succeeded decades earlier. The fire on the ground was an unholy mosaic of overburned and underburned, of fire famines and fire fiascos. Twenty-five years after the Big Blowup, the fire community received a clear, if misguided, mission that translated the idea of fire exclusion into practice. Forty years after the fire revolution, the fire community, like its primary and still-indispensable institution, the Forest Service, seemed paralyzed by inner conflicts that left it unable to fully execute any policy.

* * *

In the late afternoon on July 16, Jack Ward Thomas, John Maclean, and Steve Pyne gathered, by invitation, at a barn converted into a small conference center on the privately owned Teller Wildlife Refuge outside

Hamilton, Montana, to discuss America's fire scene. The tension of a sultry day had ended with an afternoon's outburst of dry lightning that inaugurated the region's fire season. Towering cumulonimbus then yielded to convective plumes.[1]

Retired Chief Thomas stated the obvious: the story of American fire was happening outside the barn door as the country's fire army mustered. Its infantry and armored brigades rolled along Highway 93, while overhead helicopters clattered amid the deep rumble of an occasional P2V Neptune air tanker. When the smoke rose, the fight began, as it had around the Bitterroot Valley since 1910. The three speakers brought the perspectives of science, politics, journalism, and history to bear on the question of why, but the reality was in that cavalcade of engines, hotshots, dozers, smokejumpers, helitack crews, and relic warplanes outfitted to dump retardant.

In the Bitterroot, big fires blasted Skalkaho, the Rye Creek and Burke Gulch areas, Blodgett, the East Fork, the West Fork, the Lower East Fork, and the Selway-Bitterroot Wilderness. They burned areas fired in the Big Blowup, the Pete King-Selway fires, the Sleeping Child burn; they burned bridges and signs; and they burned down boundary markers. Fires burned picnic sites and campgrounds, they burned popular hot springs and backcountry trails, they burned winter range and the radio repeaters on Sula Peak. The fires' severity worried officials about damage to watersheds, with flooding and destabilized soils to follow. But no less than infrastructure, they burned away the intellectual old growth that supported a singular suppression program. Before snows ended the season, some 2.3 million acres had burned in Idaho and Montana.[2]

Time and again, big fires in the Northern Rockies had triggered reviews of national policy, and the 2000 fires upheld that tradition. It was one thing for a prescribed fire program to falter, maybe implode lethally, but fighting fires had been for almost a century a primary duty of a Forest Service committed to custodial care of the vast lands under its stewardship. It did not have all the money or matériel it wanted or needed, but compared with what rangers had in their caches back in 1910 or 1934 or even 1967, they were immensely more powerful. Yet the fires had grown ever more mighty. They were mostly unstoppable. The firefight concentrated on those places where flames threatened towns and backcountry exurbs. The fog of battle congealed into an obscuring smoke pall that lingered for weeks in the Bitterroot Valley.

In *Young Men and Fire*, Norman Maclean had picked up and tested various conceptual tools out of the academician's cache to explain what he saw. He tried science, art, history, law, bureaucratic inquiry, and sheer storytelling. That small conclave gathered near Hamilton for policy presuppression followed his example, though they might well wonder whether their words could nudge even a little the hammer of suppression trying to strike an anvil of flames. The season ended, however, with institutional reviews that might: the National Fire Plan, a formal review of the 1995 policy, and enough studies and commentaries from enough organizations to fill NIFC's warehouse. But the enduring image came from art.

On August 6, standing on a bridge spanning the East Fork of the Bitterroot near Sula, John McColgan, a fire-behavior officer on the Alaska Type 1 IMT working in the area, took an evening photo of the river. Two elk were standing midriver with a dappled background of residual flames across a hillside behind them. The image went viral. NASA posted it. *Time* magazine ranked it with the best photos of the year. Somehow it distilled for the fire community its sense of what mattered most about the fires. There was a calm to the scene—the visual equivalent to the "still small voice" left after the fire. It spoke of fire as natural, intrinsic, and beautiful. It said that not all big fires were bad, that some fires were best left alone, that the firefight was not the apex of fire management.[3]

The 1961 fire season had bequeathed some classic images, all in black and white, notably the cauliflower smoke plume of the Sleeping Child fire boiling up like an incipient nuke, or existential scenes in which a solitary firefighter marched like a Hemingway code hero against a backdrop of obscuring smoke. But McColgan's elk photo shifted from the set-piece firefight to the fire. It showed no one, only two elk, and its flames, rather than being ravenous or devouring, were instead a source of illumination, like a battery of votive candles. The scene revealed an almost transcendent accommodation between flame and wild. That such an image could find an audience in a year noteworthy for 92,250 fires and 7.4 million acres burned—a modern record—spoke to a profound transformation in suppression itself.

* * *

That was the fire year's benign, Horacian irony. Its Juvenalian, world-gone-mad irony was the prescribed Upper Frijoles Canyon fire in New Mexico, set to restore an infirm landscape, a fire that metamorphosed into a monster, slammed the town of Los Alamos, and burned into the national lab best known for developing the atomic bomb. If the fire bust in the Northern Rockies tested one half of the fire revolution, wildfire, what became the Cerro Grande tested the other, prescribed fire. It did for prescribed burning what the Yellowstone fires had done for natural fire and the South Canyon fire had for suppression.[4]

The immediate point of ignition began at 7:20 p.m. on the evening of May 4 as a National Park Service burn crew ignited driptorches and tipped their flaming diesel-gas mix to the needles and grasses of Cerro Grande Mountain. But before that act, two earlier wildfires had demonstrated the potential risk in the Jemez Mountains. In 1977 lightning had sparked the La Mesa fire, and in 1996 an abandoned campfire kindled the Dome fire, both of which had threatened Los Alamos. The region was a tinderbox. Fuel treatments and fire-restoration projects had commenced to reduce the hazard, but like most projects, they were too little too late. Twice—once in the fall of 1992 and again in 1993—the NPS at Bandelier National Monument had tried to burn a large patch of dense woods in a side ravine of Frijoles Canyon, but both burns had smoldered out ineffectively. This time the burn plan called for a spring burn, precisely the season when, before monsoon rains, almost all the large fires of the region took off and ran with the prevailing winds. The more immediate context was four burns that had escaped in the region over the previous two weeks. The Santa Fe National Forest had a general fire ban in place.[5]

In coarse chronology, the prescribed fire proceeded with two crews of 19 firefighters. During the night one crew, exhausted, was released while the fire burned more actively than anticipated. Eventually winds pushed flames across the eastern firelines. Backup was requested from the Forest Service dispatch center in Santa Fe at 1:30 a.m., which was at that time staffed by a substitute dispatcher. If the fire had been declared a wildfire, then emergency accounts could pay for the request, but the Park Service had simply requested assistance and was told to call again in the morning. When the incident commander next called, he admitted they faced a wildfire, and a full-scale mobilization got under way. By Friday evening 45 firefighters, 2 air tankers, and 2 engines were on the lines. For reasons of safety and effectiveness, the suppression organization declined to build an underslung line below the rim and went indirect instead. That decision would also allow crews to burn out the targeted 900-acre block, but with suppression money and management. By Sunday morning, May 7, the burnout along the west flank met stiffening winds; spotting blew the fire over a dogleg on the still-unsecured Highway 4 and then into the interior of the block beyond, and finally over the slopover in an angry trajectory toward Los Alamos, directly downwind. By Monday night the fire had grown to 3,040 acres against which 330 firefighters struggled.

The fire and the firefight became an unequal contest. By Wednesday evening the fire, split into two heads, had blasted over 10,000 acres, through parts of Los Alamos and Los Alamos National Laboratory, and raced toward pueblo communities beyond. By the time it was contained on May 19, the fire had blackened 48,000 acres. It was the largest fire in recorded New Mexico history. Some 280 homes had burned. Combined costs for suppression and damages exceeded $1 billion. Interior secretary Bruce Babbitt placed a moratorium on prescribed fire throughout the national park system. If the Cerro Grande fire was an indictment on the fire revolution, it resulted in an impeachment.[6]

Norman Maclean concluded that the story of Mann Gulch was the search for a story. So at Cerro Grande, the story became the narrative of the many inquiries to find an explanatory story. The stakes were high: they would be determined by the power of whatever narrative most convincingly organized the evidence, which in turn required not only a span of events but also a theme and meaning. A decade later the narrative of Cerro Grande remains, as academics like to say, contested. The fire community might be inclined to turn to the same words it had used for South Canyon. This was just a shitty fire.

In retrospect the surface saga of Cerro Grande was an endless succession of minor glitches, operational bugs, oversights, and misreadings that finally, like a cascade of neutrons saturating a lump of U-235, set off a chain reaction and allowed a small mass of flame to explode into a conflagration. On May 11—the fire still blowing with the wind—Secretary Babbitt ordered an interagency inquiry. Its on-site investigations lasted three days. The Cerro Grande Investigation Team Report (better known as the "May 18 Report" for its release date) was announced with a full-court press of publicity in Santa Fe. The conclusion resembled those of nearly all such inquiries: the policy was sound but its execution was flawed. A second more robust investigation, the Independent Review Board, began almost on the heels of the first; it issued its report on May 26.

At the request of Senator Pete Domenici, the GAO, now a veteran analyst of the American fire scene, joined in. It published its conclusions on July 20. It also took as its premise that prescribed fire was legitimate but that the Cerro Grande disaster had exposed policy, planning, and implementation issues and that these were not restricted to the National Park Service or to Bandelier National Monument but were systemic. Planning had become more complex and required peer review, and interagency cooperation on prescribed burning did not match coordination on wildfires. The report to Congress made clear that the GAO was "not here to assign blame but to help improve the way federal land-management agencies manage future prescribed burns." Congress held two hearings in late July, one for the Senate and one for the House. Meanwhile, with a moratorium on NPS burning in place, Babbitt approached the National Academy of Public Administration to inquire into the "adequacy of the 1995 Policy" and recommend ways to improve its implementation, specifically with regard to the National Park Service.[7]

If Secretary Babbitt took the fire personally, so did the National Park Service, which voiced outrage over what it regarded as unfair treatment of the organization and its agents. It assembled a 41-page rebuttal of the investigation team's report. Superintendent Roy Weaver disputed the May 18 findings totally, though (an honorable man) he accepted final responsibility since he had signed off on the burn. Former NPS director Roger Kennedy dismissed the proceedings as a "witch hunt" and authored a book to deflect attention away from the flawed fire to the flawed policies that had allowed the American fire scene to become unmanageable. Partisans of prescribed fire and defenders of the NPS generally railed against Babbitt and what they regarded as politically motivated scapegoating. When the chips were down, he had abandoned them. Had he so chosen, Babbitt might have retorted that by accepting blame and agreeing, on behalf of the administration, to pay an obscene sum in compensation, he had taken the issue out of politics and the courts, where it would no doubt have festered like a fire smoldering in a dung pile. He had rescued the validity of prescribed fire.[8]

The NPS mustered its own Board of Inquiry, which amassed 1,600 pages of transcripts, all conducted in relative secrecy, and published its conclusions in February 2001. As with the Yellowstone crisis, its focus was less on the fire than on the first investigative report and the errors it had contained. To partisans the board corrected mistakes that had led to severe hardship on Park Service staffers who had taken necessary risks, and they resented the charge that the escaped prescribed fire had rambled into Los Alamos, not the failed backfire set by suppression crews. To critics, including Babbitt, it was a "whitewash," another exercise in misdirection. The evidence of the board remains sequestered and unavailable to the public.

Context was all. Every recalled gesture, every casual remark, every torching tree and spot fire was refracted and magnified in retrospect into portents and symbols to identify, condemn, or absolve those responsible for what happened according to whatever interpretive lens they chose to view those events. Consider, for example, the matter of other escaped burns. That another of its prescribed fires, the Outlet burn on the North Rim of Grand Canyon National Park, had only the day before slipped its leash and bolted across the North Rim, shutting the only road into and out of the park, demonstrated conclusively to apologists that winds ruled. Any fire set on those days would have escaped. Critics, however, could point to the

Outlet escape and wonder why that breakdown had not alerted colleagues at Bandelier that the same weather system would arrive the next day, that a runaway fire could only happen if they chose to light up. Like Great Aunt Augusta in *The Importance of Being Earnest*, they might liken the loss of one fire to a misfortune but regard the loss of two as carelessness. Or consider the matter of early interviews, conducted while the fire still blasted over the Pajarito Plateau. This, apologists claimed, was calculated to produce flustered comments, inchoate memories, and unfair and hasty judgments. Critics noted in reply that the failure to conduct just such interviews during the South Canyon investigation had lost vital information. So it went. Every shard of evidence could be aligned for or against.[9]

* * *

No one, however, measured Cerro Grande against the other two defining breakdowns of the new era, the fires at Yellowstone and South Canyon. Together they make a fire triangle in which each practice—suppression, natural fire, and prescribed fire—had suffered some horrific and public failure. Begin the comparisons with Yellowstone. Because of the Los Alamos National Lab, the Cerro Grande fire, too, had celebrity status. In both cases the Park Service and its apologists exploited the same rationales—that fires were inevitable, that the failures lay with bungled suppression actions, that dangerous times brought high risks, and that the costs of inaction were worse than those of doing and failing. In both instances they used errors in the initial report to deflect the subject away from the fire itself and onto the inquiry. The NPS should be applauded for attempting what others had shamefully declined to do.

But the best analogue may be South Canyon. Cerro Grande forced changes in prescribed burning comparable to those wrought by the South Canyon tragedy for fireline operations. That underslung line on Storm King Mountain might have worked, had a front not passed through; and so the slopover and burnouts at Bandelier might have been contained had a similar front not rolled across the Jemez Mountains. The real issue was accountability. Good firefighters, taking risks they thought worthwhile, had died at South Canyon, and the fire community had concluded that those risks had not been worth the sacrifice and that responsible parties should be held to higher standards. That is also what happened at Cerro Grande. In neither case had fire officers been wanton, but their judgments and in some cases executions had been flawed.

Explanations were possible, excuses were not. It was no longer sufficient for suppression to say that risks were inherent in the job, that a can-do attitude could explain or absolve reckless behavior, and that no one could be held responsible when a firefight turned ugly. So after Cerro Grande, the excuse no longer washed that prescribed fire was inherently risky and that higher risks were justified in places with higher hazards. Prescribed fire was not new, nor was its policy novel. The NPS had adopted the practice 32 years earlier. Breakdowns such as Cerro Grande could not be allowed to jeopardize whole programs or put future crews and communities at similar risk. The fire community had long lamented a double standard that granted suppression the money and freedom to maneuver that prescribed burning was denied. With Cerro Grande it finally achieved parity, even if a parity of fiasco. If there were times and places where suppression was not an option, that same reasoning applied to prescribed burning.

The fire community, if reluctantly, began to wrestle prescribed fire into the same matrix of accountability that had come after Yellowstone for natural fire and after South Canyon for suppression. Each of the fire revolution's innovations had been tested in the flames and purified by those trials. However aggrieved the NPS felt, it was Bruce Babbitt who accepted responsibility on behalf of the larger enterprise, paid out reparations, and saw his most heartfelt accomplishment, the restoration of fire, left in ash. Yet it could have been worse. As a Nature Conservancy fire officer in Florida noted sadly, if such a fire had occurred on their watch, they "would be out of business." Not the TNC fire program—TNC itself.

After Cerro Grande, a moratorium on prescribed burning was announced for the summer until reforms could be installed. The same had happened with prescribed natural fires after Yellowstone. The program would rebuild on firmer foundations, though what emerged would be vastly more costly and complicated, for it was blindingly apparent that the strategy of prescribed burning as a succession of isolated set pieces was doomed to fail. A program needed to operate across whole landscapes with plenty of buffers, interlocking checks and balances, multipartner coordination, and a shared culture of fire. Upper Frijoles Canyon was not big enough; neither was the Pajarito Plateau. The rehabilitation of the landscape had to begin at the Los Alamos borders and build out, not light up in the backcountry and let the wind drive flames to the city. The crisis threatened a sad, cruel coda to the Babbitt era.

In the end the season was redeemed by the simultaneous grand mal convulsions in the Northern Rockies and the collapse of suppression, and the year's end saw the mutual convergence of those breakdowns into the reformist agenda of the National Fire Plan. But while the long hot summer raged, the epochal seemed to segue into the Biblical. Had they time or inclination, fire officers might have tossed aside their NWCG handbooks and General Technical Reports from the Missoula lab and reached for an Old Testament prophet. "Behold," cried Isaiah (50:11), "all ye that kindle a fire, that compass yourselves about with sparks: walk in the light of your fire, and in the sparks that ye have kindled. This shall ye have of mine hand; ye shall lie down in sorrow."

CHAPTER FIVE

BLOWUP

Half the year is spent in widespread talk about the need to reintroduce fire into fire-adapted ecosystems, but the other half of the year is spent suppressing all wildfire at substantial economic cost.
—FORMER CHIEF FORESTER MIKE DOMBECK[1]

In October of 2007, a series of large wildfires ignited and burned hundreds of thousands of acres in Southern California. The fires displaced nearly one million residents, destroyed thousands of homes, and sadly took the lives of 10 people.
—INTRODUCTORY STATEMENT, CALIFORNIA FIRE SIEGE 2007[2]

Shock and Awe

The 2000 season climaxed a decade of rebuilding amid sharpened controversies, a fissured country, and worsening fires. Out of the wreckage of the late 1980s, new icons of old issues had appeared. The Endangered Species Act, not the National Wilderness Preservation Act, became the focus of environmental activism. The wildland-urban interface, not timber, became the challenge to fire protection. Fuels treatments in the interface, not prescribed burning, became the West's projects of choice for fire's restoration. The available workforce became both much smaller and vastly larger as the federal fire community shrank, professionalized, and privatized while the national fire community swelled to incorporate state agencies, municipal and volunteer fire departments, and private companies. This congregation of necessity shared problems; they only partly shared common standards, rules of engagement, or funding. Interagency agreements evolved into the National Fire Plan and by the end of the decade into a National Cohesive Strategy. Smokey Bear revised his message by replacing "forest fire" with "wildfire." Norman Maclean's young smokejumpers were channeled into novelist Nicholas Sparks's *Smokejumper* as fire whisperer. Meanwhile, a Congress that declined to pay for its foreign wars refused to pay for its war on fire. The tempo of fire management quickened, its scale swelled, and the capacity to hold this unwieldy and fractious gaggle together wore thin and threatened to snap. All the usual means of organization, or coherence in the GAO's preferred term—by agency, by policy, by region, by idea—seemed to dissolve into a frenzy of year-by-year convulsions.

All this played out amid a country shuffling itself into seemingly irreconcilable political polities, economic classes, and ethnic enclaves for which the split Supreme Court decision in the 2000 presidential election is an apt metaphor. An electorate divided between red states and blue had its counterpart in a national estate partitioned between wildlands and cities. The one pushed to reinstate fire on a landscape scale, the other to halt it at sprawl's frontier. The nation's exurbs were as fire-gerrymandered as its congressional districts. The countryside filled with subprime fires. In the 1970s the fire revolution had proceeded despite oil shocks, stagflation, Watergate, the resignation of Richard Nixon, and the final loss of the Vietnam War. In the 2000s fire seemed unable to dissociate from cascading crises, either echoing other disasters or exploited to animate them. Even chief foresters such as Mike Dombeck openly decried the schizophrenic outcome between what the fire community believed they needed to do and what they actually did. Reconciliation proved difficult: to partisans the legitimacy of the process itself seemed suspect.

What forced the issue of course were fires. Big fires—historically unprecedented burns that quickly acquired the term *megafire*—exploded not just in the usual places but in Colorado, New Mexico, and Georgia. One response was to commission studies as the era of megafire segued seamlessly into an era of metafire. A second response was to beef up suppression. Like the country after 9/11, the American fire community remilitarized, and did so along similar lines that both isolated suppression and insulated the larger society. A third was to upgrade preventive measures, to harden critical targets, in this case the I-zone. The healthy forest theme morphed into a general fuels treatment program to protect communities against catastrophic fire. The old vision of a national fire policy, advocated under the aegis of the Cold War, returned in the guise of a new national security state reconstituted as homeland security. NIFC sent incident management teams to the Twin Towers and post-Katrina New Orleans. FEMA financed some state costs for suppression; it then assumed control over the National Interagency Incident Management System and completed that program's conversion to an all-hazard emergency response format. Critics worried that the abrupt reforms threatened to abrade citizen protections with regard to environmental rights and to silently strip other budgets to cover the costs of a suppression that might not actually enhance security.

The original fire revolution, begun in the mind, had triumphed and then stumbled on the ground. The three thematic apexes of fire's management—natural fire, prescribed fire, and suppression—all suffered breakdowns. Two had been NPS operations, and one had come from the BLM, but inevitably the U.S. Forest Service took the political hits, much as its fire crews had suffered the blows in the field. Repeatedly, reviews reaffirmed policy and chastised practice. But unless notions could find expression in the field, they were bureaucratic babble. The wildland fire community did not need more policy. It needed the means to move policy to action, and for that it needed a story as much as funding. It needed a poet.

When the revolution began, the plot of the national narrative pointed to the creation of fire agencies independent of the Forest Service. Then it turned on the invention of interagency institutions amid a consensus agreement on a policy of fire by prescription. A lost decade followed, after which the project had to be reconstructed. At this point themes and regions replaced agencies as organizing conceits. Now, like separate streams of stars caught in a spiraling nebulae, institutions, ideas, and events swirled together—the Model 6 engines and the P2V and DC-10 air tankers, the Brookings consultants and the old-guard bureaucrats, the prescribed flames washing through longleaf glades and the ravenous plumes over lodgepole-clad mountains, the smoking snags and the incinerated suburbs, the GAO and the Congressional Research Service reviews, the hotshots and congressional hearings, the WFUs and the WUIs, the nimble new agencies and the recovering legacy ones, the honored dead and the living memorials. The largest fires increased a hundredfold; costs swelled by an order of magnitude. Year by year, the chronicle beat on, one topic or another dominating in any particular year but all clustering in an ever-tightening gyre.

The millennium began boldly and badly, and got bigger and worse. The American fire scene blew up.

2001: Metafire

The after-action reviews on the 2000 season continued, and when they had exhausted that year's loud and messy fires, the postmortems turned to the succeeding years, which overflowed with big and audacious fires. Sparks of insight and irrelevance blew in the winds before merging into a veritable crown fire of commentary.

Everyone even tangentially connected to fire published reports, often serially. The National Academy of Public Administration issued six papers between 2000 and 2005. The USDA Office of Inspector General reported on costs and on aircraft. After its 2000 inquiries, the GAO released 48 more reports through 2009. The Western governors held a fire summit. The International Association for Wildland Fire sponsored two fire policy convocations. The Nature Conservancy, Wilderness Society, and Yale University's Global Institute of Sustainable Forestry published analyses. The Brookings Institution prepared surveys, then brokered a Quadrennial Fire Review, the first of a series. A special blue-ribbon panel was commissioned to inquire about aircraft safety. The agencies, particularly the Forest Service, assembled task forces to address the questions posed by all the other investigations. Findings, goals, and recommendations were themselves organized by meta-analysis into more comprehensive syntheses. As the conflagrations, the costs, the casualties, and the collateral damages became more notorious, formal inquiries and consultancies rose. The jackstrawed reports became the 1,000-hour fuels of fire bureaucracy. Fire management entered its MBA era.[3]

* * *

On June 27, 2000, even before the season approached its climax, secretaries Babbitt and Glickman requested an Interagency Federal Wildland Fire Policy Review Working Group to reassess the 1995 policy and its implementation, address issues raised by the Cerro Grande reports, and "provide any other recommendations that would improve the wildland fire management programs in the two Departments." To the extent possible the same people who had written the 1995 review were reappointed. To reflect the fast-expanding range of institutions affected by wildland fire, the group added representatives from the Department of Energy, the Department of Defense, the Bureau of Reclamation, and the National Association of State Foresters. Its conclusions updated and replaced the 1995 federal fire policy.[4]

The Working Group identified six principal conclusions, nine guiding principles, and five key themes. They concluded that the 1995 policy was "still generally sound and appropriate," but with both the wildland and the interface scenes deteriorating, becoming worse and more complex, some revisions and tinkering were necessary. Implementation had been, in the understated wording of the report, "incomplete," and it could improve only with better leadership and incentives to reward the successful and penalize the laggards. Incredibly, fire plans still did not exist for all units, and many that had them had failed to integrate with land-management plans. The whole edifice needed firmer coordination, oversight, and evaluation. The once-again-affirmed policy was still not "well understood inside the fire management agencies and by the public." The "task before us—reintroducing fire—is both urgent and enormous." The public and politicians mistakenly regarded the interface as solely a federal concern, not a duty of local communities. But overall the chasm remained where it had been for decades. "The problem is not one of finding new solutions to an old problem, but of implementing known solutions." A divided society was reflected in a divided approach that left wildfire suppression as the only common practice even as fire officers, scientists, and thoughtful observers agreed that suppression, however necessary as an emergency response, only aggravated the underlying conditions.[5]

What the American fire community required was a coherent system and the capacity scaled to run it. It required more than a common policy among the usual suspects because fire had come to involve yet more federal agencies, from the Defense Department to the National Weather Service and USGS, not to mention the states, the Western Governors Association, NGOs, and local counties and municipalities. Oversight demanded not just another interagency talking shop like the NWCG but a governing body with the power to decide on priorities and allocate resources. Collective undertakings required agreement on common strategies, which meant agreed-on goals, even as each institution had to satisfy its own founding charter and distinctive tasks. Interagency collegiality required more than shared designs for shovels and fire shelters; it meant common budgetary processes, decision criteria, guidelines for execution, and evaluation procedures. It meant rendering in a way legible to outsiders the tangle of diverse and historically eccentric institutions trying to engage in a broadly collective task. The American fire community had to re-create something like what the U.S. Forest Service had achieved over the course of its first 50 years but with new parts, each with its own identity, among land agencies that had many duties besides fire management and were wary about how far to join forces.

The existing models were either too much or too little. There could be no return to the old-order Forest Service or an equivalent apparatus to synchronize the system. There could be no czar as with the Drug Enforcement Agency. No comprehensive reorganization on the model of

Homeland Security was possible. A national fire service, attractive to OMB, would destroy the achievements of the past 40 years and make wildfires more likely rather than less so because it would cleave the critical linkage between land and fire. And there was, in truth, scant desire among the participants to surrender their discretion and their distinctive élan to an invented hierarchy. Bureaucratic instincts were to avoid mistakes and do only those things for which there could be little criticism, which meant set-piece suppression. As long as Congress kept paying for those firefights, there was little incentive to change.

These concerns would not have mattered if the citizenry had not been polarized and the public lands not balkanized; if plenty of slack remained in the national estate, allowing for baffles and buffers and open rural land rather than having every nook and cranny stuffed with cabins and exurbs; and if an equally unsettled climate could allow more wet and fewer dry seasons. It would have been easier if houses were not burning every fire season and if crews did not scramble into shelters or die in brush. Torrents of money still coursed through the budget, at least when the fires came. But Congress balked. It had lavishly financed a National Fire Plan and the bills due of big fires; now it wanted to spend less and to audit what it had paid for. As the Bush administration went to war without raising bonds or extra revenue while slashing upper-income taxes, the firefight at home lost out to the forward strategies of Iraq and Afghanistan. More and more consultants, either hired or self-appointed, bawled that the national fire project was inappropriately expensive and hopelessly ineffective. In the end, Congress turned, as it first had after Yellowstone, to a grumpy GAO, which repeated its dogged demands for better accountability and the metrics to prove it.

* * *

The GAO's conclusion was as unexceptionable as it was expected. "In general implementation has been least successful when consistency and compatibility across agencies was required or when integration of fire with other disciplines was required." What stymied implementation in 2001 was what had hobbled it for decades—poor data, inadequate or inappropriate science, demographic trends that compromised the workforce, confusing terminology, conflicting values, and fumbling coordination. Of course funding had never been adequate, but it was not obvious that more money by itself would resolve what were not at heart simple problems of costs and efficiency. The fire revolution had failed to institutionalize its vision in ways that made action possible.[6]

What became the operational arm was the concurrently evolved National Fire Plan, which committed considerable funds to fuels treatments to halt catastrophic fires and to rebuild suppression capabilities. The new Congress quickly absorbed the NFP into a massive collaborative planning exercise—another practice that became obligatory during the decade—which brought together the USFS, Interior agencies, the Western Governors' Association, the NASF, the National Association of Counties, the Intertribal Timber Council, and an omnium gatherum of "interested stakeholders and experts" from the Wilderness Society to the National Cattlemen's Beef Association. The upshot was "A Collaborative Approach for Reducing Wildland Fire Risks to Communities and the Environment: 10-Year Comprehensive Strategy Implementation Plan," released in August 2001. The shared goals were to improve prevention and suppression, reduce hazardous fuels, restore fire-adapted ecosystems, and promote community assistance. The report specified a May 2002 deadline for devising a "detailed implementation plan."[7]

That started a succession of further studies and programs. The 10-year comprehensive strategy slotted slickly into an imminent Healthy Forests Initiative (HFI), enacted in August 2002; Interior and the Forest Service jointly issued a report on "Restoring Fire-Adapted Ecosystems on Federal Lands." In 2003 came an Interagency Strategy for the Implementation of Federal Wildland Fire Management Policy, a controversial expansion of the HFI into the Healthy Forests Restoration Act, and a report to Congress on local fire departments, which, whether the public believed it or not, were the first line of defense in the I-zone. The next year the Wildland Fire Leadership Council weighed in with an interagency strategy for implementation. The Western Governors' Association returned for a review, following a Forest Health Summit in 2003. Next, wildland fire use received its "implementation procedures reference guide," then the medley of fuels projects received their "cohesive strategy." Every year witnessed yet another review by yet another body, or another updated strategy for fuels, the I-zone, workforce, or coordination. Each year produced another restatement on proper terminology or a clarification on implementation. Fire officers were confused, the agencies evasive, and the public and Congress understandably baffled, and frequently alarmed and irritated.[8]

The clarifications kept coming because different people and agencies read the bland declarations of policy differently. The wrench in the gears was the operational meaning of "appropriate management response." The 2003 Implementation Strategy promulgated by the Wildland Fire Leadership Council and intended to unify practice had the effect of classifying every ignition as either a wildfire subject to suppression or as wildland fire use, which granted officers the opportunity to allow it to free-burn within broad constraints. Every fire was either one or the other and would remain so throughout its existence. The strategy collapsed an intended "spectrum of management strategies" into just two. Still, when it emerged, appropriate management response (AMR) had promised a way to expand wildland fire use, or at least parts of a fire, out of wilderness onto general wildlands. The WFU could do for the West what prescribed fire did for the Southeast.

Behind the confusion lay the irreconcilable tension between the wild and the urban. Like an endangered species, free-burning fire had to somehow thrive on a shrinking and fragmenting landscape. Urban enclaves were spreading like a runaway fractal; their protection, not fire's restoration, was driving the politics behind the National Fire Plan. As wildland fires met the interface, so agencies committed to fire's restoration had to reconcile with state and local fire departments committed solely to the protection of life and property, departments that regarded the free-burning backcountry as epidemiologists would a plague reservoir. However much ecological good long-burning fires might do in wildlands, they posed a potential threat to communities, especially those that had not adopted measures to protect themselves—had not enforced building codes, had not designed subdivisions with fire as a consideration, had not founded a fire service, had not treated developments as what they were, bits of towns with exotic landscaping.

Before the 2007 season, appropriate management response underwent another iteration. It was, in effect, a multiuse doctrine applied to fire that allowed managers to adjust their response to all or parts of the fire as deemed suitable by the land-management plan. The announced "clarification" broke down, in principle, the distinction between wildfire and wildland fire use. Everything became a *wildland fire* that should receive a response appropriate to its circumstances, with those circumstances set by the larger land-management agenda. As the mantra ran, "fire is fire," and a label once applied should not obligate operations to a particular strategy forever and always. Such decisions by classification denied the flexibility that the dismantling of the 10 a.m. policy had made possible, and not only advocates for the fire revolution but also the USDA Office of Inspector General (OIG) and the GAO argued against it. But it was difficult to achieve the kind of clarity the 10 a.m. policy had promulgated (or the hard metrics the OIG and GAO demanded) when the right decisions depended on varying circumstances, even on one fire.[9]

On June 20, 2007, the National Fire and Aviation Executive Board reissued the "clarification" by reaffirming the 2003 Implementation Strategy, specifying that "a single fire cannot be managed" as both a wildfire and a WFU concurrently, and then redefining the AMR in ways that said it could. "Beginning with the initial response to any wildland fire, decisions will reflect the goal of using available firefighting resources to manage the fire for the most effective, most efficient, and safest means available." Fire officers might be forgiven if they remained wary about what the rules of engagement actually were.[10]

Both advocates and critics saw the revision as a means to get more fire on the land. For those committed to fire's restoration, the AMR option replaced the WFU just as the WFU had replaced the PNF, and the PNF had replaced the let-burn—the parade of acronyms is an apt expression of the ceaseless confusions. For those mandated to protect communities stuffed into wildlands—however illogical or indefensible their development—the AMR seemed a way to let big fires barred from entry through the front door in through the back. Instead of resolving the controversy, the policy revisions only finessed around them or shifted the terms of discussion. It was fine, in theory, to announce that fire management would align with approved land-management plans, but it was the breakdown of consensus over those plans that had announced the revolution and kept the confusion simmering. A multiuse strategy for fire could only work on lands that allowed multiple kinds of fires. Nor were the uncertainties limited to the feds. The Society of American Foresters first declared its position on "Wildfire Fire Management" in 1989, after Yellowstone. It "subsequently renewed with revision" that statement in November 1991, August 1994, December 1994, and December 1997 before renaming the document "Wildfire Management" and revising in December 2002, and then renaming it again to "Wildland Fire Management" and revising it in June 2008. And these were self-declared professionals who had long claimed special standing on all matters pertaining to wildland fire.[11]

What should have been clear was that policy reform could not correct the underlying causes. Fire policy was not itself an informing doctrine but a symptom. Changing fire policy did not change how Americans lived on their land, which was the real driver of the fire scene. Still, with the 2001 policy, the American fire community had its vision. With the Wildland Fire Leadership Council (WFLC), it had an apparatus for governance among federal agencies. With the National Fire Plan and the 10-year collaborative strategy, it had an operational foundation for action. And with a tightly argued focus on fuels, it had a specific task and a metric for accountability. The early years concentrated on building up suppression capacity, particularly hardware such as engines, personal protective equipment, radios, and crews, while research identified landscapes of disrupted fuels and their prospective treatments along with new aggregations of "partners" and "shareholders" that tediously negotiated priorities and plans. Within a handful of years, the American fire community had much of what it had asked for, even if that brief downpour had followed a long, punishing drought.[12]

* * *

Among the proliferating voices, one agency and two terms stood out. The agency was the GAO (renamed the Government Accountability Office in 2004), which was given narrow charges for investigations but broad discretion to pursue them. It conveyed its conclusions in the measured language of economics, not just as the study of money appropriated and spent but as a moral philosophy that judged actions through incentives, preferably quantifiable. Costs and fuels were equally the currency of behavior and the index of performance. What mattered, moreover, was less this program or that program but their aggregate sum. In the GAO's drily worded judgment, the American way of fire lacked a "cohesive strategy."

The fact remained, all the studies from all the commentators identified the same suite of concerns and possible solutions. All supported what the 2001 policy review board (once again) had determined, that the crisis did not demand new solutions but the application of known ones. Most included calendars for implementation. What the GAO contributed was a gimlet-eyed focus on how outcomes measured against expectations. And while every group had its own disciplinary quirks and many sought to exploit the fire scene as a telegenic means to animate some other message, the GAO impartially held to the iron rod of its charge. So amid a swarm of texts that veered toward the urgent, if not the breathless, the GAO's library of inquiries reads with a thudding dullness that paradoxically makes their conclusions more striking.

In 2000 the GAO conducted two inquiries—one on the overall state of the American fire establishment, and one on a particular concern (Cerro Grande). That established the general pattern and tempo. In 2001 it investigated the capabilities of the federal agencies to implement the National Fire Plan while answering an inquiry from Idaho Senator Larry Craig regarding "appeals and litigation of fuel reduction projects." There were two more reports in 2002, then in each of the subsequent years, eight, four, seven (for each of three years), three, and a final three in 2009, including a summary of the decade's investigations. The Yellowstone blowout had first drawn the GAO to fire; the threat of catastrophic fires outside the wild held it. From 1999 to 2009 the GAO issued 53 reports on American fire. Those publications are the most reliable archive of documentation for the era. More than any other source they shaped the themes and language of congressional legislation. They compelled the agencies to respond. They demonstrated what policy needed in order to become practice.[13]

The gist of GAO's findings were that the federal agencies had "unquestionably improved" and had plenty of room for further improvement but that wildland fire problems had so "worsened dramatically" that the agencies were struggling to remain in place. The nation's fire environment was deteriorating faster than the nation's fire institutions could hold the line, much less enact preventive measures to gain back some of the lost ground. Along the way the GAO issued recommendations. Mostly, it worried that both emergency response and remediation were too fragmented and too costly. On many issues the agencies agreed. On some they protested that they had already recognized these concerns and had codified solutions in the form of updated policies and guidelines. The GAO replied that such documents lacked "the clarity and specificity needed by land management and firefighting officials in the field" to operate efficiently and by oversight bureaus to evaluate their achievements.[14]

The scale of the problems was staggering. Some 39 million acres of national forest (89 million by some reckonings) and, as estimated by TNC, some 110 million acres nationally, along with perhaps a tenth of national housing, were at risk. The premise behind the NFP was that restructuring fuels could abate the threat of catastrophic fire. Fuels projects could prevent fires that burned outside the range of historic conditions and thus damaged biotas, and by breaking crown fires they could shield structures and make their defense by firefighters possible. A wildland less prone to eruptive fires was one in which it was possible to fight bad fires and set good ones more effectively and cheaply. But between 2000 and 2006 the estimated cost of needed fuel treatments had increased by a third, and fire-suppression expenses and burned acres had doubled. The GAO concluded that while the agencies had genuinely improved, the threat was scaling up even more quickly. It ended its 2009 summary report by "making no new recommendations." The crisis lay with doing better—more efficiently, more broadly—what the agencies knew they had to do.[15]

Questions of efficiency ran up against the politics of a fractious citizenry and an American federalism that even in the absence of divided government favored checks and balances. To address America's disruptive pyrogeography would require a national effort on the scale of the New Deal's Civilian Conservation Corps. Even allowing for privatization and market forces, for letting contractors replace army-run camps of enrollees, nothing of the sort would happen. The country could barely enact healthcare reform; it was chimerical to imagine the nation doing something equivalent for ecosystem health. The promised fire next time was at hand.

* * *

The passion for accountability did not end with policies and position papers, nor with abstract measures of hydrocarbons thinned and dollars spent in firefights. Someone had to do those tasks. They could—would—be judged as individuals and not just as bureaucracies. They knew costs not measured in budgets.

The issue moved from the hypothetical to the existential when, in John Maclean's words, "a no-account wildfire in a remote corner of the Northwest cut off and trapped a group of firefighters, who were joined by two civilians who came out of nowhere." The fire was called Thirtymile. It began on July 9, 2001, from an abandoned picnic fire along the Chewuch River canyon. The region was already grappling with a "major rager," the Libby South fire. The Entiat Hotshots worked the Thirtymile fire through the night until relieved the next morning by a Type II district crew, the Northwest Regulars #6. In the afternoon the fire spotted uncontrollably and boiled over, and the Northwest Regulars found themselves trapped, along with two straggling tourists. They deployed shelters. Six of the crew members deployed on a scree pile. Four of those on the scree died.[16]

Since South Canyon, the agencies, particularly the Forest Service, had dedicated themselves to fireline safety. And safety seemed the one area of unquestioned success—the model, to critics, of what could be achieved in fire management with sufficient leadership and bureaucratic pressure. The Thirtymile fire challenged that record, and unlike other burnover fatalities, it had the legacy of South Canyon and a lingering sense, often bitter, that the people most responsible for that earlier tragedy had skated away without suitable blame or banishment. Again, the USFS chief declared a determination to make firefighter safety paramount. Review boards concluded that all 10 Standard Orders and 10 of the 18 Watch Out Situations had been violated or disregarded, and they specified action plans to toughen training. OSHA cited the Forest Service for two willful and three serious violations. But this time Congress also intervened.[17]

In November, Senator Maria Cantwell of Washington held hearings, decided that the causes of the two tragedies—South Canyon and Thirtymile—were "nearly identical," and sponsored legislation to demand independent investigations of Forest Service fire deaths. A year after Thirtymile blew up, Congress passed the bill, which required the Department of Agriculture's Inspector General to investigate. It is hard not to sense vindictiveness: Thirtymile would do what South Canyon had not. South Canyon had been a BLM fire, with BLM overhead, but it was manned by Forest Service crews. Thirtymile was a Forest Service affair. The legislation, ironically, applied only to the Forest Service.

In December 2006, based on the OIG report, the U.S. Attorney's office in Spokane charged the incident commander, Ellreese Daniels, with four counts of involuntary manslaughter and seven counts of lying to investigators. What further complicated the episode was the fact that Daniels was an African American and part of the cohort fast-tracked to meet agency goals for diversity. What began as a no-account wildfire inspired a big-deal after-action review that ended with criminal charges. The legal case sorely alarmed the fire community, leading the International Association for Wildland Fire to denounce it as discouraging anyone from serving in leadership positions, and indeed dozens of firefighters with supervisory qualifications did not renew their certifications. Eventually the charges were dropped or reduced, but the law remained on the books.

This was a turn toward accountability—the criminalization of poor judgment in inherently chancy settings—that neither the GAO nor the agencies foresaw or desired. It told every incident commander that a fatality might lead to felony charges. It told every prescribed burner that, as one fire-management officer put it, he went to work knowing that a wind shift might land him in jail. The agencies pushed fire officers to consider personal liability insurance. While the Federal Tort Act allowed the government to be sued even when it performed a public duty such as fire protection, it was a quantum step to download that responsibility to its agents. In wildlands there was always a degree of unpredictability and instability—that is what made them wild. No surprise then that fire officers would back off from tricky wildfires or hesitate to light a driptorch. They had to operate in a legal and political environment as overcharged with combustibles as the overgrown woods.

So even as the hearings and reports and audits continued, the Thirtymile tragedy was a reminder that not all reform was paper shuffling, that suppression was not a neutral choice, and that policies have real and potentially lethal consequences. Inevitably, the Forest Service bore the brunt of criticism on behalf of a fire community staggering from an American scene that sent shockwave after shockwave of big fires to rattle the West.

2002: Megafire

In 2002 big fires—really big fires—established themselves as a new normal. In four states they set historic records. They were not states known for monster fires.

On June 1, lightning ignited several fires that, burning together, became the Ponil Complex in northern New Mexico. The 92,522 acres burned nearly doubled the state's previous record fire, the Cerro Grande. On June 8, a Forest Service fire-prevention technician, Terry Barton, set a fire that escaped a campfire ring and raged north along the Colorado Front Range. The Hayman fire blackened 137,760 acres, destroyed 133 homes, and cost $40 million to contain. On June 18, an Apache Indian, hoping to be hired to fight fire, lit the Rodeo fire outside Cibecue, Arizona, on the White Mountain Apache Reservation. It quickly raced away, and two days later a disoriented hiker set a signal fire near Chedeski Peak that also bolted free. The two fires were close enough that they had to be treated as one; the Rodeo-Chedeski fire totaled 468,638 acres, forced the evacuation of Show Low, and cost $43.1 million. On July 12, dry-lightning storms set five fires in the Kalmiopsis Wilderness on the Siskiyou National Forest in southwestern Oregon. They burned together, eventually synthesizing into the Biscuit fire, and after burning in wilderness and inventoried roadless areas finally menaced communities along the borders until contained in mid-September. Though significant portions within the collective perimeter remained unburned, and perhaps 20 percent or more of the acreage was the result of backfires, the Biscuit racked up 499,570 acres, drew in fire personnel from throughout the United States, Mexico, and even Australia and New Zealand, and resulted in a stratospheric $155 million bill. This one fire exceeded the cost of the 1988 Yellowstone Complex.[18]

It was one thing to have a national fire crisis, as Secretary Babbitt had expressed it. It was another to face odds of holding four fire aces in one hand. Not only were these fires extraordinarily large, but they also occurred in states outside the caste renowned for big burns and in landscapes that had not historically burned with high-intensity crown fires. They were something new under the American sun, equally frightening and creepy. They were as radically novel as an emergent plague and as ancient as a Biblical one, as though both SARS and smallpox were erupting simultaneously. It did not take long for them to acquire a new label, the *megafire*.

As the name's inventors, Jerry Williams and Albert Hyde, characterized it, the megafire was an "extraordinarily high-cost, high-consequence" incident. Its impacts were "shocking to survivors, long-lasting to affected communities, and, in some ecosystems, immeasurably damaging." Megafires outstripped suppression capacities or any capability that agencies were likely to acquire. They resulted from batches of burns. They crossed land uses and political jurisdictions and hence beliefs about what they meant and how they should be addressed. They burned in wilderness and WUI, across the Philmont Boy Scout Ranch and through tribal logging berths, into old historic hamlets and new exurbs. Half were started by lightning, half by people. They rode a convective plume of public anxiety. To one group, megafires argued for more active intervention with regard to "fuels," and to another, just the opposite, to let nature's fires restore their indigenous landscapes unimpeded. But they could not be ignored.[19]

* * *

The big fire has always been the incubus of fire management. It does the most damage, runs up the highest expenses, and distorts fire plans and operations. The big burn is the black swan that unhinges the narrative, undoing what patient labor has accumulated over decades. The 10 a.m. policy was devised with the big fire in mind. But in the new millennium the question of the really big fire dominated in new ways.

Megafires constituted a veritable fire plutocracy. Of the 10,000 fires the Forest Service responded to in an average year, only 1 percent became large, and only 0.1 percent became megafires. Yet they unbalanced America's pyrogeography, accounting for 95 percent of total acres burned by wildfire and 85 percent of total suppression costs. They were a scale of fire thought banished by the massive government interventions of the 1930s. They warped the American fire scene as a similar concentration of wealth did the American economy.[20]

It was hard to avoid the use of one as a metonym for the other. The consensus perspective on how the American economy had morphed over the previous 30 years pointed to globalization, an internal shift from manufacturing to services, and policies that aimed to liberate free-ranging capital from restraints and regulation. Eerily, analogous concepts can explain the parallel evolution of the fire scene. For globalization, look to the global change suite that includes climatic warming and exotic invasives. For the leap to services, see the movement of private land from commodity production to amenities communities and of public land to wilderness and nature protection. For policy, note the determination to restore free-burning fire to something like its earlier untrammeled state while removing firefighters from heedless harm and protecting the McMansions and recreation home assets of America's privileged classes. Doctrines of laissez-faire management point to the privatization of fire protection (for those who can afford it) and a trickle-down fire ecology from the elite wild to the everyday biota on the assumption that more fire will inevitably improve the functioning of ecosystems on a landscape scale.

* * *

But it was not necessary to resort to metaphors. It was possible to explain megafires entirely within traditional concepts. First, climate. In the mid-1980s the country, particularly the West, switched into drought. Whether this was another iteration of long and erstwhile rhythms or the early symptoms of global warming did not really matter. At first it seemed to conform to a familiar three-year cycle, climaxing with the Yellowstone outbreak. Then the three years blurred into more, fire seasons lengthened, and the troughs became deeper; the 2002 drought in the Southwest, as recorded against tree-ring data, was the worst in a thousand years. Megafires were not controllable until a change in weather broke them or they ran out of fuel. The era of megafires would not end until the Long Drought ceased or the wave of fires had burned over so much that they converted the fuel regime.

Those fuels were the second consideration. It was not simply that drought had transmuted more biomass into fuel. The base biomass was greater, and it had altered in ways that made fire more likely, more vicious, and more damaging. Fires burned deeper into duff and organic soil, raced higher into the canopy, and burned wider through landscapes that were once patchy but were now uniformly cloaked in combustibles. This was the result of land use, of a century or more of everything Americans did or did not do on the land. They had shattered the ecological structure of old fire regimes and let the fragments reassemble in ways that promoted different kinds of fires and trusted that suppression could contain the new order. Montane forests such as ponderosa pine thickened with choking understories that carried fire into crowns. Explosive forests such as lodgepole pine became as homogeneous as a field of drill-planted corn; patch sizes increased by an order of magnitude. As always, fire took its character from its context. What had been seasonal nuisances, the pyric equivalent of flu season, had mutated into a virulent plague. A seeming contagion of conflagrations resulted.

Such changes had commenced in the 19th century. But to them were added contemporary factors that pushed landscapes into more explosive and damaging fires. The enthusiasm for nature protection had created special patches of pristineness called parks, wilderness, and biosphere reserves that encouraged biotas to blossom through natural means but without the pruning effects of routine fire. A concern with endangered species and old growth had promoted habitats that ironically favored eruptive burns. Such choices had altered fire regimes but did not come with a concurrent reformation in fire's management. They instead wended toward unintended consequences. Too often the fires that returned were not those that had left. Unable to manage the result, fire agencies responded by trying to suppress, which only aggravated the underlying causes.

Sprawl was the other driver of land change. Typically it replaced wild or feral lands not with incombustible cityscapes but with a fractal fringe of wooden (and often times wooden-roofed) houses that were, from a fire-behavior perspective, jackpots of fuel little different from windfall. In the East the process led less to an interface than to an intermix, as abandoned rural lands, whether farm or forest, littered the countryside with thickened patches of combustibles while emptying the woods of workers who had previously burned for hazard reduction or had doubled as firefighters. Where development was dense, fires simply propagated from house to house or from detached garage to shed and back again. Or they seized the equivalent of vacant lots. In 2005 wildfires ripped through the Long Island pine barrens and later burned on open land in Staten Island.

The fragmentation of America's landed estate had its parallel in the institutional balkanization of fire management. The character of exurban sprawl had put it beyond the pale of both urban and wildland fire protection, while its tempo outraced the capacity to install a replacement. The fire establishment had evolved to cope with urban or wildland landscapes, not places that stirred them into an omelet. No adequate mechanism existed to halt fires that blew past initial attack, and local authorities turned to the federal government in the form of a weakened Forest Service to fill the gap. Like opening a door in a room simmering with hot gases, a backdraft called megafire blew out.

Because it was not possible to reverse either drought or sprawl, the burden of fixing the problem fell to the fire agencies. The only plausible solution was to expand controlled burning, but natural fires had less room to roam than thought, and prescribed fire could not make up the difference. Even in the Southeast the amount prescribe-burned lagged seriously behind the buildup of fuels; and in the West prescription burning beyond grasslands was simply too complicated, too expensive, and too feeble. The only way to get the burned acres needed was to let wildfire—wildfire, not just natural fire—do the job.

This sentiment conflated two policies. One was the recent but unbending imperative to place safety first. After South Canyon and Thirtymile, fire supervisors were reluctant to put crews in harm's way, which meant backing off to defensible perimeters, letting multiple starts burn together, and burning out. Usefully, this approach put lots more fire on the ground. The strategy was a variant of the old confinement strategy, now born again under the doctrine of appropriate management response. Still, it had critics, some of whom thought it prescribed fire by stealth and some who viewed with horror the backfires that were often among the most severely burned portions of the fire overall.

All these factors were stirred together in ways that make it difficult to parse out their separate contributions. Many megafires burned under drought conditions that were no worse than previous episodes, but they now burned under circumstances of fuels, sprawl, endangered species and protected preserves, and firefighter safety that made them behave differently and nudged responses in directions that escalated burned acreage. What a doctrine of fire exclusion had taken away in the early years, a doctrine of fire tolerance was restoring. It just was not happening as expected or in many cases as desired. Wildfire was becoming the new prescribed fire.

* * *

What was undeniable was that burning was booming and that a minuscule fraction of fires claimed the mountainous share of resources and attention. What began as a policy of fire by prescription became a general prescription for tolerating more fire. By way of illustration, consider Arizona's current round of big fires. Before 1996 the largest burn of record was the 57,335-acre Carrizo fire in 1971, long regarded as an outrageous outlier. The next two contenders, the Verde and Castle, both half its size, had occurred in 1979. In 1977 every national forest in the state had a campaign fire, none of which exceeded 10,000 acres.

Then the megafire era laid down its own cadence. In 1996 the Lone fire burned 60,000 acres, roughly equaling the Carrizo. The fire began when an abandoned campfire set the ridgeline of Four Peaks ablaze. It was early in the season, crews were sparse, most of the mountain apart from the forested summit was chaparral or desert, and the Mazatzal Mountains made a formidable terrain. Fire officers elected to let the fire back down the slopes, with protective burning along lower-elevation roads. In 2004 lightning started a blaze in the Mazatzal Wilderness, farther north in the range. Again, this was a rugged landscape, covered with brush and scattered woods on the ridgeline, ripe with exotic grasses, and bounded by legal restrictions on what kind of suppression force might be appropriate. Officials opted to pull back to Highway 87 and backfire, leaving the bulk of the mountains to burn themselves out during the height of summer; not all the burnouts reached the main fire. The final perimeter area: 119,000 acres. A year later lightning, again acting on a desert flushed with exotic grasses thanks to winter rains, started several midsummer fires. Once more fire officers elected to herd the starts away from developed areas, permitting the separate ignitions to merge and burn more freely over desert and into the Mazatzals as the Cave Creek complex. The final reckoning: 248,000 acres. In 2011 an abandoned campfire took off in the Bear Wallow Wilderness of the Apache-Sitgreaves National Forests. The location's legal status and rugged remoteness discouraged a massive initial attack. Quickly, long-range spots appeared, while officials were reluctant to risk firefighters and believed they could hold the fire along the basin's rim. The fire blew up, and so did the backfires, and together they burned 538,000 acres. This litany does not include the Rodeo-Chedeski Fire of 2002 because it was really two fires allowed to merge. Both contributing fires were set deliberately by people, one maliciously, the other out of incompetence. The combined fire occurred, like the others, in the run-up to the summer monsoons, though it burned through the worst drought since the age of the Anasazi. The others started amid serious but not unprecedented summer dry spells. The Wallow fire burned under conditions comparable to those of the 1977 fire bust.[21]

In just 15 years the size of the largest fire had doubled four times. Climate favored large fires, but circumstances were within historic limits, and the fires burned in the season of record for all major fires. Fuels were ample, either because of drought in the mountains or rains in the desert, but both lay within historic ranges of variability. There was no abrupt metastasizing of fuels to account for the sudden doublings of burned area within a 15-year period. Something else had pushed them to such sizes, and the most obvious cause—so obvious it is easily overlooked—is simply the choices fire officers made to cope with them. What changed was fire management.

It is doubtful that Americans (certainly not its fire officers) chose to have a fire scene dominated by megafires any more than most citizens chose to replace a middle-class democracy with a Wall Street plutocracy. But that is what happened. In most of the country, certainly in the public-domain West, America's middle landscapes were shrinking and taking what might be termed the middle methods of anthropogenic burning with them. That trend was not intentional, although it reflects the decisions, both by deliberation and by default, about how people elect to live on the land and what kinds of fires they will accept. Whether or not societies get the politics they deserve, they do get the fire regimes they choose.

* * *

Even as acres burned ratcheted up, the paradox was that the nation needed far more fire, and an important fraction of high-visibility landscapes needed high-intensity fires. The core need was getting the right kind of fire regime. Bad fires had waxed, good fires had waned. Too many places were immolating from the wrong kinds of fire. Equally, too many places were rotting from lack of fire. The fire community itself contributed to the confusion by hyperventilating over jumbled statistics that showed simultaneously both record burns and too little flame.

Revolution and counterrevolution, promoting fire wholesale and preventing it through protectorates thrown around the I-zones—these were conflicting if not contradictory tasks. Those agencies that had to deal with one or the other could do so. They could erect barricades around boroughs and strengthen capacity to protect lives and property. Or they would enlarge the domains available for natural fire to burn or the institutional settings that permitted prescribed fire. Those that had to absorb both, like the Forest Service, only internalized the larger dissonance and could satisfy neither mission wholly.

In the end theoretical considerations mattered less than facts on the ground. More bad fires were burning more acres and imperiling communities. What distorted the unfolding discourse, however, were unanticipated events that caught the attention of the public or Congress. On June 17, 2002, a C-130 air tanker over a fire lost its left wing, and seconds later its right, before augering into the ground. A month later a PB4Y tanker shed a wing while on a drop run over a fire and fatally plunged to earth. Overnight the question of fire restoration was overwhelmed by a discourse about the capacity of the nation's aerial suppression forces. The public did not know about prescribed fires unless they escaped, but they saw air tankers on nearly every Action News broadcast.[22]

Worse, whether fighting fires or trying to tame them by fuels treatments, the cost of fire management had gone ballistic. This was another ground truth that went viral. It mesmerized Congress and the minds of ideologues. The megafire was a megasink for funds.

2003: Unhealthy Forests and Subprime Sprawl

On March 19, 2003, the United States and a self-styled coalition of the willing launched the Second Gulf War. The invasion of Iraq began with an overwhelming display of air power in an attempt at "shock and awe." That fall American nature unleashed its own display of shock and awe with a fiery invasion of Southern California. As the new Gulf War built on its 1991 predecessor, so the fire siege of 2003 followed the siege of 1993.

The outbreak began with three fires on October 21 and ended when the banshee Santa Ana winds subsided on November 4. There were 14 large fires in all. Thirteen had human origins, although 8 of them had uncertain causes. The Simi was a long-range spot from the Verdale fire. The Padua was an administrative creation to break up the unwieldy suppression of the Grand Prix on the San Bernardino National Forest. The Grand Prix burned for the entire duration of the siege. The Cedar fire, started from a hunter's signal fire, exceeded the 1970 Laguna fire as the largest in San Diego's history. When the smoke cleared, 750,043 acres had burned, and the flames had destroyed 3,710 homes and killed 24 persons, mostly civilian.[23]

The agencies conducted the inevitable postmortems. Two perspectives prevailed. One saw this latest uproar as simply the most recent of an ancient lineage. The South Coast had always burned: it would always burn. The scene was worse because Californians had put more of themselves in harm's way. The fires were tragic, but it was a regional event like hurricanes in Florida or tornado clusters in Oklahoma. The other perspective reflected back to the 1993 siege that seemed to announce a post-Yellowstone fire narrative in which the WUI dominated over the wild, and more savage fires had established themselves as a new normal. They noted that such outbreaks were no longer peculiar to California but in 2003 had gone global. In Australia a mammoth bushfire had rolled out of Kosciuszko National Park, over the national observatory at Mount Stromlo, and into the suburbs of Canberra. In Portugal feral fires feeding on the unraveling rural landscape had surrounded the ancient university city of Coimbra and infested its interior parks. In British Columbia, a province based on a timber economy, a "firestorm" had rampaged over the Rockies and besieged the town of Kelowna, evacuating 27,000 people. Megafires, it seemed, were the early skirmishers of a global apocalypse. What happened in California was the future of fire.

The labels and analysis mattered because they shaped possible responses. Despite the fire community's predilection to frame every event in triangles, the general trend was to bifurcate. In the South Coast these formed layers of polarities, like oppositely charged plates, across which arguments arced. The anode held that the problem was the wildland, more prone to explosive fires; the cathode argued that the city was the problem, pushing in inappropriate ways into places it did not belong. To one, the fires were the outcome of landscapes overloaded with combustibles; to the other, they were the product of shrieking winds. To one, the solution was to reduce those fuels on a landscape scale, or at least to construct fuel breaks and surround structures with defensible space. To the other, the solution was to harden houses to withstand the ember showers that typically caused their destruction. Critics answered that wholesale fuel management did no good because the winds would still drive fire across them, to which came the rebuttal that embers could be attacked by firefighters only if the overall strength of the source fire lessened in a sufficiently broad zone around the structure that crews could survive the thermal pulse of the fire front and remain to swat out the sparks.

Without conclusive evidence, the debate became scholastic, even ritualistic. The fuel scene in Southern California had worsened through fire exclusion, massive beetle kill, record drought that killed even cheatgrass, and plants weakened by overdoses of ozone. The outbursts of Santa Ana winds triggered fire avalanches. But the prime mover, the fire equivalent of the San Andreas Fault, was land development and speculation, and that had been uncontained since colonial times. There was little anyone in the fire community could do about it, however, or about the Peninsular and Transverse ranges, or about the Santa Anas, or about ignitions from malice or malpractice, whose control required a more disciplined society. The one point of vulnerability was fuel, both the natural fuels that went under the collective name of chaparral and those built fuels called houses. There might be nothing to keep fires from swelling to hideous dimensions, but it might be possible to abate the severity of their damages.

* * *

Two maps helped to nationalize the fuels problem.

One came from the Bureau of the Census, with its data milled severally by federal land agencies, regional and urban planners, conservation biologists, ecologists, and social scientists before finding its way to cartographers who plotted what fire officers knew all too keenly. Americans were once again on the move, a scattershot of exurban homesteading. It was not simply the amount of land converted to housing that mattered: it was the fragmentation of landscapes and the patches' unbounded perimeters. The more patches, the more complicated the politics. The more borders, the greater the problem of protection. More people meant more ignitions; more boundaries meant longer firelines and more protection demands per patch; more patches meant more owners, more legal jurisdictions, more conflicting values and voices. Dispersed settlement frustrated the concentration of firepower required for defense. In the West new housing construction overlaid with eerie fidelity the fire-prone landscapes. America's forests, both private and public, were "on the edge."[24]

The rough numbers accommodated both interfaces and intermixes. Researchers plotted the housing numbers against the USGS National Land Cover Data. The figures suggested that "one-tenth of the area and one-third of the housing units of the conterminous United States" were in the wildland-urban interface. The bulk of the recolonization occurred on private lands in the East; this was not, during the 1990s, regarded as a fire problem except in Florida. The most costly I-zone fires occurred in California, and these were dismissed as a quirky California pathology. But as the new millennium arrived and brought a fresh batch of census figures, the fire problem triggered a general alarm. Houses and fire-prone landscapes were mixing into a combustible cocktail and one no longer restricted to the West. Both hazard and risk went ballistic during the 2000s as funny money and subprime mortgages sloshed through the economy. The subprime landscape met the megafire.[25]

The second fuels map plotted the unnatural growth of wildland fuels. This exercise proceeded under the general rubric of forest health that had appeared as part of the conversion to ecosystem management in the early 1990s. A century or more of past practices had rendered many ecosystems vulnerable to diseases, insect infestations, and bad fires, which threatened to destabilize landscapes even further. The unhealthy ecosystems were the obverse of the unbounded interface: they challenged fire's restoration as the interface did fire protection. Catastrophic fires thus threatened both sides of the fire-management equation. What they shared were the combustibles that fueled those big burns. What fire management needed—and what overseers such as the GAO demanded—was a map of the fuel hazards akin to what geographers were doing with dispersed settlement along with a metric to measure the problem and its attempted solutions. Both requirements pointed to fuels, and that is where the National Fire Plan focused its funding.

In 2002 research, practice, and politics converged. The Bush administration announced a Healthy Forests Initiative that intended to broaden the work of the NFP. A consortium of sponsors that included the JFSP, USFS, Society of American Foresters, Colorado State University, Colorado State Forest Service, and the Western Forest Fire Research Center hosted a major conference on fire, fuel treatments, and ecological restoration to reconcile the various themes of contemporary fire management around the concept of fuel management. And the Forest Service and DOI funded a Landscape Fire and Resource Management Planning Tools Prototype Project (Landfire). It proposed "to develop the methods, tools, and protocols for producing consistent and comprehensive digital maps of current vegetation composition and condition, wildland fuel, historical fire regimes, and fire regime conditions class (FRCC) for the entire U.S. at a 30-meter spatial resolution." Put simply, it proposed to map fuels, the fire regimes they support, and their mutual change over time. Comparing historical fire regimes with contemporary regimes yielded an index of landscape health, as expressed in the kinds of fires now generated compared with those formerly common, or its departure from historic fire regimes. The upshot was a census of fire environments mapped into a geographic information system format. It showed where deviations from normal were greatest, catastrophic fires most likely, and the need for remedial action most pronounced.[26]

A parade of legislation, studies, strategies, and reviews followed. In 2003 the Wildland Fire Leadership Council signed an agreement on fuels treatments, the Healthy Forests Restoration Act added weight to the NFP, and together they allocated $2 billion for fuels treatments between 2001 and 2005. Stewardship contracts were let for "biomass services." The GAO continued its nagging demand for strategic rather than shotgun efforts. Fire research rallied with a summary of the "science basis for changing forest structure to modify wildfire behavior and severity." The Southwest Forest Health and Wildfire Prevention Act of 2004 established three institutes to study the regional problem. In 2005 Landfire published its first prototypes, mapping the central Utah mountains and the Northern Rockies. In 2006 the IAWF staged its inaugural fire behavior and fuels conference, identifying "fuels management, or how to measure success" as its theme. A massive compendium resulted. Simultaneously the Forest Service and DOI published a "Cohesive Fuels Treatment Strategy," which served for fuels what the 10-year collaborative strategy did for fire management overall. Fuels, in brief, was the organizing focus for fire research whether sponsored by the JFSP or the USFS. The institutional and conceptual apparatus appeared to be in place to cope with the combustion contagion. Then a record fire season kept the cauldron aboil.[27]

For every proposal there was a critic. Not all observers thought fuels should be the organizing principle. Biodiversity, beauty, clean air and water, solace, watersheds, silence, wildlife, historic experiences—all were values at risk, and a metric restricted to fuels dealt with none of them. The problem was not unhealthy ecosystems but sprawl. The target should not be landscape-scale slashing and burning but the "home ignition zone" that even Forest Service researchers had identified as the critical vulnerability. Where fuels needed abatement, prescribed fire, not chainsaws, should be the treatment of choice. Protection money could be better spent on local fire departments. In reply, the agencies noted that they could not control-burn in places with fuels in open riot. Besides, they declared that, in effect, they were at war, and that lands lost to savage burns surrendered all their other amenities.[28]

Other skeptics questioned not just the definition but the data behind that definition. The bins of discourse overflowed with data, but another scatter diagram only added numbers, not narratives, and the sources of those numbers were questionable. Landfire often lacked empirical evidence and resorted to "simulated historical conditions" to "characterize the departure of current landscapes using Fire Regime Condition Class (FRCC) calculations." Critics disputed those putative starting conditions and distrusted the black-box software that generated the hypotheticals, pointing to selected counterexamples that stand-replacing fires had always occurred even in ponderosa forests. Others were suspicious about an appeal to allow markets to pay for treatments and worried that the reintroduction of chainsaws and chippers would wedge open hard-won closures in a return to the bad old days. They wanted to use nature's fire to heal or at least cauterize nature's wounds.[29]

Cynics suspected that "catastrophic" fire was a manufactured crisis, like the bogus declaration by the Bush administration that Iraq had weapons of mass destruction to justify an unnecessary and ruinous invasion. That suspicion that the fire crisis was a front for scrapping environmental legislation stiffened when the Healthy Forests Restoration Act limited required NEPA analyses and restricted the use of appeals and litigation to oppose projects on the grounds that environmental jihadists were heedlessly obstructing solutions. After all, the Bush administration had systematically tried to gut many environmental laws and had appointed Mark Rey, formerly active in various forest products industries and chief architect of the detested Emergency Timber Salvage Rider Bill, as Department of Agriculture undersecretary with responsibility for the Forest Service. They saw the stripping of citizen rights of appeal as analogous to the Patriot Act's deletion of certain civil rights.

Twice, once before the Healthy Forests Restoration Act was passed and once after passage, the GAO was called on to settle the dispute. In 2001 it found no evidence of calculated obstruction by environmentalists. As measured by NEPA lawsuits, only 1 percent of projects had been appealed and none litigated. In an expanded survey in 2003, it identified appeals in 24 percent of 818 cases, or 58 percent of the 332 decisions that were potentially subject to appeal. Of these cases, 25 (3 percent) were litigated. In 74 percent of appeals the Forest Service proceeded without changes. Some 8 percent required changes before proceeding. Of the 25 litigated decisions, 19 were resolved. Protest, or at least scrutiny, had risen with program expansion and general distrust of the Bush administration, but it had hardly crippled projects. Moreover, its targets were not projects around the interface but those purportedly promoting forest health in the backcountry. Regarding the latter, even the GAO voiced suspicions. It had doubted the agencies were equipped to handle such largesse as the NFP bestowed, and it continued to harp on the need for better planning, coordination, and (inevitably) metrics. In many cases, instead of dissolving tensions, the fuels program had crystallized them.[30]

The one place of agreement was where the two maps, wildland fuels and WUI, overlapped. While this was the most politically combustible zone, it was also the one border across which all parties could shake hands instead of fists. Still, the old adage that no battle plan survives contact with the enemy was being restated, that no fire plan survives contact with nature. America's war on fire had bred a full-gauge ecological insurgency that no summer surges of hotshots and air tankers would stanch. In the early years of the fire revolution, it was possible for strategists to set the agenda. Now wildfires did.

2004: Megabucks

In 2004 the GAO directed its attention to another aspect of wildfire suppression, a crisis of "funding transfers." When its report was released in June, the fire season was just ratcheting up, and Alaska would end with the biggest fire bust on record. The season concluded with relatively fewer fires (65,461), but a staggering 8.1 million acres burned at a cost in excess of $1 billion. An interest in general accountability was reverting to a classic concern with simple accounting. Megafires were costing megabucks. Supersizing might produce economies in the fast-food industry, but big fires were rampaging through budgets and programs as wildly as they blazed over landscapes.[31]

Since 1960, 8 of the 10 worst fire seasons, as measured by acres burned or by costs, had occurred between 1994 and 2007. In the 1970s the average annual bill for wildfire-related expenditures was $420 million. In the 1980s it became $460 million, largely on the basis of the 1987 and 1988 seasons. For the 1990s it ran to $700 million. For the 2000s it doubled to $1.4 billion. Four megafires in 2002 together racked up $500 million in suppression costs. Twenty wildfires in 2006 did the same. Every indicator pointed upward. But as with the general economy, the fire bubble was set to burst. In 2007 the Forest Service spent $1.8 billion on fire control. Like Wall Street, which was not allocating invested money to the best economic use but lining its own pockets, fire suppression was siphoning off monies needed elsewhere. Although Congress eventually paid 80 percent of the transferred funds, that still left a huge hole in the budget. The agency was close to crashing. Nearly everyone agreed that the financial system underwriting wildland fire needed a systemic overhaul.[32]

The immediate panic was that paying for suppression was causing agencies to transfer funds from other programs. In the past Congress had made up the difference with supplemental appropriations, so programs could proceed without pause. Now Congress refused, nominally honoring the mechanisms it had established in 1978 but had subsequently ignored when the flames roared. Funds for wildlife, recreation, roads and trails, research, cooperative programs with states—more of everything was consumed in the convective maw of megafires. The National Fire Plan had promised to boost agency fire budgets. Instead, suppression gobbled up even those preparedness accounts. The Forest Service saw its fire program move from 10 percent of its budget to nearly half, with no end to cost inflation in sight. In 1910 Chief Henry Graves had proclaimed that fire protection was 90 percent of American forestry. Almost a century later it seemed to many observers that his forecast was becoming the future.

The deeper alarm was how expenses had blown up so swiftly and uncontrollably. When agencies had new policies, bigger budgets for fuel treatments, and had privatized much of their operation, why was the cost of fire management spiraling out of control? Congress and critics called on economists to explain the scene and propose solutions. If fuels were the metric of field operations, money was the metric of politics. Year by year commissioned reports piled up. If paper were a suppressant, those reports might have quelled the flare-up by themselves.

*   *   *

Like much else, the project had begun 10 years earlier as one of the aftershocks of the 1994 season, in this case the Forest Service's first $1 billion fire year. The agency chartered a Fire Economics Assessment Team (FEAT), which released its conclusions (as FEAR, with "Report" replacing "Team") in September 1995. It observed that real (inflation-adjusted) expenditures had not increased from 1970 to 1994, and by some measures had decreased. In 1967 the Trapper Peak fire had cost $6 million and the Sundance fire $13 million. Adjusting for inflation, those two fires translated to $107 million in firefighting costs over 74,000 acres. The obscenely expensive 2002 Biscuit fire had spent $150 million over 500,000 acres. On a per acre basis, the age of megafires was actually less costly. Still, 1994 had marked a jump in the flow of monies. FEAR determined that every factor affecting costs would escalate, "at least in the short run," and that a course correction would require investments in fire plans that "neither the Forest Service nor Congress is currently willing to make." When those causes did, in fact, rise and took nominal costs with them, suggesting that the spike marked a fiscal jump, a fire economics conference was organized in 1999 to summarize the state of understanding.[33]

That phase change applied to studies as much as to costs, and like other aspects of wildland fire, it no longer remained within the purview of the Forest Service. Congress and OMB commissioned their own inquiries. The inspector generals of the departments of Agriculture and Interior poked through the bills. The NASF weighed in on cost sharing. Even the private sector crowded into the picture. Beginning in 2000, the year of the National Fire Plan, reports ticked away with metronomic regularity. That year the USFS responded to the GAO's call for a cohesive strategy and outlined a proposed restructuring of its general fire appropriation. The National Association of State Foresters published its conclusions regarding "cost containment on large fires." Taxpayers for Common Sense released *From the Ashes*, a call for reform (or more precisely, for the Forest Service to do what its own guidelines and commissions recommended). And the Western Fire Ecology Center anted into the pot with *Money to Burn*. Everyone, it seemed, wanted more money for fire, but they wanted it better spent and thought they knew how to make that happen. The National Fire Plan distilled much of this conventional wisdom, that ounces of prevention in the form of fuels treatments would mean fewer pounds of suppression outlays.[34]

Between 2001 and 2007, the GAO conducted five inquiries specifically focused on costs. At congressional direction the DOI and USFS sponsored the National Academy for Public Administration as an independent observer, and it contributed six. The Wildland Fire Leadership Council commissioned one. The USFS, Interior, and NASF crafted a Large Fire Cost Reduction Action Plan and then folded those ideas into the 10-year collaborative strategies. The Forest Service Fire Plan Office scrutinized the impact of "large incidents." The USDA Office of Inspector General conducted an audit. The Secretary of Agriculture chartered an Independent Large Wildfire Cost Panel to examine the 2004 and 2006 budget busters. The Brookings Institution gathered retired Forest Service fire officers and other experts from outside the federal agencies to scrutinize spending. Academics published in the *Journal of Forestry*, *Forest Science*, and *Conservation Biology*. The nonprofit Thoreau Institute joined the scrum. Many tried to engage the totality of fire management, with particular attention to the trade-offs between fuel treatments, prescribed fire, Firewise community programs, and suppression. But the jaw-dropping costs went to the big fires, and that is where attention focused. Case study after case study tried to account for where the money went. The Biscuit fire, Aspen fire, Fawn Peak fire, B&B complex, Cave Creek complex, Old fire, Grand Prix fire, Padua fire, 2003 Northwest Montana area command and Northern Rockies geographic area fires—the documentation piled up, the numbers were crunched, IMTs were interviewed, and reports were duly filed.[35]

A general consensus formed that big-fire suppression costs were out of control, that money spent on suppression was less valuable than money invested in changing the conditions that made megafires possible, and that while costs had many contributing causes, solutions must be possible. The Brookings report found no evidence of fiscal malfeasance or inadequate attention to financial issues; the problems were embedded in the American way of fire. (In fact, the most expensive fires were not those that threatened exurbs but those that simmered in wilderness before escaping.) Only a few of these contributing factors were under the control of the wildland fire community, and all argued that still-higher costs would come. No single display of individual or institutional irresponsibility or single corrective reform could explain the whole. Fixing any one problem would likely squeeze the balloon and only push the expense elsewhere. Expectations about what the wildland fire community was supposed to do conflicted with how to audit the outcome. While pundits argued over which factor drove the system, the reality resembled a driverless car in which everything converged with no single set of hands on the wheel or feet on the brakes and gas.

Among the fire community, the belief swirled in the air like pine pollen that critics had no sense of what contemporary fire management was truly like. As the NASF report expressed it, "citizens, politicians, administrators and the media have no concept of reality." Yet the National Academy of Public Administration rebutted that the American taxpayer receives a "daily dose of wildland fires" and "does not know or care how or why a fire started. The taxpayer stares at the television set and sees tax dollars going up in flames." While most observers agreed that the whole apparatus of wildland fire needed reform, environmentalists argued for reform around the concept of ecosystem management, foresters and fire behavior scientists around fuel management, public administrators around tweaked policies and best practices, and the GAO around notions of accountability. While economists had long hung on the margins, publishing theories on fire financing as early as 1916, the unruly costs of megafires brought them and their methods temporarily to the fore.[36]

* * *

Most commentators agreed that the long-term solution lay with fuels management and sharper initial attack, or more broadly, with a landscape less prone to explode and with firefighting better grounded in local communities. Oddly, perhaps, this was the wildland fire version of the counterinsurgency doctrine being developed for America's other ill-begotten wars, in Iraq and Afghanistan. Suppression only aggravated the underlying conditions: real success required control over the countryside and security for its inhabitants, or in wildland terms, a more resilient ecosystem less prone to sudden disruptions, species losses, wild fluctuations, and a WUI designed to accommodate fire. But these were decadal, nation-building goals that required consistent, patient attention. It was far easier to treat fires as emergencies.

In the interim big burns and eye-popping costs poured into the gap, like gas into a vacuum. Megafires were synthesizing and distilling their surroundings. They were the visible expressions of an underlying unrest. Much as the greater threat to public health might be chronic ailments such as diabetes and hypertension, while exotic contagions such as Ebola caught the public's attention, so megafires became the visible emblems of a breakdown in public lands, which was a politically more palatable emblem of the extravagant way Americans treated their national estate. While the problems were not restricted to the Forest Service—the Interior agencies suffered from the same syndrome—as usual the USFS became the whipping boy, sometimes with cause and sometimes just from convenience.

Audits suggested that about a third of the expenses on large fires went to personnel, about 56 percent to contract services, and the remaining 11 percent to everything else. Among the oddities of firefighting was the enduring fact that machinery tended not to replace labor but only added to it. Large fires in 2000 had similar complements of personnel as in 1910. Crews became more expensive, as with labor generally over the period, but equipment costs had risen even faster, particularly as agencies appealed to private-contractor air tankers and heavy helicopters. The digital revolution only heaped up further equipment expenses. Whatever incremental improvements in productivity occurred, the increase in the size and managerial complexity of fires overwhelmed them. The situation resembled health care in that sparkling technologies, chimerical expectations, rocketing expenses, and a capacity for emergency medicine that was a marvel of the world boosted costs without a widespread improvement in national health.[37]

The other oddity, at least for many economists, was the inability of the private sector to improve services and reduce expenditures. The move to privatization had begun seriously during the Reagan administration partly out of a belief that it would improve economic efficiency and partly to reduce the size of the federal workforce. The second expectation succeeded, not only on the size of labor but in many of the goods and services the federal government had provided, such as tools and gear through the General Services Administration, fire weather forecasting through the National Weather Service, and heavy equipment that it owned. More and more of firefighting was contracted out. When the National Wildfire Suppression Association was established in 1991, it had six members; by 2013 it comprised some 200 members in 23 states and could field 130 crews along with 400 engines and logistical support. Fire aviation had its own associations. The outcome, however, refuted the assertion that costs would plummet.[38]

The perhaps unintended upshot was the creation of what critics termed a fire-industrial complex that became a lobbying force and raised costs. Even by the late 1990s, a fire shelter that GSA sold for $39.34 cost $89.98 on the open market, and a Nomex shirt priced by GSA at $43.43 went for $74.12. A short-handled shovel from a commercial vendor sold for $60 and was imported from Ireland. Heavy machinery such as dozers, lowboys, and water tenders, formerly part of the public fire cache, were now contracted at high rates and with stiff premiums for standby availability whether used or not. Perversely for those committed to free-market efficiencies, the shift from public to private goods and services had helped drive costs up. Aircraft constituted a special category. At the federal level these had long been mostly privately operated. After three fatal crashes in 2002, a blue-ribbon panel found that "the safety record of fixed-wing aircraft and helicopters used in wildland fire management is unacceptable." Because they flew on public duty, normal FAA standards did not apply. Most contractors were using legacy aircraft that were nearly impossible to maintain at safe levels with the amounts the government was willing to pay. Following the 2002 crashes, the USFS demanded that common standards apply, which stood down nearly the entire national fleet and cut the operational fleet to a third.[39]

Reviews constantly noted opportunities for minor savings in strategies and tactics and denounced the putative agency culture that regarded big fires as a blank check. Proposed solutions tended to follow the example of reforms in fireline safety. IMTs should include an auditor as they did safety officers. Fires expected to cost millions required special authorization from higher authorities; fires that cost over certain redline figures could expect after-action reviews. Egregiously expensive fires such as the Biscuit could attract the attention of the GAO or Office of the Inspector General, which functioned for fiscal matters as OSHA had come to for fireline safety. Campaigns were formulated to better educate the public and to address persistent failings in agency culture. Even seemingly trivial decisions such as where to locate fire camps, which affected how much travel was involved, or standards of mopup, which determined the size or the number of crews, could translate into serious dollars when large numbers of firefighters were mobilized. Still, the penalties for going over budget were far less severe than those for experiencing a burnover fatality. The fact remained, the system favored suppression as the least risky option, and even megabucks fires were not likely to end with agency-crippling or career-ending consequences. A Cerro Grande or Thirtymile fire were.

What the agencies needed was a way to regularize, if not standardize, decisions as it did crew categories or equipment. The preferred solution was to invest in decision-support guidelines. Since the 1970s the agencies had devised algorithms and procedures to guide baseline budget requests. For the Forest Service, this was the National Fire Management Analysis protocol (later adapted for the BLM). Always an outlier, to better satisfy its unique mission, the NPS had Firepro. But as big money flowed into preparedness (especially fuels) and suppression, congressional auditors wanted a more uniform system for all—the planning and fiscal equivalent of total mobility. This should establish the core budget adequate to cope with all but those wildfires that escaped initial attack. The 1978 reforms had argued that more accurate funding up front would reduce the emergency drafts from large fires. Now that story was about to be replayed.[40]

Actual wildfires activated a separate emergency account. To prevent mindless spending—to break the fiscal component of the 10 a.m. policy— the agencies instituted a protocol termed an escaped fire situation analysis. In a loose sense this was intended to act like a NEPA review and compel fire teams to consider alternatives, as the new policy allowed. Later, when the 1995 policy replaced earlier versions, the *escaped fire* situation analysis was replaced by a *wildland fire* situation analysis that would, in theory, expand the review process to all fire starts, not just fires that escaped. But tinkering with protocols established decades before was "cumbersome," "not scalable," and insufficiently "flexible." In 2005 the National Fire and Aviation Executive Board (NFAEB) chartered a wholesale replacement, the wildland fire decision-support system (WFDSS), with implementation scheduled for 2009. Research was called on to create a suite of decision-support tools akin to what fire behavior had produced for predicting wild and prescribed burns. Such protocols could synthesize existing science, substitute for the loss of on-the-ground fire lore, and build in cost algorithms in ways that were both consistent and legible to outsiders. What NWCG guidelines did for prescribed fire and monitoring the WFDSS could do for suppression.[41]

Whether or not the WFDSS would improve performance was unclear. But in an era in which policy allowed for many options, when the old doctrine of a unity of command was replaced by a unified command that melded fire and line officers with local officials, when fire officers had no clear expectations of what they were to do or how they would be evaluated, the project could both guide and document decisions in ways that made them transparent during postfire reviews. It was hoped it would also make the fiscal consequences of those choices more visible during the fire as well as afterward. It might expose political theater, such as air tanker runs disdainfully termed "CNN drops," done only to be seen doing something that the public expected and politicians often demanded. The immediate effect, however, was to add yet more tasks and another layer of review, which increased complexity and costs. Such reviews were unlikely to boost careers and would only add another possible threat to them.

Even among economists there was disagreement about metrics and methods. The Office of Inspector General proposed cost per acre as a simple index by which to evaluate suppression efficiency. Those more familiar with how fire behaves noted that this would perversely lead to larger fires. An expensive initial attack that held a fire to an acre would be penalized, while a strategy that spread costs over 300,000 acres would better satisfy the criteria. Behind the OIG's argument was the hope that fire officers would allow fires to expand where appropriate—would make room for more WFUs rather than instinctively suppress all ignitions—but the proposed metric said otherwise. Moreover it argued to shift costs for fires that burned onto private lands to the states. This did not reduce expenditures but only passed the buck around the players at the table and sent disruptive ripples through a wider circle.[42]

Beyond the WUI many of the values at risk also had no simple market value, and so they shifted attention to those pieces of the sprawling mosaic that could be quantified, such as costs, rather than to what those costs bought. As an organizing principle, economics introduced some perverse incentives of its own. While every economist remarked that incentives were out of line with policy, what that actually meant was that policy was complicated and confused and allowed for significant discretion by fire officers operating in an intrinsically wild setting. Paradoxically, classic economics worked best with tamed landscapes and prescribed burning, where people exercised more control over the tangled undergrowth of variables. The West, however, was moving away from the practice as too costly and cumbersome and toward big-burn suppression and big-box confinement, the first of which blew up expenditures while the second relied on a faith-based ecology that the final ledger of environmental damages would move to the black.

Neither the cloudburst of reviews nor the proposed remediations stopped the flood of suppression dollars. They did focus attention on the dramatic dissonance between what the fire community was expected to do and the tools, including fiscal, they were given to do it. Megafires made that misalignment unmistakable and unblinkingly visible.

* * *

Yet in the end, fire economists were not asked to apply their analytical firepower to the fundamentals but to hotspot fiscal flare-ups. They were asked to "contain" costs, and their megastudies had the character of an emergency response. High costs were not something separate from the whole scene that megafires were integrating. The country was unsure how to pay for wildland fire management because it was unsure what it wanted fire agencies to do, or rather it wanted them to do many competing jobs. Economics was asked to sort out what politics would not.

Commentators did what all critics do when confronted with an unfamiliar scene: they argued from the known to the unknown or, by analogy, from the world they knew to the world on fire. But no analogy fit. Though it competed with the agencies, fire was not an economic actor. Wildland fire management was not a disaster response because most fires were not disasters, and they had to integrate with land management, not just emergency relief. A city fire department could enforce codes for fire safety and nuisance abatement, and they shunned any hint of controlled burning. Wildland fire agencies could not determine land use and had to balance fire's use against its exclusion. Unlike an imposing military, fire protection had no deterrent value against its enemies. Besides, ecological nation building, not firefights, was its great challenge.

In economic terms, fire management was not an economic institution but a political one. It existed for public safety and the promotion of public assets on public lands. It could not maximize any one task: it had to serve many, few of which had measurable market values. Much like health care it was, on the whole, an example of market failure in which attempts at privatization tended to unleash rather than corral expenses. The only agreement among the inquiries was that the driving factors would worsen, and that included costs and the reluctance of Congress and the public to pay them.

The reality was that fire management had long been underfunded and had instead relied on aggressive initial attack and a favorable environment of fuels and climate to avoid long-term investments. Decades of fire exclusion had built a whopping national debt of combustibles; now the factors had turned against it in the form of crashes and blowups. The monster bills represented the higher costs of continued borrowing. There was no obvious solution that would dramatically reduce costs, only practices that would allow money to be spent more wisely. But neither was it clear that money, certainly not money for suppression, could solve the crisis. In a city every fire extinguished was a problem solved. In wildlands every fire suppressed was a problem deferred—and worsened.

No single strategy could succeed. From 2000 to 2007 every major strategy had failed at one time or another and probably at the same 2–3 percent rate. Prescribed fire had imploded at Cerro Grande, and reforms made it, in the West, prohibitively expensive. Wildland fire-use fires had occasionally blown out of their putative confinements at great distress and expense; the 2006 Warm fire, which began quietly as a burning snag that crews walked up to and could have extinguished with a canteen, escaped to burn over

58,000 acres and cost $7 million to contain. The nation's firefight was being overrun by record season after season. There was always that one fire or batch of remote snag fires that galloped beyond initial attack and wiped out what years of protection had spared. The politics of fire responded far more to those breakdowns than to modest successes. And like a recurring nightmare, the scene could repeat in perpetuity as fuels regrew, houses were rebuilt, and drought and wind returned.

* * *

In 2003 the Aspen fire burned the half of the Santa Catalina Mountains left unburned by the 2002 Bullock fire and ran through a recreational community, Summerhaven, near the top of Mount Lemmon. The main burns were too large, the terrain too rugged, the general values too sparse, and the risks too great not to back off and burn out from secure roads along the mountain flanks. The two fires consumed the Santa Catalinas. The episode seemed to encapsulate the emerging western fire scene in miniature: evacuation and point protection for high-value assets, big-box burnouts for the rest. The strategy kept firefighters safe, got acres burned, protected communities, and if lucky held costs and political uproar down.

Yet the more dramatic contrast came in viewing the scene from the north, where it was possible to see in one field of vision both Biosphere 2 (in the foreground) and the Santa Catalinas (behind). Biosphere 2, its glass domes gleaming in the desert sun, was a model of a sustainable ecology and economy, a world in which rational models worked, but it was also a world that had no tolerance for fire. The Santa Catalinas displayed a world in which full-throated wildfire was inevitable and necessary, a place whose very character as wildland frustrated social logic. As the decade evolved, the task before the wildland fire community was to reconcile those two visions, to somehow retain the character of the wild while managing through the glassy prism of disciplined reason.

2005: Centennial

The year had slightly more fires (1,292) and somewhat more acres (591,509) than the year before, and its cost quieted somewhat to barely under $1 billion. But what made the 2005 fire season historically significant was that it celebrated the centennial of the U.S. Forest Service. The agency was battered, and often instinctively cringed like a whipped dog, knowing that every gesture raised toward it would end in a slap. It was, critics and celebrants alike agreed, a shell of its former self.

But that shell resembled a heritage building, too valuable and redolent with cultural connotations to raze even if too archaic in its internal architecture and creaky in its operations to keep as a functioning institution. Over the past dozen years, it had instead had its interior ripped out and remodeled with more modern furnishings; that included having its fire code upgraded to contemporary standards. By global standards the fact that it stood at all was extraordinary. State-sponsored forestry had been one of the triumphs of colonial settlement. The British and French had pioneered the process, and forestry departments sprang up everywhere from India to Australia, from Cyprus to Mauritius, from Algeria to Cape Colony. The lesser colonizers such as the Netherlands, latecomers such as Germany, and aspiring powers such as Mexico all adopted the practice. Bureau chiefs ruled like proconsuls; they were feted and knighted. They proclaimed themselves the vanguard of global conservation. They were an institutional wave that washed deeply onto the shores of European imperialism.

With decolonization, that wave receded. In many places it vanished with the former powers; in others it became an empty edifice, little more than a heading on stationery. In settler societies the diminution took decades. Canada devolved state forestry to the provinces in 1930, leaving the Canadian Forest Service solely as a research and advisory organization; the provinces further dissolved forestry into departments of natural resources and then into bureaus of sustainability and environment. Australia's states did the same. New Zealand disestablished its Forest Service in 1984, selling off productive plantations and retaining about a quarter of the former lands for a more preservationist Department of Conservation. More and more, forestry migrated to the commercial private sector or to vague international bodies such as the FAO. It lost the high ground of modern environmentalism. Oddly, it survived best in the two Cold War antagonists, the communist Union of Soviet Socialist Republics and the capitalist United States of America. Critics pushed for its dissolution—from the left by environmentalists who wanted its lands parceled out to other agencies or brought under a new bureau, and from the right by free marketeers who pushed to privatize outright.[43]

The Forest Service was far from the organization it had been at its fiftieth anniversary. On most matters it had been turned inside out; certainly this was true for its fire program. In 1905 the national wildland fire establishment consisted of the fledgling Forest Service (with one fire guard for every 670 square miles), U.S. cavalry troops stationed in a handful of national parks, and embryonic operations in states that held public lands, such as New York and Minnesota. It was not much, and the U.S. Forest Service soon claimed preeminence in policy, practice, and political clout. For the next 50 years it amassed greater control as it sought to exclude fire as much as possible from as many landscapes as feasible. Then it watched its hegemony crumble while the old pieces, along with newly minted ones, tried to self-organize into a messy and mostly expensive pluralism.

By 2005 the Forest Service was slammed by endless metafire reports, megafire outbreaks, and megabuck invoices. Even what it had long prided itself on doing better than anyone else, fighting wildland fires, was questioned as a goal and critiqued in operations. There did not seem much to celebrate, and the festivities were relatively muted. Among its gala events were five seminars themed to its primary tasks. Inevitably, one focused on fire and was held at Boise. But perhaps the real retrospective was the open letter to Congress signed by five former chiefs that noted the self-cannibalizing costs of the fire program and the likelihood that, if unchecked by legislation, those costs could consume the agency itself, like an open-pit mine digging through the company town. Since 2002 Forest Service staff had declined 35 percent, resource specialists by 44 percent, recreation techs by 28 percent, biologists by 39 percent. The agency was unable to provide "basic stewardship." Everything was going to fire. The agency had publicized its public mission through its firefight. Now it appeared that it might crack up by that same project.[44]

The American fire establishment was ripe for reconstitution. On the occasion of its centennial, the Forest Service struggled to define what role it should have in that new order. That it was a cameo of the country made the agency's task both urgent and onerous.

* * *

The fissure widened not only between what the agency was expected to do and could do but also between what it did and the means used to evaluate it. The tension between safety and suppression became increasingly fraught. A "safety protocol review" that followed the Old and Cedar fires, two monsters of the 2003 season, concluded that "prescriptive policies are forcing fire leadership to take the risk of violating them in order to complete the mission and meet agency expectations." In other words, fire officers could not both do their job and stay within the rules.[45]

Worse, the bureaucratic reactions to the South Canyon and Thirtymile tragedies were likely aggravating the problem. Each fatality or near miss added to a morass of "check-lists, protocols, and prescriptive policies" that created a counterproductive, "rules-driven" operational environment that did not "rely on appropriate decisions and behaviors, but solely on the absence of 'bad' outcomes." One reckoning tallied 156 "ostensibly 'inviolate' wildfire suppression-related rules." And what happened in fire suppression was repeating for every other aspect of fire management. More lists, more rules, more fine-grained oversight—it was not simply that they encumbered operations but actually menaced them, and the malady was spreading to prescribed fire, wildland fire use, and aviation. People responded to the rules environment, not to the fire environment. By 2005 similar rules-driven solutions were proposed for cost containment. Auditors such as OSHA loved the rules because they established precise standards amid what was, by definition, an imprecise situation.[46]

On June 1, 2004, the Forest Service issued a briefing paper in which it proposed to move from a rules-based organization to one informed by a "defining doctrine and guiding principles." A fuller explanation followed in September from its National Fire Operations Safety Office. In effect, it presented the case for firmer goals and doctrines and more flexibility in meeting them. Behind the revised thinking was Karl Weick's concept of a high-reliability organization and, inevitably, the example of the military. A year later Tom Harbour, director of FAM, repeated a favorite quote. "Simple clear purpose and principles give rise to complex intelligent behavior. Complex rules and regulations give rise to simple stupid behavior." The next year that notion crystallized into what was dubbed the First Pulaski Conference.[47]

For six days in June, conferees nominated by each Forest Service region gathered in a closed-for-the-season ski lodge at Alta, Utah. The model was a retreat: no television, no radios, no newspapers, no rental cars, no cell phones—everyone sealed in with a late-season snowstorm. The intention was to wipe clean the bureaucratic slate and build a fire-suppression organization that could be both safe and efficient from first principles called *doctrines*. Although no one made the allusion—this was about the future, although the reference to Ed Pulaski harked back to the Big Blowup—they were returning the Forest Service to something like its origins with Gifford Pinchot's *Use Book*, which laid down the principles of national forest management in a slim volume that could fit into a hip pocket.[48]

Out of their labors came a "foundational doctrine" for fire suppression. Chief Forester Dale Bosworth formally accepted the principle behind the new doctrine along with a "roadmap" for moving doctrine into the manuals, which remained the "legal authoritative foundation." Fire and Aviation

Management instituted a new unit, Risk Management and Human Performance, dually located at NIFC and the National Advanced Fire and Resources Institute, to oversee the transformation with a goal of implementation for the 2008 fire season. Comparable conferences were held in January 2006 for fire and aviation and in July 2006 for all-hazard response. Wildland fire use would follow. Interestingly, the exercise coincided with a "redesign" for State and Private Forestry, the branch in which Fire and Aviation Management resided. Chief Bosworth accepted that "this is unprecedented and that many of the risks are unknown." But risk management was the nuclear core of the challenge.[49]

With fire seasons breaking records, costs exploding budgets, and critics hovering over the expected carcasses, the agency moved to implement quickly. Drafts for the relevant portions of the Forest Service *Manual* were targeted for 2007 with implementation for the suppression codes scheduled for early 2008. The rules would be based on principles, not replace them. Doctrine would explain how to think about events, how to recover "professional judgment" as "a legal basis to defend our actions, to reestablish our relevance as a contributor to the quality of American life, and to be consistently better prepared to meet the challenges of change." It would never be possible to register all the potential circumstances that might lead to failure when circumstances changed with each season, each gust of wind, and every ravine around the bend, yet Chief Bosworth insisted that a principles-driven approach would "not lead to anarchy." With enthusiasm running high, FAM sought authorization to use interim directives, which permitted implementation for 18 months after signing. This would allow "for time necessary to align handbooks, standards, and other agency and interagency documentation."[50]

It had been a long while since the Forest Service had led. Leading meant it would have to surrender the vestiges of the national role it had once held, for which it had more responsibility than power and whose stresses had caused internal fissures. But for a while, perhaps a historic bubble, it found itself in new circumstances. For decades it had only reacted. It had reacted to the environmental movement, to the interface issues, to tragedy fires, to perceived threats to its discretion, purposes, and funding. It had sought alliances to protect itself as much as to seize opportunities. With the refounding of fire management on doctrines, it did what it said it wanted its incident commanders to do on the fireline: lead. It sought "active collaboration with inter-departmental representation from partners and cooperators" and with key oversight groups such as OSHA and USDA's Office of General Counsel and Office of Inspector General. It wanted common agreement on what the rules of engagement would be so that evaluation could be comprehensive and fair. Three decades of interagency fusion meant it could not act in isolation, but it was big enough that if it moved, the others would have to respond. It assumed, insouciantly perhaps, that "our partners will join us in this venture."[51]

The larger fire policy remained the 1995 policy as updated in 2001. What the doctrine project intended was to mediate between those statements and field operations by replacing an unworkable welter of rules, somewhere between a Gordian knot and an Augean stable, with a slash and a sluicing. It did not wait for other agencies to respond before proceeding: it was not challenging prevailing policy per se but seeking to change its protocols and application. Still, the move would require mutual adjustments. Not all the other agencies agreed, for example, in their interpretation of the common federal policy, which mattered when fires burned across jurisdictions. In 2007, as the *Manual* rewrite was under way, clarifications were issued by the National Fire and Aviation Executive Board on what "appropriate management response" could mean.

* * *

Big costs and crises are a formula for consolidating power, and between them the National Fire Plan, the Healthy Forests Initiative, and serial megafires compelled the federal agencies to centralize their own operations and to move from interagency consultation to mutual governance. The American fire community began a slow institutional chrysalis from which, over several years, it would emerge as a new entity.

Agencies that had evolved, unit by accreted unit like beads on a string (think NPS, FWS, BIA), had to fashion hard welds under a common policy. Traditional baronies had to be combined, for purposes of fire management, into a national entity. Even the Forest Service, which to many outsiders appeared as a monolith, betrayed deep regional differences that amounted to separate satrapies. When the 10 a.m. policy no longer held them together, some new order would have to. And that was just the start. As programs and fires spilled over the borders of the public domain, the process of consolidation would have to expand to include the states, counties, municipalities, private landowners, and manifestations of a civil society such as NGOs. The interagency project had to become intergovernmental and in some respects nongovernmental. While everyone welcomed the national attention and monies, few warmed to the instruments of control that came with them. The National Park Service saw its discretion over a unique mission threatened. The Forest Service, once again, absorbed the national tensions within itself, in this case, from its regions.

The Healthy Forests Initiative and Healthy Forests Restoration Act drew the wary eye of the Office of Inspector General. In September 2006 it released its audit, concluding that the Washington Office determined funding based on "units' historical funding levels" and on "targets for number of acres to be accomplished" set in Washington, DC. There was no meaningful yardstick or reporting standard to judge whether the monies were going to communities more at risk or whether "fuels projects" were simply renaming routine prescribed burning or timber sales. As critics charged, many of the acres were far removed from the interface, and prescribed burning was boosted where it was easiest (in the Southeast), not where the need was greatest (the West). National statistics failed to parse those distinctions. As with fire policy, the HFI had tried to combine both ecosystem health and community protection, but the clear understanding in Congress was that the money would go to shield communities. The OIG recommended national guidance to prioritize projects and to ensure consistency among regions and forests. The Forest Service agreed.[52]

What happened with the fuels program happened with the other federal agencies as well. When it adopted the National Fire Plan, Interior subjected itself to the same kinds of scrutiny as the Forest Service. Yet the NPS lacked the kind of landscape inventories needed to assess the severity of fires or the outcomes of treatments; the BLM recorded information on rehabilitation outside its fire budgets. Accountability, however, demanded consistency in data and in the procedures used to prioritize projects, and the GAO especially wanted all the agencies to accept a common risk-based algorithm. Adopting uniform protocols, however, removed some of the discretion that agencies such as the NPS had traditionally enjoyed.[53]

As part of the reorganization prompted by the NFP, the agencies chartered a Wildland Fire Leadership Council in 2002 that brought together the federal land agencies. The WFLC oversaw the production of the 10-year collaborative plans to translate the NFP into a truly national project; in 2005 it updated that strategy. Meanwhile, a National Fire and Aviation Executive Board convened the agency fire directors and a representative from the states. It established a directives task group to issue briefing papers to clarify the fast-morphing programs and policies. The board also assumed responsibility for NIFC. By now NIFC had matured as the nucleus for 11 regional coordination centers that spanned the country from Puerto Rico to Hawaii to Alaska. The pyrogeography of fire management created an operational overlay atop the country's jurisdictional geography.[54]

In 2005 the Congressional Research Service issued a report that summarized the exuberant developments through 2004 and observed that congressional funding was also going to state and local governments. The next year, in registering exactly those political entities and projects to which the funds went, it tallied the pieces of an American fire scene that was no longer contained within the public domain and the agencies that oversaw it. Whatever policies were adopted and whatever funding mechanisms were enacted, they would have to embrace the full constellation of America's wildland, rural, and exurban fire institutions.[55]

In 2007 those issues were still inchoate when the National Fire and Aviation Executive Board published two memoranda. One sought yet again to clarify what "appropriate management response" actually meant in operational terms in order "to take full advantage of the flexibility afforded by existing policy." It reaffirmed that a single fire could not be managed for both protection and "resource benefits" but otherwise expanded the tactical options available. The other memo "approved and recommended the use of the Master Cooperative Wildland Fire Management and Stafford Act Response Agreement" that governed cost sharing among agencies, states, and local jurisdictions and the use of federal disaster monies for wildfire. It sought a common methodology for those bursting allocations. Like the interim directives for Forest Service doctrines, neither memo resolved the fundamentals behind the issues, but they showed where the flames were heading. The interagency model would have to become interjurisdictional, and more.[56]

* * *

No single entity or even the consortium of federal agencies could control the agenda. Wildland fire affected the atmosphere and the biosphere, upsetting climate and scrambling biodiversity. It threatened life and property in ever-expanding exurbs. It caused cracks in the woodwork of federal politics. All this translated into more researchers, more advocacy groups, and more participants on the line.

The number of entities with an interest in wildland fire climbed, matching the kinking curve of costs and burned area. NGOs added the Association for Fire Ecology (2000), primarily to promote fire research and its application, and Firefighters United for Safety, Ethics, and Environment (FUSEE; 2003), an advocacy group for more progressive fire management. Three organizations—Tall Timbers Research Station, the Association for Fire Ecology, and the International Association for Wildland Fire—sponsored regular symposia; the latter two hosted journals. Self-organizing groups to promote prescribed burning acquired sufficient critical mass in 2007 to gather into a national Coalition of Prescribed Fire Councils with the charge to facilitate all matters pertaining to prescribed fire and a "landowner's right to burn." Its local councils became, in a sense, the flipside of volunteer fire departments.[57]

The most curious institution, however, was surely The Nature Conservancy. From its Florida and prairie beachheads it had pushed far inland and was punching well above its weight. It sat on panels to review federal policy. In 2000 it joined a Fire Roundtable held at Flagstaff, which led to the idea of collaborative projects, a concept that entered into the National Fire Plan. A presentation at NIFC then crystallized the vision into a program, an agreement on Restoring Fire-Adapted Ecosystems. In 2002 this took concrete form as a system of landscape-scale collaborative projects organized as a Fire Learning Network (FLN). In 2006 that experiment matured into a 5-year Fire, Landscapes, and People program. FLN attempted for the ecosystem-restoration half of the fire equation what the interface initiatives such as Firewise would do for the other. It could bring the restoration agenda to private as well as public lands. By emphasizing cooperative ventures the FLN strategy sought to overcome the distrust that had hardened among the parties—to advance projects through functioning compromises that got work done without the threat of litigation. Collaborative gatherings would replace the courts.[58]

It was a lovely thought, but so entrenched were the suspicions, and so angry the scars, that the program needed a mutually trusted facilitator. The Nature Conservancy stepped in and provided a timely counterweight to the fears raised by the various maneuverings around the Healthy Forests Initiative to remove citizen voices. By 2010 the program had spread to 15 regional networks, 1,020 partners, 39 states, and 5 countries across 157 landscapes and 150 million acres. As always, the acres treated were far fewer than needed, but the projects created prototypes and points of positive infection, and, critically, TNC provided a model for ecologic burning, not just fuel abatement. It could flash burn excess combustibles with the best, but its metric was species and their habitat, not pounds of hydrocarbons per acre. It intervened in the name of ecological goods and services, not commodities.[59]

No one matched TNC's ability to leverage small resources to big effects. But as fires blew up as a theme, the Conservancy committed to more, even beyond the borders of the United States. It was convinced that it could not protect biodiversity without actively engaging with fire. It devoted its May/June 2001 issue of *Nature Conservancy* to the subject. Then, partnering with the World Wildlife Fund and the International Union for the Conservation of Nature, it launched a Global Fire Initiative. Ramrodding the project, Ron Myers authored a 2004 survey on *Fire, Ecosystems and People* that assessed fire "as a global conservation issue." The consensus estimate was that 82 percent of places identified as important for global conservation were at risk from "altered fire regimes." Two years later he followed with what could serve as both a prolegomenon and a manual, *Living with Fire: Sustaining Ecosystems and Livelihoods Through Integrated Fire Management.*

Because TNC was mostly dealing with landscapes inhabited by people already doing burning, there was no need to argue for prescribed fire, only to shape its application, and there was no clash over nonanthropocentric values, because the issue was not to exclude people but to regulate them. The two booklets are the clearest expression of modern thinking about fire in the literature. Because it owned the lands it managed or helped manage through conservation easements, TNC could act without every gesture becoming a cause célèbre. Because its own purposes were fairly transparent, its word was accepted. And because it was committed to practice, its recommendations were applied.[60]

TNC seemed to be everywhere, a weak gravitational force that helped hold factious alliances together and was everywhere welcome. It carried the American fire revolution, in a heavily annotated version, beyond its national borders. It demonstrated how a driptorch, with a proper fulcrum, could move the world. When the Great Recession hit, TNC suffered measurably and had to withdraw from the Global Fire Initiative. It retained its presence on a smaller scale, as always multiplying its influence through training, facilitating, technology transfer, and simply listening to those who lived on the land or cared most about it.

* * *

A good rule of thumb for wildland fire organizations is that their historical horizon is about three years: the year past, the year happening, and the year to come. The rest is ancient history or speculation. But time is relative: its density depends on the number of events, people, fires, reports, homes burned, ideas voiced, and brush cut. The mass of events since 1994, and particularly since 2000, was overwhelming. Time was compressing; the American fire community had to see over and through a thickening cone of time. Halfway through that tumultuous decade, the federal fire agencies paused to reflect on what had happened since the National Fire Plan was adopted and what the future might hold.

What emerged was the first Quadrennial Fire and Fuel Review (QFFR). It was a novel endeavor, "for the first time a unified fire management strategic vision" for the federal fire agencies. It was not policy, it had no roster of recommendations, and it did not project a budget or implementation schedule. It was, rather, an effort to look back over the past five years, sense the current conditions of the scene, and imagine how it all might look in 10 to 20 years. Specifically, it proposed to establish a baseline of data, create an "integrated strategic vision document," and devise a "vision for the future" that was connected to budget realities. It was emphatically not an official encyclical nor subject to agency imprimatur. The Brookings Institution was commissioned to oversee the enterprise. And although the National Fire and Aviation Executive Board approved it, the document did "not purport to represent any official policy or program decision by NFAEB and the federal wildland fire agencies."[61]

So while not a document of record, not pocked with bullet-point recommendations, and not charged with establishing or explaining policy, the QFFR did summarize the metafire analyses of the previous half decade, was stuffed with useful observations and suggestions, and effectively distilled the thinking that the federal agencies were both drawn to and forced into. Its vision of the future was an exercise in persistence forecasting: the near future would look pretty much like today, except worse. Its vision clustered around three "mission strategies." One, the agencies needed to allow fire "to play its natural role in ecosystem sustainability" within the given social, economic, and political context that exists. Two, they needed to treat the WUI through a strategy of resilience rather than resistance. The interfaced and intermixed enclaves needed to evolve into "fire-adapted communities" rather than rely on suppression that was too costly, too lethal, and too ineffective. Three, the American fire community needed to reconcile its land-management charter with mission creep toward all-hazard responses.[62]

These were hardly revolutionary ideas. What the QFFR brought was a clarity of expression, a sifting through data and documents and opinions that threatened to overgrow the forest of strategic imagination into impenetrability. It thinned and burned and opened vistas. What it did not do was address the widening world of fire, particularly that vaster realm of rural and wildland protection done outside the public domain. The agencies were avid to pivot away from the wildland-urban interface; the QFFR told them they could not but then noted ways they might adaptively accommodate those pressures. Not least, by modeling itself on the Department of Defense's Quadrennial Defense Review, the QFFR corroborated a growing move to make the military the exemplar for the future of the workforce that would have to enact whatever policies and programs the agencies adapted or Congress saw fit to fund.

* * *

As it celebrated its centennial, the Forest Service found itself once more, if tentatively, in the vanguard of fire management. In 1905 the USFS, along with the national parks and a handful of powerful states, defined the agenda for fire protection while forestry and the USGS furnished what limited research capabilities were available, and the country grappled with the vexing issue of paying for the services it sought. In 2005 the Forest Service, by virtue of its size and a reinvigorated sense of urgency, found itself leading reforms, this time in fire's management, while natural resource universities, the JFSP, and the USGS assisted its research endeavors. In 1906 the *Use Book* explained how rangers ought to handle fire. In 2005, after a century of rules knotted onto other rules, the agency harked back to its origins by adopting a philosophy of management by principle. Within a handful of years after the Transfer Act granted it responsibility for the national forests, the U.S. Forest Service had devised policies, found funding, developed planning protocols, undergone a trial by fire, and restructured America's pyrogeography. A century later, 50 years after the fire revolution, the USFS found itself eerily recapitulating those events.

Slowly, pulled by new monies and powers, pushed by auditors such as the GAO and OIG, drawn by the opportunities to remake degraded lands. and repulsed by fires of staggering savagery, the American fire community continued to pick up institutional shards from the rubble and glue them into something that, however ungainly, might hold fire.

2006: Where Have All the Firefighters Gone?

Another fire season, another peak, another numbing litany of numbers. The fires of 2006 roared past the old milestones. Record numbers of fires: 96,385. Record acres burned: 8,73,745. Record large fires: 1,801. Record mobilization of engines and helicopters. Record expenditures: $1,925,395,000. The 1994 season had first brushed against a $1 billion invoice. Twelve years later, after the largest buildup in wildland fire since the New Deal, that unimaginable outlay had doubled. If fires went beyond the models of fire-behavior scientists, so their funding ran off the charted models of economists.[63]

Like the U.S. military scrambling to find forces sufficient to its mission, the wildland fire community found it had fewer sources, more sinks, and greater complications. It was still striving to diversify its workforce while that workforce endured more and more scrutiny. If a burn boss flubbed a prescribed fire, if an incident commander had a fatality from a burnover, if a fire burned through $1 million or $5 million, a fire officer might be sued, audited, or jailed. Meanwhile, the size of the permanent federal workforce shrank, the proportion of that workforce involved with fire shriveled, the drive to privatize stimulated a turn to contractors, and the entangling interface put fire professionals on the line with volunteers. The fire community had more to do, fewer people to do it with, and more demands to integrate with nonfederal personnel outside the NWCG and NIFQS system. They had confusing orders that, at base, just wanted them to make the fires go away.

On October 26, a fire set by a serial arsonist in the San Jacinto Mountains, caught by Santa Ana winds in Banning Pass, abruptly exploded like a Molotov cocktail through chaparral and dispersed housing above Cabazon and wiped out a Forest Service engine crew. The five men of Engine 57 who died on the rocky slopes around a vacant home when the Esperanza fire erupted into a bilious fireball were the exurban complement to the four firefighters who had died in a wildland canyon on the Thirtymile fire. That the fire was under CalFire jurisdiction, and that the arsonist was later caught, charged with murder, and sentenced to death row, were but small compensation. These were the real fires the community faced and the real costs it paid.[64]

* * *

The question of a workforce adequate to the task before it had dogged the fire community since 1905. It took on special urgency after the multifatality fires of the 1950s and 1960s and added other concerns as the fire revolution emphasized skills beyond shovel-and-pulaski labor and as the civil rights revolution demanded greater diversity. Civil Service reforms on retirement had stripped the fire community of much of its experience. Scrutiny of plans by NEPA, operations by OSHA, and everything by OMB and Congress complicated the question of what competence meant. WFUs and WUIs were actively redefining the skills the fire community required. The reorganizations of 1998 and the adoption of the National Fire Plan made the issues more pressing.

So among the analyses and policies that spilled out as the new millennium exfoliated were a swirl of inquiries and proposals to upgrade and expand the workforce for fire. The projected hiring would boost the federal forces an average of 28 percent, with the FWS expecting a 19 percent increase and the NPS a whopping 40 percent. It made no sense to enlarge the duties of fire agencies or to embrace fire resources beyond the interagency spectrum if the people were not there to execute those tasks and engage with collaborators. It might seem odd that the community did not have a solid profile of its personnel, but then, until 1998, it really had no consistent data on its fires either. In 1999 the USFS in California conducted a survey and analyses, following a record number of unable-to-fill orders for firefighters, to determine who turned down fire assignments and why. In 2000 the Forest Service and NPS contracted with the Brookings Institution for an expanded survey of workforce demographics. The federal agencies then organized a Federal Fire Training Task Group to devise "a future strategy for workforce development for the next decade" that could serve both federal and state needs. The project merged with the NWCG Training Working Team to produce yet another of the decade's proliferating "cohesive" strategies. Those in wildland fire were unsure what a career path might look like (an old complaint). Newcomers hired through the NFP had to be initiated into the best of the old fire culture and outfitted with the skills needed to enact the requirements of a still-emerging new culture. And there was no general training program to develop "senior and mid-level federal fire officers," or, to appeal to a military analogy, there was no fire officer candidate school.[65]

Meanwhile, the workforce was seeking new allies and being forced to accept auxiliaries with very different qualifications and standards for training. The stresses came from the tectonic shift in America's pyrogeography from the pure wildlands to the interface. The exurban scene fell to states, counties, municipalities, and fire districts and looked to local and volunteer fire departments to furnish first responders. There was no way a federal fire force on the CalFire model could cover the country. Neither was it possible for local authorities to cope with the emerging plague of wildfires. As the rural countryside increasingly morphed into a landscape of exurban amenities, so the ability of locals to cope declined. Retirement communities and summer home residents were not likely to staff VFDs. Yet these agencies were "the nation's first line of defense against fire starts" in the metastasizing interface, and for all the attention lavished on federal fire fatalities, most firefighters died in vehicles and aircraft accidents, and the largest number of those killed on fire duty belonged to local companies who succumbed to medical issues such as heart attacks.[66]

Even in the 1980s the sense among federal fire authorities was that the VFDs were an important social institution and valuable to their immediate communities but not part of a national infrastructure of wildland fire. At great labor the feds found ways to coalesce around interagency standards and eventually around a common policy. But extending that aegis to thousands of local units was daunting, not just technically but politically. However much the cultures of the FWS and BLM might differ, they had far more in common than any affiliation between a BLM district in Idaho and VFDs in Texas. Yet wildfires are best controlled by rapid initial attack, and as fires flared around the interface, VFDs were that forward force.

The evolving strategy built on the 1994 report on *Fire Protection in Rural America* and its 2002 successor, *Needs Assessment of the U.S. Fire Service*, which brought into a bulging tent the U.S. Fire Administration, the NFPA, the International Association of Fire Chiefs, the National Volunteer Fire Council, and the NASF and federal agencies. The feds could not do the job—and did not want to do the job because it pried them away from the land management that was their mission and because the I-zone was gutting their budgets. The only solution was to bolster the capacity of the locals to do the task for themselves. These folks needed assistance. In 1999 the NFPA estimated that the cost of converting the nation's volunteer firefighters to paid status would run between $53 and $74 billion.[67]

That would never happen, certainly not at any rate commensurate with the growth of the WUI. But over 30 years several programs had emerged to aid rural and volunteer fire departments: the Rural Development Act (1972), the Volunteer Fire Assistance program (1975), the Cooperative Forestry Assistance Act (1978), and that old standby, the Federal Excess Personal Property (FEPP) program, which in 2002 had distributed over 70,000 pieces of equipment. The National Fire Plan revived those programs and even expanded them to involve Interior agencies. The 10-Year Comprehensive Strategy incorporated all these schemes. The U.S. Fire Administration (USFA) launched a new grant project, the Assistance to Firefighters Grant, which spilled over into wildland fire. In 2001 it had an appropriation of $100 million; in 2003 it budgeted $750 million. The programs emphasized training and equipping and the hopes that they could transfer more of the interface burden to the communities that in turn would establish Firewise standards for construction, conduct hazard abatement, and handle initial attack.

Still, the two sides of the interface divide had to coordinate if not truly cooperate. They had to operate together on fires, had to be able to communicate on shared radio systems, and had to understand how to run complementary equipment. Most VFDs could not meet federal standards, which exposed shared firefights to liability concerns. And there were always the issues of who had primary authority—who served under whom—during fire busts. The process of reconciliation would take decades.

* * *

The federal agencies were still grappling with how to "establish a unified and cohesive federal management policy" codified in manuals and practice even as policy was evolving through reforms and mutations at a rate that would make a flu virus blush. Agencies could hardly agree on common terms, much less how they might implement the "17 fire policy statements, objectives and management intents" approved by the 2003 implementation strategy. Just what is an "appropriate management response?" What, exactly, are the governing principles for a "wildland fire use for resources benefits" fire?

The NFAEB responded in 2004 by chartering a Directives Task Group to recommend ways to unclog the new regime. The Pulaski Conference perversely added to the confusion by seeking to replace explicit rules, however contradictory, with general doctrines, which had to be interpreted. Much as it sought to transfer primary responsibility for the interface to local entities, the establishment was off-loading strategic decisions to individual fire officers. The method could work, but only if the workforce was up to the task, which meant extensive reeducation—the fire-training program would also need (yet another) cohesive strategy. The National Advanced Fire and Resource Institute (NAFRI) at Tucson, the Prescribed Fire Training Center at Tallahassee, the Fire Use Training Academy at Albuquerque (later folded into NAFRI), along with state and regional fire academies for apprenticeships as a method to bolster minority recruitment, would all become more prominent even as they retooled (almost annually) to cope with the torrent of new personnel, new terms, and new reforms. The Quadrennial Fire and Fuel Review devoted its chapter 4 to "workforce preparedness and development."[68]

The collective ambition was to upgrade the existing workforce, recover personnel lost over the years, and expand to new allies and skills. The Pulaski conferees made an impassioned appeal to recover the sense of commanding importance that fire had once had for the Forest Service—an odd plea since the fire program was gobbling half the agency's budget. But the old assumptions about how fire had percolated through the agency no longer held. In the past, nearly everyone in the organization had some fire experience, and many had used seasonal firefighting as an entry position. A militia of experienced staffers could be called up during fires, and everyone more or less expected to be on call during fire season. That world was history.

Many staffers now viewed fire as a task as disruptive to their work and careers as it was to the agency's budget. They no longer responded to requests for extended service on fires. District rangers and forest supervisors often lacked the on-the-line fire experience their predecessors had acquired simply by passing through the agency. Even within the organization, fewer people understood what their policy declarations actually meant to crews charged with implementing them in the field. The QFFR accepted as "a given" that there would be no net gain in personnel. That meant enlarging the skills existing fire officers had, reactivating the militia, finding allies among sister agencies and in local fire departments, and contracting for services. In various ways, it all happened. And all of it, from the concept of doctrines to the trend toward "a more dedicated professional fire management force," looked to the military for inspiration and models.[69]

This was an old instinct. Federal firefighting had begun with the cavalry in the national parks, and for a while in the 1920s the Forest Service had requested (unsuccessfully) that army units be stationed on its lands to assist with fire control. That ambition ended with the CCC, a civilian "Tree Army," half of whose efforts went to fire protection. Many fire officers in the postwar era had served in the wars or at least in the conscripted ranks afterward. When crews took big casualties in the 1950s, the agency patterned the 10 Standard Firefighting Orders on the Marine Corps standing orders. The belief—the comfortable assumption, readily accepted—that the annual firefight in the backcountry was a moral equivalent to outright warfare came easily and often. Then it died as the military mired in Vietnam and its aftermath and as the fire revolution challenged the unquestioned supremacy of suppression. But it never ceased entirely.

Gradually, it reasserted itself as interface issues shifted attention back to firefighting and doctrines drawn from military analogues. It is here that the deeper metaphor hits pay dirt: the Forest Service, once again, was a microcosm of the country. The postwar fire community had evolved in an era of conscription in which a significant fraction of men had been drafted and the others knew they were in the pool. So it was with wildland fire. A small corps of officers and long-serving NCOs oversaw a continual stream of seasonal firefighters, typically for two to three years. But most of those in the organization, certainly before women staffed offices in large numbers, had some fire experience whether they liked it or not. They remained on call; they constituted a militia. President Jimmy Carter ended the draft in 1980, two years after the Forest Service reformed its fire policy to conclude its singular commitment to suppression, and the workforce, like the post-Vietnam military, began both a long declension and a shift in demographic composition.

By 2006 fire's workforce again resembled America's military. It had a small corps of professionals, too small for the tasks asked of it, and a reservoir of reservists from the agency militia (i.e., National Guard), which it activated at cost to other activities. There was also a growing private sector of security forces (mercenaries) for fire, consultants and contractors, that had the effect of camouflaging the real costs of operations. The general public had less and less personal involvement with the country's wars and its casualties, and even as the citizenry applauded service men and women in principle, it tended to see the conflicts as misguided in concept and perhaps fiscally ruinous. That is not a bad depiction of how the country had come to see its war on fire, or why the Forest Service in particular might turn to the military for examples of how to train. Or why, one might add, the campaign to reestablish the old agency militia was quixotic.

Before 2006 ended, the USFS saw its fire workforce both sharpened by specialization and split from its general mission. The year saw the loss of an engine company crew to an explosive cloud of flame and the partial loss of its most experienced big-fire officers to national emergencies as the incident management system, spawned from Firescope, morphed into a National Incident Management Organization. So flexible and powerful were the principles behind the ICS that teams were dispatched to coordinate the response for other major disasters. In 2003 the ICS program had been absorbed into Homeland Security as part of FEMA's National Response Plan. A year later the Forest Service reorganized its high-level fire teams by creating the National Incident Management Organization, four teams of seven members, assigned to "Command and General Staff" positions. When not on fires or other emergencies, they performed duties under Fire and Aviation Management.[70]

The trend toward a professionalization was also a kind of isolation. The Forest Service bemoaned that as fire had once been an entry job for its own workforce, now the fire corps it trained too often decamped for other agencies such as CalFire or city departments that paid better or that were less encumbered in their field operations. That churn showed the extent to which the wildland fire community was becoming a society unto itself, as members moved among agencies and identified more often with their fire colleagues than with the institutions that hired them. They had their own training, their own funding (bolstered by the NFP), and their own tribal identifications. The push for all-hazard emergency services would isolate them further. Total mobility had succeeded in allowing fire staff to move among agencies, but the drive to "professionalize" the workforce had shut down mobility within agencies. Probably, and paradoxically given the premises behind the fire revolution, their fire staff were less bonded to the agencies than when the reformation began.

* * *

As the insurgency in Iraq deteriorated, President George W. Bush refused the multilayered recommendations of the Baker-Hamilton Commission he had himself established in favor of one phase of that strategy, a troop "surge." Thanks to the Anwar Awakening, the surge helped to quell the worst of the violence. Among those alarmed by the growing unrest in the American West, there was the hope that the country might quiet the uproar, or at least get burning exurbs off the nightly news with a surge of air tankers, engines, and hotshots. Proposals even floated to create a federal fire service to be shared among the agencies, as General Services Administration did vehicles and office furniture. It might expand into an all-purpose emergency program, a kind of Coast Guard for the country's interior. It was a politically appealing idea, although there was no evidence that all-hazard emergency services could manage land any better than land agencies could provide social services after fires blew through communities.

The real revolution waiting to happen in the fire community's workforce was an overturn of generations. The maturation of the baby boom had coincided almost exactly with the 50-year cycle of the fire revolution. That generation had not embodied any particular principle other than bulk. It was neither intrinsically liberal nor conservative on political or cultural matters or on matters relating to fire. The fire revolution had begun with men older than the boomers; its full cycle would be turned by men and women younger. What the boomers did was exaggerate: they added numbers to the polarities, making compromise harder. What wildland fire needed was to redefine what it said it was doing and rewrite its story. That would not come from a generation that had lived through the promises and pitfalls. It had to come from outside. It would happen only when boomers yielded to Gen Xers and millennials.

As the monumental 2006 season segued into a 2007 season nearly as outrageous, the fire establishment continued to tinker with policy, to argue over where to invest in fuels projects, to dispatch summer surges of crews — and to wait for some extraneous event to jolt the system into another phase or just shake it into new directions.

East Meets West, 2007

Fire-planet Earth continued to divide into two grand realms of combustion. One burned surface biomass, as it had for 420 million years. The other burned fossil biomass exhumed from those ancient landscapes. These two realms competed: at any one place and time one or the other dominated. The transition between them could last for decades, however, and in such unsettled circumstances, as always, wildfire flourished.

In the United States the two realms kept parsing into three regions, and 2007 put the outcomes on full display. Although the season's tallies were half a million acres and $80 million shy of the 2006 record, from April to October explosive fires romped over the three apexes of America's institutional fire triangle—the Southeast, the Northern Rockies, and California—challenging strategies of prescribed burning, fuel abatement, appropriate management response, wildland fire use, and outright suppression. In Florida and Georgia wildfires burned 1.4 million acres. They roared over 2.75 million acres in the Northern Rockies. In California, both north and south, they burned over a million acres and took out 3,000 houses. And these were the premier sites of American firepower. Everywhere, save the Northeast, the country's pyrogeography appeared to be unraveling. Wildfire burned 648,000 acres in Oregon, 621,000 in Utah, 215,000 in Washington, and 890,000 in Nevada. For the fourth consecutive year fires had supersized and defied traditional efforts to control them—that, at least, was the prevailing perception.

Yet the story was more nuanced than that crude box score suggests. Over 9.3 million acres had burned as wildfire, but a third of that, in the Northern Rockies, came under complicated concessions to the doctrine of appropriate management response. Nationally, 3.15 million acres burned under prescription, overwhelmingly in the Southeast. And amid all the furor over semicontained wildfires, some 430,529 acres had burned as WFUs. The national total was closer to 13 million acres burned, and this figure did not include agricultural fire, especially in the Great Plains, or TNC's contribution to prescribed fire; the federal agencies were not the only fire institutions for either fire's restoration or its suppression. If there was more bad fire, there was also more good fire. Such trends were not captured by NIFC's tally of fires, acres burned, and dollars spent, nor by the demand for metrics of accountability.

Among the swarm of big events, three—one for each region—stand out as symbolically rich. The Georgia Bay complex, centered in Okefenokee Swamp, dismissed the notion that big fires were an aberration only of the West. The latest Southern California fire siege challenged every sole-cause strategy about coping with the worsening scourge. And the Meriwether fire on the Helena National Forest in Montana showed how much had changed over 40 years, or for that matter, over the 58 years since the Mann Gulch fire had hammered a geodetic marker into the bedrock of the burning West.

* * *

Acre for acre the Southeast had long had more burning than any region outside the Great Plains. Florida stood unchallenged as the premier site for controlled burning in complex landscapes. Tallahassee remained the Mecca for pilgrims wishing to learn the practice. Surrounding states followed Florida's lead.

Which made all the more stunning the outbreak of burning that began on April 16 with a downed power line south of Waycross, Georgia. Next up was a lightning fire on May 5 on Bugaboo Island in Okefenokee National Wildlife Refuge. The two burns merged with still others to become the Georgia Bay Complex. Through May the complex defiantly crunched through the refuge at almost exponential rates of growth. On May 16 burned area reached 119,000 acres; by May 22 it stood at 475,000 acres. Driven by dry winds from subtropical storm Andrea, flames had blasted south—straight as I-95—across the state line into northern Florida. A pall of smoke shrouded the region from metro Atlanta to Lake City, Florida. Twenty-one counties were declared disaster areas. Officials openly described it as a "fire siege."[1]

Georgia and Florida mobilized everything they had and called on both industry and federal forces including incident management teams. Some 3,300 firefighters from 44 states, Canada, and Puerto Rico were on the lines. This was the kind of fire expected in fast-developing California or public-land Montana, not the long-settled Southeast. Yet the same factors that underwrote big burns west of the Mississippi applied. A severe drought and record-defying relative humidity (20 percent) made everything on the land combustible, even the now-drained swamps. And there was much more to burn, thanks to public preserves like the refuge (74 percent of the total), to commercial forest land transferred to developers and exurban colonization, and to a shift from routine landscape firing to little or no burning. Fire management had not progressed at a tempo equal to the kaleidoscopic turning of land usage.

Still, a massive bulldozed firebreak protected Waycross, and fire suppression might have held the stubborn swamp fires had not Andrea, approaching from the east, sent dry wind instead of rain. A second tropical storm, Barry, finally doused the region on June 22 and drove the flames to ground. The Georgia Forestry Commission had a protection budget of $29.2 million; it spent $65.5 million. Overall suppression efforts cost $81.5 million. The crisis sparked no congressional or GAO audit. Opinions differed on whether the outbreak was a fluke or the future. To those inclined toward the apocalyptic, the convergence of changes in global climate, the national economy, and demographics, along with similar breakouts in South Carolina, suggested that the once-sequestered western fire scene was moving to new pastures.

* * *

Another fire siege, another bout of inquiries and commissions — nothing in California seemed to change except to get worse. The 2007 season, climaxed by an 11-day outbreak in October, savaged every postwar landscape in the state. Fires burned through a community in south Lake Tahoe. They burned Santa Catalina Island, gorging on native vegetation restored through the controlled removal of feral goats and invasive plants. They (again) threatened Malibu. They burned through 94,462 acres in the backcountry of the Los Padres National Forest — 70 percent in the San Rafael Wilderness and adjacent roadless areas. Beginning on October 21, with Santa Ana winds avalanching through the mountains, they powered a virtual repeat of the 2003 season, even reburning through many of the still-fresh scars. This time, though, they also ignited mountain forests, flush with conifers and houses. At 197,990 acres, the Witch fire set new records for size. Evacuations affected an estimated 910,000 residents in San Diego County alone. The fires killed 10 people outright and another seven from related causes. They destroyed 3,069 structures. Governor Arnold Schwarzenegger reconvened the Governor's Blue Ribbon Fire Commission — collected the same people — that had reviewed the 2003 fire siege.[2]

The fires testified to the profoundly interactive character of the Southern California firescape. Consider their fuels. The Zaca fire feasted on long-unburned fuels; an estimated 41 percent of the landscape had not burned since 1911 and another 46 percent not for the last 50 years. The Witch fire blasted across landscapes still recovering from fires in 2002, 2003, and 2004. The San Bernardino Mountain fires crunched through landscapes of dying conifers and rising houses, a good number of which had been expensively pretreated, though it was not the woods but the wooden structures that had carried the flames. The fuels altered by previous burns and treatments reduced damages and made some protection of houses possible, but the belief, dating back to the 1970s, that the life cycle of chaparral underwrote the South Coast's big burns and could be countered by prescribed fire was as much a casualty as were the fallen civilians.[3]

So also died the assumption that the "problem" was wind, brush, and houses. The fires burned on mountain summits and coastal plains, through wilderness and WUI, through new and old fuels, through conifers, chaparral, grass, and suburbs. Only one fire had a natural cause: the rest were started by carelessness, arson, or power lines. Southern California Edison paid $60 million in settlement fees for overloaded poles that toppled down in Santa Ana winds and sent fires into Malibu, and other lawsuits were pending for outbreaks in the San Bernardinos. The catastrophe did not originate from one cause or from one template for big fires: it was systemic. It was the product of how Southern Californians lived on their land. Increasingly, it appeared that there was little that could be done to prevent more outbreaks. Legislation promised to improve the scene for future development but left unresolved 50 years of ill-conceived sprawl across a landscape that is to fire what Key West is to hurricanes.

* * *

Because of its 25 million acres, most of it in backcountry, with a serious fraction in 15 wildernesses, the Northern Rockies had become the largest venue for that wizard's brew of evolving strategies to conjure up natural fires—PNFs, WFUs, suppression-by-confinement, and that formal ambiguity of appropriate management response. The name had proved unstable not only because the concept was morphing and its natural settings were shape-shifting, but because at its nucleus was a dilemma that had haunted the fire revolution since its opening manifestos—how to let nature's fires do nature's bidding while not threatening life, property, or the liberty to manage the national estate.

Increasingly, the problem fires were not just blowups but long-duration burns. The longer a fire lingered on the landscape, the more likely it was to blow up, the larger its perimeter if it eventually needed to be attacked, and the more varied its biological outcomes. Such fires were both the hope and horror of wildland fire management because they were more prone to do the ecological work required and to cost far more and pose higher risks to surrounding lands if they bolted beyond their intended domain.

On June 20, 2007, the NFAEB once again tried to clarify what appropriate management response meant. The policy encompassed "all of the response actions necessary to manage a wildfire or wildland fire use event for the duration of the event." A start could be suppressed, monitored, looseherded — but the choice would have to be made at the onset and then held. The next day the Northern Region of the U.S. Forest Service issued its own clarifying "guiding principles." Three weeks later national preparedness levels leaped almost overnight from a manageable two to a top-of-the-scale five as conflagrations ripped across the Sierras, the Cascades, the Great Basin, and of course the Northern Rockies.[4]

It was not possible to suppress them all. Crews, air tankers, and IMTs would be mobilized for one fire only to be dispatched to another while still en route. Of necessity fires in wilderness or roadless areas dropped in priority. The Pattengail fire on the Beaverhead-Deerlodge National Forest burned 79 days until October snows quenched it. Appropriate management response became less a manifesto for reform than a doctrine for triage, or to the minds of critics, for strategic retreat. Yet at least to some fire officers AMR provided a means to segue from a position of management from lack of resources to management by deliberate choice to improve safety, reduce costs, and enhance biotas. A review by the Lessons Learned Center documented a miasma of confusion even within the Forest Service, and more so among the agency, its cooperators, and the public. It was not enough to be given a rack of choices: fire officers needed clear decision guidelines by which to choose and by which to be evaluated.[5]

What they often had was a jumble of options, none of which was definitive. They scrounged for resources, they scrambled for suitable strategies, they improvised tactics, all beneath a Big Sky heavy with fire history. Better than most the Meriwether fire shows the outcomes. It started where the Missouri River slashes through the Big Belt range. It is storied country. Here Lewis and Clark first encountered the Rockies, a historic moment commemorated by the Gates of the Mountains Wilderness, among the first gazetted into the National Wilderness Preservation System. Dams framed both sides of the famous gorge, and around them crowded trails, campgrounds, picnic grounds, Holter Lodge, and the inevitable exurbs. The smoke plume rose within sight of Helena, the state capital.[6]

There were, initially, lots of plumes: the fire bust started on July 17–19 and did not end until September. In the Gates of the Mountains two fires were contained by initial attack while one, what became the Meriwether fire, on the ridge between Meriwether and Coulter canyons, survived despite helicopter dousings. It dropped into the canyon and smoldered. On July 23 it blew up, and it continued to blow up. On August 1 it overran another 10,000 acres. Fire officials declared breathlessly that they were dealing with "fire behavior that is off the charts!" Nature refused to stay in the black box of fire-behavior software or in the inscribed bounds of wilderness. The Meriwether fire moved up the national rankings. A Type 1 IMT was assigned. Evacuations were ordered around both lakes. Crews, engines, and bulldozers prepared to make a stand on the southeast flank along Beaver Creek Road, flanking a subdivision, where providentially fuels had been treated that spring. Before it blasted northward out of the wilderness and ended at 47,000 acres, the Meriwether rose to the number one priority fire in the nation. In that it seemed another refrain to an old song that told of big burns and big firefights in the Northern Rockies.[7]

But this time the words and tune had changed. Press releases might declare that the fire would force researchers to rewrite "the text books of how fire spreads." Instead it was operations manuals that were rewritten, whether intentionally or not. There were far too few resources to fight the fire once it bolted beyond initial attack—the entire region, the nation's suppression reserves, were overwhelmed. On August 1, with the fire engorging another 10,000 acres, the Meriwether had some 122 firefighters. Neither was there will to force a classic firefight through the ridges and ravines of the wilderness. It would be too expensive and too arduous and too damaging and too disrespectful to the values of the legally wild. The only sensible course was to defend communities where possible, to evacuate where protection was uncertain, and to let the fire work its way through the wilderness and other public lands. Whatever the manuals declaimed, this was the new reality on America's western firescapes.[8]

For a thicket of reasons, some sensible, some implausible, prescribed fire had proved too cumbersome in the West to alter conditions on the ground. The region would get its burning through wildfires like the Meriwether. Some patches would burn severely—more than anyone wished for. Other patches would hardly burn at all. And in between the seared and the spared, the creep and sweep of the long-burning fire would dapple the firescape with textured biotic sculptings of the sort the land needed. Nor could the borders of the legally wild contain such blowups. Before it died out, the Meriwether fire burned an area roughly twice the size of the wilderness.

What makes this fire especially noteworthy, however, is that among those blackened scraps of firescape was Mann Gulch, scene of the 1949 tragedy chronicled so vividly by Norman Maclean. This time no one walked or parachuted into the ravine, pulaski in hand. There were no crews strung out along the steep slopes and daydreaming or taking photos. No one subsequently scrutinized the fire's behavior. No one wrote up so much as an official narrative. The fire had scorched over the Gates of the Mountains with no more cultural consequence than had it been a herd of bighorn sheep. Fifty-eight years after the Mann Gulch tragedy, 15 years after *Young Men and Fire*, free-burning flames could pass over one of the American fire community's hallowed sites and no one thought that trying to go mano a mano with it was either prudent or required. What commanded attention were threats to the communities barnacled to its flanks. Such mixed-management fires might, in truth, be the future. The suspicion nagged that what had begun as a necessity had simply become necessary.

In 1967 some 6,767 burned acres in Glacier National Park could spark a national inquiry. Forty years later anything less than 100,000 acres was background noise. In the past dozen years everything in the American fire scene had gone on steroids. But then that was true for the country overall. In sports, Wall Street, politics, foreign policy—the 2000s had so far been a decade of evasion, excess, cheating, and misdirection that culminated in an across-the-board crash. The common image was that one sector or another of American life was poised to go up in flames. (Curiously, commentators of the financial crash behind the Great Recession couched accounts in the language of flare-ups, conflagrations, and brush fires.) For those responsible for managing the national estate, those flames were all too real.

CHAPTER SIX

BURNING OUT

The agencies have recognized that additional, strategic action is needed if they are to get ahead of the fire problem rather than simply react to it, but they have yet to take the bold steps we believe are necessary to implement such a strategic approach.
—GOVERNMENT ACCOUNTABILITY OFFICE, WILDLAND
    FIRE MANAGEMENT (2009)[1]

It's all risk and no reward.
—PAUL BOUCHER, FIRE MANAGEMENT OFFICER,
    GILA NATIONAL FOREST[2]

Recessional

Writing fire history, like fighting fire, requires anchor points. A beginning and an end, organized around a theme, are what transform a chronicle into a narrative. Stories, like asset bubbles, or blowups, cannot continue forever. The point of inflection only becomes apparent once the storyline ends.

After the 2007 season the rapidly steepening arc of a fire narrative could be seen to flex toward a terminus. In 2008, 40 years after the National Park Service formally announced a rupture in the old order, the American fire community commenced another series of inquiries, policy reforms, reorganizations, and memorials that separately and collectively proposed end points to the themes that had organized its narrative over the past half century. By 2012, 50 years after the first Tall Timbers fire conference, those changes brought what casual observers might regard as confinement to the era. The FLAME Act restructured fire financing, a National Cohesive Strategy was reassembling the kaleidoscope of institutions into something like a pattern, the 100th anniversary of the Big Blowup and the 50th anniversary of the Missoula fire lab laid down historical baselines by which to measure accomplishments, and the demographic upheaval in American society led to the reelection of Barack Obama as its first African American president.

Yet the story was not so easy to contain. Anchor points are something people create out of the landscape. Some are more natural than others, the historical equivalent of using lakes and krummholz. Others must be hacked out of the scrub of historical events and defended. As modern fire suppression entered its centennial, as the fire revolution celebrated its golden anniversary, there were plenty of examples of both. Though it might seem like an exercise in historical astrology, some patterns seemed to align amid the numbers; muted rhythms could be heard among the din of fire history. For the Forest Service, "problem fires" had come into focus and then faded in a roughly 20-year cycle. When the fire revolution was announced, it was halfway through an era of mass fire and conflagration control. From 1970 to 1989 the question of fire in wilderness dominated the agenda. Then came the wildland-urban interface. And 20 years later, restoration and resilience claimed the mantle. National policy had orbited in a roughly 30-year cycle. In 1906 the *Use Book* had spelled out how fire should be addressed. In 1935 the 10 a.m. policy was promulgated. By the mid-1960s agencies were ready to decouple that policy from their mission. In 1995 the common federal wildland fire policy was promulgated. Simply within the span of the reforms unleashed in the early 1960s, an oddly decadal pattern could be seen. The NPS had announced its new policy in 1968. The Forest Service had followed in 1978. In 1988 Yellowstone blew up. In 1998 a quiet reorganization identified the pieces that subsequently assembled into the National Fire Plan and its cohort. By 2008 the process was ready to repeat with updates.

There was no intrinsic logic why the numbers should sort themselves in these ways. But each offered ridgelines or roads from which it might be possible to contain the flaming front of events. Thematically, the story had begun with the fragmentation of Forest Service fire hegemony as each agency had pursued a fire program best suited to its charter. Then those agencies had sought accords by which to pool resources and cope with fires along their shared borders. A lost decade followed, as the fire revolution met the Reagan revolution, politics polarized, and reform sagged. In the long decade after the Yellowstone blowout, the American fire community rebuilt; themes rather than separate agencies best defined those events. With the new millennium the tempo quickened, the arc advanced year by year, topic by topic, lurch by lurch, and the heft of events rose as more and more was sucked into a common convective plume that was the American fire scene, and the fire community scrambled to keep up.

By 2010 the community could imagine a kind of thematic containment. If a new round of policy reforms and political reorganizations did not offer genuine closure, they did make it possible to sketch defensible lines in the narrative and to begin the task of blacklining them. One topic after another seemed to reach a point of inflection if not a resolution. The deeper concern, not voiced, was that America did not have a national fire problem that lent itself to a single-themed narrative. It had many fire problems, for which a cohesive narrative might prove as elusive as a cohesive strategy. Besides, narrative transitions were gateways between an endless series of paired fires. There were wildfires and tame fires, nature's fires and humanity's fires, fires that burned in living landscapes and fires that burned fossil biomass. It was not truly possible to contain them all if only because they changed character with their ever-shifting settings. It was only possible to pass between them.

Fire Is Fire, Agencies Are Agencies

The GAO had first inquired about America's fire policy in 1989, after the Yellowstone blowout. In 1999 it surveyed the scene and called for a cohesive strategy to bind the many programs, themes, and agencies together into something more than the sum of their separate parts and into an enterprise that could be held accountable. Ten years later it reinvoked that appeal.

Between those dates it issued over 50 reports on the American fire scene, each in response to a congressional inquiry. In 2009, on its own initiative, it scoped the panorama and penned a summary. The federal agencies, it concluded, had "unquestionably improved their understanding of the nation's wildland fire problem and have positioned themselves to respond to fire more effectively." Yet the conditions they addressed had worsened even faster. However much they might wish to get ahead of that deteriorating arms race, the agencies were struggling simply to react, and however sincere their desire, they were failing to take "the bold steps" necessary to cope with the yet more severe fire seasons to come. The GAO closed its books with "no new recommendations at this time."[3]

In fact, the agencies, particularly the Forest Service, were undergoing another metamorphosis. They were issuing yet more rounds of guidelines for implementing policy, both for wildlands and for the WUI. They were bringing online a fire-planning analysis program that could instill a long-sought comparability in the budgetary requests of disparate institutions. They were working to reduce the transfers of funds among programs to pay for suppression. They were mapping fuels, plotting the cancerous spread of the interface, and devising decision-support tools. And they recognized, as perhaps a supranational organ like the GAO did not, that the country was itself not cohesive, that it jostled together regions, purposes, environments, sensibilities, and traditions that viewed calls for "cohesion" as code for reestablishing by stealth the hegemony they had labored to shatter and loathed to see restored. The country would have to do with fire what it was proving increasingly unable to do with any political issue.

In the tightening gyre of improving capacity and worsening crisis, there was only so much of America's pyrogeography that the agencies could in any way control. Fire was the great synthesizer of all around it. The agencies charged with managing it had to cope with the flames without the capacity to modify what the flames were integrating.

* * *

One aspect the agencies could address was how to interpret policy. This was a further iteration in the shouting match between those who longed for the clarity once offered by the 10 a.m. policy and those who hungered for the flexibility inherent in a doctrine of appropriate management response that was never quite translated into usable rules of engagement. The 2007 guidelines for implementation issued by the NFAEB tried to assuage both sides by demanding that every fire be designated, from its first spark, either for suppression or for "resource benefits." Even the GAO and OIG scorned that compromise, and the Wildland Fire Lessons Learned Center documented the confusions it caused in the Northern Rockies fire bust. If you could treat different parts of a fire differently, why couldn't you treat the fire as a whole differently at different times of its history? After all, nature does not care how the fire started or what label it is given. Fire, so the mantra went, is fire.[4]

The NWCG organized an AMR Task Group, and based on its recommendations, the NFAEB, or Fire Executive Council as it had been renamed, agreed to modify its guidelines. The policy was fine, it concluded; it was the interpretation of spongy, multiuse language that needed sharpening. In the spring of 2008 the WFLC accepted the changes "contingent upon favorable counsel review" and identified sites for pilot programs over the 2008 season. Based on postseason reviews, the guidelines would announce (another) national change in direction. In February 2009 the Fire Executive Council approved the new "Guidance for the Implementation of Federal Wildland Fire Management Policy." It replaced the 2003 document, as amended, and came with a wildland fire decision support system, a training package, a communications plan, and a lexicon to explain good and bad usage and why certain expressions were to be airbrushed out of the scene. One of the casualties was wildland fire use, which now joined the prescribed natural fire and let-burn in the broken tool bin of fire history. Another was appropriate management response, now superseded by appropriate *strategic* response. However fire officers understood the changes, the public might be forgiven for regarding them as hair-splitting scholasticisms arguing obscure points in the theology of fire and land management. Yet those details sifted gospel from heresy, and they affected what happened on the ground.[5]

The updated guidelines addressed the wildland half of the equation. The urban half of the interface was, officials admitted, more complex. The scene involved hundreds, not to say thousands, of entities and jurisdictions; it was, according to officials, the rogue grizzly that was playing havoc with large-fire costs. And because it involved shuttling funds among the federal agencies and other entities, it was destabilizing old relationships. The Forest Service, in particular, wanted clarity about its responsibilities. Federal policy affirmed that protecting structures was the duty of states, tribes, and local departments. But because most of the fires started on public lands, because the agencies had mutual-aid agreements, because they had a duty as a neighbor, and because they had equipment that small exurbs and hamlets lacked and hence a moral obligation to help, the feds were reluctantly drawn across the border in a unseemly update of the old concept of hot pursuit. What the federal agencies wanted was to restrict those duties and liabilities. Already on the books were agreements that the feds could "assist nonfederal entities to protect the exterior of structures." They could remove brush and firewood; they could spray water or drop retardant. But their crews were not trained for structural firefighting, nor did the agencies want to convert a wildland workforce into an all-hazard emergency response service that would divorce it further from land management. They certainly did not want to bear the costs of protecting housing developments and log-slabbed McMansions from fire. Besides, research convincingly demonstrated that the home ignition zone was the critical point of vulnerability.

In 2009 the Forest Service issued new principles for "structure protection" with these concerns in mind. But, as the GAO sardonically observed, those updates did not clarify financial arrangements. The institutions with primary responsibility wanted the feds to pay as much as possible. By now Congress was supplementing up to 80 percent of the funds transferred between programs within agencies, but that still left huge budgetary craters. The same was happening among federal, state, and local entities. The national government was off-loading its decade of deficits onto the agencies, and they in turn were transferring them to governmental bodies lower in the hierarchy. The 2007 and 2008 fire sieges in Southern and Northern California, respectively, stripped $310 million from the state's reserve fund at a time of collapsing economy. Fire might be fire, but dollars were still dollars.[6]

Add those changes together, and they summed to a field strategy of big-box containment that combined burnouts with point protection. In the name of firefighter safety, cost control, and ecological benefits, fire officers would pull back to readily defensible barriers and burn out, while offering limited protection for isolated historic structures and supplemental aid to exurban enclaves. Burned acres increased. If it seemed that federal policy had arrived at a natural end point from the 10 a.m. policy, that could have been said as well in 1978 or 1995. There was no reason to believe this one would be more final than the others. Every policy had to address circumstances, and the circumstances kept changing.

* * *

The GAO's 2009 report was a retrospective summary. One reason the GAO receded from the scene was that Congress that year passed the FLAME Act, which appeared to reform the funding behind wildland fire and contained a provision that required the agencies to concoct what the GAO had long demanded, a "cohesive national strategy." As a review organ, the GAO ceded the podium to the next Quadrennial Fire Review QFR, which projected scenarios for the future.

The new-edition QFR, in fact, introduced little new. In 2005 it had forecast the future by projecting existing trends, which nearly everyone agreed on, and then congratulated itself when those proved accurate. This time it recycled the same topics, the identical themes everyone else cited; it shortened its title by dropping "fuels" in favor of a more integrated and systemic fire program; and it added two projections, that fire would have to merge further with the National Response Framework and that "fire governance" would have to advance significantly. It outlined a "core strategy" that would move beyond appropriate management response to embrace "strategic management response" as a "multi-phased approach to incident management." It wanted an "integrated fuels management portfolio" and a re-envisioned public outreach program enterprise. It projected that emergency response demands would escalate and that fire budgets would be strained. The sum of such adjustments meant a realignment, not just a reorganization. It would require novel technologies, from decision-support software to new-generation large air tankers. While the QFR thought this amounted to "shifts in mission strategy," its observations were not far from what the GAO said that year, or the Congressional Research Service, or most anyone else who was studying, analyzing, pontificating, or blogging about wildland fire.[7]

Most of the QFR's interesting observations were buried in the main text. Supplemental funding represented 27 percent of total suppression expenditures. Fire management was moving from being "event driven" to being "cost driven." The spikiness of annual burning—one year massive, two years well below average—was "part of a construct" of irregular or "asymmetric" fire, thus borrowing a striking if misplaced phrase from the misbegotten war on terror. Most fuel-reduction burning on public lands was being done by wild, not prescribed, fires—this by an order of magnitude. Disaster declarations were replacing emergency suppression funds as supplemental monies for suppression; of 74 state disaster declarations in 2008, 51 involved fire. The core problem of the interface was less the new housing construction (momentarily stalled, in any event, because of the Great Recession) but retrofitting the three decades of ill-designed existing structures. The need for "capital modernization," especially in aerial suppression, might well involve a shift from "lease-based to government-owned" aircraft. Add the complications of American federalism: "movement toward a new federal framework that pushes beyond appearances of federal fire management dominance to true intergovernmental fire management cooperation is paramount." The awkward syntax clearly reflected the tangled politics involved. Unsurprisingly, the QFR assumed all the existing trends would worsen.[8]

In the end, the QFR was less the manifesto of a revived revolution than the minutes of a fire community town hall meeting. It built on a series of specially convened panels and hundreds of interviews. It provided as much of a consensus within the federal agencies as was possible to achieve. But if only so much of the physical geography of America's fire scene lay within the control of the fire community, the same could be said of its political geography. The real power lay with Congress. When in 2009 it passed the largest reorganization of fire finances in 30 years, it followed the recurring recommendations of the GAO and required the agencies to create, within a year, that fata morgana, a cohesive national strategy. At the same time it declined to fully fund its own statute.

For such a project the federal agencies needed to regulate, though not rule, and their capability rested on the "capacity and resolve to operate as an interagency-cooperator fire service working seamlessly across jurisdictional or organizational boundaries." They had to complete the task of joining together in novel ways what the fire revolution had rent asunder. They would have to lead by example.[9]

FLAME and Fortune

Only in Washington, DC, could analysts believe that fire management was driven by policies and monies rather than by events. Without fires there would be no need for an establishment to manage them. Fires kept the political pot boiling; bad fires had more impact than good ones. So it was useful that the 2008 and 2009 seasons knocked between 4 million and 4.5 million burned acres off the average of the two preceding years and that 2010 had the lowest acres burned since 1999. Nature granted a small grace period. Call it a recession in the national economy of fire, not as great as the recession that pummeled the United States, but unlike the one emanating from Wall Street, it brought a temporary relief. And unlike the market meltdown, this one led to an attempt at serious reforms. The 2008 election unleashed a year flush with projects.

The fire scene, however, was spiraling out of control. With the National Fire Plan, Congress had doubled funding for prevention; in reply, fires doubled their acreage burned along with the costs of fighting them. Now Congress prepared to bolster suppression and reduce the ruinous practice of overdraft transfers from other programs. The outcome was, in principle, the most significant reformation in fire financing and organization since the 1978 Cooperative Forestry Act.

* * *

The Federal Land Assistance, Management, and Enhancement (FLAME) Act of 2009 addressed the concerns of the federal agencies for adequate funding and of their critics who wanted, in return, a political realignment and better programmatic auditing. Whether the FLAME Act was a legislative conclusion to over 40 years of reforms or the prologue for a new era was unclear. In reality, it perhaps was both. Or neither.

Congress specified seven elements for action. It wanted the agencies to identify the "most cost-effective means for allocating fire management budget resources." It wanted Interior and Agriculture to reinvest in the nonfire programs that had been burned away by funding transfers. It reaffirmed the doctrine of appropriate management response. It wanted the agencies to assess the level of risks to communities. It demanded that hazard-fuel reduction programs allocate funds based on priorities of risk to communities and ecosystems. It asked them to assess the impact of climate change—a forbidden topic under the Bush administration. And it wanted studies on the effects of invasive species. By and large, these were the themes that had hounded fire management for the past decade. They were topics on which the agencies themselves sought clarification.[10]

The big news was a boost in funding for emergency suppression. Congress dedicated $413 million for the Forest Service and $61 million for Interior, all to cover unanticipated spikes in firefighting. These FLAME Wildland Fire Suppression Reserve Funds would replace annoying annual requests for supplemental monies, and they would eliminate the desperate need to transfer funds from other programs. In addition, the regular suppression budget received a supplement. Altogether, suppression accounts rose to $1.855 billion, an increase of $525 million over the previous fiscal year. It was, as a fact sheet for the appropriations conference noted, "the largest non-emergency increase for wildfire suppression ever provided."[11]

With the new money came calls for further accountability. The fire program analysis software was expected to craft base budgets, adjusted to regional variations, and the wildland fire decision-support system would guide spending on big fires. Research would identify priorities for fuels treatments, anticipate bad fire seasons, and inform communities on hardening their assets and protecting themselves. Fire funding would become part of normal budgeting, not an exercise in crisis hotspotting. That was the scheme, until FLAME became another hostage to paralyzed government and Congress failed to fund to the levels it had announced, and the federal fire workforce continued to decline. Worse, no one had invited wildfire to the budget reconciliation conference. When big fires returned, so did the deficits. In 2014 the former chiefs sent another open letter to Congress.

* * *

It was easy to analyze through software, to simulate events, or to debug programs. It was harder to fly metal-fatigued airplanes to loose-herd a flaming front and to cope with bug-killed forests the size of New England. Money lubricated the gears of operations and sometimes pulled the levers of policy, but victory in the war on fire, like that for the war on cancer, could not be bought. If the agencies suffered a deficit in force with which to apply doctrine, part of the reason was that their paymasters could not agree on what they were supposed to do. The unceasing debate over aircraft distilled that quandary perfectly.

As much as anything air tankers and helicopters had symbolized the postwar mechanization and militarization of fire protection. By 1960 almost all photographic records of fire included retardant spilling out of the belly tanks of retrofitted World War II bombers and submarine hunters.

It became almost a visual cliché, the occasional absence of which suggested that suppression had somehow broken down. Air power was the totem of firepower, which also made it an aerial version of ground-based controversies. As with so much of the fire story, the recent history began in 1994 with fatalities as a C-130 lost a wing, along with three crewmen, fighting a fire near the Tehachapi Mountains in California.

The Forest Service and BLM launched a National Airtanker Study under the auspices of the National Shared Resources Group at NWCG. Already there were concerns that the fleet, operated by private vendors, was badly aging and that any reforms in policy would involve modifications in fire management's most visible expression. The study group issued its phase 1 report in March 1995 and phase 2 in November 1996. The goal was to think strategically about how large air tankers might work for initial attack and large-fire support over the coming 20 years. A review of tactics followed. As part of the 1998 reorganizations, an Interagency Airtanker Board was chartered to coordinate policy and contracts because aircraft were part of a national fleet.[12]

In the megafire year of 2002 a C-130 lost a right wing while making a retardant drop, and on another fire a PB4Y lost a left wing. The C-130 breakup was captured on video and went viral. The fatalities were starting to add up: between 1999 and 2009, 61 people died in fire aviation accidents, more than from all other fireline causes. The resolve to improve firefighter safety, now firmly grounded in the USFS, had to take to the air. The Forest Service and BLM impaneled an independent blue-ribbon commission to investigate "weaknesses and fail points in the current aviation program." Essentially it noted that the program reflected its origins in postwar opportunism (the availability of surplus military aircraft) and the emergence of quasi-autonomous federal fire agencies. Air power was added to suppression; it had not been cultivated to coevolve with it. The Forest Service and Interior agencies contracted for aircraft separately. Over time they had mingled into the semblance of a national fleet, but the current arrangement, according to the report, was unsafe, underfunded, unsustainable, and out of sync with the fast-morphing fire scene. The core concern was with large air tankers of 1,000 to 3,000-gallon capacities which, used by private contractors for fire duty, were not subject to FAA standards and inspections. In 2004 the National Transportation Safety Board sent a letter of concern to both the agencies and the FAA. But the imperative need, the panel affirmed, was "to foster cooperation and collaboration among working-level staffs, contractors, and states to raise the standards." Air operations had to improve and to integrate.[13]

By now suppression costs had become political. Aircraft could account for 20 percent or more of expenditures on fires. Prevailing wisdom assigned air support to initial attack, but more and more expenses were coming from long-duration, high-visibility fires where the practical value of retardants was debatable but its political payoff—air tankers and heavy helicopters working hotspots near the interface—seemed incontestable. Aircraft were transfiguring into aerial fire engines to protect structures. Then safety came to the fore again on August 5, 2008, when a Sikorsky S-61 on contract crashed during takeoff from a mountainous helispot near Weaverville, California, killing nine firefighters and seriously injuring four. Later, a 45-year-old air tanker bound for the same fire bust crashed on takeoff, killing its crew.[14]

An NTSB report blamed Carson Helicopters for intentionally understating the risk and the USFS and FAA for "insufficient oversight." In brief, the episode illustrated the typical American desire for firepower without high expenses, for private over government services, and for safety without government regulation. Forest Service officials regarded the accident as a "watershed event." Whether or not the FAA required compliance with norms for airworthiness, the USFS would. The immediate effect was to slash the national air tanker fleet from 44 (in 2000) to 11 (in 2011). The federal agencies had to reach beyond their usual vendors for emergency help, which further aggravated costs. Demands that contract aviation meet general FAA standards, particularly with ancient planes, further raised expenses. When the USFS suggested that upgrading the fleet would cost $2.5 billion, the formal inquiries began.[15]

What had happened to fire management generally was happening to aircraft specifically. The studies began to pile up; between 1995 and 2013, various committees issued 11. The gist was that airpower seemed self-evidently useful but that the data were inadequate to support those arguments, that various agencies relied on a nominally "national" fleet without a cohesive strategy, and that safety issues could no longer be hidden in slurry. The task forces might have been talking about fuels management or any other aspect of fire. Eventually the GAO stepped in to say just that.[16]

The recent iteration began when the OIG chastised the agency for not making "the best case for acquiring new firefighting aircraft." It wanted better cost-benefit analysis, stronger performance measures, a firmer empirical linkage between money invested in aircraft and better suppression, which is to say, fewer houses burned for fewer dollars. The agency objected to some of the concerns for reasons similar to those advanced in the cost crisis generally. Meaningful metrics were hard to come by: fire did not behave like a business. Aircraft had become indispensable because they did not, originally, have to be purchased, and the only way to continue that supplement was to transfer more from the military or to buy a fleet. No other government relied on private contractors for aerial firefighting as the United States did.[17]

To add to the complications, a controversy over the use of chemical fire retardants in wilderness and around watersheds with endangered species threatened to check the use of air tankers. Fighting fires in wildlands or along the interface edge was not the same as fighting fires in the city, where more and bigger engines could knock down more flames. The aerial fleet was a part of land management. Upgrading air tankers involved more than replacing an old tool with a newer, bigger one; the choice of tools could affect operations, and vice versa. Aircraft had to support larger strategies, and those strategies were inseparable from the general purposes of the agency's stewardship, which meant the Forest Service needed a mix of aerial platforms to do a mix of jobs. Instead, its fleet of heavy air tankers, with an average age of over 50 years, was close to shutting down.[18]

Here was another major expense at a time when suppression costs were under close scrutiny. The agencies agreed that they needed 18 to 28 large air tankers, but that contingent had to be part of a mongrel fleet including helicopters (light, medium, and heavy), observation planes, air tankers of various capacities, and even a Very Large Air Tanker or two (converted DC-10 or Boeing 747). The program could dump dollars as thick as Phos-Chek. Modern aircraft were too expensive for private vendors to purchase and convert, not without large guaranteed contracts, which defeated the arguments for privatizing. In 2012 the Wildfire Suppression Aircraft Transfer Act sought to transfer to the national fleet C-27J aircraft that the Air Force had determined might be in excess of needs, followed by Coast Guard C-130H aircraft in 2013. The Forest Service issued its "Large Air-tanker Modernization Strategy" in February 2012, specifying that the "next generation air tankers" would have a capacity of 3,000 to 5,000 gallons and would be turbine powered. Those contracts were awarded in 2013.

There was no simple solution because there was no single purpose and no cheap substitutes. Instead of breaking a deadlock over fire management and its costs, "modernizing" the fleet added to the stresses. The FLAME Act might help pay for air tanker use, but it would not finance an upgrade; those dollars had to compete with too many other claims even within the fire community. Environmental critics saw a commitment to a renewed fleet as reinforcing what had gone wrong with fire management, while suppression-based critics saw the years of indecision and harping as another blow against the core practice of their profession. The air tanker conundrum was fire management in miniature.

* * *

But if it did not modernize the fleet, FLAME did catalyze further modernizing of the federal fire establishment. As they had after the National Fire Plan, the agencies internally reconfigured. An Office of Wildland Fire Coordination replaced the BLM as Interior's banker. The governance structure for the federal agencies looked to the Wildland Fire Leadership Council for policy and to the Fire Executive Council, staffed with agency directors, for coordination. The NWCG handled implementation, meaning the working groups that helped translate concepts into mutually acceptable practices.

By now the federal fire establishment was adding institutions as agencies did -ologists. NIFC accepted a representative from the U.S. Fire Administration and a liaison from the Defense Department; WFLC included an agent from Homeland Security. A once-cozy world of federal agencies united by a shared commitment to land management had to accept more actors and more interests, many far removed from concerns over ecological integrity. With crises had come money, and with money had come more central control. Not all the agencies were pleased to see their fire autonomy constrained or to see fire funded on its own, which potentially divorced fire operations from an agency's land management mission. Nor were they pleased to put fire officers on a different career path from other managers. The autonomy fire control had once surrendered fire management was reclaiming.[19]

A deeper solution, however, had to look beyond the federal agencies and beyond Capitol Hill. The FLAME Act provided for this prospect by requiring that the secretaries of Agriculture and Interior present, within a year, a plan for a national cohesive strategy for wildland fire management. That task fell to the Wildland Fire Leadership Council, which created a smaller Wildland Fire Executive Council to oversee what promised to be a countrywide and exhausting undertaking. The project would move beyond the federal agencies to include all potential stakeholders and would make those other entities active participants in decisions.

In a nation so riven with exquisitely balanced polarities that it seemed unable to do anything other than prevent the other side from acting, the fire community proposed an omnium gatherum, a veritable constitutional convention, to demonstrate that it could do more than react, if only because fire did not take "No" for an answer.

All Lands, All Hands

The United States would never have a national fire service any more than it would nationalize forestry. The federal government could influence the states by example and grants-in-aid, and it could motivate the private sector with handouts and penalties, but it could not dictate fire policy across the country. Any national program had to involve the states. In 1911 the Weeks Act codified that realization. In aggregate the states had far more land, more people, and more money. It was the states or their agents, counties and fire districts, that had removed fire from the vernacular landscape.

The fire capacity of the states grew faster than that of the federal government. In 1950 the states spent a little more than twice as much on fire protection as the federal fire agencies did. In 1960 state expenditures exceeded the feds by a factor of five and in 1970 by a factor of six. In 2008 the National Fire Protection Association estimated the total cost of fire in America at $362 billion, or 2.5 percent of GDP. The federal wildland fire bill was one half of 1 percent of that figure. The number of houses burned in the interface was less than 1 percent of those burned elsewhere. Most firefighter fatalities came from urban and especially volunteer fire departments. The overreaching fire problem in the United States was the protection of life and property, and that task fell to state and local jurisdictions.[20]

As the interface crowded out other topics, wildland agencies had to reconcile their conception of fire with the notions of others with whom they had to deal across the wildland-urban divide. For the most part these were states or entities that operated under state charters. In many instances a private sector had emerged to supplement those services. But a private sector that wanted fire restored had also blossomed. The federal agencies had to harmonize both those who demanded fire's exclusion in the name of public safety and those who argued for fire's reinstatement in the name of the public service rendered by robust landscapes. An ideal doctrine of "all lands, all hands" required that the lands on one hand be balanced by the hands on the other.

* * *

Thanks to the Forest Service, the National Association of State Foresters had a representative on the NWCG and at NIFC. From the 1998 reforms onward, the states were actively engaged in the evolving infrastructure. The Stewardship Contracts program and National Fire Plan targeted the interface, places for which they had responsibility. The Western Governors Association campaigned for a strong voice in fuels programs, from the NFP to the Healthy Forests Initiative. The 10-year implementation strategy was designed as a collaborative platform. The states saw fire as a threat but also viewed its abatement as a source of jobs and local industry and perhaps as a running insult to the feds. The feds politely returned the gesture by observing that indifferent zoning and building codes as well as greed for development of any kind was putting those communities at risk.

Still, the WUI would burn or survive on the strength of local capabilities. By 2007 surveys had identified 51,612 vulnerable communities, of which 9 percent had community wildfire protection plans and 7 percent had adopted Firewise or equivalent practices. Apart from fuels treatments, federal aid had trained more than 64,000 firefighters from 1,854 such communities and had invested equipment in 11,595. What the Weeks Act had done for state forestry bureaus the NFP and its complement of grants-in-aid to communities were doing for the interface problem. The collaborative approach put federal money and ideas into the hands of locals, now part of a national infrastructure for fire protection. That was the "all hands" half of the doctrine.[21]

The process quickened with the 2008 Farm Bill that amended the Cooperative Forestry Assistance Act. It confirmed in law the redesign of State and Private Forestry in the USFS and required all states and territories to assess forests within their boundaries and "develop strategies to address threats and improve forest health." In fact wildfire was only one of a suite of menaces that included diseased and stunted woods, threatened species, invasive exotics, and a mountain pine beetle epidemic that was reaching the scope of the Laurentide ice sheet. These forest action plans did for the states what Landfire did generally: they allowed for a strategic assessment by which to prioritize needs regardless of ownership. By now the Forest Service had formally entered its revised policies into its *Manual*. The 2009 Collaborative Forest Landscape Restoration Program promised to expand both the scale and the character of coordinated endeavors. What unified command was to the ICS, collaboration was to preparedness projects. Together they wrote the "all lands" portion into law.[22]

But talking civilly did not mean the various parties agreed or that cohesive might mean coherent. As the scale of threat and response ratcheted up, the fundamental differences between state and federal missions endangered the very notion of collaboration as a practical outcome. The states were chartered to protect life and property; the federal agencies had to also practice ecosystem management on the public domain. Like urban departments, the states would be happy if fire vanished from the planet. They no more wanted it around citizens and economic assets than they did a drug-resistant tuberculosis. The feds, however, could not banish fire, and in fact wanted to restore it as fully as possible. So long as each side could operate in its own realm—the feds in remote wilds, the state bureaus amid rural and exurban communities—cooperation was convenient. The burs under the saddle came when the feds no longer picked up the bill for joint firefights and when they wanted to grant fire a larger role on the landscape. The states hated and distrusted anything that smacked of modified suppression or its serial euphemisms. They suspected that the 2009 guidelines for appropriate strategic response were code for fires that would linger and dawdle and eventually explode and crash onto their lands. And in one form or another they wanted the feds to pay for fire.[23]

However long and storied their cooperation, the entities were premised on fundamentally different axioms. They could not be waved away with a cavalier nod that fire knew no boundaries. In truth, fire did know borders, and people knew them even more acutely.

* * *

Consider three maps.

In 2005 the USGS published a map of large fires (over 250 acres) from 1980 to 2003. It contrasted strikingly with the map of forest fires published by the 1880 census. Over the course of a century fires in the northeast and Lake States had plummeted; so had the burned-over lands of the Southeast. Study the map carefully, however, and you will realize that the geography of big fires was largely a geography of public lands. In 2012 two other maps of American pyrogeography appeared. One, drawn from MODIS satellite data, recorded fires in speckled swaths of hotspots. The telegenic western fires that commanded media attention constituted relatively minor features: the dominant fact was a band of burning, like a Milky Way of fiery galaxies, that swept from Florida to Kansas. The two maps hardly looked like they described the same country. Simultaneously, the national Coalition of Prescribed Fire Councils released the first countywide survey of controlled burning. Organized state by state, it offered a political cartography of American fire. It bore some similarities to the MODIS image and some to the USGS map of wildfire, but the trilogy of maps complicated the story of the country's fire revolution and of what preceded it. Geography challenged history. Cartography conflicted with received narrative. Instead of a cohesive strategy, the GAO should have requested a cohesive narrative.

One explanation required only description. In the Southeast and the central Great Plains, prescribed burning—much of it on private lands—remained a vital practice. In the West, where the nation's public domain concentrated, wildland fire dominated. The Southeast had plenty of wildfire, and the West had pockets of prescribed fire, but the realms were distinct. In the Southeast prescribed fire was the preferred means to control wildfire. In the West wildland fire was becoming the primary means to do what prescribed fire could not. The contrasting maps reflected these facts, which in turn expressed two histories. The public lands dominated public discourse, and the arc of its narrative was broadly understood. A true national narrative, however, had to embrace both regions, along with others.

An all-lands, all-hands strategy, that is, had to incorporate the private sector as well as the varied public actors. It had to be intergovernmental, not just interagency; it had to account for and incorporate NGOs, landowners, and volunteer fire departments. The pyric transition, powered by industrial combustion, continued to abrade away burning by the commercial private sector, but it was resisted by a growing archipelago of private holdings, like those of The Nature Conservancy, used to advance public purposes. Meanwhile, private institutions such as Tall Timbers continued to channel public discourse, and a fire-industrial complex of consultancies and contractors sought to channel public monies. Over this swarm there was no controlling authority, nor was one sought.

In 2008 Tall Timbers sponsored a retrospective on the Yellowstone fires. The research station had long left its origins as a brash upstart behind. Like most revolutionary movements it struggled after the death of its charismatic founders; it had tried to do too much with too little, and it had assumed the role of a public agency with a private endowment. For a while it seemed to fail at what it had insisted was the only test that mattered—getting the right fire on the ground—as its home plantation began losing species and even the bobwhite quail faded. Then it righted itself and reestablished its presence as where the action was for those committed to restoring fire. The Yellowstone blowout offered a natural target of opportunity for its revival on the national stage. The fires had appeared close to the halfway mark of the fire revolution—20 years after the Green Book and 20 years before the FLAME Act. Yet when Tall Timbers convened a successor fire conference, it returned to its regional roots with the theme "The Future of Prescribed Fire." When it tried to expand into the northeast with a meeting at Cape Cod on restoring fire regimes in northern temperate ecosystems, the budgetary sequester of 2013 shut it down by denying travel to federal employees. To the extent that Tall Timbers claimed a national reach, it depended on national resources.

By the mid-1990s fire conferences abounded—too many for researchers and practitioners to attend in any one year. By now Tall Timbers was no longer a disruptive rebel but part of the establishment. Its once-startling proclamations had become mainstream. It had traded its insurgent's megaphone for a role as tribal elder. The sheer magnitude of its activities testified both to its continued vitality and to the encumbrances of adulthood. When Tall Timbers Research held its first board of trustees meeting in 1959, the minutes filled two pages. Fifty years later they ran to 224. Such was the burden of maturity.

There were others to pick up the torch, such as the Jones Ecological Research Center at Ichauway Plantation in southern Georgia. While Tall Timbers had always taken fire facts on the ground as the true measure of understanding, its own holdings were modest—4,000 acres and access to the 200-acre Wade Tract of old-growth longleaf. By contrast, The Nature Conservancy owned, or controlled through conservation easements, 119 million acres of land and 5,000 miles of rivers worldwide, all of which it actively managed; it controlled 43 percent more land than the National Park Service. It had over a million members. It held $5.64 billion worth of assets. While it could not match the Forest Service in workforce or funds to spend on collaborative endeavors, it would trump it in moral authority and the ability to muster diverse groups to a single purpose. There were few places among the 120 million acres in the United States that it estimated needed to have fire restored in which it was not a presence or at least an oracle.

And it actively managed fire. In 1962 it burned 20 acres. Fifty years later it burned 130,000. Since Yellowstone's year of fire, TNC had burned an aggregate of 1.5 million acres in more than 1,000 places. It had pioneered the use of volunteers, had found ways to amplify small numbers into large outcomes, and had served as a facilitator where public agencies were suspect. It catalyzed the fire learning networks. It oversaw ecosystem management, including fire, on many military bases. In Florida it was a bona fide member of the fire infrastructure, a partner even in suppression. Its global fire initiative had been shut down by the Great Recession, but the Conservancy had reinstated a version in cooperation with the International Union for the Conservation of Nature, the World Wildlife Fund, and the University of California, Berkeley, Center for Fire Research and Outreach.[24]

Still, no single organization could encompass the whole. The United States had an estimated billion burnable acres, some of which needed fire, some of which tolerated it, but to most of which fire would come at some time or other. As with the I-zone, the responsibility had to rest with individual landowners. But those who wanted to burn faced mounting resistance.

Individually, they could be snuffed out, one by one. The solution was to organize into voluntary cooperatives to help one another and to create a political voice. The model came from Florida (naturally) as interested parties self-organized into prescribed fire councils. The idea advanced to the state and then, in a calculated campaign, went national with assistance from the Jones Center. In 2007 the movement had sufficient momentum to form a national Coalition of Prescribed Fire Councils (CPFC). Two years later it incorporated as a nonprofit. The CPFC campaigned for legislation to guarantee a right to burn, though it found that the greatest check was sheer capacity: anything beyond a garden plot meant that fire required a landscape to operate over and neighbors to assist. The CPFC could midwife such cooperatives, furnish a template for organizing, lobby nationally, and mesh scattered interests into a functioning whole. By 2011 the movement had spread to nearly every state and was spilling over the border into Canada.[25]

In 2012, working through the NASF, the Coalition assembled the first national survey of controlled burning since 1880, a kind of statistical CAT scan of how, where, and why fire was used. Nationally, 2012 was close to a record year for wildfire—9.3 million acres. But the CPFC found that over 20 million acres had burned under prescription, the vast majority by private landowners. Despite the ever-more-restrictive constraints on open fire, more land had burned through controlled fire than with wildfire. Some 60 percent of that burning was for agriculture and 40 percent for forestry. The greatest concentration was in Florida and Georgia and along the Flint Hills of Oklahoma and Kansas. The councils were the fire-use obverse of the fire-suppressing volunteer fire departments.[26]

What had changed over the past century was the context for burning. In 1880 a landowner had to answer to no one other than his immediate neighbors, who probably also wanted to burn. Today, any burn involved a "complex web of policy, legal statutes, and liability, as well as public safety, health, and acceptance." To apply the torch was no longer strictly an individual choice. Anyone's failure affected everyone. Only "the combined problem-solving efforts of the entire fire community" could halt the erosion of what had once been an unchallenged right to fire. Yet if fire's management was no longer a government monopoly, a civil society was not robust enough to carry the torch alone.[27]

Where the private sector really flourished was in the fire-industrial complex. A small stream of consultants, contractors, and vendors, begun in earnest in the 1980s, had become a torrent. Some wealthy communities even hired private fire protection services, as they did security. The more exclusive interface was becoming fire-gated. When the shuffle between public and private sectors began, the assumption was that the costs would decline because of market pressures. Instead they inflated. The issue was not who did the work but who paid for it and why. Not everyone in the private sector wanted the right to burn; most simply wanted not to burn, and almost everyone wanted someone else to pay for the service.

Cohesive Strategies, Coherent Narratives

For many observers, the issue was policy—getting the right language or the right mix of incentives. For others, it was money. But both policies and money were only chips in the game—emblems of the real movers and shakers. The action was not policy but politics. Money was how politics mostly conveyed its decisions, which is to say, how it allocated the power to decide who would get and who would give.

On October 6–7, 2008, as the country headed into the final flurry of electoral campaigning, one of the most remarkable gatherings in American fire history convened in Emmitsburg, Pennsylvania. The Emmitsburg 13, as they styled themselves, recognized that the clash of values at the nuclear core of fire management could be resolved only by an overt bid to reform the political dynamics that ultimately underwrote decisions. Governance issues went far beyond collegial talk shops like the NWCG or superboards like the WFLC or cobbled-together collaborations like the 10-year strategy for implementation. What was needed was a fire constitution that could define rights, roles, and responsibilities. As wildland fires burst out of the backcountry, as sprawl intruded where it was not welcome, and as competing entities quarreled over what the public domain should be and what kind of fire might be tolerated on private lands, it was no longer enough to tinker with policy or tweak existing institutions. The fire revolution had spawned a loose confederation unable to cope with the common problems all members shared. The politics of confederation needed to yield to a federal system that balanced the needs of all. The participants spoke openly of drafting a kind of Federalist Papers for fire.

When the FLAME Act mandated a national cohesive strategy, a decade of discussions converged. There was the GAO's recurrent call for cohesion among the federal agencies, the QFR's futuristic projections about the two sides of the interface fusing, and the internal conversations within the Forest Service about establishing doctrines, and the demands of the Western Governors Association and the National Association of State Foresters to be integral collaborators. The feds, it was said, did not appreciate the mission of the states. The states, it was rebutted, did not understand the mixed missions of the feds. The old relationships based on cooperative agreements among fire agencies could not stand the strains of megafires, mixed purposes, and divided government.

The Emmitsburg meeting began to look like the inauguration of a self-styled constitutional convention for wildland and rural fire. A national cohesive strategy might assume many forms, but unless it had political legitimacy among all the participants, it was just another policy on paper. What emerged reflected a deeper appreciation that a cohesive strategy had to arise from a complicated, messy, laborious, compromise-laden national conversation about how the American fire scene should work.

* * *

The task went to the Wildland Fire Leadership Council, originally formed to oversee the National Fire Plan, which then charged the Wildland Fire Executive Council with oversight. The National Cohesive Wildland Fire Management Strategy was intended as a more comprehensive update that would not only look at fuels and firefighting capacity but would also synthesize all the studies that had sprouted like Morrell mushrooms after that flush of funding. It adopted as its vision, "To safely and effectively extinguish fire when needed; use fire where allowable; manage our natural resources; and as a nation, to live with fire."[28]

If that did not seem very provocative, it was only because 50 years of unremitting attempts at reform—half of fire's American century—had made such declarations seem like nostrums. Yet they were the hard residue left in the cauldron after five decades of boiling. Fire protection would survive, of course, but it would not compromise firefighter safety. Fire's use was given equal billing; it was not to be simply tolerated but to be loosed wherever "allowed." And all fire strategies would align with land management, expressed in neutral terms as "managing natural resources." Fire's management did not exist apart from its landed context.

The country would have to live with fire, not try to abolish it. A strategy of resistance had yielded to one of restoration that further yielded to one of resilience. Fire would not go away, nor did all but a few members of the wildland fire community wish it to. If the vision fell short of the pyromanticism of the early revolutionaries, if in practice it failed to measure up to even the compromised ideals that had replaced utopian hopes, it was a far cry from the simple dictum of the 10 a.m. policy. Like the country itself, America's fires had become more complex, more diverse, and far, far more powerful.

The project had three phases. Phase 1 could fashion a "national strategic framework" through elaborate forums held across the country. This spawned "A Path Toward a Cohesive Wildland Fire Strategy," vetted by the WFLC. After consultation with the Office of Management and Budget, it was submitted to Congress by the November 2010 deadline specified by the FLAME Act. In the end, two documents survived, the mandated report to Congress and a general National Cohesive Wildland Fire Management Strategy for the public. Phase 2 assessed the national scene according to three regions—Southeast, Northeast, and West, each with its own committee—and with the support of a National Science Advisory Team, intended to translate the regional goals into quantitative metrics. Its national report was released in June 2012. Phase 3 would outline actual implementation guidelines, with 5-year reviews and updates to Congress. The cohesive strategy effectively replaced the expired 10-year collaborative strategy.[29]

What did the cohesive strategy say? In keeping with the fire community's ancient rule of three, the cohesive strategy identified a trilogy of themes. One was restoring and maintaining resilient landscapes. Here was the beating heart of the reformation: to restore fire. But it also subsumed all the issues of ecosystem health, legacy fuels, and species losses and invasions that complicated fire's reintroduction, because fire needed a suitable habitat to do the biological work expected. The second theme was to create fire-adapted communities. This addressed the WUI counterreformation, and the cautious phrasing sought to shift the burden of protection from fighting fires in the adjacent wildlands to hardening the home ignition zone and from having public agencies do what individual landowners and local communities needed to do for themselves. The fire agencies could assist and catalyze: they could not substitute. The third theme was responding to wildfires. The phrasing was intended to show that public safety ruled, that fire agencies needed to fight fires better, smarter, and cheaper. But it really addressed the whole issue of how to apply strategy in the field. It was about the nature of the workforce and how it should be used. It considered "the full spectrum of fire management activities," recognizing that local, state, tribal, and federal agencies had "differences in missions." And it sought "collaborative developed methodologies" to advance the cause, even when that might mean bonding with all-hazard emergency services. Risk analysis would be the common idiom of conversation.

The project was heavy on process—top heavy, to the point of capsizing, many thought. But the process mattered more than the particulars, which were sure to change after endless reviews by OMB, Congress, and local collaborators. Despite the elaborate science nominally underpinning the assessments, this was a political project, and it had to be seen as legitimate. For the cohesive strategy to survive, it had to craft a national constituency for fire management along with a mechanism by which big agencies and small could meet and agree. It had to balance places with high-value assets and places suffused with intangibles. And it had to reconcile values for which there might be no apparent compromise. The old articles of fire's confederation had failed. Whether its proposed constitution-styled replacement could succeed was the challenge of the postrevolutionary generation.

The three strategic themes of the cohesive strategy were a fair assessment of the American fire scene. The hope was that the putative science of risk assessment would furnish a neutral basis for a national comparison of threats and needs and that politics would decide accordingly. There was little reason to believe this scenario would unfold as conceived. Risk assessment had failed spectacularly in Wall Street; why it might succeed in the even more chaotic context of wildlands marked the triumph of hope over experience.

But then politics, like fire, has its own logic. The American system works on the belief of abundance, that there is no need to ration or redistribute because there is enough on the table for everyone. When that assumption fails, the system appears cruel and unfair, and it unravels. The cohesive strategy added no new money—it could in fact be seen as a means to avoid adding money. The collective pain from wildfire would have to be very high to overcome that barrier and keep the players from squabbling and the wealthiest from simply seceding.[30]

* * *

By now the mirage that a single strategy or a single organization might encompass the whole of the American fire scene should have vanished. The nominally national cohesive strategy identified and off-loaded its charges to three regions (which eerily echoed a map of Civil War America). Each had its own character and priorities, although entities such as NIFC or NWCG could arrange for transfers. The Northeast had limited federal lands and little natural basis for fire, and thus it emphasized suppression. The Southeast, controlled predominantly by private landowners, was flush with fire and accepted prescribed burning as a core doctrine. The West, shaped by federal ownership, had a schizophrenic split between urban oases and wildlands, and it needed to reconcile suppression with natural fires. The cohesive strategy mapped a political cartography as well as a pyric one.

The West was the battleground. Here three strategies competed. One is based on resistance, one on restoration, and one on resilience. The regressive aimed to reinstate the old order. The proactive sought to get ahead of the crisis. The reactive accepted the crisis as a new normal. All were in play, like a game of rock-scissors-paper.

Begin with the resistance. This is a chimera of excluding fire, typically by suppressing it. The strategy worked in cities and places without any natural basis for burning. But where fire was integral to land management, the American fire community had learned that suppression failed ecologically, failed economically, and failed as a philosophy of land use. The capacity to fight bad fires was essential, but a firefight resembled a declaration of martial law, a callout to put down a temporary riot, not a principle by which to govern. Suppression could check bad fires (up to a point), but it did nothing to promote good ones. In naturally fire-prone places, it only destabilized landscapes. Still, it had its partisans in people who had grown up with it, treasured the clarity of its charge, and welcomed the public acclaim that made everyone in Nomex a hero. Politically, if paradoxically, it welcomed crises, which kept suppression on something like a permanent war footing and made opposition awkward.

The restoration strategy recognized that the way to cope with fire is to modify the surroundings that sustain it. It assumes effective fire management is land management: good landscapes yield good fires, and messed-up landscapes lead to messed-up fires. The strategy sought to get ahead of the crisis by creating more fire-resilient settings and communities. By selective thinning and prescribed burning it should be possible to promote the fires you want and prevent those you do not. It sought the same for communities at risk by emphasizing programs like Firewise. It bridged political boundaries through consensus projects between public and private entities. Its attraction is that the approach is rational and hopeful. The difficulty, of course, is that the approach is costly, not only in money but in political capital, and it takes patience. It took over a century to create the crisis; it could take decades to unwind it. But political will, dollars, and tolerance appear lacking. The nation's fire deficit seems less compelling than its fiscal deficit. And Congress (along with GAO and OMB) wondered aloud why the billions spent on fuel reduction had not seen the promised reduction in burned area or suppression costs. Outside a handful of states in the Southeast, prescribed fire lacks a legislative charter. Besides, if success comes it will occur on someone else's watch.

The resilience strategy accepts that the current crisis is unfixable, that events are moving faster and farther than the responses we can muster.

We can neither suppress the problem nor get ahead of it. Instead, we must respond to existing firescapes and try to contrive some useful effects out of them. In this strategy we take the fires we get, whether from lightning, careless campers, or arsonists, and work with them, even expanding them, in consideration of safety, cost, and ecological benefits. We exploit wildfires to do the work that, in the West, prescribed fire has been unable to do. We refuse to plug the gaps in the nation's fire project by throwing crews into the breach.

Instead it appeals to the doctrine of appropriate strategic response, accented in recent policy guidelines, to work with whatever ignition occurs. In some cases this may mean rapid attack and suppression; in others, point protection and a fallback to community borders. In still other situations it may mean a big-box approach in which crews retire to defensible fuelbreaks and burn out from there. The upshot is that parts will burn outside the ecological conditions to which the biota has adapted and then face tough recoveries. Other patches will burn hardly at all. A substantial swath may do more or less what a prescribed fire would have done. Unless the fire goes completely rogue, however, it should cost less, pose fewer risks, and absolve agencies of liability. That mottled landscape would become the matrix for a new regime. It would inscribe the metes and bounds for future burning. Prescribed fire would take the form of burning out operations and of fire salvage burning amid the blowouts and blotches left by quasi-managed wildfire.

* * *

A similar concern applies to narrative. A chronicle or annals is a tally of events, which might be no more than a year-by-year record. A narrative has a clear beginning and end and a thematic arc by which to span them. If the cohesive strategy is to succeed, it needs not only political legitimacy but a story that can bring together its roiling, quarreling parts, all of them informed by different values and points of view, into a coherent whole that can explain why the project makes sense. A cohesive strategy needs a coherent narrative.

There is good reason to consider the coherent strategy as a logical and aesthetic terminus, an anchor point from which to build narrative. It comes 100 years after the Weeks Act; at its chronological middle stands the inauguration of the fire revolution. It should be possible, and would seem almost self-evident, to join those end points with a narrative that could invest events with meaning and moral significance, that could separate what matters from what does not, that would make the past usable. In truth, many narratives are possible, like the constellations figured out of the stars that assist navigation.

What kind of narrative makes sense of this era? If we consider only the 50 years from the first Tall Timbers fire conference to the cohesive strategy, the simple story is one in which a singular policy applied by a hegemonic institution breaks down, fractures into a pluralism of institutions, purposes, and practices, and then gradually reassembles itself into a new whole. This assumes that the cohesive strategy is, in fact, a workable end point. Even so, two competing narratives prevail in connecting the dots from 1911 to 2011. Each interprets the story with a different outcome, and each implies a different trajectory for the future.

The first sees progress. The Forest Service put into place a fire-protection program for wildlands where none had existed. With great fortitude it developed that program, and when its hegemony began to self-destruct, the other assorted fire agencies reconstructed the project along more stable lines and began to reintroduce fire with the same conviction they had mustered to take fire out early in the century. Progress was fitful and at times errant but ever present.

So while fire management remains a work in progress, the indexes of its success are everywhere. In 1911 the pulaski tool did not exist; in 2011 fire officers were flying retardant-laden S-64s and DC-10s. In 1910 fire science consisted of maps of burned area sketched by the USGS and F. E. Clements's study of lodgepole burn forests. In 2011 whole libraries groaned under the weight of publications, dedicated labs flourished, and fire science was busy not only tracking ongoing fires but also projecting hypothetical fires over the coming decades. In 1910 rangers used stumps for lookouts and they reported fires by ground-return telephone (if they were lucky enough to be near a line). In 2011 fire officers employed nearly real-time satellite mapping, and virtually everyone had a radio or a cell phone or both. During the Big Blowup fire leadership meant William Greeley driving out to meet rangers in a buckboard. In 2011 a succession of complex IMTs cycled through the burns. At almost $1 million, some 22 percent of the agency's budget, the Big Blowup drove the Forest Service to the verge of bankruptcy. The 2011 Wallow fire alone cost $110 million, a figure nearly lost amid the serial firefights of the season, yet it constituted only 2 percent of the agency's budget that year. By any technical measure fire management was more sophisticated, more competent, better financed, and better staffed.

That large fires still burned was the outcome of exogenous forces—climate change, exurban developments, invasive pyrophytes, and a legacy of fuels. That the buildup of combustibles was the by-product of misguided zeal in the early years of fire protection was regrettable, but the error had been identified and addressed, and only an apathetic public and a laggard Congress prevented full restoration. Fire was coming back. The agencies had made a wrenching midcourse correction—that was how science worked and how a science-informed fire agency should behave. The future promised more challenges. The agencies would continue to meet them.

This is the perspective favored by most practitioners who want honor for what they have achieved and hope for what they must do in the future. It is a story that encourages. Yet it is one that places its bets on technique and technology when the real sticking points are political and cultural. Technology progresses; culture just changes.

The second narrative views the same scene but through a prism of ironic declension. All that technology, all that research, all those crews in Nomex, all those incident management teams—they have added enormously to the complexity and cost of fire management without seriously improving the quality of fire on the land. Instead, after a century of bottomless investment in money and far too many lives, we have record fires, unhinged ecosystems, and ballistic costs. What partisans see as apparent progress is not merely an illusion; it is a delusion, perhaps deadly.

In this perspective the century repeats the same story twice, with slightly different refrains. The first narrates an ironic tale of fire exclusion that ends with an ecological insurgency and the recognition that the project was misguided from its origins. The second tells an equally ironic tale of fire restoration in which, out of zealotry, failures are overlooked or dismissed with the same insouciance as in the era of fire suppression. After several decades—as many as fire suppression had to itself—the numbers show nothing like the returns promised. Outside the Southeast, prescribed fire is a niche enterprise, the fire community's equivalent of hobby farming. Letting natural fires ramble in wilderness works wonderfully until it fails, perhaps with exorbitant damages and costs. Not a few of the country's most expensive fires have been prescribed (or prescribed natural) fires that went feral. There are places where reintroducing fire—even in naturally fire-prone landscapes—worsens ecological integrity. We wanted more fire and should not be surprised if it has come in the guise of megafire by encouraging invasives. The only predictable consequence of actions is the unintended one. Besides, the deep driver of fire on Earth is not surface burning at all but the combustion of fossil biomass—this is the expression of anthropogenic fire that matters most, and it is nowhere in the agendas of fire research.

There is every reason to think another reformation will appear, and that it will end as its predecessors have. There is no point in appealing to science because science has demonstrated little capacity to improve the scene: the true fundamentals of fire are not scientific questions. They involve values, social norms, and cultural expectations as synthesized by politics. They testify to what William Faulkner called "the human heart in conflict with itself." The era of fire restoration is ending the same way its progenitor did, in irony. The history of wildland fire is no different from any other history from this era; an honest rendering would display that fact, not wave it away as unwonted negativism. The agencies are getting out the burn the same way they once got out the cut. The acres come from escape fires, and by amassing small burns into vast complexes, and by letting fires wander over landscapes while seeking to harden assets. This is a hazardous project, likely an expensive one, and a scheme founded on a faith-based ecology. No one truly knows what it will yield, only that it will not restore burning to anything like its presettlement state.

Such is the interpretation favored by intellectuals and critics, for irony has long been the official voice of modernism and is itself seemingly immune to criticism. It is a perspective skeptical of reforms, particularly promises of what will come from more research and technology. Yet while such critics can problematize, they struggle to contribute to problem solving.

But of course the end point that makes either arc possible does not exist in nature, only in the mind. The cohesive strategy could begin a reorganization so fundamental that it shapes everything that comes after it. Or it could be National Fire Plan version 2.0, or Fire Revolution 3.0, or it could be updated before the first pulaski hits the ground. It might be so savaged with compromises and exceptions by Congress that it ends up as a kind of Obamacare for wildlands. Or it might just be part of a flurry of studies, proposals, and legislation that has little effect because wildfire is reclaiming everything that has not been cleaned of combustibles and has sucked those loose papers up into its convective plume.

Considered as a political conundrum, the cohesive strategy resembles the Euro crisis that was unfolding at the same time. For both, three options exist. One is to fully integrate, to find ways to share common banes and blessings. One is to break up with the understanding that the differences among players are too extreme for them ever to reconcile. What does Finland have in common with Greece? What do CalFire and the Central Florida Prescribed Fire Council share? The third option is to plaster over the cracks with bottomless bailouts. It is far from clear how the United States will move. The likeliest choice is to do a little of each—advance cohesion where it is painless, cut separate deals that permit select actors to pursue their own agendas, and with a good bit of public bluster pay enough to keep the system from collapsing but not enough to fix it. All such outcomes are possible.

But aesthetic closure is different: the ability to seal off the seamless web of experience is what makes for art. For a cultural fire history the cohesive strategy as a defined project conveys a satisfying sense of thematic and literary completion. It has the look of symmetry and the feel of closure. It is a defensible space from which to hold a narrative and burn out.

A Billion Burnable Acres

Its fires drive the system—the fires that happen and those that do not but should. The enduring problem is to protect against the fires the country does not want and to promote those it does, or should. By 2010 wildfire was no longer a strictly western concern or prescribed fire a southern one. The U.S. Forest Service estimated that America held a billion burnable acres at risk from too little or too much fire. Its pyrogeography was a metastable alloy of restorative and ruinous burns.

But what drove the fires? The emerging consensus began with global change and legacy landscapes, or with factors pulling toward an uncertain future and with factors pushing from a now-lamented past. Among global change considerations were physical factors of climate that lengthened fire seasons, raised temperatures, lowered humidity, and encouraged extreme droughts and winds. Whatever balancing occurred, the trend seemed always to point toward harsher, longer, and more damaging settings for burning. These conditions were aided and abetted by biological ones in which indigenous species were lost or suppressed and exotics were liberated by a global circuit of diseases, by fragmenting habitats, and by plagues of insects—all of which made for brittle biotas less able to resist other insults. Legacy landscapes were the broken heritage of that process. Some reflected conditions that had built up for decades or over the course of a century, manifest particularly as an unprecedented hoard of combustible fuels. That fuel dump, cluttering over landscapes like forests lopped and left, encouraged a quantum leap in fire intensity. The bequeathed ecosystems lost the ability to survive the savage new regimes that resulted or to accommodate fire's sought-for restoration.

Behind these unintended consequences lay the unwinding mainspring of past policies and practices. Those landscapes integrated everything that had been or had not been done, and through them, so did fire. America's pyrogeography was not simply the outcome of physical factors even as integrated by biological systems. It reflected an evolution of past fires, fire regimes, and fire practices. The geography of American fire was a historical geography, and what fire scientists liked to call the "driver" were institutions, for this is how Americans had tried to broker the interaction between nature and culture. Its institutions had shaped how people had engaged fire to transform physical geography into social disaster, and they would shape how the country proposed to respond.

A century after the Big Blowup, a half century after Tall Timbers first convened a fire conference, the American fire scene remained in some respects remarkably stable. The same regions continued to burn, and the big fires broke out in the same places under comparable conditions. But that pyrogeography was undergoing a phase change in the character of its fires, in their regional distribution, and in the explanation of how and why this tipping point had been crossed. Fires swarmed, almost like bees seeking a new hive.

* * *

The new normal was partly an artifact of counting. In 1998 the federal agencies began recording the full spectrum of burning on public lands (oddly, the burned acreage from wildfire was that year the lowest in decades). A single outbreak could warp the national figures; some 80 percent of the 8.1 million acres recorded in 2004, for example, burned in Alaska. Lumping together all fires on private, state, and federal lands did not identify those distinctions.

In part, too, the numbers also reflected the persistence of old patterns. From 1970 to 2007, 12 of the 15 most catastrophic fires, as measured by insured losses, occurred in California. The 2007 season saw yet another Southern California fire siege ignited by downed power lines, arson, and accident. The 2008 season was dominated by a landscape-cloaking bust of lightning fires in Northern California, and as that episode demonstrated, even nominally mild seasons could have nationally consequential fires. The 2009 Station fire that swept the San Gabriel Mountains was the largest on record for the Angeles National Forest and prompted a congressional and GAO inquiry. What was most fascinating was that it did not occur with Santa Ana winds pushing flames down the slopes: extended burning in steep terrain and dense fuels was ample. The same could also be different.

That observation points to two trends that were redefining the fundamentals. One was the multicausal emergence of megafires. Nationally, some 97–98 percent of all fires were caught during initial attack. But the fires that escaped were less likely to be arranged along the old distribution curves. The big fires became really big. The other trend was a shift in the nation's pyrogeography. California, Florida, the Northern Rockies—these continued to define the primary scene, supplemented by the Flint Hills for agricultural burning and Alaska for free-ranging fire. But formerly quiescent regions began to creep and then explode onto the national map. Large fires rode wild in Texas and Oklahoma, previously considered immune as swaths of small towns and suburbs burned, and wildfire challenged tornadoes for the number of houses destroyed. Texas, which had seemed self-contained, as removed from the national wildfire infrastructure as from the national electricity grid, suddenly found itself claiming more and more of the nation's fire resources. In 2011 it alone experienced nearly 28,000 fires across almost 4 million acres. So, too, wildfires came to Georgia and South Carolina, spreading along the coastal plains, not yet ready to rival hurricanes for damage but conjuring up a specter like pellagra that residents had believed had vanished. Big burns came to a Southwest better known for extensive snag fires; every few years announced a new "largest fire on record." And perhaps most spectacularly, savage fires struck Utah and Colorado. These were mountain landscapes and front-range countrysides that had experienced fire only episodically. They had long been the unburned hole in the nation's cartographic donut.

The historic pattern in Colorado had been an occasional eruptive fire driven by stiff winds (like chinooks) that lasted one to two days. By the time suppression could muster its engines and crews, the fires had expired. But Colorado captured two movements that were redrawing the inherited maps. One was the outcome of excluding fire from fire-adapted ecosystems, whether those regimes had favored frequent surface burning, of which the ponderosa pine savannas that flourished along the front range were prime examples, or infrequent crown fires, exemplified by the lodgepole pine of its higher elevations. Instead of burning every five years or so, or flashing through crowns every century, fire had been shut down; the montane landscapes had missed perhaps ten or more cycles. They became hosts for every affliction of global change, thickening into often sickly woods and scrub, ideal again for mountain pine beetles and high-intensity fires.

The second movement was sprawl. The montane slopes, mingling patchy public and private lands, some protected, some developed, and some vacated, proved just as attractive for dispersed exurbanites as for beetles. By 2010 some estimates put the percentage of Colorado homes in the "red zone" at 40 percent of the total. The rate of development outpaced fire-protection capabilities. Those homeowners who undertook remedial measures could be sabotaged by neighbors who did not. The embers and flames from one indefensible structure could propagate from house to house.

For several decades fire experts had warned about the potentially lethal consequences, but no catastrophes backed up those exhortations. A bad fire was as unlikely as winning a lottery. The conditions that favored fire and those that favored development seemed out of sync. Then the stars and planets aligned: the 2002 Hayman fire ripped through the complex like an F5 tornado. It was dismissed as a freak—started by a Forest Service fire-prevention tech, for heaven's sake. But the fires kept coming. In 2012 they seized national headlines as they roared into suburbs of Fort Collins and Colorado Springs, burning a record number of houses (346) and forcing a partial evacuation of the Air Force Academy. Earlier, the Fourmile prescribed fire had run feral, blasting through exurbs and killing four people (the legislature subsequently stripped the state Forest Service of its right to burn). The next year Colorado Springs suffered through even greater losses. "We knew the risk was there," said a resident of Black Forest, one of the hammered communities. "But it hadn't burned in 100 years. We thought it never would."[31]

That fire regions—the geographic domains of big, damaging fires—were migrating was not news. In the 18th century Dark Days had obscured skies with palls of smoke that blanketed New England. In the 19th century the bad fires had moved to the Lake States and stayed for over 60 years. In the early 20th century the critical zone was the northern swath along the U.S.-Canada border between Washington and New York. Then industrialization and fire protection scrubbed the worst fires away such that by midcentury truly big fires seemed of antiquarian interest outside California, Nevada, and the Northern Rockies. By the new millennium they were back, as vicious as ever. But this time the scene of problem burning was relocating to extensive swaths of public lands and to private lands in which drought, rapid changes in land use, and inadequate institutions had abraded away the old buffering landscapes of rural America and exchanged them for combustible-rich surrogates.

Throughout, the background conditions for fire had endured. The seasons might be longer, the big fires larger and more ferocious, but fire seasons came to the same places at roughly the same times. What had changed was their interaction with American society. The sequence of regional fire prominence had followed the tidal bores and storm surges of settlement. Now that formula was apparently shifting. The algorithms had to accommodate global change, including probable shifts in climate, as it affected the basic dynamics behind the rhythms of wetting and drying and of winds that powered free-burning fire. Settlement, too, had metamorphosed from broadly agricultural purposes to urban styles that encouraged rather then removed vegetation. Still, most of the interface lay in the East, outside the patches most prone to burn explosively.

The really frightening prospect was that the rough beast of western fire might be slouching eastward to be reborn. For 30 years the threat had come from houses moving into places with abundant fire—a narrative that was commonly dismissed as an expression of regional lunacy. Now the prospect grew that the fires might be moving into the places with abundant housing. The contagion seemed set to spread to the scoffers and the smug.

* * *

Historical comparisons are tricky. Even NIFC finally posted a disclaimer in 2013 that figures before 1983 lacked provenance and might represent nothing more than quirks in recording. But the great arc of fire history over the previous 50 years had identifiable spans that tracked the changes.

Consider.

On the scale of what made a wildland fire big: in 1961, the Sleeping Child fire had blackened 28,000 acres and established its plume as the iconic image of big burns for several generations in the Northern Rockies. In 2011 wildfires burned record acreages in two states—the Miller fire in New Mexico (89,000 acres) and the Wallow fire in Arizona (538,000 acres)—while the Rockhouse stampeded over the Davis Mountains of Texas for 314,000 acres.

On the character of the intermix fire: in 1961 the Bel Air–Brentwood fire flamed through exclusive communities in the Santa Monica Mountains and was interpreted through a prism of Hollywood privilege and California exceptionalism. In 2011 the Bastrop fire outside Austin, Texas, crashed through a mixed-class landscape that included rural and recreational housing, trailer parks, a state park, and perversely overgrown woodlands.

On fire in rural landscapes: in 1961 the Harlow fire burned 43,000 acres in the foothills of the Sierra Nevada, overrunning relic towns. The 2011 Possum Kingdom Complex burned 148,000 acres and mostly new settlements in Palo Pinto County, Texas, and then experienced a series of late-season aftershocks. In 1961 every fire merited the same response: control by 10 a.m. the next day. In 2011 it was possible to apply a multiuse doctrine to each burn such that distinct portions of the same fire might be handled differently.

Such statistics, however flawed, were but a subplot. The principle storyline was not just the rise in documented burning or the transfer of that burning to a handful of malevolent monsters. The core story was the nation's growing fire famine, the continued loss of fire from places that needed it, and the inability to build new regimes on the charred landscapes left in the smoking wake of megafires. Some 39 million acres were badly out of whack on the national forests, and perhaps three times that amount on the national estate overall. They would require fire's restoration not as a one-off act of penance by vaccination-burning a plot of land but as a perpetual commitment to sustain fire regimes.

What grabbed public and political attention were burned houses, dead firefighters, and celebrities. So the 2013 season evolved into a rolling thunder of media attention as it started with incinerated suburbs outside Colorado Springs, moved to central Arizona and burned over the Granite Mountain Hotshots (killing 19), and finally swelled through the celebrated landscapes of Yosemite. Each part of that fire scene was familiar: what startled was the range of places struck and that all happened in a single season. Black Forest was an modern exurb under the responsibility of local and Colorado state agencies. Yarnell involved a city crew on a State of Arizona fire in a residual patch of a mining town. The Rim fire burned through national forest and national park with the municipal watershed for San Francisco at risk. What had once been quarantined in California was propagating elsewhere.

What was unchanged over more than 50 years was the reality of fire, a natural fact, and the need to choose, a moral one. The American fire community existed because of fires; its character was shaped by, and in turn helped shape, those fires; and by fires it would be judged.

Wallowing, 2011

In 2011 the big fires roared back. The macrostory was that Texas accounted for 45 percent of the 8.7 million acres burned. But it was not the largest and loudest fires that mattered most. While Texas unbalanced the national statistics, the finer-grained events said more about the changing character of American fire and of those who engaged it. A narrative looks for the telling details. A fire history must look for the trying fires.[1]

The country passed through two pairs of particularly revealing conflagrations. The Las Conchas and Bastrop fires showed the worst of the WUI. Both were started by downed power lines; their smoke plumes were within sight of state capitals. The Miller and Wallow fires revealed the best and worst of burning in the nation's wildlands. Both started in wilderness, though from human causes; one stayed in the Mogollon Mountains and remained invisible to the outside world, while the other bolted away amid national publicity to become the largest forest fire in Arizona's recorded history.

* * *

On June 26, amid blustering winds, a power line fell onto an aspen on a ranch in the Jemez Mountains of New Mexico. The spark almost immediately blew up into a wind-driven conflagration that within five days had become the state's largest fire on record—and then kept burning. It blazed through the Valles Caldera National Preserve, through the Santa Fe National Forest, through most of Bandelier National Monument, and over 16,000 acres of the Santa Clara Pueblo. Los Alamos was evacuated. An acre spot fire flared on the grounds of the national lab. The Las Conchas fire burned over the site of the 1977 La Mesa and 2000 Cerro Grande fires. The severe burns yielded scouring floods when the summer rains came. In the end 156,293 acres and 63 houses burned.[2]

It was déjà vu all over again, only at three times the breadth and perhaps twice the intensity. Probably only the lingering scars of the Cerro Grande fire spared Los Alamos. But any illusion that one bad burn inoculated a site blew away in the roiling, twisted plume that looked like a tornado tipped on its side. The same towns could be hit again and again. The twice-baked landscape, however, absorbed the worst of the impact. Parts of it looked nuked. Fire scientists familiar with the area, such as Craig Allen and Tom Swetnam, testified that they had never seen anything like it. If this were the new order, the future looked apocalyptic, better described by Cormac McCarthy's *The Road* than by Aldo Leopold's *Sand County Almanac*.[3]

Yet a nightmare's return might be less harrowing than a first apparition. On the Sunday of Labor Day weekend, dry winds from Tropical Storm Lee, which made landfall to the east, toppled a dead loblolly onto a power line and ignited a fire east of Austin, Texas. The call to the Bastrop VFD came in at 2:20 p.m. The first engine arrived five minutes later and instantly requested backup from the Texas Forest Service. The emergency operations center informed them that there was little to send. Other fires—the Texas Forest Service was already reinforcing nine fires in central Texas—had stripped the system of any surplus resources. After eight minutes, with the fire roaring out of control, the crews pulled back. A half hour later another falling tree started a second fire four miles away. When the flames crossed Texas Highway 21, the county officially declared the fire a disaster. Neighborhood evacuations commenced. The fire blasted into Bastrop State Park. It poured through woods and leaped the Colorado River, ranches and developments, and Highway 71.[4]

The flames burned beyond anything anyone there had witnessed before. At times they cascaded up like a colossal burnt offering, and at other times they seemed, even to fire specialists, like a mutant phenomenon that drove them to search for analogies beyond those encoded in fire behavior software. The spectacle seemed to some less like a wind-driven fire than a pyroclastic debris flow, a turbidity current of ash, embers, and gases that outran radiant heat and washed through fuels. It burned as straight as the blast of a blowtorch, and reports pointed to horizontal roll vortices that appeared to hold the pyroclastics into a ground-hugging plume much as the magnetic field of a torus contains plasma, and when the field faltered, its confined contents exploded outward, alternately constricting and collapsing the flow from something that resembled a hydraulic jet to a kind of consuming splash. By 5:00 p.m. the two fires, rushing south, merged. A third fire began from another toppled tree and later flowed into the others. By nightfall the Bastrop Complex had burned an oblong swath some 6 by 14 miles. The next day another fire took off; the winds veered and the fire front moved southwest; the flames chewed on. Then the winds died. The counterforce of engines, planes, and crews began to crest. The fire quieted, and on Wednesday it ceased to spread. By then FEMA fire assistance grants were activated. Full containment took another 23 days. Not until October 10 was the Bastrop complex officially declared out.

The reckoning included 34,356 acres and 1,645 homes but miraculously only two lives. The blowup climaxed a dismal roster of Texas records for the year—the second longest fire season, the hottest summer, the driest year since 1895, the most catastrophic losses. Some 3.9 million acres burned and over 5,600 structures. Bastrop State Park, the crown jewel of the Texas system, burned all but 250 acres of its just under 6,000 acres. The mutual-aid system that had evolved since 1996 broke down, strained far beyond its worst-case scenario; the feds rode to the rescue with $185 million in emergency assistance. The smoke plume, like a flooding river lifted out of its bed, had rushed south within eyesight of the state capitol and what was widely touted as a site of high-tech synergy. Among the most elemental expressions of nature's economy had met the next new thing in the nation's, and fire had won.

Yet the horrors could have been far worse. A red norther like this was off the scale: no system could have fully prepared for it. A decade and a half of institutional improvements had readied a far wiser and safer response, which had coped as well as possible. Officials called the evacuations correctly. VFDs that might have found themselves overrun had shrewdly pulled back and regrouped. Cleanup and rehabilitation proceeded quickly and knowledgeably. What it could not have stopped, the system accepted and then rapidly picked up afterward. Unlike the scene at Los Alamos, what terrified knowledgeable observers was not the specter of a return but of fires of a type not seen before and the prospect that they might migrate east, that the old westward surge of frontier settlement might reverse itself into an eastward migration.

* * *

The Miller fire started on April 28, 2011, an early ignition for an early-onset season. It burned in a classic Southwest setting of ponderosa pine and grassy glades, grading upslope into mixed conifer and rugged mountains, a landscape so remote that, at the recommendation of Aldo Leopold, the Forest Service had designated it the first primitive area in 1924. With the Wilderness Act it joined the founding slate of legally wild lands. The scene was ideal for natural fire, and the Gila National Forest had commenced a fire-restoration program—among the longest running in the country—in the mid-1970s. The Gila Wilderness was the regional baseline for fire.[5]

What made the Miller fire special was its anthropogenic origin. Natural fire management applied only to naturally kindled fires; the Miller was anthropogenic, exact cause unknown. It was managed as a multiple-objective burn, which involved some structure protection, some evacuations when flames moved around the Gila Cliff Dwellings National Monument, and loose-herding and burnouts from trails and ad hoc firelines. Postburn rehabilitation took the form of stream stabilizations. But because most of the landscape was wilderness, the bulk of the fire was allowed to free-burn. For partisans it was a jubilant moment. Watching the flames, one observer exulted, was like seeing an animal freed from a cage. The restored flames were the counterpart to the reinstated Mexican gray wolf. On and on the fire burned, sweeping and creeping, with mixed ecological effects. When monsoon rains finally snuffed it out (officially on July 27), the Miller fire had racked up 88,835 acres. Its claim on national firefighting resources—then strained—had been minimal. For much of the landscape it was the fourth such burn in 40 years.

This, friendly observers believed, was what the future of fire should have become. Each landscape-scale burn would help check future burns; the mosaic of patch-burning as fires torched and smoldered up and around mountains and through mixed biotas would encourage ecological resilience. The multiple reburns of Cerro Grande and Las Conchas had scoured away the capacity of the land to recover; the multiple burns on the Mogollon, rising and falling over decades, burning into one another rather than on top of one another, promoted biotic pliability, the capacity to bounce back.[6]

Even as the Miller fire still flowed through the Mogollons, its dark double boiled up across the border in the Apache-Sitgreaves National Forests. The Wallow fire started from an abandoned campfire in the Bear Wallow Wilderness on May 31 and spread to 2,800 acres through spotting. On June 1 it grew to 8,900 acres, on June 2 to 40,000, on June 3 to 100,000. It burned through Hannagan Meadow, through Alpine, through Greer, and to the outskirts of Round Valley. It burned over the old Caldwell Ranch overlooking the east fork of the Black River where in 1909 Aldo Leopold had shot a she-wolf and watched the fierce green fire in her eyes die. If the town had not halted it, the flames would have burned to the edge of the Painted Desert and Petrified Forest.[7]

At times there were as many as 3,200 personnel involved from all across the country. The high winds grounded air tankers, but there were over 30 helicopters, and as the flames threatened Eagar and Greer, a retardant-laden DC-10 arrived. The burn scar slashed across the White Mountains like a raking grizzly paw. Three Type 1 incident management teams divided up responsibilities. When the final tally was reckoned, the Wallow fire encompassed a perimeter of 538,049 acres, cost $110 million to fight, and destroyed 32 houses and 4 commercial properties. If it burned patchily, it nonetheless incinerated immense swaths of the longest contiguous forest in Arizona.

Unlike the four-month residence of the Miller fire, the Wallow burned fast and furious and expired in two weeks. It was, in classic Southwest fashion, an early-season wind-driven behemoth. With spots starting three or four miles away, there was little to contain it. Fuels, terrain, suppression, all had to wait for the winds to subside before they could steer the front's trajectory. The flames burned with the singular direction of a loosed arrow. The Bear Wallow Wilderness contained 11,000 acres. The Wallow fire that burst from it was almost 50 times larger.[8]

Probably the rugged location, the legal liabilities associated with wilderness, worries over safety, and the desire to use wildfires to do what prescribed fire had failed to do, when added together yielded a sensible caution that amounted to hesitation in initial attack. Or perhaps not; the fire was spotting within hours, and the high winds rendered useless aircraft and smokejumpers. Once it blew out of the wilderness, there was nothing to stop it, and it had everything before it. Terrain below the main wind or sheltered from its strongest gusts escaped with mild burns. Landscapes caught in their rip current burned to the rocks.

But such considerations did not stop those looking for portents and omens in the flames. Some saw the hand of global warming, though the drought was no worse than in 1977. Some saw legacy landscapes heaped with combustibles, though the flames followed the billowy contours of wind, not fuel, and the flaming front burned what looked like a bathtub ring around the Alpine Valley regardless of fuel arrays. Some saw another outbreak of the megafire plague. Some saw hobbled suppression capabilities, although a DC-10 was eventually activated (but saw no service because retardant was ineffective in the rushing winds). Others saw the complications of the interface; yet although the Wallow burned into several communities, they were not modern exurbs but small rural hamlets established over 30 years before Arizona even became a state. Such was the urge to pyromancy—to divine meaning in the flames—that the Wallow meant everything and nothing. Like all fires, it synthesized its surroundings. It was a mix of old and new. It was simply the fire the Southwest had come to have.

So, too, the fires of 2011 were, it seemed, the fires America had come to possess. They were the fires it would have to pass between.

EPILOGUE

AFTER ACTION REPORT

In 2010 the Forest Service, and its Northern Region, celebrated the centennial of the Big Blowup with far greater fanfare than they had the 50th anniversary. There were public lectures and memorials. The Museum of Northern Idaho installed an exhibit. The town of Wallace held a formal symposium in the high school gym, with Chief Forester Tom Tidwell and FAM director Tom Harbour in the audience. Former Chief Dale Boswell spoke of growing up in a Forest Service home before starting his own entry-level job on an engine crew. The culmination came at the Woodlawn Cemetery in St. Maries, where the USFS honor guard, led by a bagpiper and trailed by a pack string, marched to the site of the mass graves of the fallen and there formally dedicated a new collective headstone.

The sense of the ceremonies was more tragic than triumphant. What had changed was the new, or rather renewed, visibility of fire as fundamental to the agency's mission. But so had the meaning invested in its fire heritage. The fire abolitionism that had gradually swept over the Forest Service in the aftermath of the Big Blowup was now viewed as a regrettable error and a major source of unhealthy and ungovernable landscapes. Suppression was a young man's calling, or a young agency's. With age came understanding about the complexity of the world and the compromises that make life work. And there was the text of Norman Maclean's *Young Men and Fire*, relived at South Canyon, at Thirtymile, and at Cramer. The memorializing of the dead that began at South Canyon had spread to every mass fire-fatality site around the country, and now it came back to the font itself. The ceremonies were yet another way of affirming a commitment not to put firefighters unnecessarily in harm's way.

Yet like all remembrances, this one was an opportunity to look back over what had been done, left undone, and not done at all, and to measure the meaning of it all. Providentially, the 2010 fire season was relatively mild by contemporary standards: only 3.4 million acres burned across all jurisdictions at a cost of $1.1 billion. A vast apparatus poised to attack fire could pause and reflect. The historians among that throng might hark back to 1910. Most fire officers would contemplate the span of their own careers. Only a few of the community's elders could personally recall a life on the lines in 1960. But that 50-year span makes a convenient baseline. It was a world unrecognizable to those first holding pulaskis and driptorches in 2010.

* * *

In 1960 the U.S. Forest Service was the colossus of America's public domain and the primary vehicle for public forestry. By any measure it was the fire community's indispensable institution—the one that connected all the others, without which fire protection was a bag of misshapen marbles; in truth, there was not much else. The USFS organized and informed the whole. It controlled fire research, fire equipment development, fire training, and fire policy. It held by far the most resources in terms of crews, engines, and aircraft. It fielded the overhead teams that operated under the large fire organization. It had the authority to review its own actions. It was a benevolent hegemon. The history of wildland fire in America remained, as it had been since 1905, the history of the Forest Service. The agency accepted that charge and assumed it had achieved most of what it had set out to do.

In 1960, at the agency's request, Congress had passed the Multiple Use-Sustained Yield Act, upgrading the agency's charter for the postwar scene. But while the Forest Service conceived its land stewardship as multiple, its stewardship of fire was singular. The 10 a.m. policy gave fire protection an unambiguous standard by which to measure success and a standard that came with its own unchecked emergency account. The other federal agencies aligned with its gravitational field, like stray comets coming into orbit around a sun. So did the state foresters. So thorough was its command of the subject and so dominant its position that it seemed not only inevitable but exemplary.

But what really made the Forest Service notable was what Herbert Kaufman had identified as its salient trait: its ability to instill its values and methods in its workforce such that they knew, without being told, what to do and how to do it. They identified with the agency. They were proud, competent, and confident. And they understood that fire mattered because it challenged their mission. Fire control was not a professional career path as the civil service defined it because it did not require a forestry degree, but fire control offered the opportunity to rise through the ranks by means of diligent labor and hard-won skills. Many did. They knew fire control in their bones; they had known it all their working lives. The organization overall knew fire, too, because so many members had come into the agency through seasonal fire jobs; those members, like once-serving officers, remained on reserve duty. The agency had an internal fire militia to activate during emergencies. The alarm sounded for all; few refused the call. But the coherence Kaufman noted was selective, and so was the agency's understanding of fire. It knew fire could shape institutions as it did careers and as it did lives, but it knew fire as a formidable foe and a wily rival to its administration of the lands under its care, and it had defined itself by how well it grappled with the flames.

Fifty years later that cosmos was gone, as vanished as the world of the Aztecs. Writing about the troubled politics of democracies, Francis Fukuyama identified the Forest Service as the epitome of dysfunctional government. It could no longer meet force with an overwhelming counterforce. Little in fire was done by any agency unilaterally: everything was interagency and increasingly intergovernmental. Few projects did not involve cooperation; most demanded formal collaboration. The Joint Fire Science Program, the USGS, and even the National Institute of Standards and Technology also sponsored research, and the Forest Service was in some ways hobbled by its brick-and-mortar heritage. Fire management was no longer a government monopoly; private vendors sold equipment, set up catering services on large fires, flew aircraft, did consulting, even fielded crews and engines. Congress sought ways to banish the emergency accounts in favor of normal budgeting. FEMA disbursed emergency fire-suppression grants to states, and the U.S. Fire Administration and Homeland Security controlled the National Incident Management System. The incident command system pushed aside the large fire organization; fire teams went from 4 or 5 members to nearly 40 in its long version and expanded from fire to embrace an all-hazard model that sent them to the Twin Towers after 9/11 and to the Texas countryside in 2003 to search and retrieve debris from the disintegrated space shuttle *Columbia*.[1]

Nor was fire a stand-alone operation with an unambiguous charge. It was, in principle, inextricably intertwined with land management, which meant a fire agency had to fight fires, light fires, herd fires, and monitor fires—and prepare the land to receive fires. Fire operations spoke of management, not control. Field operations fractured because the public lands had splintered from a generic melting-pot estate governed under a doctrine of multiple use into a mosaic of special purposes and causes, often incompatible. The rough, varied middle landscape was squeezed to the point of extinction. Fire's mission was often muddled because land use was still contested. Too many fire plans remained unwritten because land management plans languished in political limbo. The idea of shifting the burden from multiple-use land to multiple-objective fire was wobbly. If fires were confused and costly, that condition largely resulted from an agency forced to act on behalf of a contradictory and cumbersome citizenry that sought, where possible, to pave over awkward differences with dollars that it was anyway increasingly unwilling to spend. But many differences could not be negotiated away or bought off with shrinking monies. Fire did not care. The agencies, but especially the Forest Service, struggled to find equilibrium. So while the mind might grasp the concept that the agency's blood enemy was potentially its best friend, when the smoke rose, the hand still reached for a pulaski. Fighting fire remained one of the few tasks the Forest Service could do that would draw more applause than catcalls.

Life on the line remained blue-collar labor with a white-collar patina. Crews were better educated, and their life experiences outside of their time in the woods were suburban. The workforce had phased into a more heavily worked professional corps and away from a citizen militia. Safety had become an organic principle that putatively trumped all other considerations. Crews no longer regularly worked night shifts on the line. On fires that escaped initial attack, they often backed off and burned out. They had personal protective equipment and communications technology undreamed of when the Chilao Hotshots roamed the San Gabriels with little more than hardhats and gloves. They had to cope with structure protection as well as with flames cresting a summit. They had an order of magnitude more mechanical muscle on call, but outside the I-zone they veered toward minimum-impact suppression tactics.

Yet as with aircraft or fuels projects, it was doubtful that firefighting had kept pace with deteriorating conditions. The diminishing workforce was not in stride with its expanding workplace. The very success of total mobility left fire folk as a guild unto themselves, devoted to technical skills and segregated from other land-management tasks, who often found it easier to communicate with counterparts in other fire organizations than with members of their own agencies who worked on watersheds or visitor services. The fire revolution in workforce had come full circle: fire remained too often an isolated program. Total mobility meant the ability to move between fire programs, not within land agencies or between fire and American culture.[2]

As the 2012 season approached, promising to match (or exceed) the record 2011 nightmare, Forest Service deputy chief Jim Hubbard circulated what became known as the "Hubbard Memo." It sought, in the name of fiscal control, to rein in the anticipated expenditures. The surest way was to reinstate for the season, albeit in muted form, the 10 a.m. policy of aggressive initial attack, or so the memo was widely interpreted. While some fires would escape, their rampage could not be blamed on the agency, and critics would not be able to harp on wishy-washy policies to restore fire. It was a bizarre piece of political theater, and it reminded at least one critic of the aborted 1991 coup attempt to restore the Soviet Union. As an informing strategy, suppression had failed. Besides, once proclaimed, emergency measures tend to continue; to reinstate the old regime even for a season tempted continuation. The memo admitted its instructions were not a sustainable practice. "Then why go there?" asked an outraged Bob Mutch.[3]

That fire officers might challenge a memo from the top was like Catholic clergy openly questioning an encyclical from the pope. And more than anything else, it illustrated what had most changed for the agency over the past 50 years. Apart from their diminished numbers, the Forest Service and its workforce were no longer of one mind and heart. The agency ranked near the bottom of government agencies in morale.

* * *

In March 2010, five months before it commemorated the firefighters who worked and died in the Big Blowup, the Forest Service was served with a fire workforce audit by the Office of Inspector General. That workforce had been in turmoil for 30 years. As with megafires and costs, the agency's staffing was approaching a crisis. Apart from its fire-specific staff, 64 percent of its workforce would be eligible to retire in 5 years and 86 percent in 10 years. The OIG determined that the Forest Service had not "taken the necessary steps to ensure it has a sufficient number of qualified staff" to do the job expected.[4]

Its corps of fire officers and NCOs had shrunk dramatically beginning around 1978. The National Fire Plan stabilized and then raised those numbers, but they still fell short of what the worsening conditions demanded. In addition to the national forests and interagency agreements, the agency was participating more actively in the National Response Framework, which further strained its most experienced fire officers. Pooling the federal fire corps helped somewhat, but big-fire complexes tended to bust out at the same time. The Forest Service had traditionally made up the difference by drafting from its militia. In 2009 approximately 70 percent of its employees (about 24,000 people) had some fire qualifications as online firefighters, helicopter managers, safety officers, or any of the 300 or so positions identified by NIFQS. The problem was that the pool of personnel was, in practice, far less than it had been. In 2008 only 9 percent of those "qualified firefighters actually took part in suppressing the agency's largest, costliest wildfires." The others stayed home.[5]

The reasons were many, but mostly, like American society generally, the Forest Service relied on the premise that "personal preferences" would coincide with agency needs, that enough of its staff would voluntarily choose fire assignments, and that neither coercion nor systematic planning would be necessary. But volunteerism was not working. Many of those surveyed offered reasons, including "someone has to stay home with the kids" and "I don't want to sleep in the dirt" and "my supervisor wants me to get my regular work done"—explanations that would have been unheard of decades earlier. Some 40 percent of those who underwent fire training actually completed the requirements to qualify, and of those who did, the overwhelming majority declined to answer the call to fire duty. Such assignments were, the audit noted, "a collateral duty" for which they were neither specially paid nor otherwise compensated. They (and their supervisors) were evaluated on their regular tasks, not their emergency fire assignments. Perversely, it was "usually against employees' best interests to make themselves available for firefighting." The upshot was a "disproportionate burden on the few who volunteer." The situation would deteriorate sharply with the upcoming mass retirements.[6]

The Forest Service accepted most of the OIG's recommendations but noted that major reforms were under way through the anticipated FLAME Act and its call for a cohesive strategy that promised to quell the impending storm. It would have to plan, and it would have to do more with less. But even among its officer corps, esprit was crumbling. Fire-management officers and incident commanders were subject to potential criminal prosecution, civil suits, and agency audits not known by previous generations. The 2001 Thirtymile fire saw an incident commander arraigned with 11 felony charges, including manslaughter, later plea-bargained to two misdemeanors, and subsequently reassigned to nonfire work and put on court-ordered probation. After that, witnesses to burnovers were advised to "lawyer up" and demand immunity before offering any testimony or participating in investigations. The dual fatalities on the 2002 Cramer fire on the Salmon-Challis National Forest led to a dismissal of the incident commander and reassignment for his bosses. Affirmative action to promote workforce diversity challenged agency staffing as much as mandated biodiversity challenged its field operations. Officers who rose through the ranks in fire might find themselves in the mud picking up after Hurricane Katrina. If this was what fire duty looked like to the permanent officer corps, it is not hard to imagine why reservists might choose to stay in their cubicles.[7]

The fire revolution, and trends in American society overall, had broken the pattern of generational succession. And the predicament was even worse than the OIG or other observers noted. The workforce itself was schismatic; the generational chasm was internalized. The agency's fire corps could not agree on what lore to hand from generation to generation or how to do it, because they could not agree on how to manage fire. Each generation saw its legacy being swept away. In 2012 that unease boiled over. The generation that restored natural fire to the wild saw in the Hubbard Memo a possible repudiation of all that their careers had struggled so mightily to achieve. Equally, the generation that had installed a fire-suppression system unrivalled on the planet saw, in megafires and big-box burnouts, the wanton destruction of what they had expended lifetimes to protect.

At the International Association of Wildland Fire's Fourth Fire Behavior and Fuels Conference in Raleigh, North Carolina, Bob Mutch, veteran of the White Cap experiment, delivered a keynote address in which he denounced the implications of the Hubbard Memo and argued for the success of restored fire in the Selway-Bitterroot Wilderness—and for the need to expand those lessons. Done right, fires could self-regulate—and thus reduce costs and threats to firefighter safety—and enhance ecological goods. At the same time a special Fire Committee of the National Association of Forest Service Retirees, headed by Al West and Mike Rogers, worked to make their voices heard within the current USFS hierarchy. They fretted over what had happened to "their work, the infrastructure, their colleagues still on duty and the community members, elected leaders who respect what they accomplished and seek them out of their retirement years to understand what is happening, why, and asking what they can do to help!"[8]

The occasion for that circular e-mail was the imminent release of a report commissioned by New Mexico congressman Steve Pearce and written by retired USFS fire officers on the damaging and costly Whitewater-Baldy Complex and Little Bear fire on the Gila National Forest in 2012. By now fire critics had taken a page out of the environmentalists' playbook and organized into NGOs such as the National Institute for the Elimination of Catastrophic Wildfire. The report promoted by Pearce was written by Roger Seewald and William Derr, and it concluded that the decisions and actions taken on the fires were in fact consistent with national policy and protocols, available resources, and actual conditions. The underlying problem was unworkable policy along with the fact that the Forest Service had been so maimed over the past 50 years that it was unable to put the land into manageable condition and to fight fires with the force necessary to stop bad burns at their origin.[9]

Both the Mutch and Seewald-Derr declarations were in effect after-action reports on an era. The scars and scabs of the fire revolution remained livid and itchy, and those who had lived through the era were unable to leave them alone. Fire policy, it seemed, might join abortion, gun control, and other irresolvable controversies from the 1960s. The OIG was correct that the future hinged on a new generation of fire officers. The existing generation, even in retirement, was unable to resolve the issues that had hounded them throughout their careers. But then few of the factors that drove the fire scene were under their control: they were the internalized quarrels of American society. The fire community was a mirror of the agencies it served, as those agencies were a cipher for the country, and as America's fires were a synthesis of American life.

* * *

It seems right that the two fires that appeared to express the best and worst of the fire revolution's legacy should happen on the same forest, because the Gila had managed the improbable task of being both at the center and at the margins of the fire revolution. So remote was it that its mountain core was closer to the Trinity test site where the first atomic bomb was exploded than it was to the nearest metropolis, Albuquerque.

Yet without much fanfare—certainly nothing like what blared out of the Sierra parks or the Selway-Bitterroot—it had pioneered an all-fires approach. For decades it had suppressed fires, or what survived the 19th-century wreckage of wretched land abuse, primarily by overgrazing. By the 1960s it exhibited a hard-core suppression organization on a California model that included an aerial fire center at Deming for air tankers and smokejumpers. Then, with a cadre of cautious doers, it adopted prescribed fire early in the 1970s, seized on the suppression options made available in the 1978 policy overhaul, experimented with natural fire, and quietly began restoring fire to what was arguably the nation's first wilderness area. Fire crews learned to fight and light at the same time. The acres added up; the fires began burning one into another. More gazetted wilderness claimed more of the mountains. The Gila acquired that most precious of creations, a fire culture.

That culture survived by being passed through generations of fire officers. Fire management and firescapes took on a shared character. As fires began interacting over time, each burning into and over the last, so did fire officers, as lore was handed down. The old-timers stayed in Silver City even in retirement. By the time the Miller fire romped insouciantly over the Mogollon Mountains, it was possible to assemble 30 years of Gila FMOs around a campfire. Even so, the suppression culture that preceded and ran through the restoration culture still endured, and it, too, stayed on site. The Whitewater-Baldy Complex threw those generations into conflict. The future depended as much on how the Gila's fire culture could absorb both as on how the Mogollons could rebound from the scouring crown fires and postburn floods.

What matters is the next generation. What matters is people like Gabe Holguin, who, when he became the Gila's fire and aviation staff officer, had come home. He had grown up in a Forest Service family on the Gila. His father, Albert, worked for the agency, which meant he answered fire calls from time to time. Better, his uncle Gary was a smokejumper stationed at Deming. Young Gabe would visit the depot and watch the training jumps. When he went to the University of New Mexico, he signed up as a chemical engineering major but quickly knew he wanted to be outdoors, so he swapped engineering for forestry at the University of Idaho and worked fire during the summers. For three years he jumped out of Grangeville, and then for two out of Missoula. He knew then that fire was what he wanted, and he knew the Gila was where he wanted to be. In 2004 he became the wilderness district's fuels specialist. He transferred to the Magdalena District on the Cibola National Forest in northern New Mexico when an FMO slot opened. In 2010 he returned to the Gila as its fire-management officer and two years later as its fire and aviation staff officer.[10]

The years had gotten tougher. The weather, benign during the early restoration program, turned nasty. The money for prescribed fire dried up. The easy burning in the recovering montane ponderosa pine belt was done. To move upslope into the disrupted mixed-conifer forest clinging like moss to the craggy stones of an extinct volcano whose cap had blown out would be complicated and edgy. The Miller fire worked because it burned in many places already treated. The Whitewater-Baldy Complex was started amid a severe drought, on 80 percent slopes, within long-unburned fir and spruce. Crews attacked the fires before being withdrawn out of safety concerns. The fires blew up, merged, and rampaged over landscapes that, over eons, had rejuvenated from episodic crown fires, though no such fires had burned within the experience of anyone alive at the time.

Weather, fuels, and an inherited past of things done and not done promised a future of larger, more ferocious flames that a new generation of fire officers would have to face. The options to either leave alone, try to stamp out, or replace on a landscape scale wild flames with tame ones were over. The prospect for returning fire to a golden age in the past, or for projecting it into an invented golden age to come, created by applied science, was gone. Fire's restoration was happening but on fire's terms, not humanity's. Gone was the illusion of resistance and control, of suppressing fire. Gone, too, was the faith in restoration or of imagined desired conditions. What remained was the hope of resilience. Gabe Holguin would play the hand he was dealt.

"Fire is fire is fire and we'll manage it."

* * *

The fires—that is what endures.

Of course they change. They change with every ridgeline and gust of wind, with the legacy of how we have lived on the land and how we seek to live on it now, with how we think protected land should look, with what institutions we invent to exploit combustion's powers and contain its dangers. Fires change with what modern science can cajole from its labs. They change with the electric arcing that bonds sky to earth and with every inspired, malevolent, stray, accidental, misplaced, or fine-crafted spark that flies from the human hand. Fire integrates it all. The kinds of fires we have synthesize the kind of world we inhabit. Yet fire and our need to engage with it endure.

Because we like to parse the world symmetrically, it is easy to imagine those fires divided into pairs of matched flames. There are many such pairings possible—wild and tame, nature's and ours, fires that burn surface biomass and fires that burn fossil fuels. There are pairs like Sleeping Child and Bel Air–Brentwood, Yellowstone and Oakland, Walhalla and the Dragon. The America that existed between 1960 and 2012 found itself burning and believed that its fires were unique, as indeed they were. Yet every era could say the same. The challenge has never been simply to decide which fires to fan and which to extinguish but how to pass between them all. That is our duty, our privilege, and our destiny—that, and how to do so with grace, courage, and humility. This was the hard lesson learned that the generations who lived through the fire revolution were handed from those who came before, though they did not recognize it as such, and it is the lesson learned they must pass on to those who will come after them.

AFTERWORD

My plan was clear, concise, and reasonable, I think. For many years I have traveled in many parts of the world. . . . Thus I discovered that I did not know my own country. I, an American writer, writing about America, was working from memory, and the memory is at best a faulty, warpy reservoir.

So it was that I determined to look again, to try to rediscover this monster land.

—JOHN STEINBECK, TRAVELS WITH CHARLEY

In late 1980 I completed the manuscript for what, 18 months later, was published as *Fire in America: A Cultural History of Wildland and Rural Fire*. That book grew out of 15 seasons of on-the-ground experience with the North Rim Longshots. The book ended its lumpy saga in the late 1970s. Over the next 30 years I pursued a global history of fire that has taken me to every continent and resulted in a suite of histories that I called Cycle of Fire. From time to time I returned to the United States to comment on major events, to write short policy-themed studies or the occasional op-ed, and to tell the story of the Great Fires of 1910.

But it also became obvious that my past years on the ground could no longer animate and inform me about the emerging future. My bred-in-the-bone understanding of American fire had spent itself as a means to explain what was happening as wildland fire and its community of nominal tenders fast marched into new country. That fact, and my deep immersion into the fire histories of Australia, Canada, Europe, Russia, and a score of other countries, had distanced me from my hard-learned intuition about how fire worked on the American landscape. Instead, I had acquired a satellite-surveillance perspective that in good academic fashion allowed me to abstract and generalize but distanced me from the particulars of the field. Too many years had passed since I had mopped up in ash, known the sour smell of old smoke, or felt the flash heat of a torching white fir. My stockpiled capital of experience had leached away. Like Steinbeck I found myself relying on the flickering flame of memory. If I wished to continue to chronicle or even comment on America's existing fire scene, I would have to re-immerse myself in the polygons and pixels of contemporary fire practice.

There are so many people to thank for the opportunity to do so. Begin with Lincoln Bramwell, chief historian of the U.S. Forest Service, and a willingness by Tom Harbour, director of Fire and Aviation Management, to finance a study (and his shrewd sense to solicit matching funding from the Department of the Interior), along with the determination of the Joint Fire Science Program to participate. Thanks to the School of Life Sciences at Arizona State University, which accommodated my calendar. Thanks to countless librarians, archivists, researchers, and fire officers who helped reeducate me. Thanks to Heidi Neeley, who assisted admirably with my literature search. Thanks to Kelly Andersson for giving a long manuscript a rapid edit under a very short deadline. Thanks to Steve LaRue for a careful edit and for his patience with my sometimes eccentric citation system. And thanks to Allyson Carter for getting the book into print.

Early readers have wondered whether the book does not focus overly much on the U.S. Forest Service. My reasons for accenting the agency are two. The first is that the emphasis reflects reality. In 1960 the USFS dominated the American fire scene, it continued to be a major player throughout the fire revolution, and it remains the only institution whose actions routinely affect all the rest. The other critical institution, the National Park Service, has sponsored its own book-length administrative history; its unique story has been told well. My charge was to survey the national panorama. And that mandated breadth of interest leads to my second reason: the narrative of fire and the Forest Service comes with an intrinsic arc, like an internal pack frame. Simply, the narrative has the structure of a three-act play. This much I knew when I began. But the history also has something else that only became apparent as my understanding matured, and this was a parallel generational theme, as one era passed, literally, to another. This, too, was a fact, but also an aesthetic possibility that offered a sense of closure too enticing to resist. It goes without saying that other literary strategies are possible. This was mine.

From the onset, I conceived the new survey as passing between two books. *Between Two Fires* would offer the play-by-play. It sketches the general history of events. It is not a chronicle of record—that would run to hundreds more pages—nor a simple extension of its predecessor but a self-contained work with its own themes and narrative arc. Its characters are ideas, institutions, and fires. It builds on my earlier work, as the activities it describes arise from past practices and policies, but it begins well before *Fire in America* ends and has its own integrity of voice and vision. The second book, *To the Last Smoke*, will seek to provide the color commentary through a sweep of essays that try to capture something of the variety, texture, and felt life of American fire and its community. That project quickly self-organized into regional groupings. The essays have been gathered into a suite of short books that the University of Arizona Press is publishing. Inevitably, throughout there is some overlap with *Fire in America* and some replaying between the new history and the regional reconnaissances, and perhaps unavoidably, some borrowing of passages all around.

I can hardly claim that *Between Two Fires* is the only narrative possible or that *To the Last Smoke* captures more than a tiny fraction of the American scene. Yet the narrative of *Between Two Fires* is a singular one, and the regional studies in *To the Last Smoke*, while suffused with a personal perspective far from the scientific stance the community believes it needs, do what no one else has attempted. Whatever the community's final judgment, they are the books I was able to write.

—Steve Pyne
Concluded in Alpine, Arizona,
amid the aftermath of the Wallow fire

ABBREVIATIONS

ADOF	Alaska Division of Forestry
AFCS	Alaskan Fire Control Service
AFS	Alaska Fire Service
AFSEEE	Association of Forest Service Employees for Environmental Ethics
AIFMC	Alaska Interagency Fire Management Council
ANCSA	Alaska Native Claims Settlement Act
ANILCA	Alaska National Interest Lands Conservation Act
AMR	appropriate management response
BIA	Bureau of Indian Affairs
BIFC	Boise Interagency Fire Center
BLM	Bureau of Land Management
CCC	Civilian Conservation Corps
CDF	California Department of Forestry
CIFFC	Canadian Interagency Forest Fire Centre
CPFC	Coalition of Prescribed Fire Councils
CWS	California Wildland System
DASP	Disaster Assistance Support Program
DESCON	Designated Control Burn System
DNR	Department of Natural Resources
DOD	Department of Defense
DOI	Department of the Interior
EFSA	escaped fire situation analysis
EPA	Environmental Protection Agency
FAM	Fire and Aviation Management

FAO	Food and Agriculture Organization of the United Nations
FCO	fire control officer
FDOF	Florida Division of Forestry
FEAR	Fire Economics Assessment Report
FEAT	Fire Economics Assessment Team
FEMA	Federal Emergency Management Agency
FEPP	Federal Excess Personal Property
FFALC	Federal Fire and Aviation Leadership Council
FFASR	Forest Fire and Atmospheric Science Research
FIRE	Flame in Real-world Environments
Firescope	Firefighting Resources of Southern California Organized for Potential Emergencies
FLAME	Federal Land Assistance, Management, and Enhancement
FLN	Fire Learning Network
FLPMA	Federal Land Policy and Management Act
FMO	fire management officer
Forplan	Forest Planning Model
FRCC	fire regime conditions class
FSEEE	Forest Service Employees for Environmental Ethics
FSU	Florida State University
FUSEE	Firefighters United for Safety, Ethics, and Environment
FWS	Fish and Wildlife Service
GIS	geographic information system
GLO	General Land Office
HFI	Healthy Forests Initiative
ICE	internal combustion engine
ICS	Incident Command System
IDNDR	International Decade for Natural Disaster Reduction
IR	infrared
ISC	Interagency Scientific Committee
ISNDR	International Strategy for Natural Disaster Reduction
IUFRO	International Union of Forest Research Organizations
JFSP	Joint Fire Science Program
LAFD	Los Angeles Fire Department

Landfire	Landscape Fire and Resource Management Planning Tools Prototype Project
LCES	lookouts, communications, escape routes, and safety zones
MACS	Multi-Agency Coordination System
MTDC	Missoula Technology and Development Center
NAFC	North American Forestry Commission
NAFRI	National Advanced Fire and Resource Institute
NAPA	National Academy of Public Administration
NAS	National Academy of Sciences
NASF	National Association of State Foresters
NBS	National Biological Survey/National Biological Service
NEPA	National Environmental Policy Act
NFAEB	National Fire and Aviation Executive Board
NFCS	National Fire Coordination Study
NFDRS	National Fire-Danger Rating System
NFP	National Fire Plan
NFPA	National Fire Protection Association
NFPCA	National Fire Prevention and Control Agency
NIFC	National Interagency Fire Center
NIFQS	National Interagency Fire-Qualification System
NIIMS	National Interagency Incident Management System
NPS	National Park Service, Park Service
NRC	National Research Council
NSF	National Science Foundation
NWCG	National Wildlife Coordinating Group
NWR	National Wildlife Refuge
O&C Lands	Oregon and California Railroad Revested Lands
OES	Office of Emergency Services
OIG	Office of Inspector General
OMB	Office of Management and Budget
PLLRC	Public Land Law Review Commission
PLTPA	Priest Lake Timber Protection Association
PNF	prescribed natural fire
QFR	Quadrennial Fire Review

QFFR	Quadrennial Fire and Fuel Review
RARE I	Roadless Area Review and Evaluation
RDA	Rural Development Act
SAFARI	Southern Africa Fire-Atmosphere Research Initiative
SCOPE	Scientific Committee on the Problems of the Environment
SES	Senior Executive Service
SWFFF	Southwest Forest Fire Fighters
SWIFCO	Southwestern Interagency Fire Council
TNC	The Nature Conservancy
TTRS	Tall Timbers Research Station
URS	United Research Services
USAID	U.S. Agency for International Development
USFA	U.S. Fire Administration
USFS	U.S. Forest Service, Forest Service
USGS	U.S. Geological Survey
VFD	Volunteer Fire Department
WFDSS	wildland fire decision support system
WFLC	Wildland Fire Leadership Council
WFU	wildland fire use
WUI	wildland-urban interface

NOTE ON SOURCES

No one appreciates better than I the gaps in this study, both the gaps in the narrative and the gaps in sources. The records in the National Archives end in the mid-1960s. Regional archives contain some material, but they are scattered and selective, not systematic. Research meant collecting printed material, published or unpublished; talking to current and former practitioners; and visiting sites relevant to American fire history over the past half century. Some agencies and places were extraordinarily helpful—Tall Timbers Research Station and The Nature Conservancy, for example. Among the federal agencies local NPS sites had materials and were willing to share them. Others were grudging or balky. The BLM was often a blank slate, the FWS was hit or miss, and the USFS was mixed—sometimes stalling and stonewalling, sometimes eager to tell their story. Apart from the Joint Fire Science Program, which has posted its records online, fire research was a placer of promised documents that never panned out. The moral is, official sponsorship does not translate into open access. Some eras (like the 1980s) are especially spotty. Caches were discovered in the desks of retiring officials—or simply vanished during frequent moves. Much documentation went missing during the transition from print to electronic communication. Some regions, such as the Lake States and Northeast, receive scant treatment in the book. There is a lot left out.

Despite these holes, there was too much to put in. The book is pretty close to the maximum of what a reader, even an avid member of the wildland fire community, might be willing to absorb. Much of the remaining evidence and insights I have poured into the companion book, a suite of regional studies, *To the Last Smoke*. All in all, they will add nearly three times as much text. They help ground the general narrative in the details of place and personality. The two undertakings are the paired wings of a single creature. My hope is that the grand narrative will give context to the essays, and the essays specificity and personality to the narrative. That strategy applies as well to their sources; there seemed no reason to repeat and pile on one book that which the other will document. Regardless, *Between Two Fires* remains in many respects more an extended interpretive essay than a chronology.

My endnotes document mostly quoted material, not background. Their main sources are published texts from periodicals, government bureaus, and books, a lot of unpublished documents, with quotations from conversations as needed. Time sifts historical records, shedding many, but also sometimes concentrating what remain. I found enough to work with. A future historian should discover more.

NOTES

Prologue: Agency on Fire

1. From a presentation by Dale Bosworth at the 1910 Fire Symposium, Wallace, Idaho, May 22, 2010 (text courtesy of Dale Bosworth).

2. Gary Snyder, "The Ark of the Sierra," in *Back on the Fire: Essays* (Berkeley, CA: Counterpoint, 2007), 9.

3. Biographical data from Mary Ellen Bosworth in Victor Geraci, comp., *The Lure of the Forest: Oral Histories from the National Forests in California* (Vallejo, CA: U.S. Department of Agriculture, Forest Service, Pacific Southwest Region, 2005); presentation by Dale Bosworth at the 1910 Fire Symposium, Wallace, Idaho, May 22, 2010; and an interview with Dale Bosworth, July 3, 2012.

4. Quote from presentation by Dale Bosworth at the 1910 Fire Symposium, Wallace, Idaho, May 22, 2010.

5. Gary Snyder, "The Ark of the Sierra," 83.

6. Numbers from Herbert Kaufman, *The Forest Ranger: A Study in Administrative Behavior* (Washington, DC: Resources for the Future, 1960), 37.

7. Kaufman, *Forest Ranger*, 48–64.

8. Kenneth B. Pomeroy, "Fire Conference," *American Forests* 66, no. 1 (January 1960): 24.

9. For a summary of rural fire defense, see Stephen J. Pyne, *Fire in America: A Cultural History of Wildland and Rural Fire* (Princeton, NJ: Princeton University Press, 1982), 450–61.

10. For a chronology of major events, see ibid., 452–53.

11. The Collins Tool Company manufactured rakes and some other implements. Forestry Suppliers sold mostly to eastern and southern customers who lacked large federal partners.

12. On the SWFFF, see C. K. Collins, "The Role of the Southwest Firefighters," *Fire Control Notes* 22, no. 2 (April 1961): 70–72. On the California inmate program, see Lloyd Thorpe, *Men to Match the Mountains* (Seattle, WA: printed by author, 1972).

13. For background history, see Pyne, *Fire in America*, and "The Interagency Idea: The Forest Protection Board," *Fire Management Notes* 40, no. 4 (1979): 13.

14. Pyne, *Fire in America*, 272–87.

15. Diane M. Smith, *The Missoula Fire Sciences Laboratory: A 50-Year Dedication to Understanding Wildlands and Fire*, General Technical Report RMRS, GTR-270 (Fort Collins, CO: U.S. Department of Agriculture, Rocky Mountain Research Station, 2012).

16. Kenneth P. Davis, *Forest Fire: Control and Use* (New York: McGraw-Hill, 1959), vii–viii.

17. The best account of how the 1910 fires influenced policy is Stephen J. Pyne, *Year of the Fires: The Story of the Great Fires of 1910* (New York: Viking, 2001).

18. Aldo Leopold, "'Piute Forestry' vs. Forest Fire Prevention," *Southwestern Magazine* (March 1920), reproduced in David E. Brown and Neil B. Carmony, eds., *Aldo Leopold's Southwest* (Albuquerque: University of New Mexico Press, 1995), 139–42.

19. Komarek quote in Ashley L. Schiff, *Fire and Water: Scientific Heresy in the Forest Service* (Cambridge, MA: Harvard University Press, 1962), 64. Raymond M. Conarro, "The Place of Fire in Southern Forestry," *Journal of Forestry* 40 (February 1942): 129–31.

20. The Florida National Forest was the product of an administrative merger by president Taft in 1911 that combined the Choctawatchee and Ocala forests. The merger was subsequently disaggregated into the Ocala, Appalachicola, and Choctawatchee forests, the latter transferred to the War Department to become Eglin Air Force Base.

21. The best introduction is Harold Biswell, *Prescribed Burning in California Wildlands Vegetation Management* (Berkeley: University of California Press, 1989), esp. 9–15.

22. On the Woodwardia fire, see Peggy Powell, "The Nation's Most Costly Fire," *American Forests* 67, no. 4 (1961): 20–21, 55–56. The fire was criminally set by a disgruntled 21-year-old Forest Service tech, later sent to prison for arson.

23. The best survey of NPS fire administration is Hal K. Rothman, *Blazing Heritage: A History of Wildland Fire in the National Parks* (Oxford: Oxford University Press, 2007).

24. Kaufman, *Forest Ranger*, 207, 206.

25. Ibid., 198, 222.

26. Michael Frome, *Whose Woods These Are: The Story of the National Forests* (Garden City, NY: Doubleday, 1962), 213. For a photo gallery of the season's fires, see "Summer of 1000 Fires," *American Forest* 66, no. 9 (September 1960): 18–20, 213–14.

27. Frome, *Whose Woods These Are*, 218–21.

28. Ibid., 325.

29. See http://www.fireleadership.gov/toolbox/leaders_meet/interviews/leaders_LynBiddison.html.

30. Letter, June 16, 1982, Lynn Biddison to chief forester, available online at http://www.fireleadership.gov/toolbox/documents/Biddison_1982_Letter.pdf. Final quote from Lynn R. Biddison, "A Historical View of Our Forest Fire Organization," *Fire Management Today* 58, no. 2 (1998): 22.

31. Merle S. Lowden, "General Functional Inspection, Region 5, August 7–25, 1961," National Archives, Record Group 95, acc. no. 72A3046, carton 169.

32. Kaufman, *Forest Ranger*, 235.

33. Frome, *Whose Woods These Are*, 340.

Three Fires, 1961

1. Arthur R. Pirsko, "California's 1961 Fire Weather Brings Near-Record Losses," Misc. Paper 70 (Berkeley, CA: Pacific Southwest Forest and Range Experiment Station, U.S. Forest Service, 1962); Earl M. Kidder, "METCALF: Never Saw Anything Like Harlow Fire in 35 Years," *Fresno Bee*, July 23, 1961. The Three Forests Interpretive Association has produced a video of the burn, *The Harlow Fire: Then and Now* (Tollhouse, California).

Interestingly, the Basin fire meanwhile romped over 17,000 hilly acres in the Sierra National Forest and became a textbook case study for blowups; see Craig C. Chandler, *Fire Behavior of the Basin Fire: Sierra National Forest, July 13–22, 1961* (Berkeley, CA: Pacific Southwest Experiment Station, U.S. Forest Service, 1961), 72.

2. Information from Robert D. Baker et al., *The National Forests of the Northern Region: Living Legacy*, FS-500 (Washington, DC: U.S. Forest Service, 1993), 189–90.

3. L. Jack Lyon, "Vegetal Development on the Sleeping Child Burn in Western Montana, 1961 to 1973," USDA Forest Service Research Paper INT-184 (Ogden, UT: Intermountain Forest and Range Experiment Station, U.S. Forest Service, 1976), 2; James E. Lotan, "Cone Serotiny: Fire Relationships in Lodgepole Pine," in *Proceedings, Tall Timbers Fire Ecology Conference 14 and Intermountain Fire Research Council, Fire and Land Management Symposium* (Tallahassee, FL: Tall Timbers Research Station, 1976), 268.

4. On the NFPA, see Rexford Wilson, "Los Angeles Conflagration of 1961: The Devil Wind and Wood Shingles," *NFPA Quarterly* (January 1962): 241–87. A good sense of the evolving crisis is conveyed in a classic documentary film, *Design for Disaster*, which was produced by the Los Angeles Fire Department. Arnold paraphrased in Carl C. Wilson, "The People Pressures in Forest Fire Control," *Proceedings, Society of American Foresters Meeting, September 27–October 1, 1964, Denver, Colorado* (Washington, DC: Society of American Foresters, 1965), 49.

5. The best accounts of the fire are Wilson, "Los Angeles Conflagration," and Capt. Harold W. Greenwood, compiler, "Bel Air–Brentwood and Santa Ynez Fires. Worst Fire in the History of Los Angeles" (official report of the Los Angeles Fire Department, 1962).

Chapter 1. Spark

1. *Proceedings, First Annual Tall Timbers Fire Ecology Conference* (Tallahassee, FL: Tall Timbers Research Station, 1962), 95.

2. John Steinbeck, *Travels with Charley: In Search of America* (1962; repr., New York: Penguin, 1997), 5, xxii, 9.

3. Michael Frome, *Whose Woods These Are: The Story of the National Forests* (Garden City, NY: Doubleday, 1962), 329, 331, 339.

4. *Proceedings, First Annual*, 95, v.

5. E. V. Komarek Sr., *A Quest for Ecological Understanding: The Secretary's Review. March 15, 1958–June 30, 1975* (Tallahassee, FL: Tall Timbers Research Station, 1977), 3.

6. Herbert L. Stoddard, *Report on Cooperative Quail Investigations, 1925–1926* (Washington, DC: Committee Representing the Quail Study Fund for Southern Georgia and Northern Florida, 1926); Simon Joseph Fraser Lovat and Edward Wilson, *The Grouse in Health and in Disease* (London: Smith, Elder, 1911).

7. Stoddard's story has been told often, but let him tell it in his own way: Herbert Stoddard Sr., *Memoirs of a Naturalist* (Norman: University of Oklahoma Press, 1969).

8. The story has been told and retold, mostly in TTRS publications, but the best summary is Komarek, *Quest for Ecological Understanding*. The station has a library that houses its archives, but the cited publication holds pretty closely to the language of the preserved documents.

9. Komarek's term *natural experiments* comes from "Fire Ecology: Grasslands and Man," in *Proceedings, Fourth Annual Tall Timbers Fire Ecology Conference* (Tallahassee, FL: Tall Timbers Research Station, 1965), 216.

10. See Stoddard quotes in *Proceedings, First Annual*, 156–57.

11. *Proceedings, First Annual*, 52.

12. The story of the first burn is found in Blane Heumann, "Reflections on 50 Years of Burning in the Nature Conservancy," *The Nature Conservancy* website, April 26, 2012, accessed October 2012, http://blog.nature.org/2012/04/26/reflections-on-50-years-of-burning-in-the-nature-conservancy/.

13. On prairie fire art, see Stephen Pyne, *Fire: Nature and Culture* (London: Reaktion, 2012), 124–29; on Leopold and fire, see Aldo Leopold, *A Sand County Almanac* (New York: Ballantine, 1990), 28–32; J. T. Curtis and M. L. Partch, "Effect of Fire on Competition Between Blue Grass and Certain Prairie Plants," *American Midland Naturalist* 39 (1949): 427–33.

14. The best quick history of the Conservancy is on its website, accessed January 2013, http://www.nature.org/about-us/vision-mission/history/index.htm.

15. "'Above All . . . Naturalness': An Inspired Report on the Parks," *Sierra Club Bulletin* (March 1963): 3.

16. United States National Park Service, *Compilation of the Administrative Policies for the National Parks and National Monuments of Scientific Significance (Natural Area Category)*, rev. ed. (Washington, DC: U.S. Government Printing Office, 1970), 100–101; hereafter, Green Book. The report was reproduced in many venues; a useful, accessible version is available in *Sierra Club Bulletin* (March 1963).

17. Ibid., 5–6, 102.

18. Ibid., 6, 9.

19. Memorandum, from Secretary of the Interior to Director, National Park Service, May 2, 1963, reproduced in Green Book, 97.

20. Hartzog quote from Tom Nichols, comments to the author, October 13, 2013.

21. Many accounts of Biswell and his career exist, but begin with his own autobiography, Harold H. Biswell, *Prescribed Burning in California Wildlands Vegetation Management* (Berkeley: University of California Press, 1989). Admirers and students

organized a Festschrift with useful memorials: David R. Weise and Robert E. Martin, technical coordinators, *The Biswell Symposium: Fire Issues and Solutions in Urban Interface and Wildland Ecosystems*, General Technical Report PSW-GTR-158 (Berkeley, CA: U.S. Department of Agriculture, Pacific Southwest Research Station, 1995).

22. For the chronology, see H. Thomas Harvey, Howard S. Shellhammer, and Ronald E. Stecker, *Giant Sequoia Ecology: Fire and Reproduction*, Scientific Monograph Series no. 12 (Washington, DC: National Park Service, 1980), xvii.

23. William C. Everhart, *The National Park Service* (Boulder, CO: Westview Press, 1983), 55. Gifford Pinchot, *Breaking New Ground*, repr. (New York: Harcourt, Brace, 1947; New York: Island Press, 1998), 44.

24. See Bruce M. Kilgore, "Origin and History of Wildland Fire Use in the U.S. National Park System," *George Wright Forum* 24, no. 3 (2007): 92–122; McLaughlin episode on p. 102. In broad form the essay follows Hal K. Rothman, *Blazing Heritage: A History of Wildland Fire in the National Parks* (Oxford: Oxford University Press, 2007), although much of Rothman's material comes from interviews with Kilgore, but the essay provides useful details and stands as the best single summary of what happened. For a condensed 30-year history of natural fire endeavors, see Gary Cones and Paul Keller, "Managing Naturally-Ignited Fire: Yesterday, Today, and Tomorrow" (Wildland Fire Lessons Learned Center, 2008).

25. Kilgore, "Origin and History," 103–4; Rothman, *Blazing Heritage*, 112–13, 115–18. I am indebted to Tom Nichols for his perceptive comments on the institutional rifts that have plagued the NPS fire program from its origins.

26. Green Book, 17–18.

27. Kilgore, "Origin and History," 102.

28. Green Book, 105, 103.

29. Ibid., 105.

30. On the idea of wilderness, I of course follow Roderick Nash, whose *Wilderness and the American Mind*, 4th ed. (New Haven, CT: Yale University Press, 2001), remains indispensable. For the legislative history, see Craig W. Allin, *The Politics of Wilderness Preservation* (Westport, CT: Greenwood Press, 1982). Quote from Linda S. Mutch and Robert W. Mutch, "Wilderness Burning: The White Cap Story," unpublished manuscript with transcriptions from recordings from the anniversary event; courtesy Robert Mutch.

31. Roderick Nash, *Wilderness and the American Mind* (New Haven, CT: Yale University Press, 1967).

32. For useful summaries, see Nash, *Wilderness* and Allin, *Politics of Wilderness Preservation*.

33. Robert Marshall, "Mountain Ablaze," *Nature* 46, no. 6 (June/July 1953), 289–92. Forest Service reprint of 1927 article. Brower quote from Rothman, *Blazing Heritage*, 119.

34. William Schwarz, comp., *Voices for the Wilderness* (New York: Ballantine, 1969); for sample fire references, see pp. 23 and 69.

35. Stewart Udall, *The Quiet Crisis* (New York: Holt, Rhinehart, and Winston, 1963), 189.

36. Many excellent accounts of the environmental movement exist; every facet is well served with its studies. Generally, I have followed Samuel P. Hays, *Health, Beauty, and Permanence: Environmental Politics in the United States, 1955–1985*

(Cambridge: Cambridge University Press, 1989), and a much distilled (and chronically expanded) version, *A History of Environmental Politics Since 1945* (Pittsburgh, PA: University of Pittsburgh Press, 2000).

37. Stewart L. Udall, "Address to the Eighth Wilderness Conference," in William Schwarz, ed., *Voices for the Wilderness* (New York: Ballantine Books, 1969), 296.

38. Public Land Law Review Commission, *One Third of the Nation's Land: A Report to the President and to the Congress by the Public Land Law Review Commission* (Washington, DC: Government Printing Office, 1970).

39. Hays, *Beauty, Health, and Permanence*.

40. Rachel Carson, *Silent Spring* (Boston: Houghton Mifflin, 1962), 22; Udall, *Quiet Crisis*, vii.

41. Clapp quote from Ashley L. Schiff, *Fire and Water: Scientific Heresy in the Forest Service* (Cambridge, MA: Harvard University Press, 1962), 13.

42. The labs soon published descriptions of their activities, and the Missoula lab has a short survey history; see *At the Southeastern Forest Experiment Station* (Asheville, NC: Southeastern Forest Experiment Station, 1961); Carl C. Wilson and James B. Davis, *Forest Fire Laboratory at Riverside and Fire Research in California: Past, Present, and Future*, Gen Tech. Report PSW-105 (Berkeley, CA: U.S. Forest Service, Pacific Southwest Forest and Range Experiment Station, 1988), and *Forest Fire Laboratory: Riverside, California* (Berkeley, CA: U.S. Forest Service, Pacific Southwest Forest and Range Experiment Station, 1963); Diane M. Smith, *The Missoula Fire Sciences Laboratory: A 50-Year Dedication to Understanding Wildlands and Fire*, General Technical Report RMRS-GTR-270 (Fort Collins, CO: U.S. Forest Service, Rocky Mountain Research Station, 2012) and Richard J. Klade, *Building a Research Legacy: The Intermountain Station 1911–1997*, General Technical Report RMRS-GTR-184 (Fort Collins, CO: U.S. Forest Service, Rocky Mountain Research Station, 2006). The labs operate under the regional research stations, most of which have their own administrative histories. Also of interest is *A Forest Fire Research Program for the Lake States* (St. Paul, MN: Lake States Forest Experiment Station, 1962), which summarizes the 1961 fire research conference in Green Bay, and James B. Craig, "Crown Fires Return to the Lake States," *American Forests* 67, no. 4 (April 1961): 18–19, 66–67.

43. See Steven H. Bullard, et al., "A 'Driving Force' in Developing the Nation's Forests: The McIntire-Stennis Cooperative Forestry Research Program," *Journal of Forestry* (April/May 2011): 141–48.

44. I follow Mark J. Schroeder et al., "Technical Development of the National Fire-Danger Rating System" (unpublished report, U.S. Forest Service), and John E. Deeming et al., *The National Fire-Danger Rating System*, U.S. Forest Service Research Paper RM-84 (Fort Collins, CO: U.S. Forest Service, Rocky Mountain Forest and Range Experiment Station, 1972), 1–2. A more general survey is available in Colin C. Hardy and Charles E. Hardy, "Fire Danger Rating in the United States of America: An Evolution Since 1916," *International Journal of Wildland Fire* 16 (2007): 217–31.

45. Quotes from Schroeder, "Technical Development," 7. Mark J. Schroeder and Charles C. Buck, *Fire Weather: A Guide for Application of Meteorological Information to Forest Fire Control Operations*, USDA Agriculture Handbook 360 (Washington, DC: Government Printing Office, 1970).

46. A quick précis of antecedents is available in Craig C. Chandler, Theodore G. Storey, and Charles D. Tangren, *Prediction of Fire Spread Following Nuclear Explosions*, Research Paper PSW-5 (Berkeley, CA: U.S. Forest Service, Pacific Southwest Forest and Range Experiment Station, 1963), 1. See also Committee on Fire Research and Fire Research Conference, *Methods of Studying Mass Fires: Second Fire Research Correlation Conference*, National Academy of Sciences–National Research Council Publication 569 (Washington, DC: Division of Engineering and Industrial Research, National Academy of Sciences, National Research Council, 1958).

47. Chandler, Storey, and Tangren, *Prediction of Fire Spread*, 1; Clive M. Countryman, *Mass Fires and Fire Behavior*, Research Paper PSW-19 (Berkeley, CA: U.S. Forest Service, Pacific Southwest Forest and Range Experiment Station,1964).

48. Countryman, *Mass Fires*. On the mass fire experiments, see Clive M. Countryman, *Project Flambeau . . . : An Investigation into Mass Fire (1964–1967), Final Report*, vol. 1 (Berkeley, CA: U.S. Forest Service, 1969). The final report included three volumes, the others on various technical matters. Regarding the NFCS, see William R. Moore, James W. Jay, and John H. Dieterich, *Defending the United States from Nuclear Fire: The Final Report of the National Fire Coordination Study* (Washington, DC: U.S. Forest Service, Division of Fire Control, 1966).

49. See William E. Towell, "Disaster Fires: A National Program for Wildfire Control," *American Forests*, July 1968, editorial.

50. See William R. Moore, "Analysis of Past Recommendations: An Analysis and Summary," in Combined Mutual Aid Analysis: National Fire Coordination Study (Washington, DC: U.S. Fire Service, Division of Fire Control, 1964).

51. Agee quote from David Carle, *Burning Questions: America's Fight with Nature's Fire* (Westport, CT: Praeger, 2002), 127.

52. Schiff, *Fire and Water*, 115.

53. Weaver quote in Carle, *Burning Questions*, 126.

54. William J. Robbins, chair, *A Report by the Advisory Committee to the National Park Service on Research* (Washington, DC: National Academy of Sciences–National Research Council, 1963), quote on p. iv; hereafter, *Robbins Report*. For background on how the report relates to the history of park management, see Richard West Sellars, *Preserving Nature in the National Parks: A History* (New Haven, CT: Yale University Press, 1997), esp. 214–33.

55. *Robbins Report*, 1, 21, 32.

56. Sellars, *Preserving Nature*, esp. 218–28.

57. Eugene P. Odum, "Concluding Remarks of the Co-Chairman," *Proceedings, Tall Timbers Fire Ecology Conference* 2, E. V. Komarek, chair (Tallahassee, FL: Tall Timbers Research Station, 1962), 178, 180.

58. Gary Gray, *Radio for the Fireline: A History of Electronic Communication in the Forest Service, 1905–1975*, FS-309 (Washington, DC. U.S. Forest Service, 1982).

59. James B. Craig, "Fire Prevention Breakthrough," *American Forests* 69, no. 3 (March 1963): 36–38.

60. My account of BLM origins derives from James Muhn and Hanson R. Stuart, *Opportunity and Challenge: The Story of BLM* (Washington, DC: Bureau of Land Management, 1988), and James R. Skillen, *The Nation's Largest Landlord: The Bureau of Land Management in the American West* (Lawrence: University Press of Kansas, 2009).

61. Background history and quote from "Progress Report: Bureau of Land Management Fire Protection Program Planning and Development" (unpublished report, BLM, 1969), 3–4. Jack F. Wilson, "From Whence They Came: A Perspective on Federal Wildland Firefighters in the Department of the Interior" (unpublished manuscript, copy given to author by Wilson), 6–7. On emergency presuppression funds, see Samuel M. Brock, "BLM Fire Control Study: Economics of Fire Control with Special Reference to BLM Protection Operations" (unpublished report, 1964), 21.

62. The legislative unfolding is nicely documented in Skillen, *Nation's Largest Landlord*, 41–46.

63. Quote from "Progress Report: Bureau of Land Management Fire Protection Program Planning and Development" (unpublished report, BLM, 1969), 4.

64. Probably the best account of Alaskan fire is Stephen J. Pyne, *Fire in America: A Cultural History of Wildland and Rural Fire* (Princeton, NJ: Princeton University Press, 1982), 497–513; and Susan K. Todd and Holly Ann Jewkes, *Wildland Fire in Alaska: A History of Organized Fire Suppression and Management in the Last Frontier*, Agricultural and Forestry Experiment Station Bulletin No. 114 (Fairbanks: University of Alaska Fairbanks, Agricultural and Forestry Experiment Station, 2006), which amplifies and extends that narrative. Together they offer a good guide to sources.

65. Charles E. Hardy, "Conflagration in Alaskan Forests 1957: A Report of Forest Fire Investigations in Interior Alaska, July 22–August 4, 1957" (unpublished report, U.S. Forest Service, Intermountain Forest and Range Experiment Station, 1957), 6.

66. A particularly helpful account of Alaskan fire at the onset of statehood is R. R. Robinson, "Forest and Range Fire Control in Alaska," *Journal of Forestry* 58 (1960): 448–53.

67. Quote from William A. Adams, "The Role of Fire in the Alaska Taiga: An Unsolved Problem" (unpublished report, BLM, Alaska State Office, Division of Fire Control, December, 1974).

68. Quote from J. H. Richardson, "Values Protected in Interior Alaska," in *Fire in the Northern Environment: A Symposium*, ed. C. W. Slaughter, Richard J. Barney, and G. M. Hansen (Portland, OR: Pacific Northwest Forest and Range Experiment Station, U.S. Forest Service, 1971), 174.

69. Adams, "Role of Fire," 15.

70. Quote from James B. Craig, "Alaska Burns," *American Forests* 75, no. 10 (October 1969): 63.

71. For details of this rapid evolution, see Brock, "BLM Fire Control Study."

72. See J. W. Jay, *Fire Studies Report: Nevada Fires: Analysis and Study* (Washington, DC: U.S. Forest Service, Division of Fire Control, 1964).

73. See Brock, "BLM Fire Control Study."

74. Chronology collated out of documents housed among National Interagency Fire Center (NIFC) historic files, in particular, Jack F. Wilson, "The Aviation Program of the Bureau of Land Management" (unpublished report, 1988); Memorandum, BLM Director-BIFC to Director, BLM, "Western Fire Coordination Center," January 21, 1971; Memorandum, Allan J. West to Regional Forester, R-5, "Lloyd Britton Detail to WO," September 29, 1982; Gordon J. Stevens, "A Chronology of BIFC Development" (unpublished report, November 1976), although the text is overwritten

with comments and objections and includes obvious errors; and Memorandum, Richard Stauber to John Chambers, "BIFC 25th Anniversary," February 22, 1990.

75. Memorandum, Merle Lowden to M. M. Nelson, "Agreements," December 18, 1968, NIFC historic files.

76. Ibid.

77. Eugene P. Odum, "Concluding Remarks of the Co-Chairman," *Proceedings, Tall Timbers Fire Ecology Conference 2*, E. V. Komarek, chair (Tallahassee, FL: Tall Timbers Research Station, 1962).

78. The best survey of America's fire provinces is the collection of essays *To the Last Smoke*, published at http://firehistory.asu.edu. Each of the regional suites has an introduction and conclusion that sketch the province's defining traits and place it within the national geography and narrative.

79. Many accounts exist of both episodes and their larger contexts, but a concise, sympathetic reading is available in Samuel T. Dana and Sally K. Fairfax, *Forest and Range Policy: Its Development in the United States*, 2nd ed. (New York: McGraw-Hill, 1980), 226–29. For a more detailed rendering within the context of the agency's rechartering, see Dennis C. LeMaster, *Decade of Change: The Remaking of Forest Service Statutory Authority during the 1970s* (Westport, CT: Greenwood Press, 1984), and for how the old and new Forest Service could split, Frederick H. Swanson, *The Bitterroot and Mr. Brandborg: Clearcutting and the Struggle for Sustainable Forestry in the Northern Rockies* (Salt Lake City: University of Utah Press, 2011).

Last Hurrahs, 1967 and 1970

1. Quote from William R. Moore, *Fire in the Northern Rockies, 1967: Some Lessons for Wartime Fire Defense*, prepared for the Office of Civil Defense, Project Order DAHC20-67-0083 (Washington, DC: U.S. Forest Service, 1967), 1. See also Neil M. Rham, *Region 1 Annual Narrative Fire Report, C.Y. 1967* (Missoula, MT: U.S. Forest Service, 1967).

2. William R Moore, Assistant Director, WO—Fire Control, "Report on Field Travel in Region 1, August 8–19 and September 3–9, 1967," RG 95, Acc. 72-A-3046, Box 168, p. 6.

3. Hal E. Anderson, *Sundance Fire: An Analysis of Fire Phenomena*, Research Paper INT-56 (Ogden, UT: Intermountain Forest and Range Experiment Station, U.S. Forest Service, 1968).

4. A cache of relevant documents, including the entire "Proceedings: Glacier Forest Fire Review," is housed in the Glacier Park Archives, Forestry (Y), Box 309, Folder 21, Forest Fire Review, 1967.

5. All quotes from documents in Glacier Park Archives, Forestry, Box 309, Folder 21: preamble from released report, untitled; Meneely from "Proceedings: Glacier Fire Review," 19; and Memorandum, Superintendent to All participants in the Glacier Fire Review, November 30, 1967.

6. Gunzel quote from "Proceedings: Glacier Forest Fire Review," 34.

7. Oddly, the main official records are the iconic photos; the best written documentation lies in newspapers, of which a composite summary is available at "Five

Fires in the Wenatchee National Forest Burn 122,000 Acres Beginning August 23, 1970," *Free Online Encyclopedia of Washington State History*, HistoryLink essay 5498.

8. Clinton B. Phillips, *California Aflame! September 22–October 4, 1970* (Sacramento: State of California, Resources Agency, Department of Conservation, Division of Forestry, 1971), 1.

9. Quotes from Carl C. Wilson, "Commingling of Urban and Forest Fires," *Fire Research Abstracts and Reviews* 13 (1971): 38; Phillips, *California Aflame!*, 1; and R. S. Alger, "The Great Oakland, Los Angeles, and San Diego Fires: September 22 to 29 1970," NOLTR 71–229 (Silver Spring, MD: Naval Ordnance Laboratory, 1971), 1.

10. Task Force on California's Wildland Fire Problem, "Recommendations to Solve California's Wildland Fire Problem," submitted to California Resources Agency, June 1972.

Chapter 2. Hotline

1. Henry W. DeBruin, "From Fire Control to Fire Management: A Major Policy Change in the Forest Service," in *Proceedings, Tall Timbers Fire Ecology Conference 14 and Intermountain Fire Research Council, Fire and Land Management Symposium* (Tallahassee, FL: Tall Timbers Research Station, 1976), 11, from a joint symposium held in 1974.

2. Craig Chandler, "Why Not Let Fires Burn" (unpublished talk, ca. 1972).

3. David Carle gives a lively, personality-centered account of the fire revolution in *Burning Questions: America's Fight with Nature's Fire* (Westport, CT: Prager, 2002).

4. Raymond Conarro, "The Place of Fire in Southern Forestry," *Journal of Forestry* 40, no. 2 (February 1942): 129–31. For an excellent distillation of the evolution of the controversy in the South, see Roland J. Diebold, "The Early History of Wildfires and Prescribed Burning," in *Prescribed Burning Symposium Proceedings* (Asheville, NC: U.S. Forest Service, Southeastern Forest Experimental Station, 1971), 11–20, and of course the many volumes of the proceedings of the Tall Timbers fire ecology conferences, especially the early conferences.

5. Merlin J. Dixon, *A Guide to Fire by Prescription* (Atlanta, GA: U.S. Forest Service, Southern Region, 1965); William R. Beaufait, *Prescribed Fire Planning in the Intermountain West*, Research Paper INT-26 (Ogden, UT: Intermountain Forest and Range Experiment Station, U.S. Forest Service, 1966); Donald T. Gordon, *Prescribed Burning in the Interior Ponderosa Pine Type of Northern California: A Preliminary Study*, Research Paper PSW-45 (Berkeley, CA: Pacific Southwest Forest and Range Experiment Station, U.S. Forest Service, 1967); Southwest Interagency Fire Council, *Guide to Prescribed Fire in the Southwest* (n.p.: Southwest Interagency Fire Council, 1968).

6. *Prescribed Burning Symposium Proceedings* (Asheville, NC: U.S. Forest Service, Southeastern Forest Experimental Station, 1971), 5.

7. Dale D. Wade and Michael C. Long, "New Legislation Aids Hazard-Reduction Burning in Florida," *Journal of Forestry* (November 1979): 725–26. For a good survey of how this legislation and its successors evolved, see Dana C. Bryan, ed., *Conference Proceedings: Environmental Regulation and Prescribed Fire: Legal and Social Challenges* (Center for Professional Development, Florida State University, 1997).

8. Louis L. Gunzel, "National Policy Change . . . Natural Prescribed Fire," *Fire Management* 35, no. 3 (Summer 1974): 6–8.

9. David D. Devet, "Descon: Utilizing Benign Wildfires to Achieve Land Management Objectives," *Tall Timbers Fire Ecology Conference* 14, 33–43.

10. For a concise but slightly different version—all versions differ in some respects—see Jack Wilson, "History of NWCG," *Fire Management Notes* 39 (Spring 1978): 13–16. For a longer version, see Jack Wilson and Jerry L. Monesmith, "The National Wildfire Coordinating Group: Then and Now" (unpublished report, NWCG files). I am grateful to the NWCG for making the full chronicle of its minutes available to me in digital format.

11. "Minutes of Meeting of NWCG. January 24–25, 1973," National Wildfire Coordinating Group files. For other particulars, see the minutes of designated meetings.

12. DeBruin, "From Fire Control to Fire Management." Self-characterizations from conversations with author in fall 1977.

13. Qualification and Certification Working Team, "National Interagency Fire Qualification System," *Fire Management Notes* 42, no. 1 (Winter 1980/81): 15–16; James B. Davis, "Building Professionalism into Forest Fire Suppression," *Journal of Forestry* (July 1979): 423–26.

14. Quotes from Dick Chase, interview by Jamie Lewis, February 19, 2007, U.S. Department of Agriculture, Forest Service Region Five History Project, pp. 23 and 20.

15. Chase, interview, pp. 35–36.

16. Victor W. Geraci and Region 5 Oral History Project Volunteers, comps., *The Unmarked Trail: Managing National Forests in a Turbulent Era: Region 5 Oral History, Volume II, 1960s to 1990s*, R5-FR-011, September 2009 (Vallejo, CA: Forest Service, Pacific Southwest Region, 2009), 153–87; quotes from pp. 161, 172.

17. Chase, interview, p. 51.

18. McGuire decision in Geraci, *Unmarked Trail*, 176; see pp. 182–87 for a brief summary of how ICS was exported.

19. A thumbnail sketch of the origins is available in Gareth C. Moon, "Welcoming Address," in Intermountain Fire Research Council, *The Role of Fire in the Intermountain West* (Missoula: Intermountain Fire Research Council and University of Montana, School of Forestry, 1970), 1.

20. Bruce Kilgore, "Research Needed for an Action Program of Restoring Fire to Giant Sequoias," in Intermountain Fire Research Council, *Role of Fire*, 176.

21. Harold H. Biswell et al., *Ponderosa Fire Management: A Task Force Evaluation of Controlled Burning in Ponderosa Pine Forests of Central Arizona*, Miscellaneous publication no. 2 (Tallahassee, FL: Tall Timbers Research Station, 1973).

22. Minutes, meeting of February 18, 1972, Tall Timbers Research Station Archives, p. 2; minutes, meeting of April 28, 1975.

23. U.S. Forest Service, *Fire in the Environment: Symposium Proceedings, May 1–5, 1972. Denver* (Washington, DC: U.S. Forest Service, 1972).

24. Harold A. Mooney and C. Eugene Conrad, technical coordinators, *Proceedings of the Symposium on the Environmental Consequences of Fire and Fuel Management in Mediterranean Ecosystems*, General Technical Report WO-3 (Washington, DC: U.S. Forest Service, 1977); Harold A. Mooney et al., technical coordinators, *Fire*

Regimes and Ecosystem Properties: Proceedings of the Conference, December 11–15, 1978, Honolulu, Hawaii, General Technical Report WO-26 (Washington, DC: U.S. Forest Service, 1981).

25. L. Jack Lyon, *Effects of Fire on Fauna*, General Technical Report WO-6 (Washington, DC: U.S. Forest Service, 1978); Carol G. Wells, *Effects of Fire on Soil*, General Technical Report WO-7 (Washington, DC: U.S. Forest Service, 1979); David V. Sandberg, *Effects of Fire on Air*, General Technical Report WO-9 (Washington, DC: U.S. Forest Service, 1979); Arthur R. Tiedemann, *Effects of Fire on Water*, General Technical Report WO-10 (Washington, DC: U.S. Forest Service, 1979); Robert Edward Martin, *Effects of Fire on Fuels*, General Technical Report WO-13 (Washington, DC: U.S. Forest Service, 1979); James E. Lotan, *Effects of Fire on Flora*, General Technical Report WO-16 (Washington, DC: U.S. Forest Service, 1981). John Hendee, George Stankey, and Robert Lucas, *Wilderness Management*, Agriculture Handbook 1365 (Washington, DC: Government Printing Office, 1978).

26. For the fire world as it appeared to Barrows, see J. S. Barrows, "Forest Fire Research for Environmental Protection," *Journal of Forestry* (January 1971): 17–20.

27. John E. Deeming et al., *National Fire-Danger Rating System*, Research Paper RM-84 (Fort Collins, CO: Rocky Mountain Forest and Range Experiment Station, U.S. Forest Service, 1972); John E. Deeming et al., *National Fire-Danger Rating System*, Research Paper RM-84, rev. ed. (Fort Collins, CO: Rocky Mountain Forest and Range Experiment Station, U.S. Forest Service, 1974); and John E. Deeming, Robert E. Burgan, and Jack D. Cohen, *The National Fire-Danger Rating System, 1978*, General Technical Report INT-39 (Ogden, UT: Intermountain Forest and Range Experiment Station, U.S. Forest Service, 1977).

28. Richard C. Rothermel, *A Mathematical Model for Predicting Fire Spread in Wildland Fuels*, Research Paper INT-115 (Ogden, UT: Intermountain Forest and Range Experiment Station, U.S. Forest Service, 1972), iii, vi.

29. Rothermel quote from "The Rothermel Fire-Spread Model: Still Running Like a Champ," *Fire Science Digest* 2 (March 2008): 2. Effects on the Riverside Lab from Dick Chase, interview, February 19, 2007, Forest Service Region Five History Project, pp. 66. On Focus, see James B. Davis and Robert L. Irwin, "FOCUS: A Computerized Approach to Fire Management Planning," *Journal of Forestry* (September 1976): 615–18.

30. Craig C. Chandler and Charles F. Roberts, "Problems and Priorities for Forest Fire Research," *Journal of Forestry* (October 1973): 626.

31. Hal E. Anderson, *Appraising Forest Fuels: A Concept*, Research Note INT-187 (Ogden, UT: Intermountain Forest and Range Experiment Station, U.S. Forest Service, 1974); James Brown, *Handbook for Inventorying Downed Woody Material*, General Technical Report GTR-INT-16 (Ogden, UT: Intermountain Forest and Range Experiment Station, U.S. Forest Service, 1974); *National Fuels Management*, Workshop, January 22–26, 1973 (unpublished handbook, U.S. Forest Service); Owen P. Cramer, ed., *Environmental Effects of Forest Residues Management in the Pacific Northwest*, General Technical Report PNW-24 (Portland, OR: Pacific Northwest Forest and Range Experiment Station, U.S. Forest Service, 1974); John M. Pierovich et al., *Forest Residues Management Guidelines for the Pacific Northwest*, General Technical Report PNW-33 (Portland, OR: Pacific Northwest Forest and Range

Experiment Station, U.S. Forest Service, 1975); David J. Parsons, "Fire and Fuel Accumulation in a Giant Sequoia Forest," *Journal of Forestry* (February 1978), 104–5.

32. Clive M. Countryman and Charles W. Philpot, *The Physical Characteristics of Chamise as a Wildland Fuel*, Research Paper PSW-66 (Berkeley, CA: Pacific Southwest Range and Experiment Station, U.S. Forest Service, 1970).

33. For an explanation of the strategy, see C. W. Philpot, "New Fire Control Strategy Developed for Chaparral," *Fire Management* 35, no. 1 (1974): 3–7, and a longer version, "Vegetative Features as Determinants of Fire Frequency and Intensity," in Mooney and Conrad, *Environmental Consequences of Fire and Fuel Management*, 12–16. Quote from Keith E. Klinger and Carl C. Wilson, "What Are We Going to Do About the Brush in Southern California?" *Fire Control Notes* 29, no. 1 (1968): 16.

34. Lisle R. Green, *Burning by Prescription in Chaparral*, General Technical Report PSW-51 (Berkeley, CA: Pacific Southwest Range and Experiment Station, U.S. Forest Service, 1981). Other information about local practices from interviews by author with former fire officers of the Angeles National Forest.

35. The southeast had its own tradition of fuelbreaks, upgraded to more modern forms during this era. Interestingly, Raymond Conarro was instrumental in the process; see Hamlin L. Williston and R. M. Conarro, "Firebreaks of Many Uses," *Fire Control Notes* 31, no. 1 (1969/70): 11–13. The northeast, too, piled on; see Von J. Johnson, "Hardwood Fuel-Breaks for Northeastern United States," *Journal of Forestry* (September 1975): 588–89.

36. Lisle E. Green and Harry E. Schimke, *Guide for Fuel-Breaks in the Sierra Nevada Mixed-Conifer Type* (Berkeley, CA: Pacific Southwest Experiment Station, U.S. Forest Service, 1971); Verdie E. White and Lisle R. Green, *Fuel-Breaks in Southern California 1958–1965* (Berkeley, CA: Pacific Southwest Experiment Station, U.S. Forest Service, 1967); Jay R. Bentley, *Conversion of Chaparral Areas to Grassland: Techniques Used in California*, Agriculture Handbook no. 328 (Washington, DC: U.S. Forest Service, 1967); and, most comprehensively, Lisle R. Green, *Fuelbreaks and Other Fuel Modification for Wildland Fire Control*, Agriculture Handbook no. 499 (Washington, DC: U.S. Forest Service, 1977). For a quick summary, see James L. Murphy, Lisle R. Green, and Jay R. Bentley, "Fuel-Breaks: Effective Aids, Not Cure-Alls," *Fire Control Notes* 28, no. 1 (1967): 4–5.

37. Among the most widely circulated accounts is Bruce M. Kilgore and George S. Briggs, "Restoring Fire to High Elevation Forests in California," *Journal of Forestry* (May 1972): 266–71.

38. SEKI archives, Kilgore accounts. Concise summary (and McLaughlin information) from Hal K. Rothman, *Blazing Heritage: A History of Wildland Fire in the National Parks* (Oxford: Oxford University Press, 2007), 106–7.

39. On Yosemite, see Rothman, *Blazing Heritage*, 115–18 (Agee quote on p. 115).

40. See "Fire in the National Parks Symposium," *Proceedings, Tall Timbers Fire Ecology Conference, 1972* (Tallahassee, FL: Tall Timbers Research Station, 1973), 339–488.

41. Quotes from Bruce M. Kilgore, "Impact of Prescribed Burning on a Sequoia–Mixed Conifer Forest," *Tall Timbers Fire Ecology Conference, 1972*, 345–76; Peter H. Schuft, "A Prescribed Burning Program for Sequoia and Kings Canyon National Parks," *Tall Timbers Fire Ecology Conference, 1972*, 77–389; and John S. McLaughlin,

"Restoring Fire to the Environment in Sequoia and Kings Canyon National Parks," *Tall Timbers Fire Ecology Conference*, 1972, 391–96, specifically 372, 386–87, 389, 394–95.

42. Rothman, *Blazing Heritage*, 123–24 (Albright quote, p. 119).

43. Ibid., 112–23.

44. A summary of the meeting is available in Memorandum, Associate Regional Director, Park System Management, Rocky Mountain Region to Superintendents, Rocky Mountain Region, Subject: Natural Fire Management, February 28, 1975.

45. Rothman, *Blazing Heritage*, 124; Green Book: United States National Park Service, *Compilation of the Administrative Policies for the National Parks and National Monuments of Scientific Significance (Natural Area Category)*, rev. ed. (Washington, DC: U.S. Government Printing Office, 1970).

46. Ibid., 124–25.

47. Ibid., 125–26.

48. Ibid., 126–27; *Fire Management*, NPS-18 (unpublished handbook, frequently amended and reissued, National Park Service), chap. 1, exhibits 1 and 2, chap. 3, p. 1

49. NPS-18, chap. 4, p. 1; chap. 5, p. 1; chap. 7, p. 2.

50. Rothman, *Blazing Heritage*, 137–40; and Teralene S. Fox, comp., *La Mesa Fire Symposium: Los Alamos, New Mexico, October 6 and 7, 1981*, LA-9236-NERP (Los Alamos National Laboratory, 1984).

51. Rothman, *Blazing Heritage*, 141–45, and my field notes from Rocky Mountain when I was sent to the park in 1983 and 1984 to write an operational fire plan.

52. David B. Butts, "Fire Management in Rocky Mountain National Park," in *Proceedings, Tall Timbers Fire Ecology Conference 14 and Intermountain Fire Research Council, Fire and Land Management Symposium* (Tallahassee, FL: Tall Timbers Research Station, 1976), 73, 69–70.

53. David B. Butts, "Case Study: The Ouzel Fire, Rocky Mountain National Park," in *Proceedings: Symposium and Workshop on Wilderness Fire*, James E. Lotan et al., technical coordinators, General Technical Report INT-182 (Ogden, UT: Intermountain Forest and Range Experiment Station, U.S. Forest Service, 1985), 248–51. "Board of Review Report for the Ouzel Fire, Rocky Mountain National Park, Colorado" (unpublished report, 1978), files, Resource Management, Rocky Mountain National Park.

54. "Board of Review Report for the Ouzel Fire, Rocky Mountain National Park, Colorado" (unpublished report, 1978), files, Resource Management, Rocky Mountain National Park; Rothman, *Blazing Heritage*, 144.

55. The theme of failure by attrition is less often addressed than successes. But lines of print do not substitute for acres burned. For an insider's, life-at-the-bottom view of the fire revolution during its early decade, see Stephen Pyne, *Fire on the Rim: A Firefighter's Season at the Grand Canyon* (Seattle: University of Washington Press, 1995).

56. James Muhn, *Opportunity and Challenge: The Story of BLM* (Washington, DC: Department of the Interior, Bureau of Land Management, 1988), esp. 210–12 for the fire program.

57. R. R. Robinson, "Fire Management Direction," in Intermountain Fire Research Council, *Role of Fire*, 143–52; William R. Moore, "Fire Management in the Northern Rockies," Intermountain Fire Research Council, *Role of Fire*, 170.

58. Unless otherwise stated, I follow Susan K. Todd and Holly Ann Jewkes, *Wildland Fire in Alaska: A History of Organized Fire Suppression and Management in the Last Frontier*, Agricultural and Forestry Experiment Station Bulletin No. 114 (Fairbanks: University of Alaska Fairbanks, 2006), 34–40. For an alternative history, for which modern Alaska is a conclusion, not as here a point of departure, see Stephen Pyne, *Fire in America* (Princeton, NJ: Princeton University Press, 1982), 507–13.

59. C. W. Slaughter et al., eds. *Fire in the Northern Environment: A Symposium* (Portland, OR: Pacific Northwest Forest and Range Experiment Station, U.S. Forest Service, 1971).

60. On the USFS role, see John A. Sandor et al., *The Forest Service in Alaska*, Alaska Region Series no. R10-28 (Juneau: U.S. Forest Service, Alaska Region, 1978).

61. George L. Turcott, "Fire Management: A Vital Factor in Land Use Planning," in George L. Turcott et al., *Fire Management in the Northern Environment: A Symposium* (Anchorage: Bureau of Land Management, Alaska State Office, 1979), 1–7; Leslie A. Viereck and Linda A. Schandelmeier, *Effects of Fire in Alaska and Adjacent Canada: A Literature Review*, BLM-Alaska Technical Report 6 (Anchorage: Bureau of Land Management, Alaska State Office, 1980).

62. Claus M. Naske and Herman E. Slotnick, *Alaska: A History of the 49th State*, 2nd ed. (Norman: University of Oklahoma Press, 1994), 224–40; Roderick Nash, *Wilderness and the American Mind*, 4th ed. (New Haven, CT: Yale University Press, 2001), 296–315.

63. Alaska Department of Education, *Applicability of Wildlands Fire-Fighting Techniques for Structural Fires* (Wasilla: Alaska Department of Education, Fire Service Training Program, 1977).

64. Stanley H. Anderson, comp., *Effects of the 1976 Seney National Wildlife Refuge Wildfire on Wildlife and Wildlife Habitat*, Resource Publication 146 (Washington, DC: Department of the Interior, Fish and Wildlife Service, 1982); Ernest Hemingway, "The Big Two-Hearted River: Part 1," in *The Short Stories of Ernest Hemingway* (New York: Scribner, 1953), 209; Roswell K. Miller, "The Keetch-Byram Drought Index and Three Fires in Upper Michigan, 1976," in *Fifth Joint Conference on Fire and Forest Meteorology* (Boston: American Meteorological Society, 1978), 63–67; Luke Popovich, "Up in Flames: Taking Heat on the Seney," *Journal of Forestry* (March 1977): 147–50.

65. Memorandum, to Regional Directors et al., from Director, U.S. Fish and Wildlife Service, Policy Update No. 13, Fire Management in the National Wildlife Refuge System, June 16, 1977.

66. "Narrative Report: Pocket Fire," Okefenokee National Wildlife Refuge (unpublished report).

67. Quotes from Art Belcher, Fire Oral History Interviews, USFWS Fire Management Office, NIFC.

68. Ibid.

69. Art Belcher, "Development of the FWS Fire Management Program: A Perspective by One Old Firedog" (unpublished report), USFWS Fire Management Office, NIFC archives.

70. See Paul S. Truesdell, "Postulates of the Prescribed Burning Program of the Bureau of Indian Affairs," in *Proceedings, Tall Timbers Fire Ecology Conference 9*

(Tallahassee, FL: Tall Timbers Research Station, 1969), 235–40; and Harry Kallander, "Controlled Burning on the Fort Apache Indian Reservation, Arizona, *Tall Timbers Fire Ecology Conference* 9, 241–49.

71. For the basic political and legal structure, I rely on Donald L. Fixico, *Bureau of Indian Affairs* (Santa Barbara, CA: Greenwood, 2012).

72. Harold H. Biswell et al., *Ponderosa Fire Management: A Task Force Evaluation of Controlled Burning in Ponderosa Pine Forests of Central Arizona*, Miscellaneous Publication no. 2 (Tallahassee, FL: Tall Timbers Research Station, 1973), 1–2.

73. On the Mesa Verde project, see Paul Broyles, "An Interagency Cooperative Effort," *Fire Management Notes* 42, no. 1 (Winter 1980–81): 3–4.

74. Biswell, *Ponderosa Fire Management*, 23.

75. Joseph F. Pechanec, "Research Needed to Guide Fire Management Direction," in Intermountain Fire Research Council, *Role of Fire*, 154.

76. Memorandum, Fire Policy and Procedure Review Committee to Merle S. Lowden, February 10, 1967.

77. Charles W. Philpot, "One Man's Opinion; or, Further Comments on a Burning Question," Intermountain Fire Research Council, *Role of Fire*, 225; Thurman H. Trosper, "Institutional Barriers in Fire Management," Intermountain Fire Research Council, *Role of Fire*, pl. 164.

78. "Fire Policy Meeting, Denver, Colorado, May 12–14, 1971," i, 1; Craig Chandler, "Why Not Let Fires Burn" (unpublished talk, ca. 1972).

79. McGuire quoted in Harold K. Steen, *The Chiefs Remember: The Forest Service, 1952–2001* (Durham, NC: Forest History Society, 2004), 41.

80. David F. Aldrich and Robert W. Mutch, "Wilderness Fires Allowed to Burn More Naturally," *Fire Control Notes* 33, no. 1 (1971): 3–5; "Ecological Interpretations of the White Cap Drainage: A Basis for Wilderness Fire Management" (unpublished report, U.S. Forest Service, 1972); Linda S. Mutch and Robert W. Mutch, "Wilderness Burning: The White Cap Story," draft manuscript (lent courtesy of Robert W. Mutch); interviews I conducted with Bob Mutch and Dave Aldrich. See also Orville L. Daniels, "Fire Management Takes Commitment," in *Proceedings, Tall Timbers Fire Ecology Conference 14 and Intermountain Fire Research Council Fire and Land Management Symposium* (Tallahassee, FL: Tall Timbers Research Station, 1976), 163–65. *Western Wildlands* ran a special fire issue in the summer of 1974 that included an article by Mutch and Daniels on the Fitz Creek experience (inexplicably spelled "Fritz" Creek).

81. A lively survey of the Gila's origins is available in Adam Burke, "Keepers of the Flame," *High Country News*, http://www.hcn.org/issues/286/15102/print_view. See also Thomas W. Swetnam and John H. Dietrich, "Fire History of Ponderosa Pine Forests in the Gila Wilderness, New Mexico," *Symposium and Workshop on Wilderness Fire*, 390–97; and "Donald R. Webb and Ronald L. Henderson, "Gila Wilderness Prescribed Fire Program," *Symposium and Workshop on Wilderness Fire*, 412–14.

82. Robert D. Gale, *Evaluation of Fire Management Activities on the National Forests*, Policy Analysis Staff Report (Washington, DC: U.S. Forest Service, 1977), 8.

83. "Fire Chiefs National Meeting. Alexandria, Virginia. October 25–November 3, 1972" (unpublished proceedings in author's possession); U.S. Forest Service, *Fire in the Environment*.

84. Gale, *Evaluation of Fire Management Activities*, 15.
85. Ibid.
86. H. Ames Harrison, "Rural Community Fire Protection Program: Two Years of Accomplishment," *Journal of Forestry* (December 1977): 780–81.
87. The title changed again in 2000 to *Fire Management Today*. Lynn R. Biddison, "USA-USSR Cooperation in Forest Fire Protection," *Fire Management* 36, no. 4 (Fall 1975): 14–15, 28.
88. National Commission on Fire Prevention and Control, *America Burning* (Washington, DC: U.S. Government Printing Office, 1973).
89. Henry W. DeBruin, "From Fire Control to Fire Management: A Major Policy Change in the Forest Service," *Tall Timber Fire Ecology Conference 14*, 11–17; and William R. Moore, "Fire, Land, and the People," *Tall Timber Fire Ecology Conference 14*, 645–54.
90. John R. McGuire, "Fire as a Force in Land Use Planning," *Tall Timber Fire Ecology Conference 14*, 439–44; and "Presentation of National Fire Management Award to Dr. Bruce M. Kilgore," ibid., 445.
91. H. P. Gibson, Lance F. Hodgin, and John L. Rich, "Evaluating National Fire Planning Methods and Measuring Effectiveness of Presuppression Expenditures" (unpublished report, U.S. Forest Service, September 1976). Gale, *Evaluation of Fire Management Activities*, 15, 53.
92. Gale, *Evaluation of Fire Management Activities*, 12.
93. Ibid., 47, 51.
94. "Fire Planning Research Status and Needs: Report of the Forest Service Fire Research Project Leaders Conference, Macon, Georgia, January 11–13, 1977"; "Revised Fire Management Policy Fact Sheet, Forest Service, USDA" (1978); Letter to AFM, Cooperative Fire Protection, and FFASR from James F. Mann, December 15, 1977, on draft policy. For explanations offered to the interested fire public, see Thomas C. Nelson, "Fire Management Policy in the National Forests: A New Era," *Journal of Forestry* (November, 1979): 723–25; James B. Davis, "A New Fire Management Policy on Forest Service Lands," *Fire Technology* 15 (February 1979): 43–50. For the final policy statement, approved at the end of 1977, see U.S. Forest Service, "National Fire Management Policy: Final Action Plan" (1977).
95. McGuire, letter of transmittal, "Evaluation of Fire Management Activities on the National Forests"; quote from Randal O'Toole, *Reforming the Fire Service: An Analysis of Federal Fire Budgets and Incentives* (n.p.: Thoreau Institute, 2002), 29. For a good in-house summary of the final round of developments, see Nelson, "Fire Management Policy."
96. Marion Clawson, *The Federal Lands Revisited* (Washington, DC: Resources for the Future, 1983), 259.
97. For a quick synopsis, see Forest History Society, "Inventory of the Gene Bernardi Papers, 1971–1991," accessed February 27, 2013, http://www.foresthistory.org/ead/Bernardi_Gene.html. For a mosaic of impressions from the California staff at the time, see Geraci, *Unmarked Trail*, 69–152.
98. The Parks Canada story offers a fascinating alternative history. For a synopsis, see Stephen J. Pyne, *Awful Splendour: A Fire History of Canada* (Vancouver: University of British Columbia Press, 2007), 444–47; for a more personal perspective,

centered on the hearth of reforms at Banff National Park, see Pyne, "Burning Banff," *ISLE (Interdisciplinary Studies in Literature and Environment)* 11, no. 1 (Summer 2004): 221–47.

99. Figures courtesy Skip Edel, National Park Service, "Fire Reporting: Summary: Number of Fires and Acres Burned by Fire Type by Year (1960–1979)," August 12, 2013.

New Normals, 1977 and 1980

1. Sources include Arnold Hartigan, "A Wrapup of the 1977 Forest Fire Season," *International Fire Chief* 43, no. 9 (November 1977): 4–5; Myron K. Lee, "Marble-Cone/Big Sur Fire: From the Command Point of View," *International Fire Chief* 43, no. 9 (November 1977): 6–8; Fred E. McBride, "Alaska Fire Season: 1977," *Fire Management Notes* 39, no. 1 (Winter 1977/78): 3–7; BLM Alaska, "Analysis of 1977 Fire Season: 10-10-77," in *La Mesa Fire Symposium: Los Alamos, New Mexico, October 6 and 7, 1981*, comp. Teralene Foxx (Los Alamos, NM: Los Alamos National Laboratory, 1984); Hal K. Rothman, *Blazing Heritage: A History of Wildland Fire in the National Parks* (Oxford: Oxford University Press, 2007), 137–41; Douglas R. Leisz and W. A. Powers, "Fire and Drought: Bad Mix for a Dry State," *Fire Management Notes* 38, no. 4 (Fall 1977): 3–7. See also remembrances recorded in Craig Allen, technical ed., *Fire Effects in Southwestern Forests: Proceedings of the Second La Mesa Fire Symposium*, General Technical Report RM-GTR-286 (Fort Collins, CO: U.S. Forest Service, 1996). On the Honda fire, Joseph N. Valencia, *Beyond Tranquillon Ridge* (Bloomington, IN: AuthorHouse, 2004).

2. Original reports destroyed in the Old fire (2009), but numerous accounts survive based on it. See, e.g., John W. Robinson, *The San Bernardinos* (Arcadia, CA: Big Santa Anita Historical Society, 1989), 107.

3. Foothill Communities Protective "Greenbelt" Program, "Report and Recommendations on the Reduction of Fire, Flood and Erosion Losses Along the Wildland/Urban Interface in the Foothills of the San Bernardino Valley (San Bernardino: Foothills Community Protective "Greenbelt" Program, 1983).

4. Albert J. Simard et al., *The Mack Lake Fire*, General Technical Report NC-83 (St. Paul, MN: U.S. Forest Service, North Central Forest Experiment Station, 1983).

5. For a summary, see Stephen J. Pyne, *Introduction to Wildland Fire: Fire Management in the United States* (New York: Wiley, 1984), 430–31.

Chapter 3. Holding

1. Quoted in the *Los Angeles Times*, December 6, 1985. She added, however, that she thought the environment was, for Reagan, a "question of benign neglect."

2. Quote in Jim Carrier, *Summer of Fire* (Salt Lake City, UT: Gibbs Smith, 1989), 42.

3. See Anne M. Burford with John Greenya, *Are You Tough Enough? An Insider's View of Washington Power Politics* (New York: McGraw-Hill, 1986), 281–82.

4. Philip Morison, review of *Fire in America*, *Scientific American* 248 (January 1983): 27–28. The precipitating article was P. Crutzen and J. Birks, "The Atmosphere After a Nuclear War: Twilight at Noon," *Ambio* 11 (1982): 114–25. A popularization

of ideas from a conference by leading scientists was published as Paul R. Ehrlich et al., *The Cold and the Dark: The World After Nuclear War* (New York: Norton, 1984). Comment on large fires from Forest History Society, *Up in Flames* (Durham, NC: The Society, 1984).

5. Randall Collins, "Fire as a Weapon and Symbol in Conflict," Academy Colloquium, "Fire in Human Evolution, Human History, and Human Society," Akademie van Wetenschappen, Amsterdam, December 15–17, 2009.

6. James E. Lotan, ed., *Computer Modeling: Its Application in Fire Management: Proceedings* (Logan, UT: Intermountain Fire Council and Utah State University, 1981), quote from p. ii.

7. Craig Chandler, "Why Not Let Fires Burn" (unpublished paper), FFASR files, p. 12.

8. Foothill Communities Protective "Greenbelt" Program, *Report and Recommendations on the Reduction of Fire, Flood and Erosion Losses Along the Wildland/ Urban Interface in the Foothills of the San Bernardino Valley* (San Bernardino, CA: Foothill Communities Protective "Greenbelt" Program, 1983); Stephen J. Pyne, "Vestal Fires and Virgin Lands: A Historical Perspective on Fire and Wilderness," in *Proceedings: Symposium and Workshop on Wilderness Fire*, James E. Lotan et al., technical coordinators, General Technical Report INT-182 (Ogden, UT: Intermountain Forest and Range Experiment Station, U.S. Forest Service, 1985), 254–62; Jim Davis and John Marker, "The Wildland/Urban Fire Problem," *Fire Command* 54, no. 10 (1986): 26–27; Jerry Laughlin and Cynthia Page, eds., *Wildfire Strikes Home! The Report of the National Wildland/Urban Fire Protection Conference* (Quincy, MA: National Fire Protection Association, 1987). For how the concept interested a broader public, see A. Richard Guth, "Wildfire Strikes Home!" *American Forests* (September/ October 1987): 54–56, 68.

9. William C. Fischer and Stephen F. Arno, comps., *Protecting People and Homes from Wildfire in the Interior West: Proceedings of the Symposium and Workshop*, General Technical Report 251 (Ogden, UT: Intermountain Forest and Range Experiment Station, U.S. Forest Service, 1988); Jim Davis and John F. Marker, "Wildland/Urban Fire Problem," *Fire Command* 54, no. 1 (January 1987): 26–27; James B. Davis, "The Wildland-Urban Interface: Paradise or Battleground?" *Journal of Forestry* 88, no. 1 (January 1990): 26–31; National Fire Protection Association, *Protection of Life and Property from Wildfire*, ANSI/NFPA 299 (Quincy, MA: NFPA, 1991); Federal Emergency Management Agency, *Report of the Operation Urban Wildfire Task Force*, FA-115 (Washington, DC: FEMA, 1992); National Association of State Foresters, *Fire Protection in Rural America: A Challenge for the Future* (Washington, DC: NASF, 1994).

10. Art Belcher, "Development of the FWS Fire Management Program: A Perspective by One Old Firedog" (unpublished manuscript, FWS Fire Staff, NIFC. Phillip E. Street, Tape 1), 13, Fire Oral History Interviews, http://training.fws.gov/ History/OralHistories.html.

11. "Fire Management: History," U.S. Fish and Wildlife Service, accessed June 17, 2011, http://www.fws.gov/fire/who_we_are/history.shtml.

12. Ronald E. Kirby, Stephen J. Lears, and Terry N. Sexson, *Fire in North American Wetland Ecosystems and Fire-Wildlife Relations: An Annotated Bibliography*, Biological Report 88(1) (Washington, DC: Fish and Wildlife Service, 1988), iii.

13. Susan K. Todd and Holly Ann Jewkes, *Wildland Fire in Alaska: A History of Organized Fire Suppression and Management in the Last Frontier*, Agricultural and Forestry Experiment Station Bulletin, no. 114 (Fairbanks: University of Alaska Fairbanks, Agricultural and Forestry Experiment Station, 2006), 41–47.

14. For an interesting account of the negotiations, see Hal K. Rothman, *Blazing Heritage* (Oxford: Oxford University Press, 2007), 148–51.

15. Rothman, *Blazing Heritage*, 150–51.

16. Quote from Todd and Jewkes, *Wildland Fire in Alaska*, 42.

17. Paraphrasing Todd and Jewkes, *Wildland Fire in Alaska*, 42, 41. For its role as a model, see Bureau of Land Management, *Alaska Interagency Fire Management Plan: Tanana/Minchumina Planning Area* (Anchorage, AK: BLM, 1982).

18. Again, the source is Todd and Jewkes, *Wildland Fire in Alaska*, 41. Herman W. Gabriel and Gerald F. Tande, *A Regional Approach to Fire History in Alaska*, BLM-Alaska Technical Report 9 (Anchorage: BLM Alaska, 1983). Larry Knapman, *Fireline Reclamation on Two Fire Sites in Interior Alaska*, BLM-Alaska Resource Management Note 1 (Anchorage, AK: BLM, 1982).

19. For the basic frame, I rely on James Muhn and Hanson R. Stuart, *Opportunity and Challenge: The Story of BLM* (Washington, DC: Bureau of Land Management, 1988), 220–69; and James R. Skillen, *The Nation's Largest Landlord: The Bureau of Land Management in the American West* (Lawrence: University of Kansas, 2009), quote from p. 102.

20. Watt quote from Skillen, *Nation's Largest Landlord*, 126; last quote from Sally K. Fairfax, "The Differences between BLM and the Forest Service," in Muhn and Stuart, *Opportunity and Challenge*, 227.

21. GAO report in Skillen, *Nation's Largest Landlord*, 135–36.

22. The best overview sources for this phase of NPS history are Rothman, *Blazing Heritage*, and Bruce M. Kilgore, "Origin and History of Wildland Fire Use in the U.S. National Park System," *George Wright Forum* 24, no. 3 (2007): 92–122.

23. Rothman, *Blazing Heritage*, 136.

24. A. Starker Leopold to Mr. Boyd Evison, Superintendent, June 9, 1983. Copy from supporting documents, NAFRI M-590 Fire in Ecosystem Management.

25. Ibid.

26. Ibid.

27. A cogent version is available in Thomas M. Bonnicksen and Edward C. Stone, "Managing Vegetation Within U.S. National Parks: A Policy Analysis," *Environmental Management* 6, no. 2 (1982): 109–22.

28. Background information on Barnes from Eric Barnes to Bruce S. Howard, January 2, 1986; quotes from Eric Barnes to Morris K. Udall, March 4, 1986. Barnes had also written to me and included copies of his letter campaign, from which I draw now. For the origins of the Christensen Report, see Norman L. Christensen et al., "Final Report: Review of Fire Management Program for Sequoia-Mixed Conifer Forests of Yosemite, Sequoia and Kings Canyon National Parks" (unpublished report, 1986), 2.

29. Background information on Barnes from Letter, Eric Barnes to Bruce S. Howard, January 2, 1986; quotes from Letter, Eric Barnes to Morris K. Udall, March 4, 1986. Barnes had also written to me and included copies of his letter campaign, from which I draw now. For the origins of the Christensen Report, see Norman L.

Christensen et al., "Final Report: Review of Fire Management Program for Sequoia-Mixed Conifer Forests of Yosemite, Sequoia and Kings Canyon National Parks" (unpublished report, 1986), 2.

30. Harold H. Biswell to Eric Barnes, March 4, 1986.

31. Christensen et al., "Final Report: Review of Fire Management Program."

32. Larry Bancroft et al., "Evolution of the Natural Fire Management Program at Sequoia and Kings Canyon National Parks," in *Symposium and Workshop on Wilderness Fire*, 178.

33. Christensen et al., "Final Report: Review of Fire Management Program," 29.

34. For an overview, see Stephen J. Pyne, *Fire in America: A Cultural History of Wildland and Rural Fire* (Princeton, NJ: Princeton University Press, 1982), 346–59. For a brief précis, see Lewis F. Southard, "The History of Cooperative Forest Fire Control and the Weeks Act," *Forest History Today* (Spring/Fall 2011): 17–20.

35. Quotes from Eliot Zimmerman, "A Historical Summary of State and Private Forestry in the U.S. Forest Service" (unpublished report 1976), 104.

36. Bureau of the Census, *Historical Statistics of the United States. Colonial Times to 1970*, Part 1, Series L (Washington, DC: Government Printing Office, 1975), 44–47, 48–55.

37. See R. A Bonninghausen, "The Florida Forest Service and Controlled Burning," *Proceedings, First Annual Tall Timbers Fire Ecology Conference* (Tallahassee, FL: Tall Timbers Research Station, 1962), 43–52. For an overview of Florida as a fire province, see Stephen Pyne, "Florida's Fire Mosaic," http://firehistory.asu.edu/florida-a-fire-mosaic/.

38. For an overview of California and fire, see Pyne, "California's Fire Complex," http://firehistory.asu.edu/californias-fire-complex/, and especially "State of Emergency," http://firehistory.asu.edu/state-of-emergency-calfire-calema-and-catastrophe-as-catalyst/.

39. Laughlin and Page, *Wildfire Strikes Home!*; Jack Cohen, "The Wildland/Urban Interface Fire Problem: A Consequence of the Fire Exclusion Paradigm," *Forest History Today* (Fall 2008): 20–26; William T. Sommers, "The Emergence of the Wildland-Urban Interface Concept," *Forest History Today* (Fall 2008): 12–18. On the NFPA, see Casey Cavanaugh Grant, "The Birth of NFPA," http://www.nfpa.org/itemDetail.asp?categoryID=500&itemID=18020&URL=About percent20NFPA/Overview/History.

40. Mark Heitlinger, with the assistance of Allen Steuter and Jane Prohaska, "Fire Management Manual," revised September 1985 (The Nature Conservancy).

41. Interviews with Ron Myers and Paula Seamon. They subsequently reviewed and approved a brief essay on their experiences, posted at http://firehistory.asu.edu/one-foot-in-the-black, from which I have drawn for this paragraph.

42. See Technical Cooperation Program, Subgroup N, Panel N3 (Thermal), "A Collection of Papers on Thermal Effects," presented at the Symposium at the Fire Service College, Dorking England. October 5–9, 1964. On Canadian-American negotiations, see Stephen J. Pyne, *Awful Splendour: A Fire History of Canada* (Vancouver: University of British Columbia Press, 2007), 340–41.

43. For an overview of USFS involvement, see Robert K. Winters, *International Forestry in the U.S. Department of Agriculture*, National Economics Division Staff

Report (Washington, DC: NED, 1980). Paul Weeden, Dennis Dube, and Robert Mutch, technical coordinators, *Proceedings: International Wildland Fire Conference, Boston, July 23–26, 1989* (Washington, DC: Fire Aviation Management, U.S. Forest Service, 1990), quote from Allan West, p. 5.

44. Stephen J. Pyne, *Burning Bush: A Fire History of Australia* (Sydney: Allen and Unwin, 1991), 357.

45. Fire Management Working Group of the North American Forest Commission, "International Cooperation in Fire Management, 1962–2008" (unpublished report, 2008).

46. Merle S. Lowden, "Fire Crisis in Brazil," *American Forests* 71 (January 1965): 42–44, 46; "From Firefighting to Revolution in Three Days," *American Forests* 71 (August 1965): 16–20, 50.

47. Robert W. Mutch, *Progress Report of the Disaster Assistance Support Program* (Washington, DC: U.S. Forest Service, 1989).

48. For a useful summary of relations at the time, see Armando González-Cában and David V. Sandberg, "Fire Management and Research Needs in Mexico," *Journal of Forestry* (August 1989): 20–23, 25–26; James C. Sorensen, "A Look at Fire Prevention in Mexico," *Fire Management Notes* 48, no. 2 (1987): 25–26.

49. The two paragraphs on the Canada/U.S. agreement are quoted, with emendations, from Pyne, *Awful Splendour*, 359–60.

50. Stewart L. Udall, *The Quiet Crisis and the Next Generation* (Salt Lake City, UT: Peregrine Smith, 1988), 262.

51. The Timber Wars have a large literature. As a summary with some historical lead in, I found useful Paul W. Hirt, *A Conspiracy of Optimism: Management of the National Forests Since World War Two* (Lincoln: University of Nebraska Press, 1994).

52. Daniel Sarewitz, "How Science Makes Environmental Controversies Worse," *Environmental Science and Policy* 7 (2004): 385–403.

53. Richard J. Barney and David J. Aldrich, *Land Management: Fire Management Policies, Directives, and Guides in the National Forest System: A Review and Commentary*, General Technical Report INT-76 (Ogden, UT: Intermountain Forest and Range Experiment Station, U.S. Forest Service, 1980); Frederick W. Bratten et al., *FOCUS: A Fire Management Planning System: Final Report*, General Technical Report PSW-49 (Berkeley, CA: Pacific and Southwest Forest and Range Experiment Station, U.S. Forest Service, 1981); Thomas J. Mills, *Integrating Fire Management Analysis into Land Management Planning*, General Technical Report PSW-74 (Berkeley, CA: Pacific and Southwest Forest and Range Experiment Station, U.S. Forest Service, 1983); David A. Seaver et al., *The Escaped Fire Situation: A Decision Analysis Approach*, Research Paper RM-244 (Fort Collins, CO: Rocky Mountain Forest and Range Experiment Station, U.S. Forest Service, 1983); Armando González-Cában et al., *Developing Fire Management Mixes for Fire Program Planning*, General Technical Report PSW-88 (Berkeley, CA: Pacific and Southwest Forest and Range Experiment Station, U.S. Forest Service, 1986); David C. Iverson and Richard M. Alston, *The Genesis of FORPLAN: A Historical and Analytical Review of Forest Service Planning Models*, General Technical Report INT-214 (Ogden, UT: Intermountain Forest and Range Experiment Station, U.S. Forest Service, 1986).

54. Chandler, "Why Not Let Fires Burn," 9–10.

55. Richard C. Rothermel, *How to Predict the Spread and Intensity of Forest and Range Fires*, General Technical Report INT-143 (Ogden, UT: Intermountain Forest and Range Experiment Station, U.S. Forest Service, 1983); Patricia L. Andrews, *BEHAVE: Fire Behavior Prediction and Fuel Modeling System: BURN Subsystem, Part 1*, General Technical Report INT-194 (Ogden, UT: Intermountain Forest and Range Experiment Station, U.S. Forest Service, 1986).

56. An excellent study of the period is William T. Sommers, "The Emergence of the Wildland-Urban Interface Concept," *Forest History Today* (Fall 2008): 12–18.

57. "User Needs/Research Planning Workshop: Fire Research to Meet User Needs: Immediate to Long Term," Missoula, Montana, April 17–19, 1984 (U.S. Forest Service Forest Fire and Atmospheric Sciences Research, 1984); Dale D. Wade to Ed Komarek, December 2, 1986, and December 18, 1896 (copies courtesy Dale Wade).

58. Northeast Forest Fire Supervisors Research Committee, "Wildland Fire Research in the Northeastern United States: A Position Paper" (July 1986), 19.

59. Sommers, "Emergence of the Wildland-Urban Interface Concept," 16.

60. Memorandum, R. Max Peterson to Deputy Chiefs, Regional Foresters, Station Directors, and Area Director, "Fire Management Activity Review: 1985 Season," January 31, 1986, pp. 1–2, 5.

61. Clinton B. Phillips and Jerry Frinecker, "The Fire Siege of 1987: Lightning Fires Devastate the Forests of California (California Department of Forestry and Fire Protection, 1988), 1–2, 21.

62. For references to the 1988 season, see notes 65–74.

63. Statistics from Kathleen M. Davis and Robert W. Mutch, "The Fires of the Greater Yellowstone Area: The Sage of a Long Hot Summer," *Western Wildlands* 15 (Summer 1989): 2–9.

64. National Park Service, "The Natural Role of Fire: A Fire Management Plan: Yellowstone National Park" (Yellowstone National Park, WY: National Park Service, Yellowstone National Park, 1972; rev., 1975). Useful background on the evolution of the plan and the subsequent fires is available in Rothman, *Blazing Heritage*, 157–81.

65. Rothman, *Blazing Heritage*, 159, 105.

66. Alston Chase, *Playing God at Yellowstone: The Destruction of America's First National Park* (Boston: Atlantic Monthly Press, 1986). For the "Yellowstone Fire Program Critique," see the summary in Yellowstone National Park, *Wildland Fire Management Plan for Yellowstone National Park* (Yellowstone National Park, 1987), 61–65.

67. I was the planner and wrote three documents: "Wildland Fire Management Plan for Yellowstone National Park," submitted August 1, 1985; "Fire Management at Yellowstone National Park: Recommendations for Future Actions," submitted August 1, 1985; "Prospectus: Yellowstone Interagency Fire Management Center," submitted July 26, 1985.

68. "Matchsticks" quote from "The Role of Fire in Yellowstone," http://www.nps.gov/yell/planyourvisit/upload/chapter1.pdf.

69. There are several good summaries of fire chronology: Davis and Mutch, "Fires of the Greater Yellowstone Area"; Greater Yellowstone Coordinating Committee, *Greater Yellowstone Area Fire Situation, 1988* (copy in the Clemson University Libraries); Richard C. Rothermel, Roberta A. Hartford, and Carolyn H. Chase, *Fire*

Growth Maps for the 1988 Greater Yellowstone Area Fires, General Technical Report INT-304 (Ogden, UT: U.S. Forest Service, Intermountain Research Station, 1994). On Canyon Creek fire, see *The 1988 Canyon Creek Fire*, R1-89-3 (Washington, DC: Forest Service, Department of Agriculture, 1989).

70. A number of studies exist regarding media coverage of the fires. The major two sources are, for magazine articles, Curt Leimbach, "Fire Proof: Social Construction of Reality and the 1988 Yellowstone Fires. An Analysis of Magazine Coverage" (thesis, University of South Florida, 1997), and for television, Conrad Smith, *Media and Apocalypse: News Coverage of the Yellowstone Fires, Exxon Valdez Oil Spill, and Loma Prieta Earthquake* (Westport, CT: Greenwood, 1992), and, in an abbreviated form, "Media Coverage of Fire Ecology in Yellowstone After 1988," in *The Ecological Implications of Fire in Greater Yellowstone: Proceedings of the Second Biennial Conference on the Greater Yellowstone Ecosystem*, ed. Jason M. Greenlee (Fairfield, WA: International Association of Wildland Fire, 1996), 23–34. The American Forestry Association's position was announced in Chris Biolgiano, "Yellowstone and the Let-Burn Policy," *American Forests* 95, no. 1/2: 21–25. Barbee quote from Robert Barbee, "Foreword," *Journal of Forestry* (December 1989): 12.

71. Bungarz quote from Rothman, *Blazing Heritage*, 161.

72. See D. Despain et al., *A Bibliography and Directory of the Yellowstone Fires of 1988* (Fairfield, WA: International Association of Wildland Fire, 1994). The most important special issues were *Western Wildlands* 15 (Summer 1989), *BioScience* 39 (November 1989), and *Journal of Forestry* (December 1989). The two best conferences are the Tall Timbers ones: *Proceedings 17th Tall Timbers Fire Ecology Conference* (Tallahassee, FL: Tall Timbers Research Station, 1991), and Ronald E. Masters, Krista E. M. Galley, and Don G. Despain, eds., *The '88 Fires: Yellowstone and Beyond: Conference Proceedings*, Miscellaneous Publication no. 16 (Tallahassee, FL: Tall Timbers Research Station, 2009). On the two primary panels, see "Interagency Final Report on Fire Management Policy, May 5, 1989," and Norman L. Christensen et al., "Ecological Consequences of the 1988 Fires in the Greater Yellowstone Area: Final Report: The Greater Yellowstone Postfire Ecological Assessment Workshop" (unpublished report submitted to Yellowstone National Park, 1989).

73. See Leimbach, "Fire Proof," 24, for "turnaround on favorable coverage." Peter Matthiessen, "Our National Parks: The Case for Burning," *New York Times Magazine*, December 11, 1988.

74. Bill Buck, "A Yellowstone Critique," *Journal of Forestry* (December 1989): 38.

75. Thomas M. Bonnicksen, "Nature vs. Man(agement)," *Journal of Forestry* (December 1989): 41–43, and Bonnicksen, "Report or Cover-up?" *Journal of Forestry* (October 1990): 26.

76. Paul Schulery and Don G. Despain, "Prescribed Burning in Yellowstone National Park: A Doubtful Proposition," *Western Wildlands* 15 (Summer 1989): 30, 1.

77. Buck, "A Yellowstone Critique," 38.

78. Ronald H. Wakimoto, "The 1988 National Fire Policy Review [abstract]," in Masters, Galley, and Despain, *The '88 Fires*, 146.

79. See Ron Wakimoto, "National Fire Management Policy," *Journal of Forestry* 15 (October 1990): 26, and Rothman, *Blazing Heritage*, 162, on Mott's decision. Wakimoto also makes the point about Yellowstone's malign influence on fire

elsewhere in "The Yellowstone Fires of 1988: Natural Process and Natural Policy," *Northwest Science* 46, no. 5 (1990): 241.

80. Thanks to Tom Nichols for accenting this internal shift within the NPS; letter to author, November 13, 2013.

81. Fire Management Policy Review Team, "Report on Fire Management Policy: December 14, 1988," forwarded by memo on December 15, 1988. "Final Report on Fire Management Policy: May 5, 1989," includes a history of the team's charter and actions. On Mott's action, see Rothman, *Blazing Heritage*, 162.

82. Christensen et al., "Ecological Consequences of the 1988 Fires," ii.

83. Ibid.

84. Norman L. Christensen et al., "Interpreting the Yellowstone Fires of 1988," *BioScience* 39 (November 1989): 685.

85. Jerry Franklin, "Toward a New Forestry," *American Forests* (November/December 1989): 37ff.; Anna Maria Gillis, "The New Forestry," *BioScience* 40 (September 1990): 558–62. The fire community still tended to refract the arguments for "an ecologically sound and socially acceptable solution" through fuels programs; see, e.g., Stephen F. Arno and James K. Brown, "Managing Fire in Our Forests: Time for a New Initiative," *Journal of Forestry* (December 1989): 44–46. A very informative survey of the civil war within forestry is Samuel P. Hays, *War in the Woods: The Rise of Ecological Forestry in America* (Pittsburgh, PA: University of Pittsburgh Press, 2007).

86. Todd Wilkinson, *Science Under Siege: The Politicians' War on Nature and Truth* (Boulder, CO: Johnson, 1998), 17–64, quote from p. 34. Additional information from interviews with Andy Stahl, director, Forest Service Employees for Environmental Ethics.

87. Wilkinson, *Science Under Siege*, 46.

88. Sharon M. Hermann, J. Larry Landers, and Ronald L. Myers, "Preface," in *Proceedings, 17th Tall Timbers Fire Ecology Conference*, ix.

89. National Fire Protection Association, *Black Tiger Fire Case Study* (Quincy, MA: NFPA, 1992).

90. See, e.g., Randal O'Toole, *Reforming the Forest Service* (n.p.: Thoreau Institute, 2000).

91. Stephen C. Nodvin and Thomas A. Waldrop, eds., *Fire and the Environment: Ecological and Cultural Perspectives*, General Technical Report SE-69 (Asheville, NC: Southeastern Forest Experiment Station, U.S. Forest Service, 1991); F. Dale Robertson, "The History of New Perspectives and Ecosystem Management," in *Ouachita and Ozark Mountains Symposium: Ecosystem Management Research*, James M. Guldin, technical comp., General Technical Report SRS-74 (Asheville, NC: Southern Research Station, U.S. Forest Service, 2004).

92. Regarding the role of slash burning in the controversy, see D. Hanley et al., eds., *The Burning Decision: Regional Perspectives on Slash*, Institute of Forest Resources Contribution no. 66 (Seattle: University of Washington Press, 1989). On Florida, see Jim Brenner and Dale Wade, "Florida's Revised Prescribed Fire Law: Protection for Responsible Burners," in *Proceedings of Fire Conference 2000: The First National Congress on Fire Ecology, Prevention, and Management*, ed. K. E. M. Galley, R. C. Klinger, and N. G. Sugihara, Miscellaneous Publication no. 13 (Tallahassee, FL: Tall Timbers Research Station, 2003), 132–36.

93. Statement of James Duffus, *Federal Fire Management: Evaluation of Changes Made After Yellowstone*, GAO/T-RCED-90-84 (Washington, DC: United States General Accounting Office, 1990), 3.

94. Ibid.

95. United States General Accounting Office, "Federal Fire Management: Limited Progress in Restarting the Prescribed Fire Program," GAO/RCED-91-42 (Washington, DC: United States General Accounting Office, 1990), 4, 6, 13, 14.

96. Alaska: Todd and Jewkes, *Wildland Fire in Alaska*, 46–47. Dude fire: "Dude Fire Staff Ride," special issue, *Fire Management Today* 62, no. 4 (Fall 2002); for a longer version, see Michael A. Johns, "The Dude Fire," version 3/10/09, http://www.fireleadership.gov/toolbox/staffride/downloads/lsr11/lsr11_Dude%20Fire_Mike_Johns_2009.pdf. Painted Cave fire: Daniel Gomes et al., *Sifting Through the Ashes: Lessons Learned from the Painted Cave Fire*, South Coast Historical Series (Santa Barbara, CA: Graduate Program in Public Historical Studies, University of California, Santa Barbara, 1993). Yosemite fires: Rothman, *Blazing Heritage*, 184–85.

97. Quote from Harold K. Steen, *The Chiefs Remember: The Forest Service, 1952–2001* (Durham, NC: Forest History Society, 2004), 98.

98. Story and quotes from Steen, *Chiefs Remember*, 100–101.

99. Bill Buck, "A Yellowstone Critique," *Journal of Forestry* (December 1989): 38–40.

100. National Research Council, *Science and the National Parks* (Washington, DC: National Academy Press, 1992).

101. Sebastian Junger, *Fire* (New York: Norton, 2001), xi; Norman Maclean, *Young Men and Fire* (Chicago: University of Chicago Press, 1992).

102. Roban Johnson, "Firefighters Go West: The Foothills Fire," *Fire Management Notes* 53/54, no. 3 (1992/93): 20–22.

103. See the retrospective review from Dan Bailey, "A Global Problem: It's All About Politics," *Wildfire* (March/April 2013): 4.

Lost Fire, 1991

1. The fire is richly documented. I found most useful Paul E. Teague, *The Oakland/Berkeley Hills Fire* (Quincy, MA: National Fire Protection Association); J. Gordon Routley, *The East Bay Hills Fire, Oakland-Berkeley, California* (Emmitsburg, MD: Federal Emergency Management Agency, U.S. Fire Administration, Fire Data Center, 1992). I should mention in particular Margaret Sullivan, *Firestorm! The Story of the 1991 East Bay Fire in Berkeley* (Berkeley, CA: City of Berkeley, 1993), and Peter Charles Hoffer, *Seven Fires: The Urban Infernos that Reshaped America* (New York: Public Affairs, 2006), which includes several chapters on the Tunnel fire.

2. Harold H. Biswell et al., *Ponderosa Fire Management: A Task Force Evaluation of Controlled Burning in Ponderosa Pine Forests of Central Arizona*, Miscellaneous Publication no. 2 (Tallahassee, FL: Tall Timbers Research Station, 1973), 1. Robert W. Mutch et al., *Forest Health in the Blue Mountains: A Management Strategy for Fire-Adapted Ecosystems*, General Technical Report PNW-GTR-310 (Portland, OR: U.S. Forest Service, Pacific Northwest Research Station, 1993), 1.

Chapter 4. Slopover

1. Harold K. Steen, ed., *Jack Ward Thomas: The Journals of a Forest Service Chief* (Seattle: University of Washington Press, 2004), 370.
2. Bruce Babbitt, "Making Peace with Wildland Fire," *Wildfire* (January 1999): 12.
3. Steward L. Udall, *The Quiet Crisis and the Next Generation* (Layton, UT: Gibbs, Smith, 1991).
4. Gary Snyder, *Back on the Fire* (Berkeley, CA: Counterpoint, 2007), 83, 12.
5. Jack Ward Thomas et al., "The Northwest Forest Plan: Origins, Components, Implementation Experience, and Suggestions for Change," *Conservation Biology* 20, no. 2 (2006): 280.
6. Norman L. Christensen et al., "The Report of the Ecological Society of America Committee on the Scientific Basis for Ecosystem Management," *Ecological Applications* 6, no. 3 (August 1996): 665–91.
7. U.S. Forest Service, "Navigating into the Future" (unpublished report, 1994), 1; Bureau of Land Management, "Ecosystem Management in the BLM: From Concept to Commitment" (unpublished report, January 1994); General Accounting Office, *Ecosystem Management: Additional Actions Needed to Adequately Test a Promising Approach*, RCED-94-111 (Washington, DC: GAO, 1994) 4; Interagency Ecosystem Management Task Force, *The Ecosystem Approach: Healthy Ecosystems and Sustainable Economies*, 3 vols. (Washington, DC: Interagency Ecosystem Management Task Force, 1995).
8. For an introduction to New Forestry, see Jerry Franklin, "Toward a New Forestry," *American Forests* 95 (November/December, 1989): 1–8, and Anna Maria Gillis, *BioScience* 40, no. 8 (September 1990): 558–62.
9. United States, State and Private Forestry, *Western Forest Health Initiative* (Washington, DC: USDA Forest Service, State and Private Forestry, 1994).
10. A good brief on Thomas's career is available in Harold K. Steen, *The Chiefs Remember: The Forest Service, 1952–2001* (Durham, NC: Forest History Society, 2004), 103–22. Not to be missed is the diary he kept: Steen, *Jack Ward Thomas*.
11. See Jack Ward Thomas, "Forest Management Approaches on the Public's Lands: Turmoil and Transition," 1992 Horace M. Albright Lecture in Conservation, College of Natural Resources, University of California, Berkeley, accessed February 13, 2014, http://www.cnr.berkeley.edu/site/lectures/albright/1992.php.
12. Quote from Steen, *Jack Ward Thomas*, 105. On the Yellowstone episode, see Rocky Barker, interview by Dave Thomas, National Park Service, "Yellowstone Fires of '88," February 15, 2008, pp. 18–19. Barker also related the story in Rocky Barker, *Scorched Earth: How the Fires of Yellowstone Changed America* (Washington, DC: Island Press, 2005), 217.
13. Quote from Steen, *Jack Ward Thomas*, 115. A chronology of postsettlement reforms is posted as "Snapshot: Fire and Aviation Management Workforce Diversity," http://www.fs.fed.us/fire/diversity/history-refs/snapshot.pdf.
14. Background career information from U.S. Forest Service, *Equal Opportunity is for Everyone. 1988 Civil Rights Report* (Washington, DC: U.S. Forest Service, 1989).
15. Steen, *Jack Ward Thomas*, 107–9.

16. For a nifty précis of Babbitt's career, see "Inside Interior: Bruce Babbitt," interview by Charles Wilkinson and Patty Limerick, Center of the American West, http://centerwest.org/wp-content/uploads/2011/01/babbit.pdf. Other information from interview with Bruce Babbitt by Stephen Pyne, Tucson, February 1, 2013.

17. Babbitt, interview.

18. Steen, *Chiefs Remember*, 109.

19. Wilkinson and Limerick, "Inside Interior," 9.

20. California, Governor's Office of Emergency Service, *The Southern California Wildfire Siege, October–November 1993* (Sacramento, CA: Governors Office of Emergency Service, 1993). For a pictorial survey, see *Inferno!: The Devastating Firestorms of October 1993 as Chronicled by the Staff of the* Orange County Register (Kansas City, MO: Andrews and McMeel, 1993).

21. Mike Davis, *Ecology of Fear: Los Angeles and the Imagination of Disaster* (New York: Metropolitan, 1998).

22. The best summary is Florida, Governor's Wildlife Response and Mitigation Review Committee, *Through the Flames: An Assessment of Florida's Wildfires of 1998: The Report of the Governor's Wildfire Response and Mitigation Review Committee* (Tallahassee, FL: The Committee, 1998).

23. Interestingly, Texas suffered from the same climatic conditions but escaped in large measure because of a massive buildup of prevention activities and prepositioned firefighting resources.

24. Statistics from National Interagency Fire Center, http://www.nifc.gov/fireInfo/fireInfo_stats_totalFires.html. Babbit quote from Bruce Babbitt, "Making Peace with Wildland Fire," *Wildfire* (January 1999): 12.

25. The best summary of the fire season overall is Woody Williams, "Pushed to the Limit," *NFPA Journal* (March/April 1995): 36–42, 47–50. The issue has several articles on the season, including one on South Canyon and another on Mann Gulch.

26. The most accessible source for the general public is John Maclean, *Fire on the Mountain: The True Story of the South Canyon Fire* (New York: William Morrow, 1999). For a short summary see Michael Isner and William Baden, "'The Whole Canyon Blew Up . . . ,'" *NFPA Journal* (March/April 1995): 51–57. The best detailed account of what happened is Bret W. Butler et al., *Fire Behavior Associated with the 1994 South Canyon Fire on Storm King Mountain, Colorado*, Research Paper RMRS-RP-9 (Oden, UT: U.S. Forest Service, Rocky Mountain Research Station, 1998). And of course there are the official reviews: *Report of the South Canyon Fire Accident Investigation Team* (n.p., 1994); "Report of the South Canyon Fire Accident Investigation Team" October 17, 1994; "Final Report of the Interagency Management Review Team: South Canyon Fire" (unpublished report, June 26, 1995); "The Occupational Safety and Health Administration's Investigation of the South Canyon Fire" (unpublished report, February 8, 1995).

27. Personal conversation; quote confirmed on July 19, 2013, by email exchange.

28. For Thomas remembrance, see note 9.

29. Steen, *Jack Ward Thomas*, 13, 96; Lavin memo on stand down for safety.

30. Interagency Management Review Team, *Final Report of the Interagency Management Review Team: South Canyon Fire* (Washington, DC: U.S. Forest Service, 1995).

31. The OSHA notice and the agencies' response are included in the Interagency Management Review Team, *Final Report*.

32. For an overview of fire on his watch, see Kelly Andersson, "An Interview with Jack Ward Thomas," *Wildland Firefighter* (1997): 4–7, 32–35 (Thomas quote p. 7), http://wildfirenews.com/fire/articles/jwt.html.

33. Steen, *Jack Ward Thomas*, 151.

34. Thomas Zimmerman, Laurie Kurth, and Mitchell Burgard, "The Howling Prescribed Natural Fire: Long-Term Effects on the Modernization of Planning and Implementation of Wildland Fire Management," in *Proceedings of 3rd Fire Behavior and Fuels Conference, October 25–29, 2010 Spokane, Washington, USA*, ed. Dale D. Wade and Mikel L. Robinson (Birmingham, AL: International Association of Wildland Fire, 2010). For background context, see also Gary Cones and Paul Keller, "Managing Naturally-Ignited Fire: Yesterday, Today, and Tomorrow" (Wildland Fire Lessons Learned Center, 2008).

35. Francis Mohr and Bob Both, "Confinement: A Suppression Response for the Future?" *Fire Management Today* 56, no. 2 (1996): 17–22.

36. "Report of the National Commission on Wildfire Disasters," *American Forests* 100, no. 9/10 (1994): 13. For the odd authorizing legislation, see *Federal Register* 61, no. 155 (Friday, August 9, 1996): 41507.

37. L. Neil Sampson and David R. Adams, eds., *Assessing Forest Ecosystem Health in the Inland West* (New York: Food Products Press, 1994); Dana C. Bryan, ed., *Conference Proceedings: Environmental Regulation and Prescribed Fire* (Tallahassee, FL: Center for Professional Development, Florida State University, 1997).

38. National Fire Protection Association, *NFPA 299: Standard for Protection of Life and Property from Wildfire* (Quincy, MA: NFPA, 1997); U.S. Operation Urban Wildfire Task Force, Federal Emergency Management Agency, U.S. Fire Administration, *Report of the Operation Urban Wildfire Task Force*, FA-115 (Emmitsburg, MD: U.S. Fire Administration, 1992); National Association of State Foresters, *Fire Protection in Rural America: A Challenge for the Future* (Washington, DC: National Association of State Foresters, 1994).

39. "Strategic Assessment of Fire Management in the USDA Forest Service" (unpublished report, January 13, 1995).

40. Cover letter, Mary Jo Lavin, William Sommers, and Steven Satterfield, January 23, 1995, to Strategic Assessment report.

41. U.S. Forest Service, Fire and Aviation Management, *Course to the Future: Positioning Fire and Aviation Management* (Washington, DC: USFS, FAM, 1995).

42. USFS, FAM, *Course to the Future*, i, 17.

43. U.S. Department of the Interior, U.S. Department of Agriculture, *Federal Wildland Fire Management: Policy and Program Review: Final Report* (Washington, DC: U.S. DOI, USDA, 1995), iii, 4.

44. Ibid., 4, 30.

45. G. Thomas Zimmerman, "Appropriate Management Responses to Wildland Fire: Options and Costs," in *Proceedings of the Symposium on Fire Economics, Planning, and Policy: Bottom Lines*, Armando González-Cabán and Philip N. Omi, technical coordinators, General Technical Report PSW-GTR-173 (Berkeley, CA: Pacific

Southwest Research Station, U.S. Forest Service, 1999), 263. Quote from U.S. DOI, USDA, *Final Report*, 30.

46. U.S. DOI, USDA, *Final Report*, 39.

47. Ibid., 23.

48. Federal Fire and Aviation Leadership Council, "Increasing Programmatic Accomplishments and Reducing Agency Differences in Prescribed Fire Management" (unpublished report, FFALC, January 10, 1996); U.S. Department of the Interior et al., *Federal Wildland Fire Management Policy and Program Review Implementation Action Plan Report* (Washington, DC: U.S. Department of the Interior, U.S. Department of Agriculture, 1996), quote from p. iv.

49. Bruce Babbitt, "Making Peace with Wildland Fire," *Wildfire* (January 1999): 12.

50. U.S. Forest Service, Fire and Aviation Management, *Fire Suppression Costs on Large Fires: A Review of the 1994 Fire Season* (Washington, DC: USFS, FAM, 1994), 1, 13.

51. Ibid., A-3.

52. Enoch Bell et al., "Fire Economics Assessment Report" (unpublished report submitted to Fire and Aviation Management, U.S. Forest Service, September 1, 1995), 3–4.

53. Ibid, 3, 8, 52, 22.

54. See also Ervin G. Schuster, David A. Cleaves, and Enoch F. Bell, *Analysis of USDA Forest Service Fire-Related Expenditures 1970–1995*, Research Paper PSW-RP-230 (Berkeley, CA: Pacific Southwest Research Station, U.S. Forest Service, 1997).

55. Interagency Management Review Team, *Final Report*, which includes a formal letter from OSHA.

56. National Wildlife Coordinating Group, "Wildland Fire Fatalities by Year" (ongoing).

57. See Wildland Firefighter Foundation website, http://www.wffoundation.org/SectionIndex.asp?SectionID=2 and http://www.nifc.gov/aboutNIFC/about_monument.html.

58. William R. Moore, "Towards the Future . . . Land, People, and Fire," *Fire Management* 35, no. 4 (Summer 1974): 5.

59. Paul, 1 Cor. 3:13.

60. Quote from Wilkinson and Limerick, "Inside Interior," 2.

61. Wilkinson and Limerick, "Inside Interior," 4.

62. For a good summaries see Diane Krahe, "The Ill-Fated NBS: A Historical Analysis of Bruce Babbitt's Vision to Overhaul Interior Science," in *Rethinking Protected Areas in a Changing World: Proceedings of the 2011 George Wright Society Conference on Parks, Protected Areas, and Cultural Sites* (Hancock, MI: George Wright Society, 2010), 160–65; and Ross W. Gorte and M. Lynne Corn, "The National Biological Survey," in *The Forest Service Budget: Trust Funds and Special Accounts*, CRS Report for Congress (Washington, DC: Congressional Research Service, Library of Congress, 1995). Babbitt's vision of the United States Geological Survey as a historical model is quoted in National Research Council, *A Biological Survey for the Nation* (Washington, DC: National Academy Press, 1993), vii.

63. Quote from H. Ronald Pulliam, "The Political Education of a Biologist," pt. 1, *Wildlife Society Bulletin* 26, no. 2 (1998): 199.

64. See Krahe, "The Ill-Fated NBS," and "A National Biological Survey: Some Issues, Concerns, and Historical Background: A Memo from the GWS Executive Office," *George Wright Forum* 10, no. 1 (1993): 2–4, which builds on the USGS historical analogy.

65. Bret W. Butler et al., *Fire Behavior Associated with the 1994 South Canyon Fire on Storm King Mountain, Colorado*, Research Paper RMRS-RP-9 (Ogden, UT: U.S. Forest Service, Rocky Mountain Research Station, 1998), preface.

66. Richard C. Rothermel, *Mann Gulch Fire: A Race That Couldn't Be Won*, General Technical Report INT-GTR-299 (Ogden, UT: U.S. Forest Service, Intermountain Research Station, 1993).

67. Butler et al., *Fire Behavior*, preface.

68. On the research, see Jack Cohen, "The Structure Ignition Assessment Model," in *The Biswell Symposium: Fire Issues and Solutions in Urban Interface and Wildland Ecosystems*, David R. Weise and Robert E. Martin, technical coordinators, General Technical Report PSW-GTR-158 (Albany, CA: U.S. Forest Service, Pacific Southwest Research Station, 1995), 85–92.

69. Kelly Andersson, "Tragedy of the Common Forest," *Oregon Daily Emerald*, 1995, http://www.wildfirenews.com/oregon/commons.html.

70. A snapshot of USFS research generally is available in Harold K. Steen, *Forest Service Research: Finding Answers to Conservation's Questions* (Durham, NC: Forest History Society, 1998), 80, 83.

71. "Joint Fire Science Program, First Annual Report, February 1999," http://permanent.access.gpo.gov/lps46819/1998/98prog.htm. See also Joint Fire Science Plan Drafting Committee, "Plan for the Joint Fire Science Plan" (unpublished document, Washington, DC, U.S. Department of the Interior, 1998).

72. Conference proceedings were published as follows: Joel S. Levine, ed., *Global Biomass Burning: Atmospheric, Climatic, and Biospheric Implications* (Cambridge, MA: MIT Press, 1991); *Biomass Burning and Global Change*, 2 vols. (Cambridge, MA: MIT Press, 1996). For the sweep of global engagements, see the Global Fire Monitoring Center, or Stephen J. Pyne, Patricia A. Andrews, and Richard L. Laven, *Introduction to Wildland Fire*, 2nd ed. (New York: Wiley, 1996), 640–44. On the ICFME, see B. J. Stocks, M. E. Alexander, and R. A. Lanoville, "Overview of the International Crown Fire Modelling Experiment (ICFME)," *Canadian Journal of Forest Research* 34 (2004): 1543–47 (the entire issue is devoted to studies that resulted from the experiments).

73. My synopsis of TNC developments derives from interviews with Paula Seamon, Ron Myers, Zach Prusak, and Walt Thomson. Useful documentation includes Mark Heitlinger with the assistance from Allen Steuter and Jane Prohaska, "Fire Management Manual," revised September 1985 (TNC); TNC, "Controlled Burning: Getting It Done" (TNC, 2009); and "A Decade of Dedicated Fire: Lake Wales Ridge Prescribed Fire Team" (TNC, 2010); "The Northeast Florida Resource Management Partnership, Draft Final Report (April 1, 2008–December 31, 2010"; "Central Florida Ecosystem Support Team, Florida Fish and Wildlife Conservation Commission, Final Report (February 1, 2010–December 31, 2010)" (January 2011); and Todd Wilkinson, "Prometheus Unbound," *Nature Conservancy* (May/June 2001): 12–20 for a survey of the national scene at the time, and in the same issue, William Stolzenburg, "Fire in the Rain Forest," 22–27 for its work in Mesoamerica.

74. "The Global Fire Monitoring Center (GFMC): Background Information," http://www.fire.uni-freiburg.de/intro/About1.html. The GFMC site hosts online versions of *IFFN*, which include an astounding library of country reports and a unique sampling of the global fire scene.

75. Will Steffen, P. J. Crutzen, and J. R. McNeill, "The Anthropocene: Are Humans Now Overwhelming the Great Forces of Nature," *Ambio* 36 (2007): 614–21.

76. Stephen J. Botti, "The National Park Service Wildland Fire Management Program," in Gonzáles-Cabán and Omi, *Proceedings of the Symposium on Fire Economics*, 7–13.

77. Jim Brenner and Dale Wade, "Florida's Revised Prescribed Fire Law: Protection for Responsible Burners," in *Proceedings of Fire Conference 2000: The First National Congress on Fire Ecology, Prevention, and Management*, ed. K. E. M. Galley, R. C. Klinger, and N. G. Sugihara, Miscellaneous Publication no. 13 (Tallahassee, FL: Tall Timbers Research Station, 2003), 132–36. For an earlier version, closer to the action, see Dale Wade and James Brenner, "Florida's Solution to Liability Issues," in *The Biswell Symposium: Fire Issues and Solutions in Urban Interface and Wildland Ecosystems*, David R. Weise and Robert E. Martin, technical coordinators, General Technical Report PSW-GTR-158 (Albany, CA: U.S. Forest Service, Pacific Southwest Research Station, 1995), 131–38.

78. Ben Jacobs, "NPS Prescribed Fire Support Modules: A Pilot Program," *Fire Management Notes* 56, no. 2 (1996): 4–6.

79. David J. Parsons, "Restoring Fire to Giant Sequoia Groves: What Have We Learned in 25 Years," in *Proceedings: Symposium on Fire in Wilderness and Park Management*, James K. Brown et al., technical coordinators, General Technical Report INT-GTR-320 (Ogden, UT: U.S. Forest Service, Intermountain Research Station, 1995), 256.

80. Quotes from Jim Bradley, "Political Considerations of Park and Wilderness Management," in Brown et al., *Proceedings: Symposium on Fire in Wilderness*, 7; Bruce Kilgore, "National Park Service Fire Policies and Programs," in Brown et al., *Proceedings: Symposium on Fire in Wilderness*, 25, 27; Robert W. Mutch, "Prescribed Fire in Wilderness: How Successful?" in Brown et al., *Proceedings: Symposium on Fire in Wilderness*, 41; Elmer J. Hurd Jr., "Fire in Wilderness and Parks: Political Issues," in Brown et al., *Proceedings: Symposium on Fire in Wilderness*, 70.

81. Arthur Carhart National Wilderness Training Center, *Wilderness Fire Planning: A Guidebook to Reestablishing the Role of Fire in Wilderness Ecosystems* (n.p.: U.S. Forest Service, Bureau of Land Management, Park Service, and Fish and Wildlife Service, 1995), iii, 1.

82. Thomas Zimmerman, Laurie Kurth, and Mitchell, Burgard, "The Howling Prescribed Natural Fire: Long-Term Effects on the Modernization of Planning and Implementation of Wildland Fire Management," in Wade and Robinson, *Proceedings of 3rd Fire Behavior and Fuels Conference*.

83. Figures from NIFC website, http://www.nifc.gov/fireInfo/fireInfo_statistics.html.

84. Weise and Martin, *Biswell Symposium*.

85. The WUI by now commanded enough attention that the *Journal of Forestry* devoted its October 1997 issue to the topic—an interesting statement on the problem as viewed by foresters. Though published later, the following analyses describe

the acceleration during the 1990s and early 2000s: Susan M. Stein et al., *National Forests on the Edge: Development Pressures on America's National Forests and Grasslands*, General Technical Report PNW-GTR-728 (Portland, OR: U.S. Forest Service, Pacific Northwest Research Station, 2007); Eric M. White and Rhonda Maaza, *A Closer Look at Forests on the Edge: Future Development on Private Forests in Three States*, General Technical Report PNW-GTR-758 (Portland, OR: U.S. Forest Service, Pacific Northwest Research Station, 2008); Roger Auch, Janis Taylor, and William Acevedo, *Urban Growth in American Cities: Glimpses of U.S. Urbanization*, USGS Circular 1252 (Denver, CO: U.S. Geological Survey, EROS Data Center, 2003).

86. For a neutral perspective on the Salvage Rider, see GAO, *Emergency Salvage Sale Program: Forest Service Met Its Target, but More Timber Could Have Been Offered for Sale*, RCED-97-53 (GAO, February 24, 1997).

87. On collaborative programs, see Courtney A. Schultz, Theresa Jedd, and Ryan D. Beam, "The Collaborative Forest Landscape Restoration Program: A History and Overview of the First Projects," *Journal of Forestry* 110, no. 7 (2012): 381–91.

88. "National Fire and Aviation Management Workforce Needs Analysis" (unpublished report, U.S. Forest Service, 1997). For the Sadler fire, see John N. Maclean, *Fire and Ashes: On the Front Lines of American Wildfire* (New York: Holt, 2003), 103–73.

89. Bruce Babbitt, "Making Peace with Wildland Fire," *Wildfire* (January 1999): 12–17. See also Babbitt, "Next Chapter in the History of Fire Fighting," Commonwealth Club of California, September 1, 1998, C-SPAN Video Library, http://www.c-spanvideo.org/program/?110977-1/next-chapter-in-the-history-of-fire-fighting.

90. Babbitt, "Making Peace," 12–17.

91. On FIRE 21 and National FFMO Conference, see the entire issue of *Fire Management Today* 58, no. 2 (Spring 1998), with special note to FAM director Mary Jo Lavin's comments on pp. 4–5. On NPS reforms, see Hal K. Rothman, *Blazing Heritage* (Oxford: Oxford University, 2007). For the Interagency Airtanker Board charter, see *Interagency Airtanker Board: Charter, Criteria, and Forms* 9857 1803 – SDTDC (U.S. Forest Service, July 1998), http://www.fs.fed.us/t-d/pubs/pdf/98571803.pdf.

92. G. Thomas Zimmerman, "Appropriate Management Responses to Wildland Fire: Options and Costs," in Gonzáles-Cabán and Omi, *Proceedings of the Symposium on Fire Economics*, 259. On NLCS, see Bruce Babbitt, "The Heart of the West: BLM's National Landscape Conservation System," in *From Conquest to Conservation: Our Public Lands Legacy*, ed. Michael P. Dombeck, Christopher A. Wood, and Jack E. Williams (Island Press, 2003), 100–102.

93. On the Sadler fire, see Maclean, *Fire and Ashes*; on Lowden Ranch, see *Lowden Ranch Prescribed Fire Review* (Lewiston, CA: Bureau of Land Management, 1999).

94. General Accounting Office, *Endangered Species Act: Impact of Species Protection Efforts on the 1993 California Fire*, GAO/RCED-94-224 (Washington, DC: GAO, 1994); *California Fire Response*, GAO/RCED-94-289R (Washington, DC: GAO, 1994); *Ecosystem Management: Additional Actions Needed to Adequately Test a Promising Approach*, GAO/RCED-94-111 (Washington, DC: GAO, 1994); *Federal Lands: Wildfire Preparedness and Suppression Expenditures for Fiscal Years 1993 Through 1997*, GAO/T-RCED-98-247 (Washington, DC: GAO, 1998); *Western*

National Forests: Catastrophic Wildfires Threaten Resources and Communities, GAO/T-RCED-98-273 (Washington, DC: GAO, 1998), 12.

95. General Accounting Office, *Western National Forests: Nearby Communities Are Increasingly Threatened by Catastrophic Wildfires*, GAO/RCED-99-79 (Washington, DC: GAO, 1999); *Western National Forests: A Cohesive Strategy Is Needed to Address Catastrophic Wildfire Threats*, GAO/RCED-99-65 (Washington, DC: GAO, 1999), quotes from pp. 7–8; *Western National Forests: Status of Forest Service's Efforts to Reduce Catastrophic Wildfire Threats*, GAO/T-RCED-99-241 (Washington, DC: GAO, 1999); *Federal Wildfire Activities: Current Strategy and Issues Needing Attention*, GAO/RCED-99-233 (Washington, DC: GAO, 1999); *Federal Wildfire Activities: Issues Needing Future Attention*, GAO/T-RCED-99-282 (Washington, DC: GAO, 1999).

96. GAO, *Cohesive Strategy*, 8–9.

97. GAO, *Status of Forest Service's Efforts*, 1.

98. U.S. Forest Service, National Management Review Team, *An Agency Strategy for Fire Management* (n.p.: USFS, 2000).

99. Strategic Overview of Large Fire Costs Team, *Policy Implications of Large Fire Management: A Strategic Assessment of Factors Influencing Costs* (Washington, DC: U.S. Forest Service, 2000), 2.

100. Ibid., 1. On the increased costs due to private servicing, see http://web.archive.org/web/20000815074639/http://www.fs.fed.us/r5/fire/1999/.

101. Lyle Laverty and Jerry Williams, *Protecting People and Sustaining Resources in Fire-Adapted Ecosystems: A Cohesive Strategy: The Forest Service Management Response to the General Accounting Office Report GAO/RECED-99-65* (n.p.: U.S. Forest Service, 2000), quoted figures from p. 43.

102. Andrew J. Patrick, *The Globalization of Wildfire: A History of the Australian/ New Zealand Deployment at the Fires of 2000* (Boise, ID: National Interagency Fire Center, 2001).

103. Ross W. Gorte, *Forest Fire Protection*, CRS Report to Congress RL30755 (Washington, DC: Congressional Research Service, Library of Congress, 2000).

104. Council on Environmental Quality, *Managing the Impact of Wildfires on Communities and the Environment: A Report to the President in Response to the Wildfires of 2000* (Washington, DC: Council on Environmental Quality, 2000), 1. A fuller "technical support document"—the source text from which the official response was crafted—is available as *Technical Support Document for Managing the Impacts of Wildfires on Communities and the Environment: A Report to the President in Response to the Wildfires of 2000* (unpublished, rev. September 1, 2000), which was used "in the preparation of the final report . . . and represents the final product from the initial work team."

Millennial Fire, 2000

1. Personal memory of the day's events. Curiously, earlier that day I had finally, after three attempts, located the Nicholson adit, the scene of Ed Pulaski's dramatic stand. The 2000 fire bust broke out as I drove from Wallace to Hamilton.

2. For a minute inventory of damages and proposed remediation on the Bitterroot National Forest, see "Bitterroot Fires 2000: An Assessment of Post-Fire Conditions

with Recovery Recommendations" (unpublished report, Bitterroot National Forest, December 2000).

3. 1 Kings 19:11–13.

4. There were many accounts, but the official accounts are four: Independent Investigation Team, "Los Alamos Prescribed Fire: Investigative Report" (May 18, 2000); "Los Alamos Prescribed Fire: Independent Review Board Report" (May 26, 2000); Barry T. Hill, *Fire Management: Lessons Learned From the Cerro Grande (Los Alamos) Fire and Actions Needed to Reduce Risk*, GAO/T-RCED-00-273 (Washington, DC: U.S. Government Accounting Office, 2000); National Park Service, Board of Inquiry, *Cerro Grande Prescribed Board of Inquiry Final Report* (Washington, DC: National Park Service, 2001). The Board's transcripts are sealed and sequestered in the Midwest Regional Office and unavailable to the public. Using the Freedom of Information Act, Tom Ribe included what unredacted portions he could get for an apologetic study, *Inferno by Committee: A History of the Cerro Grande (Los Alamos) Fire: America's Worst Prescribed Fire Disaster* (Victoria, BC: Trafford, 2010).

5. See Teralene S. Foxx, comp., *La Mesa Fire Symposium: Los Alamos, New Mexico October 6 and 7, 1981*, LA-9236-NERP (Los Alamos, NM: Los Alamos National Laboratory, 1984). A successor followed: C.D. Allen, technical ed., *Fire Effects in Southwestern Forests: Proceedings of the Second La Mesa Fire Symposium*, General Technical Report RM-GTR-286 (Fort Collins, CO: U.S. Forest Service, Rocky Mountain Forest and Range Experiment Station, 1996).

6. I have accepted the GAO account (Hill, *Fire Management*) as the most neutral.

7. Ibid., 2. National Academy of Public Administration, *Study of the Implementation of the Federal Wildland Fire Policy: Phase I Report: Perspective on Cerro Grande and Recommended Issues for Further Study* (n.p.: NAPA, 2000), quote from p. vii.

8. Roger G. Kennedy, *Wildfire and Americans: How to Save Lives, Property, and Your Tax Dollars* (New York: Hill and Wang, 2006).

9. On the Outlet fire, see Tom Pittenger, *Outlet Prescribed Fire Project, Grand Canyon National Park: Investigative Team Report* (Grand Canyon, AZ: National Park Service, 2000).

Chapter 5. Blowup

1. Michael P. Dombeck, Jack E. Williams, and Christopher A. Wood, ""Wildfire Policy and Public Lands: Integrating Scientific Understanding with Social Concerns Across Landscapes," *Conservation Biology* 18, no. 4 (August 2004): 884.

2. Ruben Grijalva, Randy Moore, and Henry Renteria, "Introductory Statement," in *California Fire Siege 2007: An Overview* (Sacramento, CA: California Department of Forestry and Fire Protection, 2008), 3.

3. Most documents will be cited later, but because this will be only reference for the other NGOs, their studies are listed here: Wilderness Society, "Wildland Fire Challenge: Protecting Communities and Restoring Ecosystems," *George Wright Forum* 22, no. 4 (2005): 32–44; Douglas C. Morton, *Assessing the Environmental, Social, and Economic Impacts of Wildfire* (New Haven, CT: Yale University, School of Forestry and Environmental Studies, Global Institute of Sustainable Forestry, 2003).

4. U.S. Department of the Interior, *Review and Update of the 1995 Federal Wildland Fire Management Policy* (Boise, ID: Interagency Federal Wildland Fire Policy Review Working Group, 2001), 4.

5. Ibid., ii–iv, 3–4.

6. Ibid., 31.

7. "A Collaborative Approach for Reducing Wildland Fire Risks to Communities and the Environment: 10-Year Comprehensive Strategy Implementation Plan," August 2001, http://forestsandrangelands.gov/resources/plan/documents/7-19-en.pdf.

8. "A Collaborative Approach for Reducing Wildland Fire Risks to Communities and the Environment: 10-Year Comprehensive Strategy Implementation Plan," May 2002, http://www.forestsandrangelands.gov/resources/plan/documents/11-23-en.pdf; "A Collaborative Approach for Reducing Wildland Fire Risks to Communities and the Environment: 10-Year Strategy Implementation Plan," December 2006, http://www.forestsandrangelands.gov/resources/plan/documents/10-yearstrategyfinal_dec2006.pdf; Department of Interior and U.S. Forest Service, "Restoring Fire-Adapted Ecosystems on Federal Lands: A Cohesive Fuel Strategy for Protecting People and Sustaining Natural Resources" (2002); Western Governors' Association Forest Health Advisory Committee, "Report to the Western Governors on the Implementation of the 10-Year Comprehensive Strategy" (November 2004). A good bibliography of the reports that really mattered is given in U.S. Department of the Interior et al., *Quadrennial Fire and Fuel Review Report* (Tucson, AZ: National Advanced Fire and Resource Institute, 2005), app. A.

9. See U.S. Government Accountability Office, *Wildland Fire Management: Lack of Clear Goals or a Strategy Hinders Federal Agencies' Efforts to Contain the Costs of Fighting Fires*, GAO-07-655 (Washington, DC: U.S. GAO, 2007).

10. For a good summary of the issues, see Firefighters United for Safety, Ethics, and Ecology, "Issue Paper: Implementing Appropriate Management Response (AMR)," http://documents.fusee.org/AMR/AMR%20Issue%20Paper.pdf?lbisphpreq=1. On the 2007 restatement, see National Fire and Aviation Executive Board, Memorandum, "Clarification of Appropriate Management Response," June 20, 2007.

11. Society of American Foresters, "Wildland Fire Management: A Position of the Society of American Foresters," June 7, 2008, http://www.safnet.org/fp/documents/wildland_fire.pdf.

12. While many thoughtful observers recognized the true situation, only a few in positions of influence voiced it. An interesting attempt is Jerry Williams, then director of FAM, "Reconciling Frictions in Policy to Sustain Fire-Dependent Ecosystems," *Fire Management Today* 65, no. 4 (Fall 2005): 4–8.

13. The best way to encapsulate the GAO's efforts is to consult its final report, which includes a chronology of its inquiries since 1999: *Wildland Fire Management: Federal Agencies Have Taken Important Steps Forward, but Additional, Strategic Action Is Needed to Capitalize on Those Steps*, GAO-09-877 (Washington, DC: General Accounting Office, 2009).

14. Ibid., 1, 32, 26.

15. Ibid., 32.

16. The study of record is John N. Maclean, *The Thirtymile Fire: A Chronicle of Bravery and Betrayal* (New York: Holt, 2007). The official report is U.S. Forest

Service, *Thirtymile Fire Investigation Report: Accident Investigation Factual Report and Management Evaluation Report: Thirtymile Fire. Chewuch River Canyon. Winthrop, Washington, July 10, 2001,* September 26, 2001, as amended October 16, 2001 (Washington, DC: USFS, 2001). Also useful is the special issue of *Fire Management Today* 62, no. 2 (Summer 2002), devoted to the fire, which includes the text of a speech by Dale Bosworth, then chief, along with the action recommendations and OSHA summary.

17. The best summaries are in *Fire Management Today* 62, no. 2 (Summer 2002).

18. The simplest overview of the 2002 fires is the two special issues of *Fire Management Today* 65, no. 1 (Winter 2005) and no. 2 (Spring 2006). The former includes several articles on the Rodeo-Chedeski fire, and the latter includes a lengthy account of the Biscuit firefight (Beth Quinn, "Monster in the Woods: The Biscuit Fire," 4–17). For a critical examination of suppression (and its backfires) as a concern, see Timothy Ingalsbee, "Collateral Damage: The Environmental Effects of Firefighting: The 2002 Biscuit Fire Suppression Actions and Impacts," May 2006, http://documents.fusee.org/SuppressionImpacts/FUSEE_Collateral_Damage_Biscuit_Fire_Report.pdf?lbisphpreq=1. The Hayman fire led to a massive inquiry published as Russell T. Graham, techn. ed., *Hayman Fire Case Study*, General Technical Report RMRS-GTR-114 (Ogden, UT: U.S. Forest Service, Rocky Mountain Research Station, 2003). Also see Jim Paxon, *The Monster Reared Its Ugly Head: The Story of the Rodeo-Chediski Fire and Fire as a Tool of Nature* (Show Low, AZ: Cedar Hill, 2007).

19. The term emerged over the course of the decade. The best distillation is Jerry T. Williams and Albert C. Hyde, "The Mega-Fire Phenomenon: Observations from a Coarse-Scale Assessment with Implications for Foresters, Land Managers, and Policy-Makers" (unpublished manuscript from Brookings Institution for U.S. Forest Service Mega-fire Project, 2009). After this section was written, *Forest Ecology and Management* published the proceedings of a conference on the topic as a special edition: *Forest Ecology and Management* 294 (April 2013).

20. Numbers from Jerry Williams, "1910 Fires: A Century Later Could It Happen Again?" (unpublished paper, Inland Empire Society of American Foresters Annual Meeting, Wallace, ID, May 20–22, 2010). The argument about a fire plutocracy was developed and published online as "Prometheus Shrugged," http://firehistory.asu.edu/prometheus-shrugged/.

21. This section comes from a brief essay published online as "Megafires, or Metastasizing Fire Regimes?" http://firehistory.asu.edu/megafires-or-metastasizing-fire-regimes/.

22. See "NTSB Recommends Rigorous Maintenance Programs for Firefighting Aircraft" (NTSB press release, April 23, 2004). For the video of the tragedy, see http://www.youtube.com/watch?v=TBcC8zqNjKk.

23. The best single source is *The Story: California Fire Siege 2003* (Sacramento, CA: U.S. Forest Service and California Department of Forestry and Fire Protection, 2004). For a heartfelt inquiry on the Cedar fire, in particular, see Robert Mutch, *FACES: The Story of the Victims of Southern California's 2003 Fire Siege* (Tucson, AZ: Wildland Fire Lessons Learned Center, 2007). For other, more formal reviews, see California Office of Emergency Services, *2003 Southern California Fires: After Action Report June 17, 2004* (n.p.: Governor's Office of Emergency Services, 2004);

U.S. Forest Service, Pacific Southwest Region, *The Story One Year Later 2004: An After Action Review*, R5-PR-015 (n.p.: USFS, Pacific Southwest Region, 2005); William F. Maxfield, *San Bernardino County Fire Chiefs' Association Lessons Learned Report: Fire Storm 2003 "Old Fire,"* http://www.firescope.org/training/aars/2003/2003-old-fire-lessons-learned-report.pdf.

24. For a useful sample of the studies and their range of disciplines, see V. C. Radeloff et al., "The Wildland-Urban Interface in the United States," *Ecological Applications* 15, no. 3 (June 2005): 799–805; Roger Auch, Janis Taylor, and William Acevedo, *Urban Growth in American Cities: Glimpses of U.S. Urbanization*, Circular 1252 (Reston, VA: U.S. Geological Survey, 2004); Susan Stein et al., *National Forests on the Edge: Development Pressures on America's National Forests and Grasslands*, General Technical Report PNW-GTR-728 (Portland, OR: U.S. Forest Service, Pacific Northwest Research Station, 2007).

25. Radeloff et al., "Wildland-Urban Interface," 802.

26. The best summary of developments is U.S. Department of the Interior and U.S. Forest Service, *Protecting People and Natural Resources: A Cohesive Fuels Treatment Strategy*, February 2006, 1–3, http://www.forestsandrangelands.gov/resources/documents/CFTS_03-03-06.pdf. Philip N. Omi and Linda A. Joyce, technical eds., *Fire, Fuel Treatments, and Ecological Restoration: Conference Proceedings: April 16–18, 2002*, Proceedings RMRS-P-29 (Fort Collins, CO: U.S. Forest Service, Rocky Mountain Research Station, 2003); Matthew G. Rollins and Christine K. Frame, techn. eds., *The LANDFIRE Prototype Project: Nationally Consistent and Locally Relevant Geospatial Data for Wildland Fire Management*, General Technical Report RMRS-GTR-175 (Fort Collins, CO: U.S. Forest Service, Rocky Mountain Research Station, 2006), 1.

For an intelligent summary of the criticisms, see Jacqueline Vaughn and Hanna J. Cortner, *George W. Bush's Healthy Forests: Reframing the Environmental Debate* (Boulder: University Press of Colorado, 2005). For a popular summary of the conflict, see Paul Trachtman, "Fire Fight," *Smithsonian* (August 2003): 42–48, 51–52. To capture the full-throated outrage the administration inspired, see Carl Pope and Paul Rauber, *Strategic Ignorance: Why the Bush Administration Is Recklessly Destroying a Century of Environmental Progress* (San Francisco: Sierra Club, 2004).

27. Russell T. Graham, Sarah McCaffrey, and Theresa B. Jain, *Science Basis for Changing Forest Structure to Modify Wildfire Behavior and Severity*, General Technical Report RMRS-GTR-120 (Fort Collins, CO: U.S. Forest Service, Rocky Mountain Research Station, 2004); Patricia L. Andrews and Bret W. Butler, comps., *Fuels Management: How to Measure Success: Conference Proceedings*, Proceedings RMRS-P-41 (Fort Collins, CO: U.S. Forest Service, Rocky Mountain Research Station, 2006); U.S. Department of the Interior and U.S. Forest Service, *Protecting People and Natural Resources*; David L. Peterson et al., *A Consumer Guide: Tools to Manage Vegetation and Fuels*, General Technical Report PNW-GTR-690 (Portland, OR: U.S. Forest Service, Pacific Northwest Research Station, 2007).

28. For an analysis of the debate through media, see Jayne Fingerman Johnson et al., "U.S. Policy Response to the Fuels Management Problem: An Analysis of the Public Debate About the Healthy Forests Initiative and the Healthy Forests Restoration Act," in Andrews and Butler, *Fuels Management*, 59–66.

On the home ignition zone, see J. D. Cohen, "Preventing Disaster: Home Ignitability in the Wildland-Urban Interface," *Journal of Forestry* 98 (2000): 15–21; "The Wildland-Urban Interface Problem: A Consequence of the Fire Exclusion Paradigm," *Forest History Today* (Fall 2008): 20–26.

29. Quote from Sarah Pratt, Lisa Holsinger, and Robert E. Keane, "Using Simulation Modeling to Assess Historical Reference Conditions for Vegetation and Fire Regimes for the LANDFIRE Prototype Project," in Rollins and Frame, *LANDFIRE Prototype Project*, 277.

30. General Accounting Office, Letter to the Honorable Larry Craig, "Forest Service: Appeals and Litigation of Fuel Reduction Projects," August 31, 2001, GA001-1114R; General Accounting Office, *Forest Service: Information on Appeals and Litigation Involving Fuels Reduction Activities*, GAO-04-52 (Washington, DC: GAO, 2003). On its running criticism of the escalated fuels program, begin with General Accounting Office, *The National Fire Plan: Federal Agencies Are Not Organized to Effectively and Efficiently Implement the Plan*, GAO-01-1022T (Washington, DC: GAO, 2001). Follow the evolution through the inquiries listed in the 2009 summary, Government Accountability Office, *Wildland Fire Management: Federal Agencies Have Taken Important Steps Forward, but Additional, Strategic Action Is Needed to Capitalize on Those Steps*, GAO-09-877 (Washington, DC: GAO, 2009).

31. A good summary of fire funding is Ross Gorte, "The Rising Cost of Wildfire Protection" (research paper, Headwater Economics, June 2013).

32. Numbers from U.S. Forest Service, FAM, updated April 15, 2013. A very deft summary is available in Timothy Ingalsbee, *Getting Burned: A Taxpayer's Guide to Wildfire Suppression Costs* (Eugene, OR: Firefighters United for Safety, Ethics, and Ecology, 2010), 6–7.

33. Ervin G. Schuster, David A. Cleaves, and Enoch F. Bell, *Analysis of USDA Forest Service Fire-Related Expenditures 1970–1995*, Research Paper PSW-RP-230 (Berkeley, CA: U.S. Forest Service, Pacific Southwest Research Station, 1997); Enoch Bell et al., "Fire Economics Assessment Report" (unpublished report to FAM, September 1, 1995), quote from p. 52; Armando González-Cabán and Philip N., technical coordinators, *Proceedings of the Symposium on Fire Economics, Planning, and Policy: Bottom Lines*, General Technical Report PSW-GTR-173 (Berkeley, CA: U.S. Forest Service, Pacific Southwest Research Station, 1999). Inflation adjustments from Bureau of Labor Statistics inflation calculator.

34. U.S. Forest Service, Fire and Aviation Management, briefing paper, July 13, 2000; National Association of State Foresters, "Cost Containment on Large Fires: Efficient Utilization of Wildland Fire Suppression Resources" (Washington, DC: NASF and U.S. Forest Service, State and Private Forestry, 2000); Jonathan Oppenheimer, *From the Ashes: Reducing the Harmful Effects and Rising Costs of Western Wildfires* (Washington, DC: Taxpayers for Common Sense, 2000); Timothy Ingalsbee, *Money to Burn: The Economics of Fire and Fuels Management: Part One: Fire Suppression* (Washington, DC: American Lands Alliance, 2000).

35. A remarkably good bibliography is available in Ingalsbee, *Getting Burned*. A summary of GAO reports themed on costs is included in Government Accountability Office, *Wildland Fire Management: Actions by Federal Agencies and Congress Could Mitigate Rising Fire Costs and Their Effects on Other Agency Programs*, GAO-09-444T

(Washington, DC: GAO, 2009). On the National Academy of Public Administration, see Frank Fairbanks, *Wildfire Suppression: Strategies for Containing Costs: Background and Research* (Washington, DC: National Academy of Public Administration, 2002), quote from p. 1; National Academy for Public Administration, *Containing Wildland Fire Costs: Improving Equipment and Services Acquisition* (Washington, DC: NAPA, 2003); *Containing Wildland Fire Costs: Enhancing Hazard Mitigation Capacity* (Washington, DC: NAPA, 2004). For an early announcement within the fire community, see the special issue on "Reducing Costs on Large Fires," *Fire Management Today* 61, no. 3 (Summer 2001). As a sample of academic publications, see Krista M. Gebert, David E. Calkin, and Jonathan Yoder, "Estimating Suppression Expenditures for Individual Large Fires," *Western Journal of Applied Forestry* 22, no. 3 (2007): 188–96; Geoffrey H. Donovan and Thomas C. Brown, "An Alternative Incentive Structure for Wildfire Management on National Forest Land," *Forest Science* 51, no. 5 (2005): 387–95. On the USFS Fire Plan Office study, see "Consolidation of 2003 National and Regional Large Incident Strategic Assessment and Oversight Reviews Key Findings" (September 22, 2003). Among NGO contributions three stand out: Oppenheimer, *From the Ashes*; Ingalsbee, *Money to Burn*; and Randal O'Toole, *Reforming the Fire Service: An Analysis of Federal Fire Budgets and Incentives* (n.p.: Thoreau Institute, 2000). On the cadence of reports and replies, see U.S. Forest Service, U.S. Department of the Interior, and National Association of State Foresters, "Large Fire Cost Reduction Action Plan" (March, 2003); Wildland Fire Leadership Council, *Large Fire Suppression Costs: Strategies for Cost Management* (n.p.: Strategic Issues Panel on Fire Suppression Costs, 2004); Secretary of Agriculture Independent Cost-Control Review Panel, "FY 2004 Large Cost Wildfires: Report" (March 23, 2005); USDA, Office of Inspector General, Western Region, "Audit Report: Forest Service Large Fire Suppression Costs," Report No. 08601-44-SF (November 2006). On case studies, see Alex E. Dunn, "The Old, Grand Prix, and Padua Wildfires: How Much Did These Fires *Really* Cost?" (unpublished report, U.S. Forest Service); "Consolidation of 2003 National and Regional Large Incident Strategic Assessment and Oversight Review Key Findings" (U.S. Forest Service, September 22, 2003), http:// http://www.forestsandrangelands.gov/resources/reports/documents/2003/403-417-en.pdf, which included summaries from five fires; Tonto and Prescott National Forests, "Regional Large Fire Cost Review: Cave Creek Complex: Final Assessment Report" (August 15, 2005). A kind of summary of IMTs and their use exists in the U.S. Forest Service and U.S. Department of the Interior, "The National Interagency Complex Incident Management Organization Study" (2004), http://www.wflccenter.org/news_pdf/80_pdf.pdf.

36. NASF, "Cost Containment on Large Fires," 7; Fairbanks, *Wildfire Suppression*, 1.

37. Figures from Schuster, Cleaves, and Bell, *Analysis of USDA Forest Service Fire-Related Expenditures*.

38. On National Wildfire Suppression Association, see http://www.nwsa.publishpath.com/nwsa-history.

39. Numbers from Richard J. Mangan, "Equipment Standardization Reduces Costs on Wildland Fires," *Fire Management Today* 61, no. 3 (Summer 2001): 13. For a spirited critique of the privatization movement, see Ingalsbee, *Money to Burn*, 7–8. The shovel example comes from personal experience. Bettina Boxall and Julie

Cart wrote a five-part series of investigative pieces in 2008 for the *Los Angeles Times* on the fire-industrial complex, which helped advertise that term: "As Wildfires Get Wilder, the Costs of Fighting Them Are Untamed," July 27; "Air Tanker Drops in Wildfires Are Often Just for Show," July 29; "A Santa Barbara Area Canyon's Residents Are Among Many Californians Living in Harm's Way in Fire-Prone Areas," July 31; "Beige Plague," August 2; "'Stay or Go' Policy puts Australian Families on Front Lines of Firefighting," August 3.

On aircraft, see *Federal Aerial Firefighting: Assessing Safety and Effectiveness: Blue Ribbon Panel Report to the Chief, USDA Forest Service and Director, USDI Bureau of Land Management* (Washington, DC: U.S. Forest Service, U.S. Department of the Interior, Bureau of Land Management, 2002), quote from p. i. Reduction in air tanker fleet from USDA Office of Inspector General, Western Region, "Audit Report: Forest Service's Replacement Plan for Firefighting Aerial Resources," Report no. 08601-53-SF (unpublished report, July 2009), p. i.

40. For an early interagency effort, see National Fire Plan Coordinators, "Developing an Interagency Landscape-Scale Fire Planning Analysis and Budget Tool" (unpublished report, 2001).

41. Appropriately for an online program, the best description is at http://wfdss.usgs.gov/wfdss/WFDSS_About.shtml.

42. For a spirited critique, see Mike Dubrasich, "An Open Letter to the US Senate Regarding Fire Suppression Costs," February 7, 2007.

43. See Stephen Pyne, "Forestry and Forest Governance: A Brief Interpretive History" (research paper for World Bank, submitted June 2009).

44. See Statement of R. Max Peterson, F. Dale Robertson, Jack Ward Thomas, Michael P. Dombeck, and Dale N. Bosworth, Retired Chiefs of the Forest Service, "On the FY2008 Appropriation for the U.S. Forest Service" (April 2007), and the subsequent Letter from Max Peterson, Dale Robertson, Jack Ward Thomas, Michael P. Dombeck, and Dale N. Bosworth, March 24, 2008, to Nick J. Rahall II, Chairman, Committee on Natural Resources, U.S. House of Representatives in support of FLAME Act, reproduced at http://pinchot.org/news/439.

45. Forest Service National Fire Operations Safety Information Briefing Paper, June 1, 2004.

46. Paul Keller, "The First Pulaski Conference: How We Did It" (final draft report, U.S. Forest Service, 2005), and for a general audience, "A First Step Toward Improved Fireline Safety and Efficiency" *Fire Management Today* 66, no. 2 (Spring 2006): 6.

47. "Defining Doctrine and Guiding Principles of Wildland Fire Suppression in the USDA Forest Service," September 1, 2004. Harbour quote in Edward D. Hollenshead, Mark Smith, Franklin O. Carroll, and Paul Keller, *Fire Suppression: Foundational Doctrine* (Washington, DC: U.S. Forest Service, 2005), 1.

48. Hollenshead et al., *Fire Suppression*.

49. The source documentation is available at http://www.fs.fed.us/fire/doctrine/dialogues.html. For a useful chronology of the early years, see "Doctrine Dialogue, USFS FAM, June 01, 2007, at http://www.fs.fed.us/fire/doctrine/dialogue/dialogue.vol5.htm. Also pertinent for the spirit of the undertaking is "Defining Doctrine for Wildland Fire Suppression in the USDA Forest Service," April 1, 2006 (unpublished

paper, U.S. Forest Service). On the redesign, see Lincoln Bramwell, *Forest Management for All: State and Private Forestry in the U.S. Forest Service* (Durham, NC: Forest History Society, 2013), 85. Bosworth quote from Dale N. Bosworth, Chief Forester, to regional foresters, station directors, area director, IITF director, WO staff, "Fire Suppression Foundational Doctrine," January 26, 2006, http://www.fs.fed.us/fire/doctrine/implementation/source_materials/2006-chiefs_acceptance.pdf.

50. "Professional judgment" quote from "Defining Doctrine for Wildland Fire Suppression," 3. Bosworth quote from Bosworth, "Fire Suppression Foundational Doctrine." "For time necessary" quote from "Doctrine Dialogue, USFS FAM, June 01, 2007."

51. "Doctrine Dialogue, USFS FAM, June 1, 2007."

52. U.S. Department of Agriculture, Office of Inspector General, Southeast Region, *Audit Report: Implementation of the Healthy Forests Initiative*, Report no. 08601-6-AT (Washington, DC: USDA, Office of Inspector General, Southeast Region, 2006), i–ii.

53. See U.S. General Accounting Office, *Major Management Challenges and Program Risks: Department of the Interior*, GAO-03-104 (Washington, DC: GAO, 2003); *Wildland Fires: Forest Service and BLM Need Better Information and a Systematic Approach for Assessing the Risks of Environmental Effects*, GAO-04-705 (Washington, DC: GAO 2004); *Wildland Fire Rehabilitation and Restoration: Forest Service and BLM Could Benefit from Improved Information on Status of Needed Work*, GAO-06-670 (Washington, DC: GAO 2006).

54. The record of the WFLC is archived online at http://forestsandrangelands.gov/leadership/archive.

55. Ross W. Gorte, *Wildfire Protection in the 108th Congress*, Major Studies and Issue Briefs of the Congressional Research Service, 05-RS-22024 (Washington, DC: Congressional Research Service, Library of Congress, 2005): *Wildfire Protection Funding*, CRS Report for Congress, RS21544 (Washington, DC: Congressional Research Service, Library of Congress, 2006).

56. National Fire and Aviation Executive Board, memorandum, "Clarification of Appropriate Management Response," June 20, 2007, and "Master Cooperative Wildland Fire Management and Stafford Act Response Agreement," January 25, 2007.

57. All appeared since the Internet arrived and so have their primary presence online. See Association for Fire Emergency, http://fireecology.org/about/history/; Firefighters United for Safety, Ethics, and Ecology, http://www.fusee.org/about/mission.html; Coalition of Prescribed Fire Councils, http://www.prescribedfire.net/about-us.

58. Again, the program is mostly documented online. Begin at http://www.conservationgateway.org/ConservationPractices/FireLandscapes/FireLearningNetwork/Pages/fire-learning-network.aspx.

59. Numbers from "Fire Learning Network, 2002–2011," https://www.conservationgateway.org/Documents/FLP_FLN_Summary_v4Sept11.pdf.

60. Ron L. Myers, "Fire, Ecosystems and People: A Preliminary Assessment of Fire as a Global Conservation Issue" (The Nature Conservancy, 2004), https://www.conservationgateway.org/Files/Pages/Global_Fire_Assessment.aspx, and *Living with Fire: Sustaining Ecosystems and Livelihoods Through Integrated Fire Management* (The Nature Conservancy Global Fire Initiative, 2006).

61. U.S. Department of the Interior et al., *Quadrennial Fire and Fuel Report* (Tucson, AZ: National Advanced Fire Resource Institute, 2005), http://www.forestsandrangelands.gov/strategy/documents/foundational/qffr_final_report_20050719.pdf.

62. Ibid., 3–4.

63. National Interagency Coordination Center, *2006 Season Summary and Statistics* (unpublished report), esp. p. 66. Tabular data is available at http://www.predictiveservices.nifc.gov/intelligence/2006_statssumm/fires_acres.pdf.

64. As usual, the best summary account comes from John N. Maclean, *The Esperanza Fire: Arson, Murder and the Agony of Engine 57* (Berkeley, CA: Counterpoint, 2013).

65. The Brookings Institution study, conducted by Al Hyde, was titled *Where Have All the Firefighters Gone?* (Brookings Institution, 2006). Its data was released to the agencies and reported in general terms to the NWCG. Federal Fire Training Task Group, *Federal Fire Training Strategy: Training and Development for the Next Generation of Federal Wildland Fire Managers*, Version 2.0 (2002), quotes from pp. 3–5; numbers on expected increasing in staff from p. 22.

66. National Association of State Foresters, Core Team, *The Changing Role and Needs of Local, Rural, and Volunteer Fire Departments in the Wildland-Urban Interface: Recommended Actions for Implementing the 10-Year Comprehensive Strategy: An Assessment and Report to Congress* (Washington, DC: The Core Team, 2003), 3.

67. Ibid., 13.

68. NFAEB, Directives Task Group, "Implementation Strategy for the Implementation of the Federal Wildland Fire Policy" (2004), https://www.nifc.gov/policies/policies_documents/GIFWFMP.pdf; NAFRI, http://www.nafri.gov; U.S. Department of the Interior et al., *Quadrennial Fire*, 34–39.

69. Pulaski Conference; U.S. Department of the Interior et al., *Quadrennial Fire*, 37, 34.

70. See http://www.fs.usda.gov/nimo.

East Meets West, 2007

1. Primary documentation from Georgia Forestry Commission, "The Historic 2007 Georgia Wildfire: Learning from the Past, Planning for the Future" (unpublished report, 2008).

2. My survey relies primarily on two documents: CalFire, U.S. Forest Service, Pacific Southwest Region, and California Office of Emergency Services, *California Fire Siege 2007: An Overview* (Sacramento: California Department of Forestry and Fire Protection), and Jon E. Keeley et al., "The 2007 Southern California Wildfires: Lessons in Complexity," *Journal of Forestry* (September 2009): 287–96. My own interpretation of the Southern California scene is available at http://firehistory.asu.edu/californias-fire-complex.

3. Numbers from Keeley et al., "2007 Southern California Wildfires," 291.

4. NFAEB, Memorandum, June 20, 2007, "Clarification of Appropriate Management Response"; U.S. Forest Service, Region One, "Appropriate Management Response Summary for the Northern Rockies" (final version, July 21, 2007).

5. Information Collection Team, "Initial Impressions Report: Appropriate Management Response: Northern Rockies 2007" (Wildland Fire Lessons Learned Center, 2007), http://www.wildfirelessons.net/communities/resources/viewincident?Document Key=979383ac-7dc7-4a50-ac78-75928892d597, and for a condensed version, Josh McDaniel, "Changing Face of Fire Management in the Northern Rockies" (Wildland Fire Lessons Learned Center, Winter 2007), https://higherlogicdownload.s3.amazon aws.com/WILDFIRELESSONS/AMR.pdf?AWSAccessKeyId=AKIAJH5D4I4F WRALBOUA&Expires=1427236875&Signature=4rF%2B%2F3kZuYrt7U 8Wkjyr43HrtfQ%3D.

6. Information from personal inspection of fire sites and discussions with Helena National Forest personnel, June 2012.

7. Quote from press release cited in WildlandFirefighter.com, August 1, 2007, accessed February 17, 2014, http://wlfhotlist.com/threads/1274-MT-HNF-Meriwether.

8. Ibid.

Chapter 6. Burning Out

1. Government Accountability Office, *Wildland Fire Management: Federal Agencies Have Taken Important Steps Forward, but Additional, Strategic Action Is Needed to Capitalize on Those Steps*, GAO-09-877 (Washington, DC: GAO, 2009), 32.

2. Quote from Adam Burke, "Keepers of the Flame," *High Country News*, November 8, 2004, http://www.hcn.org/issues/286/15102/print_view.

3. Robin M. Nazzaro, *Wildland Fire Management: Lack of Clear Goals or a Strategy Hinders Federal Agencies' Efforts to Contain the Costs of Fighting Fires*, GAO-07-655 (Washington, DC: Government Accountability Office, 2007); GAO, *Federal Agencies Have Taken Important Steps*, 32.

4. GAO, *Federal Agencies Have Taken Important Steps*, 16.

5. For the background, see AMR Task Group, "Modifying Guidance for Implementation of Federal Wildland Fire Policy. Briefing Paper," to Fire Executive Council, June 13, 2008; Fire Executive Council, "Guidance for the Implementation of Federal Wildland Fire Management Policy" (February 13, 2009), which includes a thumbnail history since 2001; Wildland Fire Leadership Council, memorandum, "Modification of Federal Wildland Fire Management Policy Guidance," May 2, 2008; FEC, "Modification of Federal Wildland Fire Management Policy Guidance: Communication Plan" (July 18, 2008); National Wildlife Coordinating Group chair, memorandum, "Terminology Updates Resulting from Release of the 'Guidance for the Implementation of Federal Wildland Fire Management Policy (2009),'" April 30, 2010; and for a PowerPoint presentation on the changes, Dick Bahr, chair, National Wildlife Coordinating Group Fire Policy Committee, "Federal Wildland Fire Management Policy: Update on Implementation and Terminology" (2009).

6. See GAO, *Federal Agencies Have Taken Important Steps*, 28.

7. Fire Executive Council and NASF Forest Fire Protection Committee, "Quadrennial Fire Review" (January 2009), iii–iv.

8. Ibid., vi, 1, 2, 6, 37, 29.

9. Ibid., vi.

10. The Federal Land Assistance, Management and Enhancement Act of 2000, Report to Congress (2000).

11. "FLAME ACT of 2009. Fact Sheet FY 2010 Interior & Environment Appropriations Conference."

12. U.S. Forest Service and Department of Interior, "National Study of Airtankers to Support Initial Attack and Large Fire Suppression, Final Report Phase 1 (March, 1995) and Final Report Phase 2 (November, 1996)," and "National Study of Tactical Aerial Resource Management to Support Initial Attack and Large Fire Suppression" (April 1998). USFS et al., "Interagency Airtanker Board. Charter, Criteria, and Forms" *5700 Aviation Management* (July 1998).

13. Blue Ribbon Panel on Aerial Firefighting, *Federal Aerial Firefighting: Assessing Safety and Effectiveness: Blue Ribbon Panel Report to the Chief* (Washington, DC: U.S. Department of Agriculture, Forest Service, U.S. Department of the Interior, Bureau of Land Management, 2002), i–vii, quote from p. i. Statistics on fatalities vary by source. I have used the NWCG, *Historical Wildland Firefighter Fatality Reports*, listed by year, cause, and state, http://www.nifc.gov/safety/safety_documents/year.pdf; the Associated Aerial Firefighters organization (http://airtanker.org) tallies a lower figure, 53.

14. National Transportation Safety Board, *Aircraft Accident Report: Crash During Takeoff of Carson Helicopters, Inc., Firefighting Helicopter Under Contract to the U.S. Forest Service, Sikorsky S-61N, N612AZ, Near Weaverville, California, August 5, 2008*, NTSB/AAR-10/06 (Washington, DC: NTSB, 2010), A-04-29-33.

15. Ibid.

16. Bill Gabbert, "GAO Air Tanker Report Says More Data and Planning Needed," http://fireaviation.com/2013/08/21/gao-air-tanker-report-says-more-data-and-planning-needed/. U.S. Government Accountability Office, *Wildland Fire Management. Improvements Needed in Information, Collaboration, and Planning to Enhance Federal Fire Aviation Program Success*, GAO-13-684, August 2013, http://purl.fdlp.gov/GOP/gpo40657.

17. The best distillation of the USFS perspective is "Large Airtanker Modernization Strategy" February 10, 2012, http://www.fs.fed.us/fire/aviation/airtanker_modernization_strategy.pdf. On the difficulty of determining cost benefits, see Matthew P. Thompson et al., "Airtankers and Wildfire Management in the US Forest Service: Examining Data Availability and Exploring Usage and Cost Trends," *International Journal of Wildland Fire* 22, no. 2 (2012): 223–33, http://dx.doi.org/10.1071/WF11041.

18. The retardant controversy can be accessed through the USFS website, http://www.fs.fed.us/fire/retardant/index.html. The legal maneuvering began with a 2006 notice by the USFS that it intended to conduct an environmental analysis in response to complaints, released in October 2007, and then invalidated by a district court in July 2010.

19. Good summaries of federal organization at the time are available in NWCG, "Overview of Wildland Fire Organizations" (January 2009) and DOI Office of Inspector General, "Wildland Fire Management Overview," ER-IS-MOA-0010-2009 Wildlands Fire Inspection (May 2009).

20. U.S. Bureau of the Census, *Historical Statistics of the United States: Colonial Times to 1970*, Part 1 (Washington, DC: U.S. Department of Commerce, Bureau of the Census, 1975), series L 44–47. John R. Hall Jr., *The Total Cost of Fire in the United States* (Quincy, MA: National Fire Protection Association, 2011). For a sample of how states regarded flexible interpretations of "appropriate management response," see Montana Department of Natural Resources and Conservation, "Appropriate Management Response (AMR) Policies: A State Perspective" (unpublished position paper, February 2008). The paper argues in particular that "it is critical that agencies debunk the popular criticism that AMR is analogous to 'let burn,' 'wildfire use,' or 'prescribed natural fire'" (unpublished position paper).

21. Wildland Fire Leadership Council, "A Collaborative Approach for Reducing Wildland Fire Risks to Communities and the Environment. Monitoring and Performance Report. Executive Summary," 2007, http://www.forestsandrangelands.gov/leadership/archive/documents/wflc_monitoring_perfrpt_execsummary.pdf.

22. Lincoln Bramwell, *Forest Management for All: State and Private Forestry in the U.S. Forest Service* (Durham, NC: Forest History Society, 2013), 85–87. On CFLRP, see http://www.fs.fed.us/restoration/CFLRP.

23. For a good summary of these attitudes, see Dan Smith, "State Forestry Agency Perspectives Regarding 2009 Federal Wildfire Policy Implementation" (briefing paper from the National Association of State Forester prepared by the Forest Fire Protection Committee, July 12, 2010).

24. Blane Heumann, "Reflections on 50 Years of Burning in The Nature Conservancy," http://blog.nature.org/2012/04/reflections.on-50-years-of-burning-in-the-nature-conservancy. *U.S. Fire Learning Network Field Guide* (TNC, 2009), https://www.conservationgateway.org/Documents/USFLN_Field_Guide_2009_0.pdf, and 2010 addendum, https://www.conservationgateway.org/Files/Pages/us-fire-learning-network-aspx50.aspx.

25. The best summary is the coalition's website, http://www.prescribedfire.net.

26. Mark A. Melvin, "2012 National Prescribed Fire Use Survey Report," Technical Report 01-12, Coalition of Prescribed Fire Councils, 2012, http://www.stateforesters.org/sites/default/files/publication-documents/2012_National_Prescribed_Fire_Survey.pdf.

27. Ibid., 19.

28. The principal cache of record is the WFLC-overseen website: http://www.forestsandrangelands.gov/strategy/. The foundational document is *A National Cohesive Wildland Fire Management Strategy* (2010).

29. The primary documents are U.S. Forest Service, *A National Cohesive Wildland Fire Management Strategy* (Washington, DC: Wildland Fire Leadership Council, 2010); *The Federal Land Assistance, Management and Enhancement Act of 2009: Report to Congress* (Washington, DC: Wildland Fire Leadership Council, 2011); *A National Cohesive Wildland Fire Management Strategy: Phase II National Report* (Washington, DC: Wildland Fire Leadership Council, 2012); and the regional reports: *A National Cohesive Wildland Fire Strategy: Southeastern Regional Risk Assessment* (Washington, DC: Wildland Fire Leadership Council, 2012); *Northeast Regional Risk Analysis Report* (Washington, DC: Wildland Fire Leadership Council, 2012); and *Western Regional Science-Based Risk Analysis Report* (Washington, DC:

Wildland Fire Leadership Council, 2012), all released on September 30, 2011. For the scientific component, see *A Comparative Risk Assessment Framework for Wildland Fire Management: The 2010 Cohesive Strategy Science Report*, Gen. Tech. Report RMRS-GTR-262 (U.S. Forest Service, 2011).

30. See the classic study by David Potter, *People of Plenty: Economic Abundance and the American Character* (Chicago: University of Chicago Press, 1958).

31. Quote from Jack Healy, "In Colorado, Nature Takes a Fiery Toll Despite a Community's Efforts to Prepare," *New York Times*, June 14, 2013.

Wallowing, 2011

1. On the 2011 season in Texas, see Texas A&M Forest Service, "2011 Texas Wildfires: Common Denominators of Home Destruction" (2013), http://texasforest service.tamu.edu/uploadedFiles/FRP/New_-_Mitigation/Safety_Tips/2011%20 Texas%20Wildfires.pdf.

2. Las Conchas fire information at InciWeb, June 2, 2013, http://inciweb.nwcg .gov/incident/2385, and "The Las Conchas Fire," Bandelier National Monument, http://www.nps.gov/band/learn/nature/lasconchas.htm.

3. Allin and Swetnam quotes from personal conversations and talks at professional meetings.

4. I relied on information collected on a research trip in April 2012.

5. Basic data on the Miller fire from InciWeb, updated July 27, 2011, and final map, http://inciweb.nwcg.gov/incident/map/2207/29/.

6. See, e.g., Zachary A. Holden, Penelope Morgan, and Andrew T. Hudak, "Burn Severity of Areas Reburned by Wildfires in the Gila National Forest, New Mexico, USA," *Fire Ecology* 6, no. 3 (2010): 77–85.

7. Wallow fire statistics from InciWeb report, final fire progression map, http:// inciweb.nwcg.gov/incident/maps/2262/1/.

8. Basic information from InciWeb, updated July 21, 2011, and personal observations, http://inciweb.nwcg.gov/incident/2262/.

Epilogue: After Action Report

1. Francis Fukuyama, "America in Decay," *Foreign Affairs*, August 20, 2014, http://www.foreignaffairs.com/articles/141729/francis-fukuyama/america-in-decay.

2. One way to track this transformation is to follow the literature of firefighting as memoir. The corpus grew slowly until Maclean opened the floodgates. Perhaps the most useful comparison is between my *Fire on the Rim: A Firefighter's Season at Grand Canyon* (New York: Weidenfeld and Nicolson, 1989), which describes the life of a smokechaser from 1967 to 1981, and Matthew Desmond, *On the Fireline: Living and Dying with Wildland Firefighters* (Chicago: University of Chicago Press, 2007), which combines personal experience with sociological analysis for a more contemporary look. The most widely circulated is probably Murry A. Taylor, *Jumping Fire: A Smokejumper's Memoir of Fighting Wildfire* (New York: Harcourt, 2000).

3. James E. Hubbard, memo, May 25, 2012, "2012 Wildfire Guidance." Mutch's comments were circulated by e-mail but then formalized into "Framing Our Fire

Story to Promote Sustainable Policies and Practices," *Wildfire* (May/June 2012): 12–16.

4. USDA Office of Inspector General, "Forest Service's Firefighting Succession Planning Process" (audit report 08601-54-SF, March 2010), 1.

5. Ibid., 3.

6. Ibid., 2, 3, 26.

7. On the Cramer fire, see Linda Donoghue and George Jackson, *Cramer Fire Fatalities: North Fork Ranger District, Salmon-Challis National Forest, Region 4, Salmon, ID, July 22, 2003*, 0351-2M48-MTDC (Missoula, MT: Forest Service Technology and Development Program, 2003); K. R. Close, "Fire Behavior vs. Human Behavior: Why the Lessons from Cramer Matter," in *Eighth International Wildland Firefighter Safety Summit: Human Factors: 10 Years Later*, ed. B. W. Butler and M. E. Alexander (Missoula, MT: International Association of Wildland Fire, 2005), http://www.iawfonline.org/summit/2005.php. For the legal consequences of this improved scrutiny, see John Maclean, *The Esperanza Fire* (Berkeley, CA: Counterpoint, 2013).

8. Bob W. Mutch, "Framing Our Fire Story to Promote Sustainable Policies and Practices," *Wildfire* (May/June 2013): 12–16; e-mail memo, Darrel L. Kenops, to NSFSR Members, April 2, 2013.

9. Bill Derr and Roger Seewald, "Congressional Review of Several Fires in New Mexico During May and Early June, 2012: Seewald Report" (2013), http://pearce.house.gov/sites/pearce.house.gov/files/1%20Seewald%20Report.pdf; William A. Derr, "Wildfire Review Report: Whitewater-Baldy Complex and Little Bear Fires: New Mexico: May thru June 2012," http://documents.fusee.org/AMR/Derr%20Fire%20Report.pdf?lbisphpreq=1. There were other spontaneous reports from former FMOs or residents of the area, all lamenting the losses of what they had known in their youth or over their careers; see, e.g., Allen Campbell, "Whitewater Fire, a Lasting Legacy," http://pearce.house.gov/sites/pearce.house.gov/files/WHITEWATER,_the_legacy_PDF.pdf.

10. Interviews with Gabe Holguin, November 13, 2013, and January 17, 2014. I am grateful for his willingness to share his career and thoughts.

INDEX

Advisory Board on Wildlife Management. *See* Leopold Report
Aerospace Corporation, 114–15
Agee, James, 70, 136
Agency Strategy for Fire Management, 343–44
aircraft: accidents, 376, 388; origins, 11; upgrades, 426–30
Alaska: 1950s and 1960s, 80–89, 95, 99, 106, 123, 135, 144, 146; 1970s, 147–51, 191–98, 227–29, 231, 244, 250–51, 285, 292, 322, 324, 328, 333–34, 350, 382, 398, 447–48; 1980s, 191–94. *See also* Alaska (state); *and names of federal agencies*
Alaska (state), 78, 148–50, 192–93
Alaska fires (1957), 78
Alaska Fire Service (AFS), 21, 151, 191, 193
Alaska National Interest Lands Conservation Act (1980), 79, 99, 150–51, 166, 191, 195–96, 244
Albini, Frank, 125, 317
Albright, Horace, 137, 200, 232
Aldrich, Dave, 136, 162, 222, 269 fig. 10, 277
Allen, Craig, 453
all-hazard emergency service model, 86, 211, 359, 396, 402, 409–10, 422, 439, 459
all-lands, all-hands doctrine, 431, 434

America Burning, 453
American Forestry Association, 68, 79, 204, 298, 307
Amicarella, L. A., 169, 283
AMR. *See* appropriate management response
Anderson, Hal, 68, 94, 124, 126–27
Andrews, Patricia, 225
appropriate management response (AMR), 302, 334, 340, 363–65, 374, 397, 399, 401, 407, 411, 414–15, 421–23, 425
appropriate strategic response, 422, 442
Arnold, Keith, 15, 31
A-Rock fire (1990), 250–51
Arthur Carhart National Wilderness Training Center, 333
Aspen fire (2003), 274 fig. 20, 385, 392
Association for Fire Ecology, 399
Association of Forest Service Employees for Environmental Ethics (FSEEE), 246, 276
Australia, 18, 39, 68, 121, 145, 186, 216–17, 220, 322, 339, 345, 370, 377, 393, 467
automated lightning detection system, 78, 146, 149

Babbitt, Bruce, 271 fig. 15, 275, 280, 285–86, 290, 294, 302, 304, 313–15, 318, 339, 345–46, 352–53, 355, 360, 370

525

Bandelier National Monument, 142, 167, 178, 202, 290, 345–46, 351–52, 354, 452
Barbee, Robert, 49, 136, 232, 234, 268 fig. 9
Barnes, Eric, 199
Barrows, Jack, 122–23
Bastrop fire (2011), 450, 452–54
Battlement Creek fire (1976), 147
Beadle, Henry, 37, 120
Behave, 225
Bel Air–Brentwood fire (1961), 29, 31–32, 68, 450, 466
Belcher, Art, 154–55
Bernardi, Gene, 164, 171
Bernardi v. Butz, 171
BIA. *See* Bureau of Indian Affairs
Biddison, Lynn, 24–25, 87, 104
BIFC. *See* Boise Interagency Fire Center
Big Blowup, 5, 7, 9, 12, 18, 23, 26, 30, 58, 91, 95, 173, 178, 230–31, 234–35, 255, 290, 306–8, 311, 327, 345, 348–49, 355, 418, 443, 447, 461, 467; centennial celebration of, 457–58
Big Burn (fossil fuels), 325–27
Big Cypress National Preserve, 197, 329
big fire costs, 304–6, 371, 385
Big Tree. *See* sequoia
Biosphere II, 274 fig. 20, 392
Biscuit fire (2002), 370, 384–85, 388
Biswell, Harold, 21–22, 37, 46–49, 70, 88, 119, 121, 136, 199–200, 247, 268 fig. 9, 276, 329, 335
Bjornsen, Bob, 111
Black Tiger fire (1989), 247
BLM. *See* Bureau of Land Management
Blue Mountains (Oregon), 252, 260–62
Boise Interagency Fire Center (BIFC), 80–84, 108, 111, 113–14, 116–17, 140–43, 145–47, 153–54, 157, 164, 178–79, 190–91, 193, 195–96, 205, 213, 219, 235, 255, 286
Bonnicksen, Thomas, 199, 237

Bonninghausen, R. A., 39, 338
Bosworth, Dale, 3, 5–7, 267 fig. 7, 305, 396
Bosworth, Irwin and Mary Ellen, 5–6
Botti, Stephen, 328
Boucher, Paul, 418
Broken Arrow fire (1985), 199
Brookings Institution, 343, 359–60, 385, 402, 405
Brower, David, 56
Brown, James, 127
Buck, Bill, 237–38, 253
Bungarz, Denny, 235
Bureau of Indian Affairs (BIA), 20–21, 38, 75, 108, 110, 113, 144–45, 156–58, 192, 242, 397; 1960s and 1970s, 155–58. *See also* Weaver, Harold
Bureau of Land Management (BLM): early years, 74–78; 1970s, 144–47; 1980s, 194–96; role in BIFC, 81–84. *See also* Alaska; Boise Interagency Fire Center
Burford, Ann M., 182–83, 221
Burgan, Bob, 125
Bush (the elder) administration, 226, 242, 248, 251, 276
Bush (the younger) administration, 362, 380–82, 425
Butts, David, 140, 142, 233
Byram, George, 15, 67

CalFire. *See* California Department of Forestry
California: diaspora of fire officers, 25–26, 87–88, 210–11; fire disaster plan, 86, 97–98, 114, 116, 209–10; impact on national system, 25–26, 31, 85–89, 116–17, 128–29, 136, 171, 188, 210–11, 271 fig. 16, 287, 289, 306, 447–48. *See also* California Department of Forestry; Northern California; Southern California
California Department of Forestry, 6, 114, 116, 131, 209–10, 230, 258–59, 307, 340–41, 404–5, 409, 445

Canada, 18, 66, 38, 118, 121, 163, 175, 186, 193, 216–20, 226, 234–35, 296, 322–23, 393, 412, 436, 449, 467
Canada/United States Forest Fire Fighting Assistance Agreement, 219–20
Cantwell, Senator Maria, 369
Canyon Creek fire (1988), 231–32, 234
Carrizo fire (1971), 158
Cave Creek complex (2005), 375
CCC. *See* Civilian Conservation Corps
Cerro Grande fire (2000), 345, 350–56
chamise, as model fuel, 129–31
Chandler, Craig, 99, 161, 188, 223, 225
Chapman, H. H., 19–20, 37, 118
Chase, Alston, 232, 237
Chase, Dick, 115–16
Christensen, Norman L., 199–200, 202, 235, 239, 243, 246, 270 fig. 12, 277, 279
Christensen Committee: 1985, 199–200, 202; 1988, 235, 239, 243; 1996, 279
Civil Defense Act, 10. *See also* Office of Civil Defense
Civilian Conservation Corps (CCC), 11–13, 24, 75, 77, 102, 112, 140, 209, 307, 368, 408
civil rights, 50, 63, 90
civil society, for fire, 5, 35–36, 41, 100, 118, 211, 244, 246, 248, 251–52, 276, 279, 324, 328, 397, 436; international, 324–25. *See also* The Nature Conservancy
Clarke-McNary Act (1924), 9–10, 21, 167, 203, 205, 207, 209
Clawson, Marion, 170
Clean Air Act, 45, 61, 90, 99, 208
Cliff, Edward, 83
Clinton administration, 278, 282, 286, 318, 340–41, 346
Coalition of Prescribed Fire Councils, 272 fig. 18, 400, 433, 436
Coffman, John, 22
Cohen, Jack, 318

Collaborative Forest Landscape Restoration Program, 432
Conarro, Raymond, 20, 102
consent decree, 171, 221, 246, 253–54, 283–84
Cooperative Fire Protection Program, 167, 203
Cooperative Forestry Assistance Act, 169, 425, 432
Countryman, Clive, 129, 131–32
Covington, Wally, 270 fig. 13, 280, 285, 339
Cramer fire (2002), 457, 462
Crutzen, Paul, 321, 325
Curtis, John, 20, 40

Daniels, Ellreese, 369
Daniels, Orville, 162, 269 fig. 10
Davis, Kenneth, 15, 22, 121, 186
Davis, Mike, 288
DeBonis, Jeff, 245–46
DeBruin, Henry, 99, 108–10, 165
Decker fire, 21
Deeming, John, 123
Defense Department, 73, 225, 254, 302, 361, 430
demographics, 61–63. *See also* workforce
Department of the Interior: 1960s, 74–80; 1970s, 144–58, 430; 1980s and 1990s, 190–202; Interior Fire Coordination Committee, 251, 430. *See also individual agencies within Interior*
Derr, William, 463–64
Designated Control Burn System (DESCON), 106, 166
Despain, Don, 182, 238, 241
Dickenson, Russell, 186, 192
Dieterich, Jack, 68
Disaster Assistance Support Program (DASP), 217–18
Divison of State and Private Forestry (USFS), 9, 13, 108, 111, 169, 203, 224, 283, 300, 396, 432
doctrines, as strategy, 395–97

Dombeck, Mike, 284, 294, 296, 340, 357–58
Dome fire (1996), 351
duBois, Coert, 128, 215
Dude Creek fire (1990), 250

Eastern Fire Management project, 226
ecosystem management, as doctrine, 253, 261, 270 fig. 14, 278–83, 298, 306, 314, 318–19, 338, 341, 345, 379, 386
Ecosystem Research Group, 211
El Cariso Hotshots, 88
Elko fires (1964), 80–81
Emmitsburg 13, 437
Endangered Species Act (1973), 45, 54–55, 90, 99, 133, 144, 151, 191, 194, 244, 249, 278, 314, 318, 357
Environmental Regulation and Prescribed Fire conference, 298
equipment development, 458; centers, 30, 65, 73, 316
Esperanza fire (2006), 404
Everglades National Park, 20, 44, 50, 85, 102, 135–36, 140, 197, 208, 232, 320, 329
Everhardt, Gary, 135, 138–41
Everhardt, William, 48
Evison, Boyd, 198
exurban fire. *See* wildland-urban interface

fatality fires: aircraft, 426–28; Battlement Creek fire (1976), 147; Cramer fire (2002), 457; Dude Creek fire (1990), 250; Esperanza fire (2006), 404; Great Fires of 1910, centennial, 457; Honda fire (1977), 178; at Merritt Island NWR (1981), 155; in 1950s and 1960s, 88; Pocket fire (1979), 154; South Canyon fire (1994), 254, 290, 293–97, 299–300, 306–8, 311–12, 316–17, 335, 337, 339, 344, 350, 352, 354–55, 368–69, 374, 394, 457; Thirtymile fire (2001), 368–70

Faulkner, William, 445
Federal Aviation Administration, 427–28
Federal Emergency Management Agency (FEMA), 117, 189, 225, 258, 283, 299, 302, 358, 409, 454, 459
Federal Excess Personal Property (FEPP) program, 10, 203, 205, 406
Federal Land Assistance, Management, and Enhancement (FLAME) Act of 2009. *See* FLAME Act
Federal Land Policy and Management Act (FLPMA), 146, 148, 194
Federal Wildland Fire Management Policy (1995), 297, 302–4, 306, 339–40, 350, 353, 360, 360, 389, 397, 419; implementation of federal policy, 421–23. *See also* Federal Wildland Fire Management Policy and Program Review
Federal Wildland Fire Management Policy and Program Review, 300, 302–4. *See also* Federal Wildland Fire Management Policy
FEMA. *See* Federal Emergency Management Agency
Final Report of the Interagency Management Review Team (1995), 295, 307
fire, contrast to logging, 281
fire, generational themes, 310–12. *See also* workforce
fire, globalization, 321–27
fire, privatization, 387–88
fire by prescription, 108, 110, 120, 122, 127, 137, 147, 149, 166, 175, 180–81, 359, 374; as concept, 102–3. *See also* prescribed fire; prescribed natural fire
Fire by Prescription Symposium, 120
fire conferences: general, 120–23; international, 322. *See also names of specific conferences*
fire costs. *See* big fire costs; fire economics; fire funding
fire councils, 117–18

fire counterrevolution, 182–85, 220–22, 228–30
fire-danger rating, 14–15, 26, 66, 69, 77, 109, 123, 125–26, 216, 220. *See also* National Fire-Danger Rating System
fire economics, 167–69, 304–6, 382–92; and big fires, 292. *See also* fire funding
Fire Economics Assessment Report (FEAR), 305–6
Fire Economics Assessment Team (FEAT), 384
Fire Fighter Retirement Program, 165, 171, 174, 253, 284, 404
Firefighters United for Safety, Ethics, and Environment (FUSEE), 399
Firefighting Resources of Southern California Organized for Potential Emergencies. *See* Firescope
fire funding, 167–69, 304–6, 382–92, 426. *See also* fire economics
fire globalization, 215–20, 252
fire history, as narrative, 443–45
Fire in America, 184
Fire in the Environment Conference, 163
Fire in the Northern Environment conference, 148
fire labs, 13, 15, 39, 65–66, 68, 94, 104–8, 114–16, 122–26, 129–30, 134, 169, 179, 189, 226, 233, 269 fig. 11, 316–18, 356, 419. *See also* Macon lab; Missoula lab; Riverside lab
Fire, Landscapes, and People program, 400
Fire Learning Network (FLN), 400, 435
Fire Management Notes, 164, 284
fire policy, cycle of, 419. *See also* Federal Wildland Fire Management Policy; fire by prescription; 10 a.m. policy
fire plutocracy, 371
Firepro, 197–98, 388
fire research. *See* fire conferences; fire labs; research

fire revolution, 33–92; and counterrevolution, 182–85, 220–22, 228–30; revival (revolution 2.0), 277–87; stall in, 172–76; summary, 172–73
fires. *See* fire seasons; *and names of specific fires*
Firescope, 69, 88, 108, 113–16, 125–26, 160, 178–79, 188, 195, 210, 409
fire seasons: 1961, 28–32; 1967, 93–96, 204; 1970, 93, 96–98, 269 fig. 11; 1977, 177–78; 1980, 178–81; 1985, 229–30; 1987, 230–31; 1988, 230–41, 290–91; 1990, 250; 1993, 287–90, 299; 1994, 289–90, 292–97, 345, 403; 1996, 290, 351; 1999, 290; 2000, 290–91, 343–45, 348–56; 2003, 291, 377, 394, 377–78; 2005, 392; 2006, 403; 2007, 357; 2010, 457–58; 2011, 450; 2012, 449–50; 2013, 451
fire strategies, American West, 441–42
fire study tours, 164–65, 216
fire suppression: 1960s, as example, 24–25; 1970s, 169–70; 1980s, 187–89, 226–28; 1990s, 335–38. *See also* Boise Interagency Fire Center; 10 a.m. policy; total mobility
fire textbooks, 15, 121, 186. See also *Forest Fire: Control and Use*
fire training, 12, 69, 88, 125, 127, 146, 217–18, 330–31, 333, 339, 396, 405, 407, 462
fire treaties, 218–20, 323–24
Fire Use Training Academy, 407
Firewise, 212–13, 336, 385, 400, 406, 432, 441
Fish and Wildlife Service (FWS): 1970s, 151–55; 1980s, 190–92
Flagstaff model, 280–82
FLAME Act, 418, 423, 425–26
Flathead fire (1967), 95–96
Flint Hills, 20, 41, 89, 107, 186, 213, 330, 436, 448
Florida: as fire province, 85, 87, 206–9; influence on The Nature

Conservancy, 213–15, 324; role in fire revolution, 35–41, 102–3; role in prescribed fire, 19–20, 249, 298, 328–30; significant fires in, 155, 212, 229, 290–91, 412–13
Florida Division of Forestry (FDOF), 20, 39, 112, 207–10, 212, 338
Florida Forest Service. *See* Florida Division of Forestry
Focus, 222
Fons, Wallace, 15, 67, 124
Foothills fire (1992), 254–55, 309
Forest and Rangeland Renewable Resources Planning Act, 164, 166–67
Forest Fire and Atmospheric Science Research (FFASR). *See* research; U.S. Forest Service
Forest Fire: Control and Use (1959), 15, 121, 186
forest health, 119, 260–62, 270 fig. 13, 279–82, 298, 341, 363, 379, 380, 382, 432. *See also* Blue Mountains; Flagstaff model
Forest History Society, 184
Forest Planning Model. *See* Forplan
The Forest Ranger (Kaufman), 8, 22, 27, 71, 76, 171, 343, 458–59. *See also* Kaufman, Herbert
Forest Service Ethics and Course to the Future, 279
Forest Service *Manual*, 161, 168, 169, 173, 223, 396–97, 432
Forest Stewardship Program, 339, 380, 431
Forplan, 187, 223
Foss, Phillip, 76
fossil fuel combustion, 325–27
Frandsen, William, 124
Franklin, Jerry, 244, 280
Frome, Michael, 23–24, 27, 33–34, 186
Frye, Jack, 152–53
fuels: as concept, 126–28; and fuel-breaks, 130–32; as models, 125–30; as metric, 280

Gale, Robert, 168
Gale Report, 164, 168
Gallagher Peak fire (1980), 180–81
Gates of the Mountains Wilderness, 415–17
General Land Office (GLO), 74–75
Gila National Forest, 55, 63, 106, 162, 329, 418, 454–55, 463–65
Gisborne, Harry, 15, 122–23
Glacier Wall fire (1967), 92, 95–96
Gleason, Paul, 307
Global Fire Initiative, 324, 401, 435
Global Fire Monitoring Center, 324
globalization, 321–27
global warming, 220, 228, 240, 322–23, 326, 372, 456
Goldammer, Johann G., 324
Government Accountability Organization, 195, 236, 239, 242, 249, 252, 276, 279–81, 332, 341–46, 352, 358–60, 362, 364, 366–67, 369, 379–80, 382, 384–86, 388, 398, 403, 413, 418, 420–24, 428, 433, 437, 441, 447
Granite fire (1994), 296, 301, 333, 451
Granite Mountain Hotshots, 451. *See also* hotshot crews
Graves, Henry, 5, 383
Great Fires of 1910. *See* Big Blowup
Greeley, William, 5, 14, 443
Green Book (NPS), 49–52, 139, 231, 434
Gunzel, Les, 96, 105

Hansen, James, 322, 240
Harbour, Tom, 271 fig. 16, 395, 457, 468
Hardy, Charles, 21
Harlow fire (1961), 28–29, 450
Hartesveldt, Richard, 47
Hartzog, George, 46, 72, 95
Hawkins Bill, 207
Hayman fire (2002), 370, 449
Hazardous Fuels Reduction Program, 337

Healthy Forests Initiative (HFI), 363, 380, 397–98, 400, 432
Healthy Forests Restoration Act, 363, 380–82, 398
Heitlinger, Mark, 213
Helen Allison Savanna Preserve, 40–41, 267 fig. 6
Hemingway, Ernest, 152
Hirsch, Stanley, 68, 115
Hoberg's Resort, 47, 70, 88, 119
Hochdoerffer fire (1996), 304, 339
Holguin, Gabe, 465–66
home ignition zone, 318, 381
Homeland Security, 362, 409, 430, 459
Honda fire (1977), 178
hotshot crews, 12, 24, 25, 88, 97, 114, 147, 193, 197–98, 254, 271 fig. 15, 271 fig. 16, 285, 293–94, 307, 324, 368, 451, 460. *See also names of individual crews*
Howling fire (1994), 296–97, 301, 333–34
Hubbard Memo, 460–61, 463
Hurd, Elmer, Jr., 353
Hyde, Albert, 371

Inaja fire (1956), 88
Incident Command System, 115–16, 206, 218, 255, 258, 409, 432, 459
Independent Review Board, May 26 report of Cerro Grande fire, 352
Interagency Airtanker Board, 339, 427
Interagency Ecosystem Management Task Force, 280
Interagency Fire Management Policy Review Team, 235, 239, 242–44, 246, 249, 251
Interagency Strategy for the Implementation of Federal Wildland Fire Management Policy, 363
Interior Fire Coordination Committee, 251
Intermountain Experiment Station, 21, 123, 159. *See also* Missoula lab

Intermountain Fire Research Council, 117, 119, 159, 165, 187
International Association for Wildland Fire, 252, 255, 276, 323, 360, 369, 399, 406, 463
International Union for the Conservation of Nature, 401, 435
International Union of Forest Research Organizations, 164, 215–16
interstate fire compacts, 203–4
I-zone. *See* wildland-urban interface

Joint Fire Science Program (JFSP), 320–21, 339, 380–81, 403, 459, 468
Joseph W. Jones Center for Ecological Research, 252, 435
Junger, Sebastian, 254, 276, 309

Kallander, Harry, 20, 119
Kaufman, Herbert, 8, 22, 27, 63, 71, 76, 171, 343, 458–58. See also *The Forest Ranger*
Kenai fires (1969), 78–79, 82, 153
Kennedy, Roger, 353
Kilgore, Bruce, 49–50, 117, 136–37, 166, 187, 199, 233, 268 fig. 9, 276, 332
Komarek, E. V., 20, 22, 33, 36–39, 46, 62, 119–21, 148, 246, 261, 268 fig. 8, 276, 329
Komarek, Roy, 119

Laguna fire (1970), 97
La Mesa fire (1977), 142, 177, 351, 452
Land and Water Conservation Fund, 59–60
Landfire, 69, 380–81, 432
Landscape Fire and Resource Management Planning Tools Prototype Project. *See* Landfire
Large Airtanker Modernization Strategy, 429
large fire organization, 4, 11, 24, 458–59
Las Conchas fire (2011), 452, 455

Lassie, 64
Laverty, Lyle, 344
Lavin, Mary Jo, 254, 283–84, 294, 300
Lawrence, Don, 40, 267
LCES, 307
Leonard, Brad, 242
Leopold, Aldo, 19, 40–41, 43, 55–56, 118, 279, 453–55. See also *A Sand County Almanac*
Leopold, A. Starker, 41, 46–48, 72, 136, 198. *See also,* Leopold Report
Leopold Report, 41–42, 45, 47–49, 51, 58, 71–72, 74, 76, 90, 96, 135, 143, 196, 231–32, 243, 253, 279; influence on NPS, 41–43
let-burn, 49, 103, 105, 136–37, 149, 233, 330, 334, 365, 422
light-burning, 18–19, 44, 47, 49, 56, 86, 102, 261
Little, Silas, 121
Living with Fire, 401
Lone fire (1996), 374
lookouts, communications, escape routes, and safety zones. *See* LCES
Loop fire (1966), 88
Los Alamos (New Mexico), 177, 290, 316, 345, 350–51, 353–55, 452–54
Los Angeles County Fire Department, 32, 116, 131, 210, 257
Los Angeles Fire Department, 32, 115
Lowden, Merle, 26, 83, 87–88, 159, 163, 217
Lowden Ranch fire (1999), 332, 340

Mack Lake fire (1980), 179–81
Maclean, John, 293, 344, 348, 368
Maclean, Norman, 254, 276, 285, 293, 308–12, 317, 344, 349, 352, 358, 417, 457. *See also* Mann Gulch; *Young Men and Fire*
Macon lab, 15, 39, 65–66, 124, 134, 169, 226, 316, 318
Malibu fire (1993), 287–89
Mann Gulch (Montana), 254, 293, 307–8, 310–12, 317, 352, 412, 417.
See also Maclean, Norman; *Young Men and Fire*
Marble-Cone fire (1977), 177
Master Cooperative Wildland Fire Management and Stafford Act Response Agreement, 399
Mathiesssen, Peter, 236–37
McArdle, Richard, 27, 55, 307
McGuire, John, 106, 116, 160–62, 165–69, 171
McLaughlin, John, 48, 50–51, 136
McSweeney-McNary Act (1928), 14
Mees, Romaine, 115
megafire, 7, 35, 231, 345, 358, 370–77, 379, 383–86, 390, 394, 397, 427, 438, 447, 451, 456, 463; definition of, 371
Meriwether fire (2007), 416–17
metafire, 343, 359–70
Mexico, 163, 217–18, 239, 323–24, 345, 370, 393
Michigan Department of Natural Resources, 152–53
Midnight Sun Hotshots, 193, 271 fig. 15, 285. *See also* hotshot crews
militia, fire, 6, 11, 112, 171, 205–6, 284, 344, 407–9, 459–61
Millar, Dick, 115
Miller fire (2011), 450, 452, 454–56, 465
Minor, Vicki, 308
Missoula lab, 13, 15, 65, 68, 94, 122–26, 179, 226, 233, 269 fig. 11, 316–18, 419
Missoula Technology and Development Center (MTDC), 73, 316
Moccasin fire (1972), 140
Montague, Dick, 115
Moore, William "Bud," 52, 68, 88, 93–94, 147, 162–63, 165, 269 fig. 10, 310
Mormon Lake Hotshots, 147
Morton, Rogers, 108, 251
Mott, William Penn, 239, 242
Muir, John, 55–56, 200

Multi-Agency Coordination System (MACS), 115–16, 178
multiple use: as concept, 31, 55, 100, 159, 276; in BLM, 76, 150. *See also,* Multiple-Use Sustained-Yield Act
Multiple Use-Sustained Yield Act, 8, 45, 76, 90–91
Mutch, Bob, 136, 162, 233, 269 fig. 10, 277, 333, 461, 463–64
Myers, Ron, 20, 214, 324, 401

narratives, of American fire, 443–46; ironic, 444–45; progressive, 443–44
NASA, 170, 240, 273 fig. 19, 322, 350
NASF. *See* National Association of State Foresters
National Academy of Public Administration (NAPA), 353, 360, 386
National Academy of Sciences (NAS), 14, 67, 71, 314, 318
National Advanced Fire and Resource Institute (NAFRI), 407
National Advanced Resource and Technology Center, 88, 125, 127, 146, 218, 333
National Airtanker Study, 427
National Association of State Foresters (NASF), 75, 116, 189, 203–5, 216, 242, 299–300, 343–44, 360, 363, 384–86, 398, 406, 431, 436–37
National Biological Service. *See* National Biological Survey
National Biological Survey (NBS), 314–16, 320
National Cohesive Wildland Fire Strategy, 62, 69, 341–42, 344, 346, 357, 363, 366, 384, 418, 420, 430, 433, 437–40, 442–43, 445–46, 462
National Commission on Fire Prevention and Control, 109, 165
National Commission on the Wildland/Urban Interface Problem, 189
National Commission on Wildfire Disasters, 189, 297–98

National Environmental Policy Act (NEPA), 45–46, 58, 61, 90, 99, 146, 223, 243–44, 284, 295, 381–82, 389, 404
National Fire and Aviation Executive Board (NFAEB), 304, 365, 389, 397–99, 402, 407, 415, 421
National Fire Coordination Study (NFCS), 68–69, 72, 88, 92–93, 115, 162, 204
National Fire-Danger Rating System (NFDRS), 67–68, 70, 94, 123–27, 129, 143, 148, 152, 161, 164, 180–81, 193, 232, 257
National Fire Defense Plan, 69
National Fire Effects Workshop, 122
National Fire Management Analysis, 388
National Fire Plan (NFP), 69, 168, 213, 346–47, 350, 356–57, 362–66, 380, 383–84, 397–98, 400–401, 404, 406, 419, 425, 430–31, 438, 445, 461
National Fire Prevention and Control Agency (NFPCA), 109, 165. *See also* U.S. Fire Administration
National Fire Protection Association (NFPA), 14, 31, 116, 189, 212–13, 216, 242, 247, 258, 293, 299, 336, 343, 406, 431
National Fire Research Planning Conference, 169
National Fuel Inventory System, 127, 161
National Interagency Fire-Qualification System (NIFQS), 69, 112–13, 139, 152, 179, 253, 275, 285, 404, 462
National Interagency Incident Management System (NIIMS), 116, 358
National Park Service (NPS): in Alaska, 191–93; 1960s, 38–72, 74–75, 86, 88, 92, 95–96; 1970s, 134–44; 1980s, 196–202; research within, 71–72; response to Cerro Grande fire, 353–54. *See also* Leopold Report;

NPS-18; Robbins Report; *and individual parks*
National Research Council, 67, 216, 253, 314
National Response Plan, 409, 461
National Science Foundation (NSF), 72, 121, 164
National Transportation Safety Board, 427–28
National Wilderness Preservation System, 52, 54, 90, 415. *See also* Wilderness Act
National Wildfire Coordinating Group (NWCG), 108–13, 116–17, 139–40, 145–46, 152, 157, 187, 190, 193, 205, 275, 284, 304, 311, 319, 324, 339, 356, 361, 389, 404–5, 421, 427, 430–31, 437, 440
National Wildfire Suppression Association, 387
National Wildland-Urban Interface Fire Protection Initiative, 247, 299
natural fire, 72, 85–88, 100, 130, 136–38, 143, 160, 170, 174–75, 181, 191, 196–97, 200–202, 207, 223, 227, 232–35, 241, 248, 253, 156, 288, 305, 329–30, 332–35, 343, 350, 354–55, 359, 373–74, 376, 414, 440, 444, 454–55, 463–64. *See also* let-burning; prescribed natural fire; wildland fire use
The Nature Conservancy (TNC), 39–41, 87, 213–15, 220, 244, 246, 252, 267 fig. 6, 276, 324–25, 328, 331, 338, 342, 355, 360, 367, 400–401, 411, 434–35
Navigating into the Future roundtable, 279
Needs Assessment of the U.S. Forest Service, 406
NEPA. *See* National Environmental Policy Act
New Deal. *See* Roosevelt administration
New Forestry, 212, 244–45, 248, 280–82
New Perspectives, 248, 251, 253

NFDRS. *See* National Fire-Danger Rating System
NFP. *See* National Fire Plan
NFPA. *See* National Fire Protection Association
Nichols, Tom, 271 fig. 16
North American Forestry Commission (NAFC), 121, 163, 217–18
Northeastern States Forest Fire Protection Compact (1949), 9, 203, 218
Northern California, 5, 47, 86, 103, 230, 423, 447
Northern Forest Fire Laboratory. *See* Missoula lab
Northern Rockies, 5, 7, 28, 30, 56, 85, 88, 92, 95, 128, 141, 170, 181, 231, 234, 265 fig. 4, 290, 329, 340, 345, 347, 348–50, 355, 380, 385, 411, 414–16, 421, 448–50
Northwest Forest Plan, 222, 278, 282, 284
NPS. *See* National Park Service
NPS-18, 140–43, 169, 172, 197, 202, 232–33, 238, 332
nuclear winter, 184, 220, 225, 321
NWCG. *See* National Wildlife Coordinating Group

Occupational Safety and Health Administration, 242, 295–97, 307, 342, 368, 388, 395–96, 404
Odum, Eugene, 72
Office of Civil Defense, 14, 67, 204, 225. *See also* Federal Emergency Management Agency
Office of Inspector General (OIG), 360, 364, 369, 384–85, 388–90, 396, 398, 403, 421, 428, 461–64
Office of Management and Budget (OMB), 167, 170, 197, 221, 251, 362, 384, 404, 439–41
Office of Wildland Fire Coordination, 430
Old Faithful, 189, 230, 232, 234, 246, 283, 289, 294. *See also* Yellowstone National Park

Operation FireStop, 15, 131
Ordway, Katharine, 41, 231
Oregon and California (O&C) Railroad Revested Lands, 79–80, 190, 247, 284
O'Toole, Randall, 247
Ouzel fire (1978), 142–44, 233, 240–41

Pachenec, Joseph, 159
Pacific Southwest Experiment Station, 47–48, 131, 136, 164
Palm Coast fire (1985), 208, 212, 229
Panorama fire (1980), 178–79, 181, 188
Parks Canada, 175
Philpot, Charles, 129, 131–32, 159, 225–26, 242, 302
Pinchot, Gifford, 5, 18, 34, 48, 79, 395. See also *Use Book*
PNF. *See* prescribed natural fire
Pocket fire (1979), 154
Ponil complex (2002), 450
Prescribed Burning Act of 1990, 208
Prescribed Burning Symposium, 103–4
prescribed fire: as concept, 20; big-box strategy, 442; early history, 19–21, 44; Florida exemplar, 20, 85, 87, 208–9, 329; guidebooks for, 187; maps of, 272 fig. 17, 272 fig. 18; in The Nature Conservancy, 40–41, 213–14; 1960s, 36–38, 46–56, 88, 103–6; 1970s, 118–22, 135–36; 1980s, 330–31; as policy, 102–3, 174; problems with, 174, 179–80, 199–201, 223–24, 255, 330–32, 340–41, 345. *See also* fire by prescription; prescribed natural fire
prescribed fire support modules, 331
Prescribed Fire Training Center, 330, 407
prescribed natural fire (PNF), 113, 125–26, 128, 138, 142, 158, 162, 170, 173, 180–81, 192, 196, 201–2, 223, 227, 232–34, 237–39, 242–43, 249–51, 253, 255, 296–97, 311, 328–29, 334, 355, 365, 414, 422, 444; as concept, 105–8. *See also* fire by prescription; prescribed fire; wildland fire use
Prineville Hotshots, 293
privatization, 387–88
problem fires, cycle of, 419
Project Flambeau, 129
Project Skyfire, 65, 123
Protecting People and Sustaining Resources in Fire-Adapted Ecosystems: A Cohesive Strategy, 344
Public Land Law Review Commission (PLLRC), 60, 62, 156, 160
Pulaski Conference, 395, 407
Pulliam, Ron, 314–15
Pungo Lake fire (1981), 155
Putnam, Ted, 295, 316
pyrogeography: alignment with society, 327–28, 373–76; American fire regions, 449–50; fuels maps, 378–82; maps of, 433–37

Quadrennial Fire [and Fuel] Review, 360; 2005, 402–3, 407; 2009, 423–24
The Quiet Crisis (1963), 43, 58, 247, 277

Rainbow Series, 122–23, 187
Rasmussen, Boyd, 83, 145
Rattlesnake fire (1953), 88
Reagan administration, 183, 185, 189–90, 194–95, 198, 201–2, 220–22
Reciprocal Protection Act (1955), 9–10
red card. *See* National Interagency Fire-Qualification System
Red Skies of Montana, 6, 122
regions, in fire history, 85–86
research, fire: crisis in 1980s, 224–26; fire labs, 13, 15, 39, 65–66, 68, 94, 104–8, 114–16, 122–26, 129–30, 134, 169, 179, 189, 226, 233, 269 fig. 11, 316–18, 356, 419; Interior, 71–72, 314–16; international, 322–23; Joint Fire Science Program, 320–21; role in fire revolution, 122–34; U.S. Forest Service, 13–16, 189, 224–26, 316–18

resilience strategy, 441–42, 466
resistance strategy, 441, 466
restoration strategy, 441, 466
Richardson, J. H., 79
Rim fire (2013), 451
Riverside lab, 15, 65, 68, 104, 108, 114, 115–16, 126, 129–30, 179, 189, 226, 316, 318
Roadless Area Review and Evaluation, 160, 166
Robbins, William J., 71. *See also* Robbins Report
Robbins Report, 71–72, 74, 135, 253, 315
Robertson, F. Dale, 248, 251, 253, 282
Robinson, Laird, 310
Robinson, Roger R., 78, 82–83, 146–47, 193, 310
Rockhouse fire (2011), 450
Rocky Mountain National Park, 137, 140, 142–44, 233
Rodeo-Chediski fire (2002), 370, 375
Rogers, Mike, 463
Roosevelt administration, 13, 101–2, 183, 346, 368, 403
Roscommon Equipment Development Center, 65, 73
Rothermel, Richard, 68, 124–26, 225, 235, 269 fig. 11, 277, 317
Rothermel fire behavior model, 125–28, 132–34, 164, 187, 225, 235, 319
Rothman, Hal, 139, 143
Rural Community Development Act, 148, 164, 406
Rural Fire Defense, 68
Russia, 66, 77, 164–65, 322–24, 339, 393, 461

Saddle Mountain fire (1960), 24
Sadler fire (1999), 337, 340
safety, 307; aircraft, 376, 426–28. *See also* fatality fires; fire seasons, 1994
Sagebrush Rebellion, 146, 194–95, 221
Sampson, Neil, 298

A Sand County Almanac (Leopold), 43, 453. *See also* Leopold, Aldo
San Dimas Equipment Development Center, 30, 70
Satterfield, Steven, 300
Schechter, Claudia, 302
Schiff, Ashley, 70–71
Schimke, Harry, 136
Schuft, Peter, 136–37
Scientific Committee on the Problems of the Environment (SCOPE), 121–22, 186, 216
Seamon, Paula, 214
Seewald, Roger, 463–64
Selway-Bitterroot Wilderness, 106, 136, 160, 162, 329, 349, 463–64. *See also* White Cap Project
Seney fires (1976), 152–54
sequoia, 47–49, 86–87, 127–28, 136–37, 196, 199–200, 202, 251, 279
Sequoia-Kings Canyon National Park, 46–49, 64, 88, 106, 136, 139, 166, 197–99, 232–33, 235, 241, 252, 271 fig. 16, 332
Show, S. B., 15, 215
Sierra Club, 34, 41, 44, 56, 59, 100, 254, 277
Sierra Nevada parks, 46, 135–36, 140–41, 197, 232, 464. *See also* Sequoia-Kings Canyon National Park; Yosemite National Park
Silcox, Ferdinand Augustus "Gus," 13, 20
Sleeping Child fire (1961), 28–31, 264 fig. 2, 349–50, 450, 466
smokejumpers, 4, 21, 57, 73, 77–78, 81–82, 95, 122, 149, 196, 254, 293–94, 307–8, 310, 317, 324, 349, 358, 456, 464–65
Smokey Bear, 4, 10, 134, 203, 292, 357
Snyder, Gary, 5, 7, 277–78
Society of American Foresters, 20, 119, 121, 148, 212, 248, 365, 380
Sommers, William, 226, 300
South Canyon fire, 254, 290, 293–97, 299–300, 306–8, 311–12, 316–17,

335, 337, 339, 344, 350, 352, 354–55, 368–69, 374, 394, 457
South Coast. *See* Southern California
Southern California, 7, 12, 24–25, 31, 86–88, 97, 113, 116, 129–30, 141, 179, 187, 256, 259, 287–88, 299, 357, 377–78, 412–14, 447
Southern California Edison, 414
Southern Forest Fire Lab. *See* Macon lab
Southwest Forest Fire Fighters (SWFFF), 12, 114, 147, 149
sprawl, 185. *See* wildland-urban interface
Starvation Creek fire (1994), 296
states (American): fire management by, 84–92; fire programs, 202–11; impact of fire revolution, 205–6. *See also* California; Florida; National Association of State Foresters
Stegner, Wallace, 8, 313
Steinbeck, John, 33–34, 467
Steuter, Al, 213
Stoddard, Herbert, Sr., 19, 22, 36–39, 46, 56, 120, 268 fig. 8, 276
Strange, Tex, 112
Strategic Assessment of Fire Management in the USDA Forest Service, 300
Strategic Overview of Large Fire Costs Team, 344
Street, Phillip, 190
Sula complex (2000), 265 fig. 4, 349–50
Sundance fire (1967), 93–95, 97, 384
Swetnam, Tom, 162, 453

Tall Timbers fire ecology conferences, 33, 35–40, 46, 48, 58, 65, 70, 72, 88, 102, 110, 118–22, 136–37, 148, 165, 186–87, 212, 216, 236, 256, 276, 304, 328, 338, 342, 418, 434, 443, 447
Tall Timbers Research Station (TTRS), 35–40, 46, 48, 51, 65, 70, 72, 85, 90, 103–4, 118–20, 152, 157, 186, 196, 212, 214, 236, 244, 246, 248, 252, 268 fig. 8, 276, 304, 328, 338, 399, 434–35, 447; task force on ponderosa pine management, 119, 157–58, 261
Tallahassee (Florida), 20, 35, 39, 85, 87, 119, 208, 330, 338–39, 407, 412
Task Force on California's Wildland Fire Problem, 97, 114
10-acre policy, 127, 163–64, 167–68
10 a.m. policy, 13–14, 20, 26, 48, 84, 94, 101, 106, 119, 127, 159, 161, 162–64, 167–69, 223, 227, 229, 276, 311, 364, 371, 389, 397, 419, 421, 423, 438, 450, 458, 461; exceptions to, 162–63
10 Standard Fire Fighting Orders, 88, 307, 325, 368, 408
Texas fires (2011), 450
Thirtymile fire (2001), 368–70
Thomas, Jack Ward, 244, 270 fig. 14, 275, 282–84, 286, 294–96, 301, 319, 348–49
Tidwell, Tom, 457
Timber Salvage Rider, 336, 382
TNC. *See* The Nature Conservancy
total mobility, as concept, 108–11
Trapper Peak fire (1967), 93–94, 384
Truesdale, Denny, 304
TTRS. *See* Tall Timbers Research Station
Tunnel fire (1991), 256–60, 299, 335
Turcott, George, 149

Udall, Stewart, 41, 43, 45, 58, 61, 71, 76, 78, 95, 145, 183, 221, 247, 277
U.N. Food and Agriculture Organization (FAO), 10, 121, 163, 215–17, 220, 323–24, 393
University of California, Berkeley, 5, 21, 41, 46–48, 70, 72, 136, 198, 282, 435
U.S. Agency for International Development (USAID), 216–18, 324

Use Book (Pinchot), 395, 403, 419. *See also* Pinchot, Gifford
U.S. Fire Administration (USFA), 160, 165, 189, 212–13, 336, 406, 430, 459
U.S. Forest Service (USFS): centennial, 7, 392–94, 403; contrast between 1960 and 2010, 458–61; Division of State and Private Forestry, 9, 13, 108, 111, 169, 203, 224, 283, 300, 396, 432; international forestry, 215, 248; 1960s, 7–27, 22–28, 30, 35, 38–39, 41, 44, 47–48, 50, 54; 1970s, 158–72; 1978 reforms, 169–70, 215, 227, 346, 425; 1980s, 220–28; 1990s, 277–78, 282–85. *See also* research; workforce; *and individual interagency institutions, fires, and seasons*
U.S. Geological Survey (USGS), 272 fig. 17, 313–15, 321, 361, 379, 403, 433, 433, 459
U.S.-Mexico fire treaty, 218

van Gelder, Randy, 115
van Wagtendonk, Jan, 136, 268 fig. 9, 276
volunteer fire departments, 10, 69, 206, 276, 308, 357, 400, 405–6, 431, 434, 436, 453–54

Wakimoto, Ron, 238
Wallow fire (2011), 375, 443, 450, 452–56
Walsh Ditch fire (1976), 152–53
Waterfalls Canyon fire (1972), 138, 140, 142, 240
Weaver, Harold, 20, 22, 37, 48–49, 70, 118–19, 121, 155, 157
Weaver, Roy, 353
Weeks Act, 9, 91, 167, 209, 431–32, 442
Weick, Karl, 395
Wenatchee fires (1970), 96–98
West, Al, 463
Western Forest Fire Lab. *See* Riverside lab

Western Forest Health Initiative, 280, 298
Western Governors Conference, 242, 285, 360–63, 431, 437
WFU. *See* wildland fire use
Whitaker's Forest, 48, 88, 136
White Cap Project, 162, 166, 233, 269 fig. 10, 463
Whitewater-Baldy complex (2012), 463, 465
Whose Woods These Are (1960), 23, 27, 33, 186
Wilderness Act (1994), 39, 45, 52–55, 58–59, 61–63, 90, 99, 166, 191, 454
Wilderness Fire Planning, 333
Wilderness Fire Symposium (1983), 201
Wildfire Strikes Home! conference, 212
Wildfire Suppression Aircraft Transfer Act, 429
wildland fire decision support system (WFDSS), 389
Wildland Firefighter Memorial, 309
Wildland Firefighters Foundation, 308–309
Wildland Fire Leadership Council (WFLC), 304, 363, 365, 380, 385, 398, 430, 438
Wildland Fire Lessons Learned Center, 415, 421
wildland fire use (WFU), 297, 334, 339, 363–64, 391, 395–96, 407, 411, 415, 422, 419. *See also* prescribed natural fire
wildland-urban interface (WUI), 87, 98, 188–89, 206, 211–14, 250, 255–56, 259, 276, 289, 291, 300, 302–5, 313, 316, 318, 326, 335–37, 339–40, 342, 357–59, 363, 371, 376, 379, 382, 386, 390, 402, 404, 406, 414, 419–20, 431–32, 435, 452, 460
Williams, Jerry, 344, 371
Wilson, Carl, 97
Wilson, Jack, 75–76, 111, 145
Witch fire (2007), 413

Women in Fire Management conference, 246
Woodwardia fire (1959), 21
Worff, Bill, 269 fig. 10
workforce, fire: impact of civil rights, 63, 171–72, 174, 221, 246, 253–54, 283–84, 403–10, 461–64; 1960s, 11–12; relation to demographics 61–64. *See also* Fire Fighter Retirement Program
World Wildlife Fund, 324, 435, 401
Wright, Henry, 119, 186
WUI. *See* wildland-urban interface

Yarnell Hill fire (2013), 451
Yellowstone fires: 1988, 231–43, 246–47, 249, 251–54, 256, 265 fig. 3, 266 fig. 9, 270 fig. 12, 276–77, 283, 286, 289–90, 292, 294, 296, 313, 322, 332–34, 339–41, 347, 350, 353–55, 362, 365, 367, 370, 372, 377, 419–20, 434–35, 466
Yellowstone National Park, 22, 41, 43, 106, 136–37, 140, 173, 182, 188, 197, 216, 226, 229, 231–43, 246–47, 251–54, 256, 313. *See also* Old Faithful
Yosemite National Park, 46–47, 49, 88, 105–6, 136, 157, 232, 250–51, 268 fig. 9, 451
Young Men and Fire, 254, 285, 293–94, 308, 310–12, 317, 344, 349, 417, 457. *See also* Maclean, Norman; Mann Gulch

Zahniser, Howard, 62, 277
Zimmerman, Eliot, 204
Zimmerman, Tom, 303, 340

ABOUT THE AUTHOR

Stephen J. Pyne is a regents professor in the School of Life Sciences, Arizona State University. He is the author of over a score of books, most of them on fire, and especially fire's history. He has written big-screen fire histories for the United States (up through the 1970s), Australia, Canada, Europe (including Russia), and Earth overall. In his earlier life, he spent 15 seasons with the North Rim Longshots at Grand Canyon National Park and another three seasons writing fire plans for the National Park Service.